Mathematical
Methods *in*
Chemical
Engineering

Mathematical Methods *in* Chemical Engineering

V. G. JENSON
Senior Lecturer
Department of Chemical Engineering
The University, Birmingham, England

and

G. V. JEFFREYS
Professor & Head
Department of Chemical Engineering
University of Aston, Birmingham, England

SECOND EDITION

ACADEMIC PRESS

A Harcourt Science and Technology Company

San Diego   San Francisco   New York   Boston
London   Sydney   Tokyo

Copyright © 1977 by ACADEMIC PRESS
Ninth printing 2000

Academic Press
A Harcourt Science and Technology Company
Harcourt Place, 32 Jamestown Road, London NW1 7BY, UK
http://www.academicpress.com

Academic Press
A Harcourt Science and Technology Company
525 B Street, Suite 1900, San Diego, California 92101-4495, USA
http://www.academicpress.com

ISBN 0-12-384456-8

A catalogue record for this book is available from the British Library

Library of Congress Catalog Number: 77-77362

Printed and bound in Great Britain by Antony Rowe Ltd, Eastbourne

00 01 02 03 04 05 IB 14 13 12 11 10 9

PREFACE TO THE SECOND EDITION

The first edition of this book was published in 1963 when both authors were lecturers in the Department of Chemical Engineering at Birmingham University and the contents were directed to the undergraduate and postgraduate courses offered in the department. At that time, the courses presented in most university engineering departments were analytically based and only simple numerical methods for use with desk calculators were introduced. Powerful digital computers are now freely available which provide alternative methods for solving the more complicated problems, especially those in optimization studies. In addition S.I. units have been adopted almost universally and these developments have made a revision of the first edition desirable.

In this second edition, chapters 1 to 10 have been condensed slightly but the treatment of boundary layer theory has been introduced in Chapter 7. Chapter 11 has been completely rewritten with an emphasis on computer methods without restricting the presentation to a particular computer language. This has been achieved by the inclusion of logic flow diagrams wherever possible both here and in the subsequent chapters of the text. It is hoped that this will encourage readers to write programs in whatever language they prefer. Chapter 12 has been modified to prepare the reader for some of the optimization techniques described in a completely revised Chapter 13, and finally additional problems for the reader to attempt are offered together with references to their solution.

Throughout the text we have tried to adhere to the S.I. system of units but on a number of occasions we have found the "second" inconvenient and have expressed time in minutes or hours. We are assured that this liberty is permissible.

Birmingham
June, 1977   V. G. J. and G. V. J.

PREFACE TO THE FIRST EDITION

The development of new processes in the chemical industry is becoming more complex and increasingly expensive. If the research and development of the process can be carried out with confidence, the ultimate design will be more exact, and therefore the plant will operate more economically. In all facets of such a project, mathematics, which is the language of the quantitative, plays a vital role. Therefore training in mathematical methods is of the utmost importance to chemical engineers. The present text has been written with these ideas in mind, and we would emphasize that our aim is to present mathematics in a form suitable for the engineer rather than to teach engineering to mathematicians. To the pure mathematician an elegant proof is an end in itself, but to the chemical engineer it is merely a means to an end. Consequently this book only sketches demonstrations of the validity of theorems, to encourage the reader to have more confidence in the use of the technique for the solution of engineering problems. In addition, an attempt has been made to sort out the useful from the trivial and flamboyant of the wide variety of mathematical techniques available.

The material presented in this book is based on various undergraduate and post graduate courses given in the Chemical Engineering Department of the University of Birmingham. Many of the worked examples have been selected from research work carried out in the department, and these are supplemented with problems taken from the chemical engineering literature. Some chapters of the book (notably Chapters 4 and 5) are almost entirely mathematical, but wherever possible the text has been illustrated with chemical engineering applications.

The book was written when both authors were lecturers in the Chemical Engineering Department of the University of Birmingham. It is hoped that the text will encourage chemical engineers to make greater use of mathematics in the solution of their problems.

We wish to express our thanks to Professor J. T. Davies for initiating the venture, and to Professor F. H. Garner, and Professor S. R. M. Ellis for their interest and advice in the preparation of the text.

Birmingham
July, 1963   V. G. J. and G. V. J.

CONTENTS

Chapter 4

COMPLEX ALGEBRA

Chapter 5

FUNCTIONS AND DEFINITE INTEGRALS

Chapter 6

THE LAPLACE TRANSFORMATION

Chapter 7

VECTOR ANALYSIS

Chapter 8

PARTIAL DIFFERENTIATION AND PARTIAL DIFFERENTIAL EQUATIONS

Chapter 9

FINITE DIFFERENCES

Chapter 13

OPTIMIZATION

Chapter I

THE MATHEMATICAL STATEMENT OF THE PROBLEM

1.1. INTRODUCTION

NEARLY all applied science consists of performing experiments and interpreting the results. This may be done quantitatively by taking accurate measurements of the system variables which are subsequently analysed and correlated, or qualitatively by investigating the general behaviour of the system in terms of one variable influencing another. The first method is always desirable, and if a quantitative investigation is to be attempted it is better to introduce the mathematical principles at the earliest possible stage, since they may influence the course of the investigation. This is done by looking for an idealized mathematical model of the system.

The second step is the collection of all relevant physical information in the form of conservation laws and rate equations. The conservation laws of chemical engineering are material balances, heat balances, and other energy balances; whilst the rate equations express the relationship between flow rate and driving force in the fields of fluid flow, heat transfer, and diffusion of matter. These are then applied to the model, and the result should be a mathematical equation which describes the system.

The type of equation (algebraic, differential, finite difference, etc.) will depend upon both the system under investigation, and the detail of its model. For a particular system, if the model is simple the equation may be elementary, whereas if the model is more refined the equation will be a more difficult type. The appropriate mathematical techniques are then applied to this equation and a result is obtained. This mathematical result must then be interpreted through the original model in order to give it a physical significance.

In this chapter, only the simplest problems will be considered and the ideas of simple models will be introduced. The more complicated models will be introduced throughout the book in the chapters dealing with the particular mathematical techniques which are needed for the completion of the solutions.

1.2. REPRESENTATION OF THE PROBLEM

A simple example of the application of these ideas will be given first. The apparatus shown diagrammatically in Fig. 1.1 is to be used for the continuous extraction of benzoic acid from toluene, using water as the extracting solvent. The two streams are fed into a tank A where they are stirred vigorously, and the mixture is then pumped into tank B where it is allowed to settle into two layers. The upper toluene layer and the lower water layer are removed separately, and the problem is to find what proportion of the benzoic acid has passed into the solvent phase.

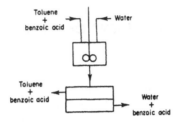

FIG. 1.1. Single-stage mixer settler

The problem is idealized in Fig. 1.2 where the two tanks have been combined into a single stage. The various streams have been labelled and two material balances have already been used, (a) conservation of toluene, and (b) conservation of water. These flow rates have been expressed on a solute free basis to simplify the analysis. The concentration of benzoic acid in each stream has also been stated and this completes the mathematical model.

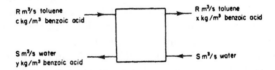

FIG. 1.2. Idealized single-stage solvent extraction

So far, it has been assumed that all flow rates are steady, and that toluene and water are immiscible. Further assumptions are now made that the feed concentration c remains constant, and that the mixer is so efficient that the two streams leaving the stage are always in equilibrium with one another. This last fact can be expressed mathematically by

$$y = mx \qquad (1.1)$$

where m is the distribution coefficient.

The equation is now derived from the model by writing down a mass balance for benzoic acid.

Input of benzoic acid $= Rc$ kg/s

Output of benzoic acid $= Rx + Sy$ kg/s

Since benzoic acid must leave at the same rate as it enters,

$$Rc = Rx + Sy \qquad (1.2)$$

The pair of equations (1.1) and (1.2) contain four known quantities (R, S, c, m) and two unknown quantities (x, y), and they can be solved for the unknowns as follows.

$$Rc = Rx + mSx$$

$$\therefore \quad x = \frac{Rc}{R + mS}, \quad y = \frac{mRc}{R + mS} \tag{1.3}$$

Therefore, the proportion of benzoic acid extracted is

$$\frac{Sy}{Rc} = \frac{mS}{R + mS} \tag{1.4}$$

As a numerical example, if $S = 12R$, $m = 1/8$, and $c = 1.0$; then $x = 0.4$, $y = 0.05$, and the proportion of acid extracted is 60%.

At this stage, it can be seen that even in this simple problem, two dimensionless groups which are characteristic of the problem have arisen quite naturally as a result of the investigation. Putting

$$\alpha = R/mS \tag{1.5}$$

and $E = Sy/Rc$, equation (1.4) becomes

$$E = 1/(\alpha + 1) \tag{1.6}$$

That is, the proportion extracted is governed solely by the value of the dimensionless group α.

1.3. Solvent Extraction in Two Stages

The above example will now be reconsidered, but two stages will be used for the extraction of benzoic acid instead of one stage. Each stage still consists of two tanks, a mixer and a settler, with counter-current flow through the stages. The idealized flow system is shown in Fig. 1.3 where the

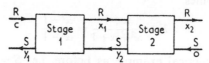

FIG. 1.3.   Idealized two-stage extraction

symbols have the same meaning as in the previous example, and the different concentrations in a particular phase are distinguished by suffixes. In accordance with chemical engineering practice, the suffix denotes the number of the stage from which the stream is leaving. The assumptions which were made above are made again, and equation (1.1) is still valid for each stage separately, giving

$$y_1 = mx_1, \quad y_2 = mx_2 \tag{1.7}$$

A benzoic acid mass balance is now taken for each stage.

	Stage 1	*Stage* 2
Input of acid (kg/s)	$Rc + Sy_2$	Rx_1
Output of acid (kg/s)	$Rx_1 + Sy_1$	$Rx_2 + Sy_2$

The fact that benzoic acid must enter and leave a stage at the same rate gives the two equations

$$Rc + Sy_2 = Rx_1 + Sy_1$$

$$Rx_1 = Rx_2 + Sy_2$$

Using equations (1.7) to eliminate x_1 and y_2, the inter-stage concentrations,

$$Rc + mSx_2 = (Ry_1/m) + Sy_1 \tag{1.8}$$

and

$$Ry_1/m = Rx_2 + mSx_2 \tag{1.9}$$

Eliminating y_1 between equations (1.8) and (1.9),

$$R(Rc + mSx_2) = (R + mS)(Rx_2 + mSx_2)$$

$$\therefore \quad R^2 c = x_2(R^2 + mRS + m^2 S^2)$$

$$\therefore \quad x_2 = \frac{R^2 c}{R^2 + mRS + m^2 S^2} \tag{1.10}$$

$$\therefore \quad y_1 = \frac{mRc(R + mS)}{R^2 + mRS + m^2 S^2} \tag{1.11}$$

Again, the proportion of benzoic acid extracted is

$$\frac{Sy_1}{Rc} = \frac{mS(R + mS)}{R^2 + mRS + m^2 S^2} \tag{1.12}$$

Introducing the dimensionless groups E and α again from equation (1.5), equation (1.12) becomes

$$E = \frac{\alpha + 1}{\alpha^2 + \alpha + 1} = \frac{\alpha^2 - 1}{\alpha^3 - 1} \tag{1.13}$$

Using the same numerical example as before, i.e. $S = 12R$, $m = 1/8$, and $c = 1\cdot0$; then $x_2 = 0\cdot21$, $y_1 = 0\cdot066$, and the proportion extracted is 79%.

A greater degree of extraction has thus been obtained with two stages than with one stage, everything else being the same.

1.4. SOLVENT EXTRACTION IN N STAGES

This improved extraction leads to the consideration of more than two stages in the extraction system. The algebraic treatment was quite simple for one stage, only requiring the solution of two equations in two unknowns.

The application to two stages involved the solution of four equations in four unknowns, and following the same procedure for N stages, it would be necessary to solve $2N$ equations. This is too laborious and would require an individual solution for every integer value of N, and more advanced mathematical techniques are obviously needed to reduce the work. One method using matrix algebra will be illustrated in Chapter 12, but a second method, using finite differences and anticipating the contents of Chapter 9, will be used here.

The general arrangement is as shown in Fig. 1.4 where the flow rates of the two streams are still denoted by R and S, and the benzoic acid concentrations by x and y. The suffix notation is again used to distinguish

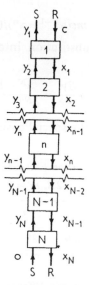

FIG. 1.4. Idealized N-stage extraction

between the different states of each stream, the suffix denoting the stage which the stream has just left. This time, a benzoic acid material balance is applied to the general stage n of the system.

$$\text{Input of acid (kg/s)} = Rx_{n-1} + Sy_{n+1}$$

$$\text{Output of acid (kg/s)} = Rx_n + Sy_n$$

Since entry rate and exit rate must be equal,

$$Rx_{n-1} + Sy_{n+1} = Rx_n + Sy_n \tag{1.14}$$

The distribution coefficient equation (1.1) is still valid for any value of n.

$$\therefore \quad y_n = mx_n$$

and equation (1.14) becomes

$$Rx_{n-1} + mSx_{n+1} = (R+mS)x_n$$

Introducing α again from equation (1.5),

$$\alpha x_{n-1} + x_{n+1} = (\alpha+1)x_n$$

or in standard form,

$$x_{n+1} - (\alpha+1)x_n + \alpha x_{n-1} = 0 \qquad (1.15)$$

This is a second order linear finite difference equation and the method of solution will be discussed in Chapter 9. The solution is quoted here for completeness, viz.

$$y_n = mx_n = mc(\alpha^{N+1} - \alpha^n)/(\alpha^{N+1} - 1) \qquad (1.16)$$

and this may be verified by substitution into equation (1.15). The proportion extracted, E, is given by

$$E = \frac{Sy_1}{Rc} = \frac{\alpha^N - 1}{\alpha^{N+1} - 1} \qquad (1.17)$$

which gives equation (1.13) for the special case $N = 2$.

TABLE 1.1. *Proportion Extracted in N Stages*
($S = 12R$, $m = 1/8$)

N	1	2	3	5	10
E (%)	60·0	78·9	87·7	95·2	99·4

Table 1.1 gives a few values of E for different values of N for the particular system considered. This shows how the proportion extracted increases with the number of stages, and indicates that more than ten stages are likely to be wasteful whereas one stage gives a poor degree of extraction. This type of problem will be continued in Chapter 13, where the most economical number of stages will be determined by financial considerations.

1.5. SIMPLE WATER STILL WITH PREHEATED FEED

Figure 1.5 illustrates a distillation apparatus consisting of a boiler B with a constant level device C, fed with the condenser cooling water. The steam is condensed in A and collected in the receiver D. Some of the latent heat of evaporation is returned to the boiler by preheating the feed. Denoting the condenser feed rate by F kg/s and the temperature by T_0 °C, the exit water temperature by T °C, the excess water over-flow rate by

W kg/s and the distillation rate by G kg/s, the performance of the apparatus can be investigated.

Input of water to the still (kg/s) $= F$

Output of water from the still (kg/s) $= W$

Output of steam from the still (kg/s) $= G$

$$\therefore \quad F = W + G \tag{1.18}$$

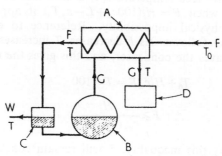

FIG. 1.5. Water still with heat exchanger

If heat is supplied to the boiler at a rate H J/s, the latent heat of evaporation of water is L J/kg, its heat capacity is c_v J/kg °C, and 0 °C is taken as the datum temperature, a heat balance can be taken over the boiler.

Heat input (J/s) $= H + (F - W) c_v T$

Heat output (J/s) $= 100 c_v G + LG$

$$\therefore \quad H + G c_v T = (100 c_v + L) G \tag{1.19}$$

where equation (1.18) has been used to eliminate W. Equation (1.19) contains two unknown quantities, G and T, and another equation is needed to complete the description. This is obtained from a heat balance over the condenser.

Heat gained by cooling water (J/s) $= F c_v (T - T_0)$

Heat lost by condensing steam (J/s) $= G(L + 100 c_v - c_v T)$

$$\therefore \quad F c_v (T - T_0) = G(L + 100 c_v - c_v T) \tag{1.20}$$

In deriving equation (1.20) it has been assumed that the distillate is cooled to the exit water temperature.

Eliminating G between (1.19) and (1.20),

$$F c_v (T - T_0) = H$$

$$\therefore \quad T = T_0 + H/c_v F \tag{1.21}$$

From (1.19) and (1.21),

$$H = G(100c_v + L - c_v T_0 - H/F)$$

$$\therefore \quad G = \frac{FH}{F(100c_v + L - c_v T_0) - H} \tag{1.22}$$

This equation gives the rate of distillation in terms of the heat input and the temperature and flow rate of the cooling water.

If an attempt is made to interpret equation (1.22) for a constant heat supply rate and constant feed temperature, it is seen that as F is decreased, G increases. In fact when $F = H/(100c_v + L - c_v T_0)$ it appears that G is infinite, which is a physical impossibility. Reference to equation (1.21) resolves the difficulty as follows. As F decreases, T increases; but T cannot exceed 100 °C as it leaves the condenser, and this gives the restriction

$$T_0 + H/c_v F = T \leqslant 100$$

$$\therefore \quad F \geqslant \frac{H}{c_v(100 - T_0)} \tag{1.23}$$

If F does not satisfy this inequality, T will remain constant at 100 and equation (1.19) becomes

$$H = LG$$

$$\therefore \quad G = H/L$$

The temperature restriction has a further influence in that the amount of steam produced by the boiler exceeds the capacity of the condenser. Denoting the rate of collection of distillate by D kg/s, equation (1.20) becomes

$$Fc_v(100 - T_0) = DL$$

$$\therefore \quad D = c_v(100 - T_0) F/L \tag{1.24}$$

The complete solution is thus given by equations (1.22), (1.23), and (1.24), i.e.

$$D = c_v(100 - T_0) F/L \quad \text{for } F \leqslant H/c_v(100 - T_0)$$

and

$$D = \frac{FH}{F(100c_v + L - c_v T_0) - H} \quad \text{for } F \geqslant H/c_v(100 - T_0)$$

and this solution is illustrated by the continuous line in Fig. 1.6. This shows that when $F = H/c_v(100 - T_0)$, the rate of collection of distillate is a maximum at a value H/L.

The above analysis has been made without reference to heat losses, and an attempt to allow for these would lead to a much more complicated model. A qualitative investigation would suggest that heat losses in the feed line to the boiler would be detrimental, and more serious at low values of F; but heat losses in the vapour line can have two effects. If F is greater than the optimum value, losses from the vapour line will be detrimental, but if F is smaller than the optimum value, heat losses from the vapour line will actually increase the yield. On the basis of these considerations, the second dotted line has been drawn in Fig. 1.6.

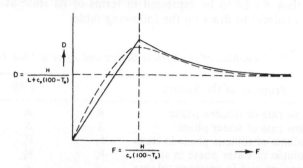

FIG. 1.6. Variation of distillation rate with feed rate

1.6. UNSTEADY STATE OPERATION

In the examples considered so far, the system has been in a steady state, allowing the material entering the system to be equated to the material leaving the system and this has always given algebraic equations. In unsteady state problems, however, time enters as a variable and some properties of the system become functions of time. In the application of conservation laws it is no longer true that the rate of entry of material will equal the rate of exit, since an allowance must be made for material accumulating within the system. The general conservation law now becomes

$$\text{INPUT} - \text{OUTPUT} = \text{ACCUMULATION} \qquad (1.25)$$

The example of Section 1.2 will now be reconsidered as it starts from the following situation. Assume that the single stage contains V_1 m³ of toluene, V_2 m³ of water, and no benzoic acid. Assume, as before, that the mixer is so efficient that the compositions of the two liquid streams are in equilibrium at all times, and, in addition, that a stream leaving the stage is of the same composition as that phase in the stage. This is all illustrated in Fig. 1.7, which shows the state of the system at a general time θ, where x and y are now functions of time.

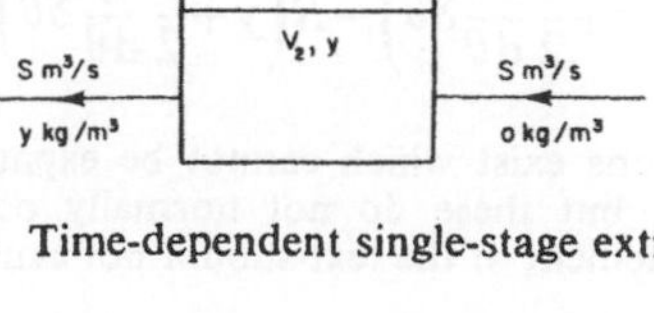

FIG. 1.7. Time-dependent single-stage extraction

Since the flow rates of water and toluene are constant, V_1 and V_2 will remain constant. The conditions are always changing, and so the material balance must be applied during a small time increment $\delta\theta$. Any function of θ can be expanded by Taylor's theorem,† and this allows the state of the system at a time $\theta+\delta\theta$ to be expressed in terms of its state at time θ. In this case, it is helpful to draw up the following table.

TABLE 1.2. *The Condition of the System Before and After a Time Increment*

Property of the System	θ	$\theta+\delta\theta$
Flow rate of toluene phase	R	R
Flow rate of water phase	S	S
Volume of toluene phase in stage	V_1	V_1
Volume of water phase in stage	V_2	V_2
Conc. of acid in entrance toluene	c	c
Conc. of acid in exit toluene	x	$x+\dfrac{dx}{d\theta}\delta\theta$
Conc. of acid in exit water	y	$y+\dfrac{dy}{d\theta}\delta\theta$
Amount of acid in toluene layer	V_1x	$V_1x+V_1\dfrac{dx}{d\theta}\delta\theta$
Amount of acid in water layer	V_2y	$V_2y+V_2\dfrac{dy}{d\theta}\delta\theta$

During a time interval $\delta\theta$,

$$\text{Input of acid} = Rc\,\delta\theta$$

$$\text{Output of acid} = R\left(x+\frac{1}{2}\frac{dx}{d\theta}\delta\theta\right)\delta\theta+S\left(y+\frac{1}{2}\frac{dy}{d\theta}\delta\theta\right)\delta\theta$$

$$\text{Accumulation} = V_1\frac{dx}{d\theta}\delta\theta+V_2\frac{dy}{d\theta}\delta\theta$$

Since input − output = accumulation,

$$Rc\,\delta\theta-R\left(x+\frac{1}{2}\frac{dx}{d\theta}\delta\theta\right)\delta\theta-S\left(y+\frac{1}{2}\frac{dy}{d\theta}\delta\theta\right)\delta\theta = V_1\frac{dx}{d\theta}\delta\theta+V_2\frac{dy}{d\theta}\delta\theta$$

$$\therefore\quad Rc-R\left(x+\frac{1}{2}\frac{dx}{d\theta}\delta\theta\right)-S\left(y+\frac{1}{2}\frac{dy}{d\theta}\delta\theta\right) = V_1\frac{dx}{d\theta}+V_2\frac{dy}{d\theta}$$

† Mathematical functions exist which cannot be expanded by Taylor's theorem as shown in Section 3.3.6, but these do not normally occur in chemical engineering applications, and the statement in the text should not cause any difficulty.

Taking the limit as $\delta\theta \to 0$,

$$Rc - Rx - Sy = V_1\frac{dx}{d\theta} + V_2\frac{dy}{d\theta} \qquad (1.26)$$

In stating the output, the arithmetic mean of the concentrations was taken, but the later calculations show that this was not necessary; the same equation (1.26) would have been obtained by using the concentrations at time θ instead. It should be noted, however, that the accumulation term arises solely because the concentrations change during the time interval; thus these changes cannot be disregarded completely.

Only the first two terms were taken in the Taylor series for x and y, and these are all that are necessary since the further terms which contain a factor $(\delta\theta)^2$ will disappear when the limit $\delta\theta \to 0$ is taken.

Equation (1.26) must now be solved in conjunction with the equilibrium relationship (1.1). Eliminating y,

$$Rc - Rx - mSx = V_1\frac{dx}{d\theta} + mV_2\frac{dx}{d\theta}$$

Rearranging,

$$\frac{dx}{Rc - (R+mS)x} = \frac{d\theta}{V_1 + mV_2}$$

and integrating,

$$\frac{-\ln[Rc - (R+mS)x]}{R+mS} = \frac{\theta}{V_1 + mV_2} + A \qquad (1.27)$$

The constant of integration A can be evaluated by using the given initial state of the system that the stage contains no benzoic acid at zero time; i.e. when

$$\theta = 0, \quad x = 0 \qquad (1.28)$$

$$\therefore \quad \frac{-\ln Rc}{R+mS} = A$$

$$\therefore \quad \ln Rc - \ln[Rc - (R+mS)x] = (R+mS)\,\theta/(V_1 + mV_2)$$

or in exponential form:

$$x = \frac{Rc}{R+mS}\left[1 - \exp\left(-\frac{R+mS}{V_1 + mV_2}\theta\right)\right] \qquad (1.29)$$

This equation satisfies the condition $x = 0$, at $\theta = 0$, satisfies the differential equation (1.26) for any positive value of θ, and when $\theta \to \infty$, $x \to Rc/(R+mS)$. This is the steady state solution which has already been obtained in equation (1.3).

1.7. Salt Accumulation in a Stirred Tank

Two further points are illustrated by the following example. A tank contains 2 m³ of water. A stream of brine containing 20 kg/m³ of salt is fed into the tank at a rate of 0·02 m³/s. Liquid flows from the tank at a rate of 0·01 m³/s. If the tank is well agitated, what is the salt concentration in the tank when the tank contains 4 m³ of brine?

1.7.1. Simple Treatment

Firstly, the following simple model of this system can be set up. Liquid enters at a faster rate than it leaves, so that liquid accumulates in the tank at a rate of 0·01 m³/s. The increase in tank contents is $4-2 = 2$ m³, and hence the operation will last for 200 s.

During this 200 s, 4 m³ of brine enter carrying 80 kg of salt and 2 m³ of brine leave. Assuming that the final concentration in the tank is X, and that the concentration of the outlet stream increases linearly with time, the following material balance can be established.

$$\text{Input of salt during 200 s(kg)} \qquad = 80$$

$$\text{Output of salt during 200 s(kg)} \qquad = 2(X/2)$$

$$\text{Accumulation of salt in the tank (kg)} = 4X$$

Using equation (1.25),

$$80 - X = 4X$$

$$\therefore \quad X = 16 \text{ kg/m}^3 \tag{1.30}$$

This is a very simple model of the system which yields an answer by elementary algebra. It does, however, contain a further assumption beyond those given in the question; the assumption concerns the time variation of the outlet concentration. This additional assumption need not be made if a more detailed model is taken in a similar manner to that in the previous section.

1.7.2. More Detailed Treatment

The state of the system at a general time θ is shown in Fig. 1.8, where both V and x are functions of θ. Again, a table can be constructed showing the state of the system at time θ and at time $\theta + \delta\theta$.

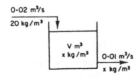

Fig. 1.8. Accumulation of salt in surge tank

TABLE 1.3. *The Condition of the System Before and After a Time Increment*

Property of the System	θ	$\theta + \delta\theta$
Brine input rate (m³/s)	0·02	0·02
Input salt concentration (kg/m³)	20	20
Brine output rate (m³/s)	0·01	0·01
Output salt concentration (kg/m³)	x	$x + \dfrac{dx}{d\theta}\delta\theta$
Volume of liquid in tank (m³)	V	$V + \dfrac{dV}{d\theta}\delta\theta$
Salt content of tank (kg)	Vx	$\left(V + \dfrac{dV}{d\theta}\delta\theta\right)\left(x + \dfrac{dx}{d\theta}\delta\theta\right)$

During a time interval $\delta\theta$,

$$\text{Input of brine} \quad = 0{\cdot}02\delta\theta$$

$$\text{Output of brine} \quad = 0{\cdot}01\delta\theta$$

$$\text{Accumulation of brine} = \left(V + \frac{dV}{d\theta}\delta\theta\right) - V$$

$$\therefore \quad 0{\cdot}02\delta\theta - 0{\cdot}01\delta\theta = \frac{dV}{d\theta}\delta\theta$$

$$\therefore \quad 0{\cdot}01 = \frac{dV}{d\theta} \tag{1.31}$$

A material balance must also be satisfied by the salt, viz.

$$\text{Input of salt} \quad = 0{\cdot}02 \times 20\delta\theta = 0{\cdot}4\delta\theta$$

$$\text{Output of salt} \quad = 0{\cdot}01x\,\delta\theta$$

$$\text{Accumulation of salt} = \left(V + \frac{dV}{d\theta}\delta\theta\right)\left(x + \frac{dx}{d\theta}\delta\theta\right) - Vx$$

Since input − output = accumulation,

$$0{\cdot}4\delta\theta - 0{\cdot}01x\,\delta\theta = Vx + x\frac{dV}{d\theta}\delta\theta + V\frac{dx}{d\theta}\delta\theta + \frac{dV}{d\theta}\frac{dx}{d\theta}(\delta\theta)^2 - Vx$$

$$\therefore \quad 0{\cdot}4 - 0{\cdot}01x = x\frac{dV}{d\theta} + V\frac{dx}{d\theta} + \frac{dV}{d\theta}\frac{dx}{d\theta}\delta\theta \tag{1.32}$$

Taking the limit $\delta\theta \to 0$, the last term in equation (1.32) vanishes. In deriving this equation, the salt concentration in the outlet stream at time θ has been used in the output term; since, as in Section 1.6, any correction to this contains a factor $\delta\theta$ which will lead to its elimination when the limit $\delta\theta \to 0$ is taken. The first important point is now evident; the accumulation consists of two terms. The one term is due to a volume change, the other is due to a concentration change, and both are essential in the solution.

Equations (1.31) and (1.32) must now be solved for x and V as functions of θ. Equation (1.31) can be solved immediately;

$$V = A + 0 \cdot 01\theta$$

but when $\theta = 0$, $V = 2$

$$\therefore \quad V = 2 + 0 \cdot 01\theta \tag{1.33}$$

Substituting equations (1.31) and (1.33) into (1.32),

$$40 - x = x + (200 + \theta)\frac{dx}{d\theta}$$

Rearranging,

$$\frac{dx}{40 - 2x} = \frac{d\theta}{200 + \theta}$$

and integrating,

$$-\tfrac{1}{2}\ln(40 - 2x) = \ln(200 + \theta) + B$$

where B is the constant of integration. But when

$$\theta = 0, \quad x = 0 \tag{1.34}$$

$$\therefore \quad -\tfrac{1}{2}\ln 40 = \ln 200 + B$$

Eliminating B and combining the logarithms,

$$-\tfrac{1}{2}\ln(1 - 0 \cdot 05x) = \ln(1 + 0 \cdot 005\theta)$$

$$\therefore \quad 1 - 0 \cdot 05x = (1 + 0 \cdot 005\theta)^{-2}$$

$$\therefore \quad x = 20 - 20(1 + 0 \cdot 005\theta)^{-2} \tag{1.35}$$

Equation (1.35) gives the salt concentration in the tank at any time, and equation (1.33) gives the volume of brine in the tank. From (1.33) it is seen that $V = 4$ when $\theta = 200$, as in Section 1.7.1, and therefore from (1.35)

$$X = 20 - 20(1 + 1)^{-2}$$

$$\therefore \quad X = 15 \text{ kg/m}^3 \tag{1.36}$$

Comparison of this result with the previous result (1.30) shows that the first simple model has a significant error of 6·7% due to the additional assumption of a linear variation of x with θ; an assumption which is shown to be incorrect by equation (1.35).

The two treatments of this example illustrate the important point that the solution of a problem depends upon the choice of model. A simple model needs many assumptions and yields an approximate answer quickly, whereas a more complicated model needs fewer assumptions and yields a more accurate answer by more advanced mathematical techniques. It is evident that the most accurate result will be obtained by making the minimum number of assumptions consistent with obtaining a tractable mathematical problem. In this way the mathematical techniques which are available to an investigator control the applicability of any theoretical prediction made by him. It is therefore desirable that the widest selection of mathematical techniques should be at his disposal.

Alternatively, the availability of digital computers and a knowledge of numerical techniques enable more complex models to be used to obtain numerical solutions to specific problems. This approach however means that it is more difficult to make predictions about the general behaviour of a system.

1.8. Radial Heat Transfer through a Cylindrical Conductor

In the last two problems the properties of the system were functions of time. There are many problems of a similar nature, where the properties of the system are functions of position instead of time, and the following example illustrates the application of the method to this type of problem.

Two concentric cylindrical metallic shells are separated by a solid material. If the two metal surfaces are maintained at different constant temperatures, what is the steady state temperature distribution within the separating material?

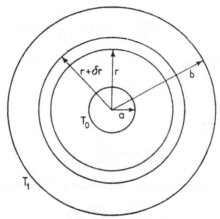

Fig. 1.9. Radial heat flow through cylindrical conductor

This is a steady state problem, but some of the properties of the system depend upon the position at which they are measured. In this case, the temperature (T) and the heat flow rate per unit area (Q) are both functions of the radius (r), and the heat balance must be related to a space interval between r and $r + \delta r$ as shown in Fig. 1.9. Following the same procedure as before, but considering variations of r instead of θ, Table 1.4 can be compiled

TABLE 1.4. *The Condition of the System on Either Side of a Space Increment*

Property of the System	r	$r + \delta r$
Temperature	T	$T + \dfrac{dT}{dr}\,\delta r$
Heat transfer area/unit length	$2\pi r$	$2\pi(r + \delta r)$
Radial heat flux density	Q	$Q + \dfrac{dQ}{dr}\,\delta r$
Total radial heat flow	$2\pi r Q$	$2\pi(r + \delta r)\left(Q + \dfrac{dQ}{dr}\,\delta r\right)$

Considering the element of the thickness δr,

$$\text{Heat input to inner surface} \quad = 2\pi r Q$$

$$\text{Heat output from outer surface} = 2\pi(r + \delta r)\left(Q + \frac{dQ}{dr}\,\delta r\right)$$

$$\text{Accumulation of heat} \quad = 0$$

Since input − output = accumulation,

$$2\pi r Q - 2\pi(r + \delta r)\left(Q + \frac{dQ}{dr}\,\delta r\right) = 0$$

$$\therefore \quad rQ - rQ - r\frac{dQ}{dr}\,\delta r - Q\,\delta r - \frac{dQ}{dr}(\delta r)^2 = 0$$

Cancelling out rQ, dividing by δr, and taking the limit $\delta r \to 0$,

$$r\frac{dQ}{dr} + Q = 0 \qquad\qquad (1.37)$$

But Q is related to T by

$$Q = -k\frac{dT}{dr} \qquad\qquad (1.38)$$

where k is the thermal conductivity. Equation (1.37) becomes

$$-kr\frac{d^2 T}{dr^2} - k\frac{dT}{dr} = 0$$

or

$$r\frac{d^2 T}{dr^2} + \frac{dT}{dr} = 0 \qquad\qquad (1.39)$$

since k is constant. Referring to Table 1.4, the change in heat flow rate across the element gives rise to two terms in this example, one due to a change in temperature gradient and the other due to a change in the area of conduction.

The problem can be solved by treating (1.39) as a homogeneous second order differential equation (see Section 2.4.3), or by solving equations (1.37) and (1.38) in succession. Equation (1.37) gives

$$\frac{dQ}{Q} = -\frac{dr}{r}$$

$$\therefore \quad \ln Q = -\ln r + \ln A$$

or

$$Q = A/R$$

where $\ln A$ is chosen for the constant of integration. Putting this value of Q into equation (1.38),

$$\frac{A}{r} = -k\frac{dT}{dr}$$

$$\therefore \quad \frac{dr}{r} = -\frac{k\,dT}{A}$$

$$\therefore \quad \ln r = -(kT/A) + B \tag{1.40}$$

Equation (1.40) involves two constants of integration, A and B, and these can be evaluated from the fixed boundary temperatures.

At
$$r = a, \quad T = T_0 \Big\}$$

and at
$$r = b, \quad T = T_1 \Big\} \tag{1.41}$$

$$\therefore \quad \ln a = -(kT_0/A) + B$$

and
$$\ln b = -(kT_1/A) + B$$

$$\ln a - \ln b = k(T_1 - T_0)/A$$

$$\therefore \quad A = k(T_1 - T_0)/(\ln a - \ln b)$$

$$\therefore \quad B = (T_1 \ln a - T_0 \ln b)/(T_1 - T_0)$$

Putting these values of A and B into equation (1.40),

$$(T_1 - T_0)\ln r = (\ln b - \ln a)T + T_1 \ln a - T_0 \ln b$$

$$= (T - T_0)(\ln b - \ln a) + (T_1 - T_0)\ln a$$

$$\therefore \quad \frac{T - T_0}{T_1 - T_0} = \frac{\ln r - \ln a}{\ln b - \ln a} \tag{1.42}$$

1.9. Heating a Closed Kettle

The problems considered so far have only used material or heat balances in the derivation of the equations. In the following example, the equation is derived from a rate equation.

A closed kettle of total surface area A m² is heated through this surface by condensing steam at temperature T_s °C. The kettle is charged with M kg of liquid of specific heat C J/kg °C at a temperature of T_0 °C. If the process is controlled by a heat transfer coefficient h W/m² °C, how does the temperature of the liquid vary with time?

Consider a time interval $\delta\theta$.

$$\text{Heat input (J)} \quad = hA(T_s - T)\,\delta\theta$$

$$\text{Heat output (J)} \quad = 0$$

$$\text{Accumulation (J)} = MC\frac{\mathrm{d}T}{\mathrm{d}\theta}\,\delta\theta$$

Since input − output = accumulation,

$$hA(T_s - T)\,\delta\theta = MC\frac{\mathrm{d}T}{\mathrm{d}\theta}\,\delta\theta \tag{1.43}$$

Rearranging,

$$\frac{\mathrm{d}T}{T_s - T} = \frac{hA\,\mathrm{d}\theta}{MC}$$

and integrating,

$$-\ln(T_s - T) = (hA/MC)\,\theta + B \tag{1.44}$$

B can be evaluated by using the known fact that

$$\text{when} \quad \theta = 0, \quad T = T_0 \tag{1.45}$$

$$\therefore \quad -\ln(T_s - T_0) = B$$

$$\therefore \quad \ln\frac{T_s - T}{T_s - T_0} = -\frac{Ah}{MC}\,\theta \tag{1.46}$$

In this problem, equation (1.43) could have been written down immediately from the definition of the heat transfer coefficient, thus avoiding the necessity of taking a heat balance.

1.10. Dependent and Independent Variables, Parameters

Any solution of a differential equation is an algebraic equation involving symbols. These symbols fall into three classes, (a) independent variables, (b) dependent variables, and (c) parameters.

1.10.1. *Independent Variables*

These are quantities describing the system which can be varied by choice during a particular experiment independently of one another. Examples are time and coordinate variables.

1.10.2. *Dependent Variables*

These are properties of the system which change when the independent variables are altered in value. There is no direct control over a dependent variable during an experiment. The relationship between independent and dependent variables is one of cause and effect; the independent variable measures the cause and the dependent variable measures the effect of a particular action. Examples of dependent variables are temperature, concentration, and efficiency.

1.10.3. *Parameters*

This is by far the largest group, consisting mainly of the characteristic properties of the apparatus and the physical properties of the materials. The group contains all properties which remain constant during an individual experiment, but since different constant values can be taken by a property during different experiments, the correct term for them is "parameters". Examples are overall dimensions of the apparatus, flow rates, heat transfer coefficients, thermal conductivity, specific heat, density, and initial or boundary values of the dependent variables.

Referring to Section 1.6, the solution to the problem of starting a single stage liquid–liquid extraction system is given by equations (1.1) and (1.29).

$$y = mx \tag{1.1}$$

$$x = \frac{Rc}{R+mS}\left[1 - \exp\left(-\frac{R+mS}{V_1+mV_2}\theta\right)\right] \tag{1.29}$$

and by eliminating x,

$$y = \frac{mRc}{R+mS}\left[1 - \exp\left(-\frac{R+mS}{V_1+mV_2}\theta\right)\right] \tag{1.47}$$

The symbols can be classified as follows:

Independent variable θ

Dependent variables x, y

Parameters m, R, S, c, V_1, V_2

The parameters are all fixed in value during an experiment, but can be varied between experiments for comparison purposes. θ is the only variable whose value can be altered during an experiment, x and y vary during an experiment and between experiments; their values depend upon the choice of values for both the parameters and the independent variable.

The main use of this classification is in the interpretation of the differentiation process. A dependent variable is usually differentiated with respect to an independent variable, and occasionally with respect to a parameter. By differentiating either (1.29) or (1.47) with respect to θ, an expression giving the rate of change of concentration with time can be obtained, but differentiating (1.1) with respect to the dependent variable x, gives $dy/dx = m$, an almost useless piece of information which does not throw any light on the behaviour of the apparatus.

So far, the problems considered in Sections 1.2, 1.3, 1.4, and 1.5 have been equations between parameters since no independent variables were involved, and hence no dependent variables could be present. Each experiment found a set of related constant values, the only variation could be between experiments when some of the parameters may be altered in value.

The problems of Sections 1.6, 1.7, and 1.9 involved time as an independent variable, and in Section 1.8, a radial coordinate was the single independent variable. These problems all gave rise to ordinary differential equations. When more than one independent variable is needed to describe a system, the usual result is a partial differential equation, and this type of problem will be dealt with in Chapter 8.

1.11. BOUNDARY CONDITIONS

An ordinary differential equation usually arises in any problem which involves a single independent variable. The general solution of this differential equation will contain arbitrary constants of integration, the number of constants being equal to the order of the differential equation (see Chapter 2). To complete the solution of a particular problem, these arbitrary constants have to be evaluated.

In formulating the equation, the conservation law (1.25) is applied to an infinitesimal increment of the independent variable, and this yields a differential equation. There is usually a restriction on the range of values which the independent variable can take and this range describes the scope of the problem. Special conditions are placed on the dependent variable at these end points of the range of the independent variable. These are naturally called "boundary conditions", and are used to evaluate the arbitrary constants in the general solution of the differential equation.

These conditions have already been used in the examples considered earlier in this chapter. In Section 1.8, a second order differential equation (1.39) was derived, and its solution (1.40) contains two arbitrary constants A and B. The independent variable in this problem is r, and the boundaries are given by $r = a$ and $r = b$. The boundary conditions are given by equations (1.41) in the form of restrictions on the dependent variable (T) at the boundaries defined by values of the independent variable (r).

In the other examples in Section 1.6, 1.7, and 1.9, θ was the independent variable, all of the equations were first order, and each solution only contained one arbitrary constant. When time is the independent variable, the boundary condition is frequently called the "initial condition". This is because it specifies the starting state of the apparatus.

The three most frequently used boundary conditions in heat transfer are:

(1)　Boundary at a fixed temperature, $T = T_0$.
(2)　Constant heat flow rate through the boundary, $dT/dx = A$.

(2a) Boundary thermally insulated, $dT/dx = 0$.

(3) Boundary cools to the surroundings through a film resistance described by a heat transfer coefficient, $k\,dT/dx = h(T - T_0)$. k is the thermal conductivity, h is the heat transfer coefficient, and T_0 is the temperature of the surroundings.

Boundary conditions in problems involving partial differential equations will be described when they arise in Chapter 8.

1.12. Sign Conventions

Most sign difficulties are a result of trying to think too deeply about them, thus causing a state of complete confusion. When setting up the model the temptation to anticipate the solution of the problem is very strong but it must be resisted. The formulation of equations can be reduced to a set of rules which can be systematically applied and although it is difficult to frame such a set of rules to cover the wide variety of problems which can arise, an attempt has been made in the next section. The slavish observance of rules is not to be generally recommended, but it is best in the first instance to do just this because it completely eliminates any possibility of anticipating properties of the solution, and ensures that the signs are correct. The resulting equation can frequently be subjected to a physical sign check after it has been established.

All terms must be treated as positive during the formulation, and negative signs will only occur due to two causes.

(a) The first cause is the negative sign in the general conservation law (1.25) viz.

$$input - output = accumulation$$

(b) The second cause is by the definition of the rate equations governing heat transfer, mass transfer, and fluid flow. This will be illustrated below for the case of heat transfer.

When an independent variable is defined for a system, an origin and positive direction have to be included in the definition. This has not been a serious problem so far, since when time is the independent variable, its direction is the natural one of successive events and its origin can be taken as the instant when the initial conditions are established. The radial co-ordinate (r), the other independent variable which has occurred, is measured outwards from the axis of the body of revolution, the axis being the origin. In other cases it is always necessary to specify the origin of the independent variable and also its positive direction.

Turning to heat flow, the quantity of heat (Q) conducted past a point through unit cross-sectional area in unit time is counted as positive if it flows in the same direction as the positive direction of the independent variable (coordinate), and negative if it flows in the negative direction of the coordinate. According to Fourier's Law, the rate of conduction of heat is proportional to the temperature gradient, and the direction of the flow is down the temperature gradient as shown in Fig. 1.10 where the two possible cases are shown. On the left-hand side of the diagram, the temperature gradient is positive and as a result, the heat flows from right to left and is negative. The reverse is true in the right-hand side of the diagram where the

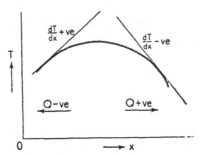

FIG. 1.10. Sign convention in heat transfer

temperature gradient is negative and the heat flow positive. In both cases, the equation can be written in the consistent form

$$Q = -k \frac{dT}{dx} \tag{1.48}$$

where k is the conductivity. This equation has already been used (1.38), but in that application the negative sign was of no importance.

For mass transfer, material diffuses down a concentration gradient, and for fluids, the flow is down a pressure gradient. By analogy with the Fourier Law, these rate equations also contain a negative sign in the definition.

The treatment of chemical reaction terms should present no difficulty if the following working principle is adopted. A single chemical reaction serves to remove reactants and supply products. The products are thus put into the system and should be included with the input terms whilst the reactants, being removed from the system, should be included with the output terms. In a more complicated reacting system each reaction should be considered separately with each reversible reaction divided into two individual reactions with the products of one becoming the reactants of the other.

In conclusion, if the rules which have been illustrated in this chapter are followed, and the solution is not anticipated, the signs will take care of themselves. This can be illustrated by repeating the example of Section 1.7 with a different boundary condition. Instead of the tank containing 2 m³ of water initially, let it contain 2 m³ of brine of concentration 25 kg/m³. The only physical difference that this makes is that the brine concentration will now decrease with time instead of increase. Mathematically, only equation (1.34) need be altered. The list of properties in Table 1.3 will be unaltered, and the change in salt concentration with time is still adequately described by $dx/d\theta$ without artificially changing its sign. In this new problem, $dx/d\theta$

is now a negative quantity since x decreases with time and it would be a mistake to say that the output salt concentration at $\theta + \delta\theta$ is $x - (dx/d\theta)\,\delta\theta$. A further reason why this is a mistake can be seen from Taylor's theorem which is the origin of this term; the sign is governed by the sign of $\delta\theta$ which is positive. Table 1.3 is therefore still true for this modified problem, and the solution is identical until equation (1.34) is reached, and this now becomes:

$$\text{when} \quad \theta = 0, \quad x = 25 \tag{1.49}$$

$$\therefore \quad -\tfrac{1}{2}\ln(-10) = \ln 200 + B$$

Here, the logarithm of a negative number is involved due to integrating $(40 - 2x)^{-1}$ in an unconventional manner. This need not give rise to any difficulty if $\ln(-1)$ is treated as an imaginary quantity by analogy with $\sqrt{(-1)}$, and B is eliminated as before by subtraction.

$$-\tfrac{1}{2}\ln\frac{40 - 2x}{-10} = \ln\frac{200 + \theta}{200}$$

$$\therefore \quad \ln\frac{x - 20}{5} = -2\ln(1 + 0.005\theta)$$

$$\therefore \quad x = 20 + 5(1 + 0.005\theta)^{-2}$$

This is the correct solution to this modified problem, and it illustrates how signs can be allowed to take care of themselves, and how the different boundary conditions select appropriate particular solutions from the general solution of the differential equation.

1.13. Summary of the Method of Formulation

(1) Pick a mathematical model of the process by making assumptions about its ideal behaviour.

(2) Define dependent variables to measure the properties being investigated.

(3) Define one or more independent variables in terms of which the dependent variables can be expressed.

(4) Define the parameters of the system which particularize the general relationship between the dependent and independent variables.

(5) Establish the state of the system at a typical point defined in terms of the independent variables.

(6) Take an infinitesimal increment in each of the independent variables, and establish the state of the system at neighbouring points in terms of the state at the typical point by using Taylor's theorem.

(7) Apply the relevant conservation laws and rate equations to the element of the system defined by (6).

(8) Cancel out terms where possible, including the incremental intervals.

(9) Take the limit as the size of the increments tends to zero.

(10) Establish the state of the system at all boundaries defined by special values of the independent variables.

These rules include the possibility that there may be more than one independent variable, resulting in the formulation of a partial differential equation. The treatment of such problems will be given in Chapter 8.

Although only the first two terms of Taylor's theorem are usually needed in practice, the infinite series of terms is effectively included, since the later terms involve incremental quantities raised to high powers and these disappear when the incremental quantities tend to zero.

The solution given by the analysis is related to the model of the system being investigated, and not to the system itself. The assumptions made in order to derive an equation for the model may or may not be valid, and if the result does not agree with the experiment, then the model is probably wrong and not the mathematics. As shown in Section 1.7, a more complicated model gives a more accurate result, but frequently, if the model is too detailed, the equation derived defies rigorous mathematical analysis. This difficulty can be overcome by using a numerical method to find an approximate solution to the difficult equation as shown in Chapter 11, or by simplifying the model with a further assumption. It must be realized, however, that as the number of assumptions increases, the likelihood of the result agreeing with practice decreases.

Tables 1.2, 1.3, and 1.4 have been compiled in this chapter to facilitate the description of the method of formulating equations. These tables should be compiled mentally and throughout the rest of the book they will be omitted. If any reader finds difficulty in formulating the equations in the rest of the book, it is suggested that similar tables should be constructed.

Chapter 2

ORDINARY DIFFERENTIAL EQUATIONS

2.1. Introduction

AN equation relating a dependent variable to one or more independent variables by means of its differential coefficients with respect to the independent variables is called a differential equation. If there is only one independent variable the equation is said to be an "ordinary differential equation". If there are two or more independent variables and the equation contains differential coefficients with respect to each of these, the equation is said to be a "partial differential equation". Thus

$$\frac{d^3y}{dx^3} - \frac{d^2y}{dx^2} + 5\left(\frac{dy}{dx}\right)^2 + 2x^3\frac{dy}{dx} + 4y = 4e^x\cos x \qquad (2.1)$$

is an ordinary differential equation; whereas

$$\rho C_p\frac{\partial T}{\partial \theta} = k\left(\frac{\partial^2 T}{\partial x^2} + \frac{\partial^2 T}{\partial y^2} + \frac{\partial^2 T}{\partial z^2}\right) \qquad (2.2)$$

is a partial differential equation. Equation (2.2) contains the partial differential coefficients of T with respect to each of the independent variables θ, x, y, and z.

Partial differential equations will be dealt with in Chapter 8, whilst in this chapter attention will be confined to ordinary differential equations.

2.2. Order and Degree

The order of a differential equation is equal to the order of the highest differential coefficient that it contains. Thus in equation (2.1) the order of the highest differential coefficient is three and therefore equation (2.1) is a third order differential equation. The degree of a differential equation is the highest power of the highest order differential coefficient that the equation contains after it has been rationalized. Equation (2.1) is of the first degree, even though it contains the first order differential coefficient raised to the second power and the cube of the independent variable. However as the first order differential coefficient is squared, equation (2.1) is referred to as a third order first degree non-linear ordinary differential equation. It is non-linear because the equation contains the term $(dy/dx)^2$ and generally differential equations are said to be non-linear if any products exist between the dependent variable and its derivatives, or between the derivatives

25

themselves. Thus the following equations are non-linear.

$$\frac{dy}{dx} + y^2 = \sin x \tag{2.3}$$

$$\left[\frac{d^2 y}{dx^2}\right]^2 + \frac{dy}{dx} + y = 0 \tag{2.4}$$

Equation (2.3) is non-linear because the dependent variable y appears as y^2, whereas equation (2.4) is non-linear because it is of the second degree.

The standard methods of solving the simplest types of ordinary differential equations that occur very frequently in chemical engineering will be considered in the present chapter, whereas more difficult types of ordinary differential equations will be considered later.

2.3. First Order Differential Equations

First order equations contain no derivatives higher than the first, but relate the first order differential coefficient to the two variables. Unfortunately there is no general method of solution of these equations because of their different degrees of complexity, but it is possible to classify the different types of first order equations so that each can be solved by a particular procedure. These are
(a) Exact equations.
(b) Equations in which the variables can be separated.
(c) Homogeneous equations.
(d) Equations solvable by an integrating factor.
Of these (d) is the most important and consequently (a), (b) and (c) will be treated briefly.

2.3.1. *Exact Equations*

Occasionally, a differential equation can be expressed in the form

$$df(x, y) = 0 \tag{2.5}$$

where $df(x, y)$ is an exact differential coefficient of some function. When this is so, the solution is obviously,

$$f(x, y) = C \tag{2.6}$$

where C is an arbitrary constant which must be evaluated from the boundary conditions of the problem. Frequently, it is possible to show that $df(x, y)$ is an exact differential coefficient by inspection. Thus in the equation,

$$x\frac{dy}{dx} + y = 0 \tag{2.7}$$

the left-hand side is immediately seen to be the exact differential coefficient of xy with respect to x, and therefore the solution is

$$xy = C$$

However, consider the equation,

$$x^3 - y\sin x + (\cos x + 2y)\frac{dy}{dx} = 0 \qquad (2.8)$$

Only an experienced mathematician would recognize equation (2.8) to be an exact equation, but the reader familiar with the elementary concepts of partial differentiation could easily verify that equation (2.8) is an exact equation as follows. If $\phi = \phi(x,y)$, equation (8.16) shows that

$$d\phi = \frac{\partial \phi}{\partial x} dx + \frac{\partial \phi}{\partial y} dy \qquad (2.9)$$

Comparing this with equation (2.8),

$$\frac{\partial \phi}{\partial x} = x^3 - y\sin x$$

and

$$\frac{\partial \phi}{\partial y} = \cos x + 2y$$

The most general function which can be differentiated partially with respect to x to give the first of these expressions is

$$\phi = \tfrac{1}{4}x^4 + y\cos x + f(y) \qquad (2.10)$$

Differentiating equation (2.10) partially with respect to y gives

$$\frac{\partial \phi}{\partial y} = \cos x + f'(y) = \cos x + 2y$$

$$\therefore \quad f(y) = y^2 + \text{constant}$$

The solution of equation (2.8) is therefore

$$x^4 + 4y\cos x + 4y^2 = C \qquad (2.11)$$

2.3.2. *Equations in which the Variables can be Separated*

In the most simple first order differential equations, the independent variable and its differential can be separated from the dependent variable and its differential by the equality sign, using nothing more than the normal processes of elementary algebra. When this is possible, solution of the equation follows by straightforward integration of the rearranged equation. The example in Section 1.6 was of this type and as the procedure of solving is elementary, no further discussion of these equations will be made. It will suffice to state that they frequently appear in engineering problems.

2.3.3. *Homogeneous Equations*

A differential equation of the type,

$$\frac{dy}{dx} = f\left(\frac{y}{x}\right) \tag{2.12}$$

is termed a homogeneous differential equation of the first order. Such an equation can be solved by making the substitution $y = vx$ and thereafter integrating the transformed equation. Thus let

$$y = vx \tag{2.13}$$

and therefore

$$\frac{dy}{dx} = v + x\frac{dv}{dx} \tag{2.14}$$

Substitution in equation (2.12), rearranging and integrating, gives

$$\ln x = \int \frac{dv}{f(v) - v} + C \tag{2.15}$$

Equation (2.15) is the solution of the general homogeneous equation (2.12). It should be pointed out, however, that the integral in equation (2.15) is not always expressible in terms of elementary functions. Thus the final solution after substituting for v may not be very useful in practice even though it is correct.

Homogeneous differential equations of first order arise in batch chemical reactor analyses when the materials undergo chemical reaction by two or more simultaneous second order reaction paths. A typical example is the halogenation of a hydrocarbon and the method of treating such a problem is illustrated in the example below.

Example 1. Liquid benzene is to be chlorinated batchwise by sparging chlorine gas into a reaction kettle containing the benzene. If the reactor contains such an efficient agitator that all the chlorine which enters the reactor undergoes chemical reaction, and only the hydrogen chloride gas liberated escapes from the vessel, estimate how much chlorine must be added to give the maximum yield of monochlorbenzene. The reaction can be assumed to take place isothermally at 55 °C when the ratios of the specific reaction rate constants are

$$\frac{k_1}{k_2} = 8{\cdot}0 \quad \text{and} \quad \frac{k_2}{k_3} = 30{\cdot}0$$

as published by Macmullen (*Chem. Eng. Prog.* **44**, 183, 1948). k_1 refers to reaction I, k_2 to reaction II, and k_3 to reaction III below.

$$C_6H_6 + Cl_2 \longrightarrow C_6H_5Cl + HCl \qquad \text{I}$$

$$C_6H_5Cl + Cl_2 \longrightarrow C_6H_4Cl_2 + HCl \qquad \text{II}$$

$$C_6H_4Cl_2 + Cl_2 \longrightarrow C_6H_3Cl_3 + HCl \qquad \text{III}$$

Solution

Take a basis of 1·0 mole of benzene fed to the reactor and introduce the following variables to represent the state of the system at time θ,

$$p = \text{moles of chlorine present}$$

$$q = \text{moles of benzene present}$$

$$r = \text{moles of monochlorbenzene present}$$

$$s = \text{moles of dichlorbenzene present}$$

$$t = \text{moles of trichlorbenzene present}$$

Then
$$q+r+s+t = 1 \qquad\qquad\qquad\qquad \text{IV}$$

and the total amount of chlorine consumed is given by

$$y = r+2s+3t \qquad\qquad\qquad\qquad \text{V}$$

Now assume that the reaction volume V is constant. Therefore, the rate of the first reaction of benzene is given by $R_1 = k_1 pq$ which is a rate of removal of benzene and a rate of introduction of monochlorbenzene to the system. The accumulation of benzene is $V\,dq/d\theta$ hence the material balances for the four aromatics can be determined thus

$$0 - k_1 pq = V\frac{dq}{d\theta} \qquad\qquad\qquad\qquad \text{VI}$$

$$k_1 pq - k_2 pr = V\frac{dr}{d\theta} \qquad\qquad\qquad\qquad \text{VII}$$

$$k_2 pr - k_3 ps = V\frac{ds}{d\theta} \qquad\qquad\qquad\qquad \text{VIII}$$

$$k_3 ps = V\frac{dt}{d\theta} \qquad\qquad\qquad\qquad \text{IX}$$

The time θ can be eliminated as a variable by dividing equations VII to IX by equation VI. The first ratio gives

$$\frac{dr}{dq} = \frac{k_2 r}{k_1 q} - 1 = \frac{r-8q}{8q} \qquad\qquad\qquad\qquad \text{X}$$

This is now a first order homogeneous differential equation which can be solved by the substitution technique.

Let
$$r = vq \qquad\qquad\qquad\qquad \text{XI}$$

and therefore
$$\frac{dr}{dq} = v + q\frac{dv}{dq} \qquad\qquad\qquad\qquad \text{XII}$$

Substitution of XI and XII into X and integration of the result gives

$$\ln q = \ln K - 8/7 \ln (7v + 8) \qquad\qquad \text{XIII}$$

where K is the constant of integration. Removing the logarithms,

$$q = K\left(\frac{7r}{q} + 8\right)^{-8/7} \qquad\qquad \text{XIV}$$

But when $\theta = 0$, $q = 1$ and $r = 0$;

$$\therefore \quad K = 8^{8/7}$$

and after rearrangement,

$$r = 8(q^{1/8} - q)/7 \qquad\qquad \text{XV}$$

Similarly, from equations VI and VIII it can be shown that

$$\frac{\mathrm{d}s}{\mathrm{d}q} = \frac{s}{240q} - \frac{r}{8q} \qquad\qquad \text{XVI}$$

When r has been eliminated by using equation XV, equation XVI becomes a first order differential equation which can be solved by the integrating factor method given in Section 2.3.4. The result is

$$s = \frac{240}{7 \times 29 \times 239}(29q - 239q^{1/8} + 210q^{1/240}) \qquad\qquad \text{XVII}$$

For any value of q it is possible to determine the corresponding values of r and s from equations XV and XVII, and then the value of t from equation IV. Similarly by means of equation V the total quantity of chlorine gas consumed is determined. These last stages are arithmetical and are consequently not reported in detail; but the results of the calculations are summarized in graphical form in Fig. 2.1 in which the mole fraction of each halogenated hydrocarbon is plotted against the total amount of chlorine consumed. The amount of chlorine required for the maximum production of monochlorbenzene is from Fig. 2.1 equal to 1·0 mole of chlorine per mole of benzene charged.

2.3.4. *Equations Solved by Integrating Factors*

There are many equations which can be solved by a variety of integrating factors, but the most useful integrating factor is the one used to integrate the first order linear differential equation,

$$\frac{\mathrm{d}y}{\mathrm{d}x} + Py = Q \qquad\qquad (2.16)$$

where P and Q are functions of x only.

The basis of the solution of equations of this type is that there exists a factor by which equation (2.16) can be multiplied so that the left-hand side becomes a complete differential coefficient. The factor is called "the integrating factor". Assume that the integrating factor R is a function of x only, and multiply equation (2.16) by R, thus

$$R\frac{dy}{dx} + RPy = RQ \tag{2.17}$$

in which the left-hand side of equation (2.17) is the complete differential coefficient of some product, say Ry. Then

$$\frac{d}{dx}(Ry) = R\frac{dy}{dx} + y\frac{dR}{dx} \tag{2.18}$$

Equation (2.18) represents the left-hand side of equation (2.17) if

$$\frac{dR}{dx} = PR \tag{2.19}$$

The variables R and x can be separated in this equation, and the solution is

$$R = \exp\left(\int P \, dx\right) \tag{2.20}$$

That is, the integrating factor for the solution of the first order linear differential equation (2.16) is $\exp(\int P \, dx)$.

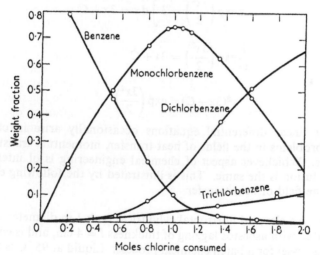

FIG. 2.1. Rate of reaction of benzene

Occasionally, a first order, first degree differential equation has to be solved in which P and Q are functions of both the dependent and the independent variable. When this arises it is often possible to linearize the equation by changing the dependent variable. Thus, consider the following example

Example 2. Solve

$$xy - \frac{dy}{dx} = y^4 \exp\left(\frac{-3x^2}{2}\right)$$

or
$$\frac{x}{y^3} - \frac{1}{y^4}\frac{dy}{dx} = \exp\left(\frac{-3x^2}{2}\right) \qquad\qquad \text{I}$$

Let $z = 1/y^3$ thus changing the dependent variable. Then

$$\frac{dz}{dy} = -\frac{3}{y^4} \quad \text{and} \quad \frac{dz}{dx} = -\frac{3}{y^4}\frac{dy}{dx} \qquad\qquad \text{II}$$

Substitution of II into I gives

$$3xz + \frac{dz}{dx} = 3\exp\left(\frac{-3x^2}{2}\right) \qquad\qquad \text{III}$$

Equation III is a linear first order differential equation which has the integrating factor $\exp\left(\int 3x\,dx\right)$ according to equation (2.20). Therefore

$$\exp\left(\frac{3x^2}{2}\right)\frac{dz}{dx} + 3xz\exp\left(\frac{3x^2}{2}\right) = 3 \qquad\qquad \text{IV}$$

Integrating,

$$z\exp\left(\frac{3x^2}{2}\right) = 3\int dx = 3x + C \qquad\qquad \text{V}$$

$$\therefore \quad \frac{1}{y^3}\exp\left(\frac{3x^2}{2}\right) = 3x + C$$

$$\therefore \quad y^3(3x + C) = \exp\left(\frac{3x^2}{2}\right) \qquad\qquad \text{VI}$$

First order linear differential equations occasionally arise in chemical engineering problems in the fields of heat transfer, momentum transfer, and mass transfer. Whichever aspect of chemical engineering is of interest the method of solution is the same. This is illustrated by the following example taken from the field of heat transfer.

Example 3. An elevated horizontal cylindrical tank 1 m diameter and 2 m long is insulated with asbestos lagging of thickness $l = 4$ cm, and is employed as a maturing vessel for a batch chemical process. Liquid at 95 °C is charged into the tank and allowed to mature over 5 days. If the data below applies, calculate the final temperature of the liquid and give a plot of the liquid temperature as a function of time.

Liquid film coefficient of heat transfer (h_1)   = 150 W/m² °C
Thermal conductivity of asbestos (k)   = 0·2 W/m °C
Surface coefficient of heat transfer by convection and
  radiation (h_2)   = 10 W/m² °C

Density of liquid (ρ) $= 10^3 \text{ kg/m}^3$

Heat capacity of liquid (s) $= 2{,}500 \text{ J/kg} \,^\circ\text{C}$

Atmospheric temperature (t) which can be assumed to vary according to the relation $t = 10 + 10\cos(\pi\theta/12)$, where θ is the time in hours

Atmospheric temperature at time of charging $= 20\,^\circ\text{C}$

Heat loss through supports is negligible. The thermal capacity of the lagging can be ignored

Solution

Area of tank (A) $= (\pi \times 1 \times 2) + 2(\tfrac{1}{4}\pi \times 1^2) = 2{\cdot}5\pi \text{ m}^2$

Rate of heat loss by liquid $= h_1 A(T - T_w)$

Rate of heat loss through lagging $= \dfrac{kA}{l}(T_w - T_s)$

where T represents the bulk liquid temperature,

T_w represents the inside wall temperature of the tank,

T_s represents the outside surface temperature of the lagging

Rate of heat loss from the exposed surface of the lagging $= h_2 A(T_s - t)$

Now,

$$\begin{bmatrix} \text{Rate of heat} \\ \text{loss from} \\ \text{liquid} \end{bmatrix} = \begin{bmatrix} \text{Rate of heat} \\ \text{transfer through} \\ \text{lagging} \end{bmatrix} = \begin{bmatrix} \text{Rate of heat} \\ \text{loss from} \\ \text{surface} \end{bmatrix} \qquad \text{I}$$

$$\therefore \quad h_1 A(T - T_w) = \frac{kA}{l}(T_w - T_s) = h_2 A(T_s - t) \qquad \text{II}$$

Rearranging the first part of this equation to give

$$T_w\left(h_1 + \frac{k}{l}\right) = h_1 T + \frac{k}{l} T_s \qquad \text{III}$$

and substituting into the last part of equation II gives

$$T_s = t + \left(\frac{k h_1}{h_1 h_2 l + h_1 k + h_2 k}\right)(T - t) \qquad \text{IV}$$

or $\qquad\qquad\qquad T_s = 0{\cdot}326 T + 0{\cdot}674 t \qquad\qquad\qquad \text{V}$

using the numerical values given earlier.

Considering the thermal equilibrium of the liquid,

Input rate $= 0$

Output rate $= h_2 A(T_s - t)$

Accumulation rate $= V\rho s \dfrac{dT}{d\theta}$

$$\therefore \quad -h_2 A(T_s - t) = V\rho s \frac{dT}{d\theta} \qquad \text{VI}$$

Using the numerical values given above, remembering a factor of 3600 because θ is measured in hours, and substituting for T_a from equation V,

$$\frac{dT}{d\theta} = -0.072(0.326T + 0.674t - t) \qquad\qquad \text{VII}$$

or in standard form,

$$\frac{dT}{d\theta} + 0.0235T = 0.0235t = 0.235 + 0.235\cos(\pi\theta/12) \qquad \text{VIII}$$

Equation VIII is a first order linear differential equation which can be solved by means of the integrating factor $e^{0.0235\theta}$. The solution is

$$Te^{0.0235\theta} = 0.235 \int e^{0.0235\theta}\,d\theta + 0.235 \int e^{0.0235\theta}\cos(\pi\theta/12)\,d\theta \qquad \text{IX}$$

The second term on the right-hand side can be integrated by parts to give

$$\frac{0.235\,e^{0.0235\theta}}{(0.0235)^2 + (0.262)^2}(0.0235\cos 0.262\theta + 0.262\sin 0.262\theta)$$

Hence, the complete solution of equation VIII is

$$T = 10 + 0.08\cos 0.262\theta + 0.89\sin 0.262\theta + Ke^{-0.0235\theta} \qquad \text{X}$$

with the boundary condition that at $\theta = 0$, $T = 95$.

$$\therefore \quad 95 = 10 + 0.08 + K$$

$$\therefore \quad K = 84.92$$

Therefore, the final solution is

$$T = 10 + 0.08\cos 0.262\theta + 0.89\sin 0.262\theta + 84.92\,e^{-0.0235\theta} \qquad \text{XI}$$

Inspection of equation XI shows that the second term can only affect T by $\pm 0.08\,^\circ\text{C}$ at most. Similarly, the third term will only contribute $\pm 0.89\,^\circ\text{C}$ at most. Both of these terms are negligible from a practical point of view, consequently it is justifiable to express equation XI as

$$T = 10 + 85\,e^{-0.0235\theta} \qquad\qquad \text{XII}$$

Figure 2.2 gives a plot of T vs θ in accordance with equation XII and after 120 h, $T = 15\,^\circ\text{C}$. At this temperature level, the term $0.89\sin 0.262\theta$ is not negligible but at $\theta = 120$, $\sin 0.262\theta = 0$.

2.4. SECOND ORDER DIFFERENTIAL EQUATIONS

As already stated, a second order differential equation is one in which the highest order differential coefficient appearing is the second. However, the equation may contain other terms involving first order differential coefficients

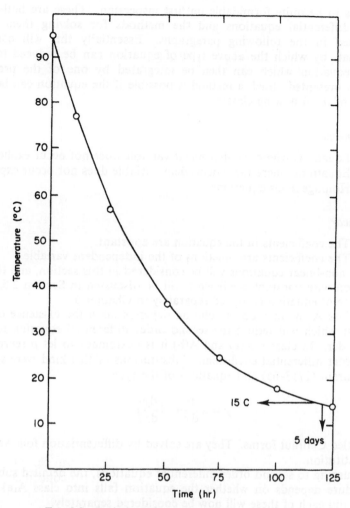

FIG. 2.2. Variation of tank temperature with time

in addition to functions of the independent and dependent variables. Furthermore, the equation may be linear or non-linear. Thus,

$$\frac{d^2 y}{dx^2} = f(x) \tag{2.21}$$

is a simple second order differential equation easily solved by two successive integrations. On the other hand, the equation,

$$y\frac{d^2 y}{dx^2} + 1 = \left(\frac{dy}{dx}\right)^2 \tag{2.22}$$

appears to be quite formidable on first inspection. These are both second order differential equations and the methods for solving them will be described in the following paragraphs. Essentially this will consist of proposals by which the above type of equation can be reduced to a first order equation which can then be integrated by one of the procedures already presented. Such a method is possible if the equation can be placed in one of the following classes.

A. Non-linear

(a) Equations where the dependent variable does not occur explicitly.
(b) Equations where the independent variable does not occur explicitly.
(c) Homogeneous equations.

B. Linear

(a) The coefficients in the equation are constant.
(b) The coefficients are functions of the independent variable.

The non-linear equations will be considered in this section, and the linear equation with constant coefficients will be discussed in Section 2.5, but the last type of equation is treated separately in Chapter 3.

In class A the method of solution depends upon the existence of a substituent which will reduce the second order differential equation to one of first order. In classes A(a) and A(b) it is customary to let p represent the first order differential coefficient. Substitutions of this kind were suggested by Clairaut (1713–65) and equations of the type,

$$y = x\frac{dy}{dx} + f\left(\frac{dy}{dx}\right) \tag{2.23}$$

are called Clairaut forms. They are solved by differentiation followed by the p substitution.

Returning to second order differential equations, the detailed substituting procedure depends on whether the equation falls into class A(a) or class A(b), and each of these will now be considered separately.

2.4.1. Equations where the Dependent Variable does not occur Explicitly

When the p substitution is made in this case, the second derivative of y is replaced by the first derivative of p thus eliminating y completely and producing a first order differential equation in p and x. The following example will clarify the technique.

Example 1. Solve

$$\frac{d^2y}{dx^2} + x\frac{dy}{dx} = ax \qquad\qquad\text{I}$$

Let

$$p = \frac{dy}{dx}$$

and therefore

$$\frac{dp}{dx} = \frac{d^2 y}{dx^2}$$

Substitution into equation I gives

$$\frac{dp}{dx} + xp = ax \qquad \qquad \text{II}$$

Equation II is a first order linear differential equation which can be solved by using the integrating factor $\exp(\tfrac{1}{2}x^2)$.

$$\therefore \quad p \exp(\tfrac{1}{2}x^2) = a \int x \exp(\tfrac{1}{2}x^2)\, dx$$

$$= a \exp(\tfrac{1}{2}x^2) + C$$

$$\therefore \quad p = a + C \exp(-\tfrac{1}{2}x^2) = \frac{dy}{dx}$$

$$\therefore \quad y = ax + C \int \exp(-\tfrac{1}{2}x^2)\, dx \qquad \qquad \text{III}$$

$$\therefore \quad y = ax + A \operatorname{erf}\left(\frac{x}{\sqrt{2}}\right) + B \qquad \qquad \text{IV}$$

where A and B are the two constants of integration. The integration of the last term in equation III involving the "error function" will be treated in Chapter 5. Here, it is sufficient to quote the result.

2.4.2. *Equations where the Independent Variable does not occur Explicitly*

In this case, the same substitution of p for the first derivative is made, but the substitution for the second derivative of y is obtained as follows.

Let

$$p = \frac{dy}{dx} \qquad \qquad (2.24)$$

$$\therefore \quad \frac{d^2 y}{dx^2} = \frac{dp}{dx} = \frac{dp}{dy}\frac{dy}{dx} = p\frac{dp}{dy} \qquad \qquad (2.25)$$

Hence, when the independent variable does not appear explicitly, make the substitutions as given by equations (2.24) and (2.25). The following example will make the procedure quite clear.

Example 2. Solve

$$y\frac{d^2 y}{dx^2} + 1 = \left(\frac{dy}{dx}\right)^2$$

Substitution of equations (2.24) and (2.25) gives

$$yp\frac{dp}{dy}+1 = p^2 \qquad\qquad\text{I}$$

Separating the variables,

$$\frac{dy}{y} = \frac{p\,dp}{p^2-1}$$

Integrating,

$$\ln y + \ln a = \tfrac{1}{2}\ln(p^2-1) \qquad\qquad\text{II}$$
$$\therefore\quad a^2 y^2 = p^2 - 1$$
$$\therefore\quad p = \frac{dy}{dx} = \sqrt{(a^2 y^2 + 1)} \qquad\qquad\text{III}$$
$$\therefore\quad x = \int\frac{dy}{\sqrt{(a^2 y^2 + 1)}}$$

This is a standard integral, and therefore the solution of equation III is

$$x = (1/a)\sinh^{-1}(ay) + b$$

or

$$ay = \sinh(ax+c)$$

where a and c are the two constants of integration.

2.4.3. Homogeneous Equations

In Section 2.3.3, the first order homogeneous differential equation was expressed in the form,

$$\frac{dy}{dx} = f\left(\frac{y}{x}\right) \qquad\qquad (2.12)$$

Regarding x and y as of the same dimensions, equation (2.12) is an equation between dimensionless groups. The corresponding dimensionless group containing the second differential coefficient is $(x\,d^2 y/dx^2)$. In general, the dimensionless group containing the nth coefficient is

$$x^{n-1}\frac{d^n y}{dx^n}$$

The second order homogeneous differential equation can be expressed in a form analogous to equation (2.12), viz.

$$x\frac{d^2 y}{dx^2} = f_1\left(\frac{y}{x},\frac{dy}{dx}\right) \qquad\qquad (2.26)$$

The substitution for a first order homogeneous equation is

$$y = vx \qquad\qquad (2.13)$$

If this substitution is made again,

$$\frac{dy}{dx} = v + x\frac{dv}{dx} \qquad (2.14)$$

as before, and

$$\frac{d^2y}{dx^2} = 2\frac{dv}{dx} + x\frac{d^2v}{dx^2} \qquad (2.27)$$

Substituting these expressions into equation (2.26) gives

$$2x\frac{dv}{dx} + x^2\frac{d^2v}{dx^2} = f_1\left(v, v + x\frac{dv}{dx}\right) \qquad (2.28)$$

In equation (2.28), a factor x^2 is associated with the second differential coefficient of v, a factor x with the first differential coefficient of v on both sides of the equation, and v occurs alone. Equation (2.28) can therefore be written in the alternative equivalent form,

$$x^2\frac{d^2v}{dx^2} = f_2\left(v, x\frac{dv}{dx}\right) \qquad (2.29)$$

It is to be noted that if equation (2.26) is multiplied by x, the left-hand side assumes the same form as the left-hand side of equation (2.29). However, the right-hand sides only take the same form if the right-hand side of equation (2.26) is a linear homogeneous function of its two arguments. A short cut from equation (2.26) to (2.29) is therefore possible if the given differential equation is both homogeneous and linear.

Equation (2.29) is solved by using the second substitution

$$x = e^t \quad \text{or} \quad t = \ln x \qquad (2.30)$$

$$\therefore \quad \frac{dv}{dx} = \frac{dv}{dt}\frac{dt}{dx} = \frac{1}{x}\frac{dv}{dt} \qquad (2.31)$$

i.e.

$$x\frac{dv}{dx} = \frac{dv}{dt} \qquad (2.32)$$

Differentiating equation (2.31) again with respect to x,

$$\frac{d^2v}{dx^2} = -\frac{1}{x^2}\frac{dv}{dt} + \frac{1}{x}\frac{d}{dx}\left(\frac{dv}{dt}\right)$$

$$= -\frac{1}{x^2}\frac{dv}{dt} + \frac{1}{x}\frac{dt}{dx}\frac{d}{dt}\left(\frac{dv}{dt}\right)$$

$$= -\frac{1}{x^2}\frac{dv}{dt} + \frac{1}{x^2}\frac{d^2v}{dt^2}$$

e.

$$x^2\frac{d^2v}{dx^2} = \frac{d^2v}{dt^2} - \frac{dv}{dt} \qquad (2.33)$$

Putting equations (2.32) and (2.33) into equation (2.29) gives

$$\frac{d^2 v}{dt^2} - \frac{dv}{dt} = f_2\left(v, \frac{dv}{dt}\right) \tag{2.34}$$

The original differential equation (2.26) has now been reduced to a form (2.34) where the independent variable (t) does not occur explicitly, and it can be solved by the method given in Section 2.4.2. The actual procedure is illustrated in the following example.

Example 3. Solve

$$2x^2 y \frac{d^2 y}{dx^2} + y^2 = x^2 \left(\frac{dy}{dx}\right)^2 \qquad \text{I}$$

This equation is homogeneous since it can be expressed in the form (2.26) after division by $2xy$. Use the first substitution $y = vx$ and its differentiated forms (2.14) and (2.27).

$$\therefore \quad 2vx^3\left(2\frac{dv}{dx} + x\frac{d^2 v}{dx^2}\right) + v^2 x^2 = x^2\left(v + x\frac{dv}{dx}\right)^2 \qquad \text{II}$$

Simplifying into the form of equation (2.29),

$$2vx^2 \frac{d^2 v}{dx^2} + 2vx\frac{dv}{dx} = x^2\left(\frac{dv}{dx}\right)^2 \qquad \text{III}$$

Now making the second substitution $x = e^t$ and simplifying,

$$2v\frac{d^2 v}{dt^2} = \left(\frac{dv}{dt}\right)^2 \qquad \text{IV}$$

Equation IV does not contain the independent variable explicitly, and it can be solved as suggested in Section 2.4.2. Put $p = dv/dt$, and obtain

$$2vp\frac{dp}{dv} = p^2 \qquad \text{V}$$

This has two solutions, viz.

$$2v\frac{dp}{dv} = p \quad \text{or} \quad p = 0$$

The second solution, called a "singular solution", yields the result,

$$y = Ax \qquad \text{VI}$$

which undoubtedly satisfies the original equation I. The first solution can be integrated twice to give

$$y = x(B \ln x + C)^2 \qquad \text{VII}$$

By choosing the arbitrary constants in equation VII to be $B = 0$, $C = \sqrt{A}$, equation VI is seen to be included in equation VII as a special case. Equation VII is therefore the complete solution of equation I.

The value of the substitutions given in Sections 2.4 is rather limited, because in many engineering applications the transformed equation cannot be integrated. Such a difficulty was encountered by Marshall and Pigford[†] in their problem on the dynamics of a laminar jet. The authors of this text also met the same difficulty in attempting to predict the temperature profile along a graphite electrode by using the above substitution. The particular problem had to be solved by the method to be given in Chapter 11. However, the problem is presented here to illustrate the type of difficulty that can arise and to show that an approximate yet valuable solution can still be obtained by elementary procedures.

Example 4. A graphite electrode 15 cm in diameter passes through a furnace wall into a water cooler which takes the form of a water sleeve. The length of the electrode between the outside of the furnace wall and its entry into the cooling jacket is 30 cm; and as a safety precaution the electrode is insulated thermally and electrically in this section, so that the outside surface temperature of the insulation does not exceed 50 °C. If the lagging is of uniform thickness and the mean overall coefficient of heat transfer from the electrode to the surrounding atmosphere is taken to be 1·7 W/°C m² of surface of electrode; and the temperature of the electrode just outside the furnace is 1500 °C, estimate the duty of the water cooler if the temperature of the electrode at the entrance to the cooler is to be 150 °C. The following additional information is given.

Surrounding temperature $\qquad = 20\,°C$

Thermal conductivity of graphite $= k_T = k_0 - \alpha T$ W/m °C

where $k_0 = 152\cdot6$ and $\alpha = 0\cdot056$.[‡] (These numerical values are calculated as an example in Chapter 10.) The temperature of the electrode may be assumed uniform at any cross-section.

Solution

The sectional area of the electrode $(A) = 0\cdot0177$ m².

Referring to Fig. 2.3 and using the symbols stated there; a heat balance over the length of electrode δx at a distance x from the furnace is

Input $\qquad = -k_T A \dfrac{dT}{dx}$ $\qquad\qquad$ I

Output $\qquad = -k_T A \dfrac{dT}{dx} + \dfrac{d}{dx}\left(-k_T A \dfrac{dT}{dx}\right)\delta x + \pi D U (T - T_0)\,\delta x$ $\qquad$ II

Accumulation $= 0$

$$\therefore \quad \frac{d}{dx}\left(k_T \frac{dT}{dx}\right)\delta x = \frac{\pi D U}{A}(T - T_0)\,\delta x \qquad\qquad \text{III}$$

† Marshall and Pigford, "Differential Equations in Chemical Engineering", p. 76. University of Delaware, Newark, Delaware (1947).
‡ *I.C.T.* **5**, 86 (1937). McGraw-Hill Book Co. Ltd., New York.

where $U =$ overall heat transfer coefficient from the electrode to the surroundings and $D =$ electrode diameter.

Substituting for k_T and simplifying by dividing throughout by $A\,\delta x$ and letting $\pi DU/A = \beta$, equation III becomes

$$(k_0 - \alpha T)\frac{d^2 T}{dx^2} - \alpha\left(\frac{dT}{dx}\right)^2 - \beta(T - T_0) = 0 \qquad \text{IV}$$

Equation IV is a second order non-linear differential equation which does not contain the independent variable explicitly. Thus let

$$p = \frac{dT}{dx} \quad \text{and} \quad \frac{d^2 T}{dx^2} = p\frac{dp}{dT}$$

Equation IV then becomes

$$(k_0 - \alpha T)\,p\,\frac{dp}{dT} - \alpha p^2 - \beta(T - T_0) = 0 \qquad \text{V}$$

Inspection of equation V shows that the variables p and T cannot be separated, and one might be tempted to introduce simplifying assumptions

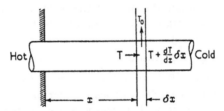

Fig. 2.3. Heat transfer through electrode

at this stage; or attempt to solve equation IV by some other method. However, the way in which p occurs in equation V suggests the substitution $p^2 = z$. Putting $y = (T - T_0)$ also gives

$$[(k_0 - \alpha T_0) - \alpha y]\frac{dz}{dy} - 2\alpha z - 2\beta y = 0 \qquad \text{VI}$$

which is a first order differential equation that can be solved by the integrating factor method. The integrating factor is

$$\exp\left(-\int\frac{2\alpha\,dy}{k_0 - \alpha T_0 - \alpha y}\right) = (k_0 - \alpha T_0 - \alpha y)^2$$

Substituting the integrating factor into equation VI and converting back to the original variables gives

$$x = \int\frac{(k_0 - \alpha T)\,dT}{\sqrt{[C + \beta(k_0 - \alpha T)(T - T_0)^2 - \frac{2}{3}\alpha\beta(T - T_0)^3]}} \qquad \text{VII}$$

Equation VII cannot be integrated analytically because of the cubic polynomial under the square root sign, or graphically because the arbitrary constant C is completely unknown. Furthermore, there is no way of evaluating C other than perhaps by trial and error which would not be very accurate. Hence in order to obtain an estimate of the heat load on the cooler it will be assumed that the heat lost to the atmosphere per m² of electrode surface (Q) is constant, and equal to 1250 W/m². Equation III then simplifies to

$$\frac{d}{dx}\left[(k_0 - \alpha T)\frac{dT}{dx}\right]\delta x = \frac{\pi D Q}{A}\delta x \qquad \text{VIII}$$

or

$$(k_0 - \alpha T)\frac{d^2 T}{dx^2} - \alpha\left(\frac{dT}{dx}\right)^2 = K \qquad \text{IX}$$

where $K = \pi D Q/A$.

The solution of equation IX is obtainable by the substitution

$$p = \frac{dT}{dx} \quad \text{and} \quad p\frac{dp}{dT} = \frac{d^2 T}{dx^2}$$

which gives

$$p^2 = \frac{B}{\alpha(k_0 - \alpha T)^2} - \frac{K}{\alpha} \qquad \text{X}$$

where B is the constant of integration. Reducing the power of p to unity, and choosing the negative root because the temperature falls as distance x increases,

$$\frac{dT}{dx} = -\sqrt{\left[\frac{B}{\alpha(k_0 - \alpha T)^2} - \frac{K}{\alpha}\right]} \qquad \text{XI}$$

Equation XI can be integrated with the aid of the substitution $z = (k_0 - \alpha T)^2$. The result is

$$T = \frac{k_0}{\alpha} - \frac{1}{\alpha}\sqrt{\left[\frac{B}{K} - \alpha K(x + C)^2\right]} \qquad \text{XII}$$

where C is the second arbitrary constant. The boundary conditions at the ends of the electrode give
at $x = 0$,

$$1500 = \frac{k_0}{\alpha} - \frac{1}{\alpha}\sqrt{\left(\frac{B}{K} - K\alpha C^2\right)} \qquad \text{XIII}$$

and at $x = 0.3$,

$$150 = \frac{k_0}{\alpha} - \frac{1}{\alpha}\sqrt{\left[\frac{B}{K} - K\alpha(0.3 + C)^2\right]} \qquad \text{XIV}$$

It is more convenient to determine how B and C occur in the required solution, rather than solve these equations directly. The duty of the water cooler is the rate at which heat arrives at $x = 0.3$, which is given by

$$H = -(k_0 - 150\alpha) A \frac{dT}{dx}\Big|_{x=0.3}$$

$$= \frac{-(k_0 - 150\alpha) AK(0.3 + C)}{\sqrt{[(B/K) - K\alpha(0.3 + C)^2]}}$$

by differentiating equation XII and putting $x = 0.3$. Substituting from equation XIV for the denominator of this expression,

$$H = -AK(0.3 + C) \qquad\qquad \text{XV}$$

Solving equations XIII and XIV for C gives

$$C = -0.15 - 4500(k_0 - 825\alpha)/K$$

Substituting into XV gives

$$H = A[4500(k_0 - 825\alpha) - 0.15K] \qquad\qquad \text{XVI}$$

$$= 8400 \text{ W}$$

and this is the duty of the water cooler.

2.5. LINEAR DIFFERENTIAL EQUATIONS

This type of equation is frequently encountered in most chemical engineering fields of study, ranging from heat, mass, and momentum transfer to applied chemical reaction kinetics. It is one of the most important types of equation to be found in applied science, and therefore considerable study has been devoted to finding solutions to the many different forms it may take.

The general linear differential equation of the nth order having constant coefficients may be written

$$P_0 \frac{d^n y}{dx^n} + P_1 \frac{d^{n-1} y}{dx^{n-1}} + \dots + P_{n-1} \frac{dy}{dx} + P_n y = \phi(x) \qquad (2.35)$$

where $\phi(x)$ is any function of x.

The treatment to be described below for the second order equation applies equally well to linear equations of any other order. However, in the case of a higher order equation, the mathematical manipulations may be cumbersome as the reader will soon appreciate, but since no new principles are involved, further discussions will be restricted to the second order case.

2.5.1. *Second Order Linear Differential Equations*

The general equation can be expressed in the form,

$$P \frac{d^2 y}{dx^2} + Q \frac{dy}{dx} + Ry = \phi(x) \qquad (2.36)$$

where P, Q, and R are the constant coefficients.

Let the dependent variable y be replaced by the sum of two new variables u and v, i.e.

$$y = u + v \qquad (2.37)$$

Substitution in (2.36) gives

$$\left[P\frac{d^2 u}{dx^2} + Q\frac{du}{dx} + Ru \right] + \left[P\frac{d^2 v}{dx^2} + Q\frac{dv}{dx} + Rv \right] = \phi(x) \qquad (2.38)$$

If v is a particular solution of the original differential equation, then

$$P\frac{d^2 v}{dx^2} + Q\frac{dv}{dx} + Rv = \phi(x) \qquad (2.39)$$

and by comparison with equation (2.38) it can be seen that

$$P\frac{d^2 u}{dx^2} + Q\frac{du}{dx} + Ru = 0 \qquad (2.40)$$

That is, u satisfies equation (2.40) which is the same as equation (2.36) with $\phi(x)$ replaced by zero. Therefore, since y is the general solution of equation (2.36) and is the sum of u and v, where v is a particular solution of the original differential equation, u must be the complement of the particular solution which must be added in order to get the general solution. That is, the general solution of the linear differential equation will be the sum of a "complementary function" and a "particular integral". The problem resolves itself into one of finding some specific solution that will satisfy the original equation—if necessary by trial and error, and finding the complete solution of the same equation assuming that the right-hand side is zero. The general solution will be the sum of these two solutions and will contain two arbitrary constants which can be evaluated from the boundary conditions of the specific problem being considered.

The above remarks on the general approach to the solution of the second order linear differential equation apply to any linear differential equation irrespective of its order. However, the number of arbitrary constants that will have to be evaluated will equal the order of the equation. That is, the general solution of an nth order differential equation will be the sum of a complementary function and a particular integral and will contain n arbitrary constants. The arbitrary constants will arise in the complementary function which will therefore have n terms. These points should be borne in mind when solving a linear differential equation.

2.5.2. The Complementary Function

Referring back to the second order differential equation (2.36), the problem is to find a solution of the equation when $\phi(x)$ is zero. That is a solution of the equation,

$$P\frac{d^2 y}{dx^2} + Q\frac{dy}{dx} + Ry = 0 \qquad (2.41)$$

Initially let it be assumed that the solution is

$$y = A_m e^{mx} \tag{2.42}$$

Then

$$\frac{dy}{dx} = A_m m e^{mx}$$

and

$$\frac{d^2 y}{dx^2} = A_m m^2 e^{mx}$$

Substitution of the above expressions for the first and second differential coefficients into equation (2.41) and simplifying gives

$$A_m e^{mx}(Pm^2 + Qm + R) = 0 \tag{2.43}$$

Equation (2.43) will be satisfied if either A_m or the term within the brackets is zero. That is if

$$Pm^2 + Qm + R = 0 \tag{2.44}$$

A_m cannot be zero by definition in equation (2.42) and therefore the values of m determined by equation (2.44) will give solutions of the form (2.42) which satisfy the "reduced equation" (2.41). Equation (2.44) is termed the "auxiliary equation". It will have two roots which could be unequal or equal, real or complex (see Chapter 4), and these will be considered individually.

(a) *Unequal Roots to Auxiliary Equation.* Let the roots of equation (2.44) be distinct and of values m_1 and m_2. Therefore, the solution of equation (2.41) will be

$$y = A_1 e^{m_1 x} \quad \text{or} \quad y = A_2 e^{m_2 x}$$

and the most general solution of the reduced equation (2.41) will be

$$y = A_1 e^{m_1 x} + A_2 e^{m_2 x} \tag{2.45}$$

Equation (2.45) is the "complementary function". It has two terms and two arbitrary constants as would be expected for a second order linear differential equation. If m_1 and m_2 are complex it is customary to replace the complex exponential functions with their equivalent trigonometric forms so that the re-defined arbitrary constants are also real. The third example below illustrates this point.

(b) *Equal Roots to Auxiliary Equation.* If the roots of the auxiliary equation are equal, that is if $m_1 = m_2 = m$, the auxiliary equation can be written

$$Pm^2 + Qm + R = 0 \tag{2.46}$$

and the solution of the reduced equation is

$$y = A e^{mx} \tag{2.47}$$

But as this only contains one term and one arbitrary constant it is not the complete complementary function of a second order equation. Therefore let

$$y = V e^{mx} \tag{2.48}$$

be a second trial solution in which V is a function of x. Then

$$\frac{dy}{dx} = e^{mx}\frac{dV}{dx} + mVe^{mx}$$

and

$$\frac{d^2 y}{dx^2} = e^{mx}\frac{d^2 V}{dx^2} + 2me^{mx}\frac{dV}{dx} + m^2 Ve^{mx}$$

Substitution of these expressions for the first and second differential coefficients into equation (2.41), using (2.46) and simplifying gives

$$\frac{d^2 V}{dx^2} = 0$$

Double integration of this gives

$$V = Cx + D$$

and the complementary function becomes

$$y = (Cx + D)e^{mx} \tag{2.49}$$

The above solution contains two terms and two arbitrary constants. Since the arbitrary constant D can be made equal to A in equation (2.47), it can be seen that the previous solution (2.47) is included in the later solution (2.49). Equation (2.49) is therefore acceptable as the complete complementary function when the auxiliary equation has equal roots.

Example 1. Solve

$$\frac{d^2 y}{dx^2} - 5\frac{dy}{dx} + 6y = 0 \qquad\qquad \text{I}$$

The auxiliary equation is

$$m^2 - 5m + 6 = 0 \qquad\qquad \text{II}$$

and the roots of this are

$$m_1 = 2, \quad m_2 = 3$$

The solution to equation I is therefore

$$y = Ae^{2x} + Be^{3x} \qquad\qquad \text{III}$$

where A and B are the two arbitrary constants.

Example 2. Solve

$$\frac{d^2 y}{dx^2} + 6\frac{dy}{dx} + 9y = 0 \qquad\qquad \text{I}$$

The auxiliary equation is

$$m^2 + 6m + 9 = 0 \qquad\qquad \text{II}$$

and the roots of this are

$$m_1 = m_2 = -3$$

The solution of equation I is therefore

$$y = (A + Bx)e^{-3x} \qquad\qquad \text{III}$$

where A and B are the two arbitrary constants.

Example 3. Solve

$$\frac{d^2 y}{dx^2} - 4\frac{dy}{dx} + 5y = 0 \qquad\qquad \text{I}$$

The auxiliary equation is

$$m^2 - 4m + 5 = 0 \qquad\qquad \text{II}$$

and the roots are

$$m = 2 \pm i$$

Therefore the complementary function is

$$y = A\exp[(2+i)x] + B\exp[(2-i)x] \qquad\qquad \text{III}$$

or the solution may be written as

$$y = e^{2x}(E\cos x + F\sin x) \qquad\qquad \text{IV}$$

using the relationships between exponential and trigonometric functions.

In conclusion it will be appreciated that irrespective of the nature of the roots, the complementary function takes the form:

Either
$$y = Ae^{m_1 x} + Be^{m_2 x} \qquad\qquad (2.50)$$

for distinct roots whether real or complex.

Or
$$y = (A + Bx)e^{mx} \qquad\qquad (2.51)$$

for equal roots, which must always be real. In all forms the expression will have arbitrary constants equal in number to the order of the differential equation.

2.5.3. *Particular Integrals*

To obtain the complete solution of the linear differential equation, it is necessary to find some particular solution of the equation,

$$P\frac{d^2 y}{dx^2} + Q\frac{dy}{dx} + Ry = \phi(x) \qquad\qquad (2.36)$$

where $\phi(x)$ may be a constant or a function of the independent variable x. If $\phi(x)$ is zero the complementary function is the complete solution and therefore nothing further is required. However, in most problems encountered, $\phi(x)$ is not zero and a particular integral will have to be found in order to complete the solution.

There are many general methods of determining the particular integral. Of these only two will be discussed here, as these two are capable of solving most problems arising in applied science. They may be termed.

(a) The method of undetermined coefficients.

(b) The method of inverse operators.

Method (a) is confined to linear equations with constant coefficients, and particular forms of the function $\phi(x)$. Method (b) is of more general applicability. Each of the methods will be discussed below.

2.5.4. *Particular Integrals by the Method of Undetermined Coefficients*

This is essentially a trial and error procedure, and whilst it is easy to apply, some experience is required before the form of the particular integral can be recognized. It is most unlikely that the chemical engineer will have had this experience and consequently the following rules which choose the form of the particular integral are submitted as a guide.

(i) $\phi(x)$ *is constant*, say C. Then obviously a particular integral of equation (2.36) is

$$y = C/R \tag{2.52}$$

because all the differential coefficients will be zero. The complete solution will therefore be

$$y = A\,e^{m_1 x} + B\,e^{m_2 x} + C/R \tag{2.53}$$

This rule applies to any order linear differential equation with constant coefficients.

(ii) $\phi(x)$ *is a polynomial of the form* $a_0 + a_1 x + a_2 x^2 + \ldots a_n x^n$; where all the coefficients $a_0, a_1, \ldots, a_n$ are constants.

The form of the particular integral will also be a rational integral function of the independent variable x because only functions of x whose differential coefficients are positive integral powers of the independent variable are themselves integral powers of x. Consequently, the form of the particular integral will be

$$y = \alpha_0 + \alpha_1 x + \alpha_2 x^2 + \ldots + \alpha_n x^n \tag{2.54}$$

Furthermore, the degree of the particular integral must equal the degree of x in $\phi(x)$, otherwise the degree of x on each side of the differential equation will be unequal and this is impossible in a linear differential equation with constant coefficients.

The solution procedure is therefore to assume a polynomial of the type shown in equation (2.54) and obtain the first, second, etc., differential coefficients of this expression. These differential expressions must then be substituted into the original differential equation and the coefficients α_0 to α_n evaluated by equating coefficients. The following example will make the procedure clear.

Example 4. Solve

$$\frac{d^2 y}{dx^2} - 4\frac{dy}{dx} + 4y = 4x + 8x^3 \qquad\qquad \text{I}$$

Because the degree of $\phi(x)$ is 3, let

$$y = p + qx + rx^2 + sx^3 \qquad\qquad \text{II}$$

in order to evaluate the particular integral. Then

$$\frac{dy}{dx} = q + 2rx + 3sx^2 \qquad\qquad \text{III}$$

and

$$\frac{d^2 y}{dx^2} = 2r + 6sx \qquad\qquad \text{IV}$$

Substitution of these expressions for the differential coefficients into I gives

$$(2r + 6sx) - 4(q + 2rx + 3sx^2) + 4(p + qx + rx^2 + sx^3) = 4x + 8x^3 \qquad \text{V}$$

Equating coefficients of equal powers of x,

$$2r - 4q + 4p = 0$$

$$6s - 8r + 4q = 4$$

$$4r - 12s = 0 \quad \text{or} \quad r = 3s$$

$$4s = 8 \quad \text{or} \quad s = 2$$

$$\therefore \quad r = 6, \quad q = 10, \quad \text{and} \quad p = 7$$

$$\therefore \quad y_p = 7 + 10x + 6x^2 + 2x^3 \qquad\qquad \text{VI}$$

The complementary function for this case of equal roots is

$$y_c = (A + Bx)e^{2x} \qquad\qquad \text{VII}$$

and the complete solution is

$$y = y_c + y_p = (A + Bx)e^{2x} + 7 + 10x + 6x^2 + 2x^3 \qquad \text{VIII}$$

(iii) $\phi(x)$ *is the form* Te^{rx}; where T and r are constants. The particular integral will also be of exponential form because all the differential coefficients of e^{rx} will be multiples of e^{rx}. Therefore, if the particular integral is assumed to be

$$y_p = \alpha e^{rx} \qquad\qquad (2.55)$$

The first differential coefficient will be $\alpha r e^{rx}$, and the second differential coefficient will be $\alpha r^2 e^{rx}$. Hence substitution into the original differential equation (2.36) gives

$$(Pr^2 + Qr + R)\alpha e^{rx} = Te^{rx} \qquad\qquad (2.56)$$

or

$$\alpha = \frac{T}{Pr^2 + Qr + R} \qquad\qquad (2.57)$$

Evaluation of α from the known constants enables the particular integral to be determined.

(iv) $\phi(x)$ *is of the form* $G \sin nx + H \cos nx$; where G and H are constants, either of which could be zero, whilst n is a non-zero constant. The form of the particular integral will also be a trigonometrical function of the same type because all the differential coefficients will be multiples of $\sin nx$ or $\cos nx$. Therefore assume the particular integral to be

$$y_p = L \sin nx + M \cos nx \tag{2.58}$$

The first differential coefficient of equation (2.58) will be

$$\frac{dy}{dx} = nL \cos nx - nM \sin nx$$

and the second differential coefficient

$$\frac{d^2 y}{dx^2} = -n^2(L \sin nx + M \cos nx)$$

Substitution in the original differential equation gives

$$L = \frac{(R - n^2 P) G + nQH}{(R - n^2 P)^2 + n^2 Q^2} \tag{2.59}$$

$$M = \frac{(R - n^2 P) H - nQG}{(R - n^2 P)^2 + n^2 Q^2} \tag{2.60}$$

Equations (2.59) and (2.60) enable the constants L and M to be evaluated and thus the particular integral can be determined.

The above four forms of the particular integral that would be obtained from each type of expression of $\phi(x)$ are summarized in Table 2.1 overleaf.

(v) *Modified procedure when a term in the particular integral duplicates a term in the complementary function.* In each of the above forms of the particular integral, it is possible for one or more terms to be identical with one or more terms in the complementary function. When this occurs the general procedure is to multiply the assumed form of the particular integral by the independent variable, which will destroy the similarity of the terms. However, it is possible that this multiplication by the independent variable will make other terms in the particular integral and complementary function identical. If this should happen the assumed form of the particular integral should be multiplied by the square of the independent variable, and if similarities still exist the assumed form should be multiplied by the third power of the independent variable and so on until no identical terms remain between the complementary function and the particular integral.

Identical terms in both parts of the solution of the differential equation can arise in many ways. For instance, when $\phi(x)$ is a constant, the first trial form for the particular integral will be a constant according to Table 2.1. If, in addition, one of the roots of the auxiliary equation is zero one term in the complementary function will also be a constant. Hence the modified

TABLE 2.1. *Normal Forms of Particular Integrals*

Right-hand Side of Differential Equation $\phi(x)$	Particular Integral	Coefficient Values
A constant C	A constant K	$K = \dfrac{C}{R}$
A polynomial $a_0 + a_1 x + \ldots + a_n x^n$	A polynomial $\alpha_0 + \alpha_1 x + \ldots + \alpha_n x^n$	To be determined by substituting in the differential equation and equating coefficients
An exponential $T e^{rx}$	An exponential αe^{rx}	$\alpha = \dfrac{T}{Pr^2 + Qr + R}$
Trigonometrical $G \sin nx + H \cos nx$	Trigonometrical $L \sin nx + M \cos nx$	$L = \dfrac{(R - n^2 P) G + nQH}{(R - n^2 P)^2 + n^2 Q^2}$ $M = \dfrac{(R - n^2 P) H - nQG}{(R - n^2 P)^2 + n^2 Q^2}$

P, Q, and R refer to the coefficients in the differential equation (2.36).

trial form of the particular integral should be a constant multiplied by the independent variable, in which case the similarity will have been removed.

Example 5. Solve

$$3\frac{d^2 y}{dx^2} - 6\frac{dy}{dx} = 18 \qquad\qquad \text{I}$$

The auxiliary equation is

$$3m^2 - 6m = 0 \qquad\qquad \text{II}$$

and the roots are

$$m = 0 \quad \text{and} \quad m = 2$$

The complementary function is

$$y_c = A e^{m_1 x} + B e^{m_2 x} = A + B e^{2x} \qquad\qquad \text{III}$$

Since one term in the complementary function is a constant, the particular integral cannot be a constant even though the right-hand side of equation I contains a constant. Therefore let the form of the particular integral be

$$y = Cx \qquad\qquad \text{IV}$$

Then

$$\frac{dy}{dx} = C$$

and

$$\frac{d^2 y}{dx^2} = 0$$

Substitution in equation I gives

$$3(0) + 6(C) = 18 \qquad \text{V}$$

or

$$C = 3$$

and the particular integral is $3x$. Therefore the complete solution is

$$y = 3x + A + Be^{2x} \qquad \text{VI}$$

The above complication can arise also when a root of the auxiliary equation is equal to the coefficient of the independent variable in the exponential term on the right-hand side of the equation. In this case too, an example will make the procedure clear. Therefore consider

Example 6. Solve

$$3\frac{d^2y}{dx^2} + 10\frac{dy}{dx} - 8y = 7e^{-4x} \qquad \text{I}$$

The auxiliary equation is

$$3m^2 + 10m - 8 = (3m - 2)(m + 4) = 0 \qquad \text{II}$$

$$\therefore \quad m = 2/3 \quad \text{or} \quad -4$$

and the complementary function is

$$y_c = A\,e^{2x/3} + Be^{-4x} \qquad \text{III}$$

The second term in equation III is of the same form as that given in Table 2.1 for the particular integral of equation I. Therefore, take the modified form

$$y = Cxe^{-4x} \qquad \text{IV}$$

for the particular integral.

$$\therefore \quad \frac{dy}{dx} = (1 - 4x)\,Ce^{-4x}$$

and

$$\frac{d^2y}{dx^2} = (16x - 8)\,Ce^{-4x}$$

Substitution in equation I and solving for C gives $C = -\frac{1}{2}$, and the complete solution of equation I is

$$y = y_c + y_p = A\,e^{2x/3} + Be^{-4x} - \tfrac{1}{2}xe^{-4x} \qquad \text{V}$$

Another way in which the above complication can appear, is when the roots of the auxiliary equation are both equal to the value m, the coefficient in the exponent in a term of the form Ce^{mx}. Thus for an auxiliary equation with equal roots the complementary function will be $(A + Bx)e^{mx}$ where the first term is similar to the expression Ce^{mx}. Hence the proposed form of the particular integral would be Ex^2e^{mx}, since multiplying by x alone would give a form similar to the second term of the complementary function. E would be evaluated in a manner similar to that illustrated in the above examples.

Finally, when the roots of the auxiliary equation are complex and the right-hand side of the differential equation contains the terms

$$(A \sin nx + B \cos nx) e^{px} \tag{2.61}$$

the modified trial form of the particular integral will be

$$x(\alpha \sin nx + \beta \cos nx) e^{px} \tag{2.62}$$

and the evaluation of α and β is as above.

2.5.5. *Particular Integrals by the Method of Inverse Operators*

The symbol (dy/dx) representing the differential coefficient signifies that the operation of differentiation of y has been carried out with respect to the independent variable x. In fact if there is no doubt whatever of the independent variable, it is acceptable to write simply "Dy" for the differential coefficient implying that the differentiation process has been carried out. Hence it is convenient to refer to the symbol "D" as the "differential operator", and prefixing it to a variable means that a differentiation has been carried out with respect to an obvious independent variable. The letter "D" alone has no significance and must be placed in front of the dependent variable to signify the differential coefficient. That is,

$$Dy = dy/dx$$

Similarly

$$D(Dy) = D^2 y = d^2 y/dx^2$$

and

$$D(D^2 y) = D^3 y = d^3 y/dx^3$$

so that

$$D^n y = d^n y/dx^n$$

It should be noted that

$$(Dy)^2 = (dy/dx)^2$$

without ambiguity.

The above symbolic representation in terms of the differential operator can be extended to expressions involving different order differential coefficients. Thus the expression,

$$\frac{d^2 y}{dx^2} + 3 \frac{dy}{dx} + 2y \tag{2.63}$$

may be written

$$D^2 y + 3Dy + 2y = (D^2 + 3D + 2) y \tag{2.64}$$

and the latter may also be factorized,

$$(D+1)(D+2) y \tag{2.65}$$

Hence it appears that the differential operator D can be treated as an ordinary algebraic quantity with certain limitations.

The three basic laws of algebra are the following.

(a) *The Distributive Law.* This states that

$$A(B+C) = AB+AC \tag{2.66}$$

which also applies to the differential operator D, viz.

$$D(u+v+w) = Du+Dv+Dw \tag{2.67}$$

In fact the distributive law was applied to the operator in equation (2.64).

(b) *The Commutative Law.* This states that

$$AB = BA \tag{2.68}$$

which does not in general apply to the differential operator D. It is conventional to consider that the differential operator differentiates every term which follows it, and not the terms preceding it. Transferring a term past the operator thus alters the value of the expression. For example,

$$Dxy \neq xDy$$

but it is true that the expression (2.65) can be given the equivalent forms,

$$(D+1)(D+2)y = (D+2)(D+1)y$$

The operator will thus commute with itself but not with variables.

(c) *The Associative Law.* This states that

$$(AB)C = A(BC) \tag{2.69}$$

which does not in general apply to the differential operator D.

The following two equations illustrate how far the law is obeyed.

$$D(Dy) = (DD)y$$

but
$$D(xy) = (Dx)y+xDy \tag{2.70}$$

In the first term on the right-hand side of equation (2.70) the closing of the brackets cuts y off from the operator thus destroying the simple equality.

The basic laws of algebra thus apply to the pure operators, but the relative order of operators and variables must be maintained. Also, the influence of the operator must not be hindered by brackets as in equation (2.70) unless this is specifically desired. Consequently these properties can be utilized particularly in the solution of linear differential equations thus simplifying the solving procedure. Some of these properties will be illustrated in order to familiarize the reader with the techniques.

Application of the Differential Operator to Exponentials. Let the differential coefficient of e^{px} be required. This expression could be written

$$D e^{px} = p e^{px} \tag{2.71}$$

Also for the second differential coefficient,

$$D^2 e^{px} = p^2 e^{px} \qquad (2.72)$$

and for the nth differential coefficient,

$$D^n e^{px} = p^n e^{px} \qquad (2.73)$$

In general, if $f(D)$ is some polynomial of the differential operator, it is possible to write

$$f(D) e^{px} = f(p) e^{px} \qquad (2.74)$$

that is, the coefficient of x replaces D in the function. This is quite obvious, for if $f(D)$ is expanded and then each term in the expansion multiplied by e^{px}, each of these new terms will be similar to the left-hand side of equation (2.73) and the differentiated form similar to the right-hand side. Specifically, if in equation (2.64) $y = e^{px}$, then the expression can be written

$$(D^2 + 3D + 2) e^{px} = (p^2 + 3p + 2) e^{px} \qquad (2.75)$$

The above concept can be extended to more complex functions involving exponentials. For example, if the differential coefficient of $y e^{px}$ is required, then

$$D(y e^{px}) = e^{px} Dy + y D e^{px} = e^{px}(D+p) y \qquad (2.76)$$

Similarly the second differential coefficient would be

$$D^2(y e^{px}) = D e^{px}(D+p) y = e^{px} . D(D+p) y + (D+p) y . D(e^{px})$$

$$= e^{px}(D+p)(D+p) y = e^{px}(D+p)^2 y \qquad (2.77)$$

Similarly it follows by comparison of equations (2.76) and (2.77) that

$$D^n(y e^{px}) = e^{px}(D+p)^n y \qquad (2.78)$$

or in even more general terms, if $f(D)$ is a polynomial in D,

$$f(D)(y e^{px}) = e^{px} f(D+p) y \qquad (2.79)$$

The above property of the differential operator involving exponentials is most important as will be seen later.

Finally, the operation performed on the exponential function in equations (2.71) to (2.79) can be extended to trigonometrical functions by using the complex relationships between exponential and trigonometrical functions to be given in Chapter 4. Let the nth differential coefficient of $\sin px$ be required. Now

$$D^n(\sin px) = D^n \operatorname{Im} e^{ipx} \qquad (2.80)$$

$$= \operatorname{Im} D^n e^{ipx}$$

$$= \operatorname{Im} (ip)^n e^{ipx} \qquad (2.81)$$

where "Im" represents the imaginary part of the function which follows it.

The evaluation of equation (2.81) depends upon whether n is odd or even. Put $n = 2m$ to make n even.

$$\therefore \quad (ip)^n = (-p^2)^m$$

and because $$e^{ipx} = \cos x + i \sin x$$

$$\therefore \quad D^{2m}(\sin px) = (-p^2)^m \sin px \tag{2.82}$$

Similarly, for n odd,

$$D^{2m+1}(\sin px) = (-p^2)^m p \cos px \tag{2.83}$$

The corresponding formulae for differentiating $\cos px$, which can be derived in the same way are

$$D^{2m}(\cos px) = (-p^2)^m \cos px \tag{2.84}$$

$$D^{2m+1}(\cos px) = -(-p^2)^m p \sin px \tag{2.85}$$

The principles involved in equations (2.80) to (2.85) can be extended to operations on more complicated trigonometrical functions by considering real or imaginary parts as shown. This is left to the initiative of the reader in attempting the examples at the end of the book.

The Inverse Operator. The operator D signifies differentiation, i.e.

$$D\left[\int f(x)\,dx\right] = f(x) \tag{2.86}$$

so that $$\int f(x)\,dx = D^{-1}f(x) \tag{2.87}$$

which suggests that the reciprocal of the operator D placed before a function implies integration of that function with respect to an obvious independent variable. Thus D^{-1} is the "inverse operator" and is an integrating operator. Because of its relationship to the differential operator D it would be expected that it can be treated as an algebraic quantity in exactly the same manner as D. This will now be considered.

Equations (2.74) and (2.79) can be extended to an infinite series of positive powers of D, so that if the function of D can be expanded in ascending powers of D, the equations still apply.

Example 7. Solve

$$\frac{dy}{dx} - 4y = e^{2x} \qquad\qquad \text{I}$$

$$\therefore \quad (D-4)y = e^{2x} \qquad\qquad \text{II}$$

Transferring the operator to the other side of the equation,

$$y = \frac{1}{D-4}e^{2x} \qquad\qquad \text{III}$$

$$= \frac{-1}{4(1-\frac{1}{4}D)}e^{2x}$$

$$\therefore \quad y = -\tfrac{1}{4}[1+(\tfrac{1}{4}D)+(\tfrac{1}{4}D)^2+(\tfrac{1}{4}D)^3+\dots]e^{2x} \qquad\qquad \text{IV}$$

using the binomial expansion. Performing the operations on e^{2x} gives

$$y = -\tfrac{1}{4}e^{2x}[1+\tfrac{1}{2}+(\tfrac{1}{2})^2+(\tfrac{1}{2})^3+\dots] \qquad\qquad \text{V}$$

The series in the brackets is a geometrical progression whose sum to infinity is 2.

$$\therefore \quad y = -\tfrac{1}{2}e^{2x} \qquad\qquad \text{VI}$$

is the particular integral of equation I.

Because the inverse operator in equation III can be expanded in a series of positive powers of D, it satisfies the condition stated above. Thus, the general property given in equation (2.74) could have been applied much earlier to equation III, viz.

$$y = \frac{1}{D-4}e^{2x} = \frac{1}{2-4}e^{2x} = -\tfrac{1}{2}e^{2x}$$

as before.

In the above example, equation (2.74) has been used in the form,

$$\frac{1}{f(D)}e^{px} = \frac{1}{f(p)}e^{px} \qquad\qquad (2.88)$$

However, if $f(p) = 0$, $e^{px}/f(p)$ is infinite and reference has to be made to the more general equation (2.79) in order to resolve this difficulty. When $f(p) = 0$, $(D-p)$ must be a factor of $f(D)$. Hence

$$f(D) = (D-p)^n \, \phi(D) \qquad\qquad (2.89)$$

where n is the lowest integer which makes $\phi(p)$ finite and not zero. Then

$$\frac{1}{f(D)}e^{px} = \frac{1}{(D-p)^n \, \phi(D)}e^{px}$$

$$= \frac{1}{(D-p)^n} \frac{e^{px}}{\phi(p)}$$

by applying equation (2.74). Application of equation (2.79) with $y = 1$ gives

$$\frac{1}{f(D)}e^{px} = \frac{e^{px}}{\phi(p)} \frac{1}{D^n} 1 \qquad\qquad (2.90)$$

D is raised to a negative power in equation (2.90), and this implies integration.

$$\therefore \quad \frac{1}{f(D)} e^{px} = \frac{e^{px} x^n}{\phi(p) n!} \tag{2.91}$$

It is unnecessary to include the arbitrary constants arising in the integrations since these are included in the complementary function.

Example 8. Solve

$$\frac{d^2 y}{dx^2} - 8\frac{dy}{dx} + 16y = 6x\,e^{4x} \qquad\qquad \text{I}$$

In terms of the differential operator, equation I becomes

$$(D^2 - 8D + 16)\,y = (D-4)^2\,y = 6x\,e^{4x} \qquad\qquad \text{II}$$

and the particular integral is

$$y_p = \frac{6}{(D-4)^2}\, x\,e^{4x}$$

$$= 6\,e^{4x}\,D^{-2}\,x$$

$$= 3\,e^{4x}\,D^{-1}\,x^2$$

$$= x^3\,e^{4x} \qquad\qquad \text{III}$$

The complementary function is

$$y_c = (A + Bx)\,e^{4x} \qquad\qquad \text{IV}$$

and the complete solution is

$$y = y_c + y_p = (A + Bx + x^3)\,e^{4x} \qquad\qquad \text{V}$$

Application of the Inverse Operator to Trigonometrical Functions. Trigonometrical functions can be written as the real or imaginary parts of e^{ipx} in which case they can be treated in that form by the inverse operator. The analytical procedure to follow would be very similar to that given above for real exponential functions combined with the principles presented in equations (2.80) to (2.85). Therefore no further discussion need be given to this particular topic.

Application of the Inverse Operator to Polynomial Functions. The inverse operator can be made to operate on the polynomial function $\phi(x)$ by expanding the function of the inverse operator in terms of D by the usual binomial series for negative exponents. The method is illustrated by the following example.

Example 9. Solve

$$\frac{d^2 y}{dx^2} - \frac{dy}{dx} - 6y = 4x^3 + 3x^2 \qquad\qquad \text{I}$$

Introducing the differential operator, equation I becomes

$$(D^2 - D - 6)y = (D-3)(D+2)y = 4x^3 + 3x^2 \qquad \text{II}$$

and the particular integral is

$$y_p = \frac{1}{(D-3)(D+2)}(4x^3 + 3x^2) = -\frac{1}{5}\left[\frac{1}{(3-D)} + \frac{1}{(2+D)}\right](4x^3 + 3x^2) \qquad \text{III}$$

Expanding each term of the partial fractions by the binomial theorem gives

$$y_p = -\frac{1}{5}\left[\left(\frac{1}{3} + \frac{D}{9} + \frac{D^2}{27} + \frac{D^3}{81} + \frac{D^4}{243} + \cdots\right)\right.$$

$$\left. + \left(\frac{1}{2} - \frac{D}{4} + \frac{D^2}{8} - \frac{D^3}{16} + \frac{D^4}{32} - \cdots\right)\right](4x^3 + 3x^2) \qquad \text{IV}$$

$$= -\frac{1}{5}\left[\frac{5}{6} - \frac{5}{36}D + \frac{35}{216}D^2 - \frac{65}{1296}D^3 + \frac{275}{7776}D^4 - \cdots\right](4x^3 + 3x^2) \qquad \text{V}$$

$$= -\frac{4x^3 + 3x^2}{6} + \frac{12x^2 + 6x}{36} - \frac{7(24x+6)}{216} + \frac{13 \times 24}{1296} - 0\ldots \qquad \text{VI}$$

$$= -\frac{144x^3 + 36x^2 + 132x - 10}{216}$$

The complementary function is

$$y_c = A\,e^{3x} + B\,e^{-2x} \qquad \text{VII}$$

and the complete solution is

$$y = y_c + y_p = A\,e^{3x} + Be^{-2x} - (72x^3 + 18x^2 + 66x - 5)/108 \qquad \text{VIII}$$

In the above paragraphs, methods of finding the particular integral and the complementary function have been presented for a second order linear differential equation. These methods are quite general and are equally applicable to any order linear differential equation. The second order equation was chosen simply for convenience in order to bring out the salient principles in the solution of these equations. The same treatment would have to be given to a third, fourth or higher order equation.

2.5.6. *Illustrative Problems*

Linear differential equations arise very frequently in the solution of chemical engineering problems. Their appearance is not surprising when one considers that chemical engineers are chiefly concerned with the prediction of rates of transfer of heat, momentum, and material. Consequently the method of solving such problems will now be illustrated by two examples taken from the various important fields of the subject.

Problem 1. *Simultaneous Diffusion and Chemical Reaction in a Tubular Reactor*

A tubular chemical reactor of length L and $1 \cdot 0 \text{ m}^2$ in cross section is employed to carry out a first order chemical reaction in which a material A is converted to a product B. The chemical reaction can be represented,

$$A \longrightarrow B$$

and the specific reaction rate constant is $k \text{ s}^{-1}$. If the feed rate is $u \text{ m}^3/\text{s}$, the feed concentration of A is c_0, and the diffusivity of A is assumed to be constant at $D \text{ m}^2/\text{s}$, determine the concentration of A as a function of length along the reactor. It may be assumed that there is no volume change during the reaction, and that steady state conditions are established.

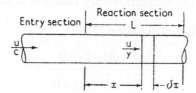

FIG. 2.4. Tubular reactor analysis

Solution

Take a coordinate x to specify the distance of any point from the inlet of the reactor section, let c denote the variable concentration of A in the entry section $(x < 0)$, and y denote the concentration of A in the reactor section $(x > 0)$, as shown in Fig. 2.4. The concentration will vary in the entry section due to diffusion, but will not vary in the section following the reactor.†

A material balance can be taken over the element of length δx at a distance x from the inlet.

	x	$x + \delta x$
Bulk flow of A	uy	$uy + u \dfrac{\mathrm{d}y}{\mathrm{d}x} \delta x$
Diffusion of A	$-D \dfrac{\mathrm{d}y}{\mathrm{d}x}$	$-D \dfrac{\mathrm{d}y}{\mathrm{d}x} + \dfrac{\mathrm{d}}{\mathrm{d}x}\left(-D \dfrac{\mathrm{d}y}{\mathrm{d}x}\right) \delta x$

The accumulation in this case is zero.

Rate of removal of A by reaction $= ky\delta x$

since the reactor is of unit cross-sectional area. Thus,

$$uy - D\frac{\mathrm{d}y}{\mathrm{d}x} - \left(uy + u\frac{\mathrm{d}y}{\mathrm{d}x}\delta x\right) - \left[-D\frac{\mathrm{d}y}{\mathrm{d}x} + \frac{\mathrm{d}}{\mathrm{d}x}\left(-D\frac{\mathrm{d}y}{\mathrm{d}x}\right)\delta x\right] = ky\,\delta x \qquad \text{I}$$

† Wehner, J. F. and Wilhelm, R. H. *Chem. Eng. Sci.* **6**, 89 (1956).

Simplifying, dividing by δx, and rearranging,

$$\therefore \quad D\frac{d^2 y}{dx^2} - u\frac{dy}{dx} - ky = 0 \qquad \text{II}$$

Similarly, the material balance in the entry section gives

$$D\frac{d^2 c}{dx^2} - u\frac{dc}{dx} = 0 \qquad \text{III}$$

which can also be obtained from II by removing the reaction term. Equations II and III are both second order linear differential equations with zero right-hand side, thus the complementary function is the complete solution in both cases.

The auxiliary equation of equation II is

$$Dm^2 - um - k = 0 \qquad \text{IV}$$

$$\therefore \quad m = u(1 \pm a)/2D \qquad \text{V}$$

where

$$a = \sqrt{(1 + 4kD/u^2)} \qquad \text{VI}$$

Therefore, the solutions of equations II and III are

$$y = A\exp\left[\frac{ux}{2D}(1+a)\right] + B\exp\left[\frac{ux}{2D}(1-a)\right] \qquad \text{VII}$$

and

$$c = \alpha + \beta\exp(ux/D) \qquad \text{VIII}$$

which contain four arbitrary constants, A, B, α, and β. The four boundary conditions are

$$\text{at } x = -\infty, \quad c = c_0 \qquad \text{IX}$$

$$\text{at } x = 0, \quad c = y \qquad \text{X}$$

$$\text{at } x = 0, \quad \frac{dc}{dx} = \frac{dy}{dx} \qquad \text{XI}$$

$$\text{at } x = L, \quad \frac{dy}{dx} = 0 \qquad \text{XII}$$

The first condition specifies the state of the feed stream, and the second ensures continuity of composition. The third condition, taken with equation X, is necessary to conserve material at the boundary assuming that the diffusivities are equal in both sections. The final condition forbids diffusion out of the reactor and is necessary as a conservation law for the section following the reactor. A full argument supporting the last condition is given in the reference quoted at the start of the problem. Equations IX to XII give respectively,

$$\alpha = c_0 \qquad \text{XIII}$$

$$\alpha + \beta = A + B \qquad \text{XIV}$$

$$2\beta = A(1+a) + B(1-a) \qquad \text{XV}$$

$$A(1+a)\exp\left[\frac{uL}{2D}(1+a)\right] + B(1-a)\exp\left[\frac{uL}{2D}(1-a)\right] = 0 \qquad \text{XVI}$$

Eliminating α and β from XIII, XIV and XV gives

$$2c_0 = A(1-a) + B(1+a) \qquad \text{XVII}$$

Solving XVI and XVII for A and B gives

$$A = \frac{2c_0(a-1)}{K}\exp\left(-\frac{uLa}{2D}\right) \qquad \text{XVIII}$$

$$B = \frac{2c_0(a+1)}{K}\exp\left(\frac{uLa}{2D}\right) \qquad \text{XIX}$$

where $\qquad K = (a+1)^2\exp(uLa/2D) - (a-1)^2\exp(-uLa/2D) \qquad$ XX

Putting these values of A and B into equation VII gives the final result.

$$\frac{y}{c_0} = \frac{2}{K}\exp\left(\frac{ux}{2D}\right)\left\{(a+1)\exp\left[\frac{ua}{2D}(L-x)\right] + (a-1)\exp\left[-\frac{ua}{2D}(L-x)\right]\right\} \quad \text{XXI}$$

From equation XXI, if diffusion is neglected, the first order piston type flow equation results thus,

$$\frac{c_0 - y}{c_0} = 1 - \exp(-kx/u) \qquad \text{XXII}$$

after a rather involved application of L'Hôpital's rule as $D \to 0$ (see Section 3.3.7).

Problem 2. The Continuous Hydrolysis of Tallow in a Spray Column

1·017 kg/s of a tallow fat mixed with 0·286 kg/s of high pressure hot water is fed into the base of a spray column operated at a temperature of 232 °C and a pressure of 4·14 MN/m². 0·519 kg/s of water at the same temperature and pressure is sprayed into the top of the column and descends in the form of droplets through the rising fat phase. Glycerine is generated in the fat phase by the hydrolysis reaction and is extracted by the descending water so that 0·701 kg/s of final extract containing 12·16% glycerine is withdrawn continuously from the column base. Simultaneously 1·121 kg/s of fatty acid raffinate containing 0·24% glycerine leaves the top of the column. If the effective height of the column is 22 m and the diameter 0·66 m the glycerine equivalent in the entering tallow 8·53% and the distribution ratio of glycerine between the water and the fat phase at the column temperature and pressure is 10·32, estimate the concentration of glycerine in each phase as a function of column height. Also find out what fraction of the tower height is required principally for the chemical reaction. The hydrolysis reaction is pseudo first order and the specific reaction rate constant is 0·0028 s⁻¹.

Solution

Consider Fig. 2.5 which illustrates the flows of extract and raffinate in the hydrolyser column. L represents kg/s of raffinate and G kg/s of extract

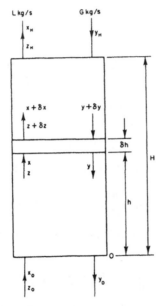

Fig. 2.5. Continuously operating fat-hydrolysing column under steady-state conditions

whilst the appropriate suffix signifies the position in the hydrolyser column. In addition to the symbols given in Fig. 2.5, let

x = weight fraction of glycerine in raffinate
y = weight fraction of glycerine in extract
y^* = weight fraction of glycerine in extract in equilibrium with x
z = weight fraction of hydrolysable fat in raffinate
S = sectional area of tower (m²)
a = interfacial area per unit volume of tower (m⁻¹)
K = overall mass transfer coefficient expressed in terms of extract compositions (kg/m² s)
m = distribution ratio
k = specific reaction rate constant (s⁻¹)
ρ = mass of fat per unit volume of column (kg/m³)

h = distance coordinate from base of column (m)
w = kg fat per kg glycerine
H = effective height of column (m)

Consider the changes occurring in the element of column of height δh. These are

(a) glycerine transferred from fat to water phase,

$$KaS(y^*-y)\,\delta h \qquad\qquad\qquad \text{I}$$

(b) rate of destruction of fat by hydrolysis,

$$k\rho Sz\,\delta h$$

and therefore the rate of production of glycerine is

$$k\rho Sz\,\delta h/w \qquad\qquad\qquad \text{II}$$

Then a glycerine balance over the element δh is by Fig. 2.5,

$$Lx+\frac{(k\rho Sz\,\delta h)}{w}-L\!\left(x+\frac{dx}{dh}\,\delta h\right)=Gy-G\!\left(y+\frac{dy}{dh}\,\delta h\right)=KaS(y^*-y)\,\delta h \qquad \text{III}$$

A glycerine balance between the element and the base of the tower is

$$\frac{Lz_0}{w}+Gy=Lx+\frac{Lz}{w}+Gy_0 \qquad\qquad\qquad \text{IV}$$

Finally, the glycerine equilibrium between the phases is

$$y^*=mx \qquad\qquad\qquad \text{V}$$

From the last two parts of equation III and equation V it can be shown that

$$KaSmx=KaSy-G\frac{dy}{dh} \qquad\qquad\qquad \text{VI}$$

and substitution of IV in the element glycerine balance III,

$$k\rho S\!\left[\frac{z_0}{w}+\frac{G}{L}(y-y_0)-x\right]\delta h-L\frac{dx}{dh}\,\delta h=-G\frac{dy}{dh}\,\delta h \qquad \text{VII}$$

Multiply equation VII by $(KaSm/LG)$ and substitute for x from VI,

$$\frac{k\rho S^2\,Ka}{LG}\!\left[\frac{mz_0}{w}+\frac{mG}{L}(y-y_0)\right]-\frac{k\rho S}{L}\!\left(\frac{KaS}{G}y-\frac{dy}{dh}\right)$$

$$-\left(\frac{KaS}{G}\frac{dy}{dh}-\frac{d^2y}{dh^2}\right)+\frac{KaSm}{L}\frac{dy}{dh}=0 \qquad \text{VIII}$$

Equation VIII contains the following groups of constants which can be denoted by the following parameters,

$$r=\frac{mG}{L},\quad p=\frac{k\rho S}{L},\quad q=\frac{KaS}{G}(r-1) \qquad \text{IX}$$

Substitution of the above groups into VIII gives on rearranging,

$$\frac{d^2 y}{dh^2} + (p+q)\frac{dy}{dh} + pqy = \frac{pq}{r-1}\left(ry_0 - \frac{mz_0}{w}\right) \qquad \text{X}$$

Equation X is a second order linear differential equation with constant coefficients.

The auxiliary equation is

$$m_1{}^2 + (p+q)m_1 + pq = 0 \qquad \text{XI}$$

The roots of XI are

$$m_1 = -p \quad \text{or} \quad -q$$

and therefore the complementary function is

$$y_c = A e^{-ph} + B e^{-qh} \qquad \text{XII}$$

The right-hand side of equation X is a constant and therefore the particular integral is a constant and equal to the right-hand side of equation X divided by the coefficient of y, as given in Section 2.5.4.

$$\therefore \quad y_p = \frac{ry_0 - mz_0/w}{r-1} = C \qquad \text{XIII}$$

and the complete solution is

$$y = A e^{-ph} + B e^{-qh} + \frac{ry_0 - mz_0/w}{r-1} \qquad \text{XIV}$$

where A and B are arbitrary constants to be evaluated from the following boundary conditions.

At
$$h = 0, \quad x = 0$$

and at
$$h = H, \quad y = 0$$

From equation VI,

$$mx = y - \frac{r-1}{q}\frac{dy}{dh}$$

and from equation XIV,

$$\frac{dy}{dh} = -pA e^{-ph} - qB e^{-qh} \qquad \text{XV}$$

Therefore at
$$h = 0$$

$$A + B + \left(\frac{r-1}{q}\right)(pA + qB) + C = 0 \qquad \text{XVI}$$

and at
$$h = H$$

$$A e^{-pH} + B e^{-qH} + C = 0 \qquad \text{XVII}$$

Equations XVI and XVII are two simultaneous equations in A and B. The arbitrary constants are best evaluated from these two equations by defining a further constant. Thus let

$$v = \frac{q+rp-p}{q} = 1 + \frac{k\rho G}{KaL} \qquad \text{XVIII}$$

Inserting this constant into equations XVI and XVII and solving gives

$$B(v\,e^{-qH} - r\,e^{-pH}) = C(e^{-pH} - v) \qquad \text{XIX}$$

and

$$A(v\,e^{-qH} - r\,e^{-pH}) = C(r - e^{-qH}) \qquad \text{XX}$$

Substituting these values of A and B into equation XIV,

$$y(v\,e^{-qH} - r\,e^{-pH}) = C[(r - e^{-qH})\,e^{-ph} + (e^{-pH} - v)\,e^{-qh} + v\,e^{-qH} - r\,e^{-pH}] \qquad \text{XXI}$$

and expressing y_0 from C in terms of the other variables and using the fact that $y = y_0$ at $h = 0$,

$$y_0 = \frac{mz_0}{w(r - e^{-qH})}\left[1 - \left(\frac{r-1}{r-v}\right)e^{-pH} + \left(\frac{v-1}{r-v}\right)e^{-qH}\right] \qquad \text{XXII}$$

Substituting the above value of y_0 into equation XIII and then into equation XXI gives

$$y = \frac{mz_0}{w(r-v)}\left[e^{-ph} + \left(\frac{e^{-pH} - v}{r - e^{-qH}}\right)e^{-qh} + \left(\frac{v\,e^{-qH} - r\,e^{-pH}}{r - e^{-qH}}\right)\right] \qquad \text{XXIII}$$

Equation XXIII gives the weight fraction of glycerine in the extract phase as a function of column height h.

Allowing for the solubility of water in tallow, taking mean flow rates, and using the data given in the problem gives

$$L = 1\cdot076, \quad G = 0\cdot474, \quad y_0 = 0\cdot188$$

$$\therefore \quad r = 4\cdot544, \quad p = 0\cdot650, \quad q = 2\cdot56Ka, \quad v = 1 + \frac{0\cdot90}{Ka}$$

Solving equation XXIII for Ka with $H = 2\cdot2$ m gives the value of the mass transfer coefficient as

$$Ka = 0\cdot0632 \text{ kg glycerine per sec per m}^3 \qquad \text{XXIV}$$

With the value of $Ka = 0\cdot0632$, values of y, y^* and z can be determined as functions of column height with the aid of equations XXIII, VI, V, IV. The results of the arithmetical calculations are shown in graphical form in

Fig. 2.6. Figure 2.6 shows that the chemical reaction is virtually complete in the bottom 9 m of the column, or 40% of the column.

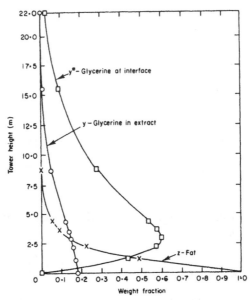

FIG. 2.6. Concentration distributions in hydrolysing column

2.6. SIMULTANEOUS DIFFERENTIAL EQUATIONS

These are groups of differential equations containing more than one dependent variable but only one independent variable, and in these equations all the derivatives of the different dependent variables are with respect to the one independent variable. This distinguishes an equation which is a member of a group of simultaneous differential equations, from a partial differential equation which will contain more than one independent variable and consequently will also contain partial derivatives. All the derivatives in the simultaneous differential equations will be total derivatives.

The basis of the solution of simultaneous differential equations is algebraic elimination of the variables until only one differential equation relating two of the variables remains. The final differential equation is solved in the ordinary way by one of the conventional methods. If the equation is linear with constant coefficients, one of the methods already presented in this chapter will be suitable, otherwise a method to be given later must be employed.

The variables to be eliminated depend on the set of equations, and the ultimate information required from the mathematical analysis. The variable may be the independent variable or one of the dependent variables. These possibilities will now be considered.

2.6.1. *Elimination of the Independent Variable*

Consider the pair of simultaneous differential equations,

$$\frac{dx}{d\theta} = f_1(x, y) \tag{2.92}$$

and

$$\frac{dy}{d\theta} = f_2(x, y) \tag{2.93}$$

in which x and y are the dependent variables and θ is the independent variable. $f_1(x, y)$ and $f_2(x, y)$ are two functions of x and y. Equations of this kind can be solved by eliminating the independent variable by dividing one equation by the other, say equation (2.93) by equation (2.92) to give

$$\frac{dy}{dx} = \frac{f_2(x, y)}{f_1(x, y)} \tag{2.94}$$

which is an ordinary differential equation in x and y, and would be solved as such.

Example 1 solved in paragraph 2.3.3 was really one of the above type. Elimination of the time gave a homogeneous equation which was solved by standard solution methods for homogeneous equations.

The above method of eliminating the independent variable is restricted to first order first degree equations in which the independent variable appears only in the derivative. Consequently its scope is very limited.

2.6.2. *Elimination of One or More Dependent Variables*

This approach has considerably wider application, however it becomes very involved with equations of high order and therefore it would be better to make use of matrices if there are many equations. The matrix methods of solving large sets of simultaneous equations are presented in Chapter 12, whereas the following paragraphs will be restricted to the simpler situations. Thus, the independent variable can be eliminated by

(a) Taking advantage of the algebraic properties of the differential operator.

(b) Systematic elimination.

(a) *By Use of the Differential Operator.* It has been pointed out that the operator D obeys the fundamental laws of algebra and therefore can be used in the elimination process. The method is best illustrated by an example.

Example 1. Solve

$$(D^2 + D - 6)y + (D^2 + 6D + 9)z = 0 \qquad\qquad \text{I}$$

and

$$(D^2 + 3D - 10)y + (D^2 - 3D + 2)z = 0 \qquad\qquad \text{II}$$

Equations I and II can be written

$$(D + 3)(D - 2)y + (D + 3)^2 z = 0 \qquad\qquad \text{III}$$

and

$$(D - 2)(D + 5)y + (D - 2)(D - 1)z = 0 \qquad\qquad \text{IV}$$

Multiply III by $(D+5)$ and IV by $(D+3)$ to give

$$(D+3)(D+5)(D-2)y+(D+5)(D+3)^2 z = 0 \qquad \text{V}$$

and $\qquad (D+3)(D+5)(D-2)y+(D+3)(D-2)(D-1)z = 0 \qquad \text{VI}$

Elimination of y and simplifying the remainder gives

$$(11D+13)(D+3)z = 0 \qquad \text{VII}$$

whose solution is

$$z = A e^{-13x/11} + B e^{-3x} \qquad \text{VIII}$$

where A and B are arbitrary constants.

Substitution of z from equation VIII into equation I gives

$$(D^2+D-6)y = Ee^{-13x/11} \qquad \text{IX}$$

where $\qquad E = -\left[\left(\dfrac{13}{11}\right)^2 - \left(\dfrac{6 \times 13}{11}\right) + 9\right] A$

The terms involving Be^{-3x} add up to zero.

The particular integral of y is

$$y_p = \frac{1}{D^2+D-6} Ee^{-13x/11} = \frac{E}{5}\left(\frac{1}{D-2} - \frac{1}{D+3}\right)e^{-13x/11} \qquad \text{X}$$

$$= Ge^{-13x/11} \qquad \text{XI}$$

where $\qquad G = -\dfrac{121}{700} E = \dfrac{4}{7} A$

The complementary function is

$$y_c = He^{2x} + Je^{-3x} \qquad \text{XII}$$

Substituting equations VIII and XII into equation II shows that $J = 2B$. The general solution is thus

$$y = \tfrac{4}{7} A e^{-13x/11} + 2Be^{-3x} + He^{2x}$$
$$z = A e^{-13x/11} + B e^{-3x}$$

(b) *Systematic Elimination.* This term is proposed for the elimination process which reduces the amount of algebra involved to a minimum. Essentially the method consists of setting up a table indicating the number of times and in what form each variable appears in the simultaneous equations. Then the variable that appears in the simplest manner is first eliminated by making the appropriate substitution. Following this the second least frequent variable is removed from the remaining equations by the appropriate substitution, and the substitution process is continued until only one equation remains. Generally terms which appear as derivatives are left to the final steps of the substitution process. The following example will demonstrate the technique.

Example 2.† 1·25 kg/s of sulphuric acid (heat capacity 1500 J/kg °C) is to be cooled in a two-stage countercurrent cooler of the following type. Hot

† Suggested by W. M. Crooks, Department of Chemical Engineering, University of Birmingham.

acid at 174 °C is fed to a tank where it is well stirred in contact with cooling coils. The continuous discharge from this tank at 88 °C flows to a second stirred tank and leaves at 45 °C. Cooling water at 20 °C flows into the coil of the second tank and thence to the coil of the first tank. The water is at 80 °C as it leaves the coil of the hot acid tank. To what temperatures would the contents of each tank rise if due to trouble in the supply, the cooling water suddenly stopped for 1 h?

On restoration of the water supply, water is put on the system at the rate of 1·25 kg/s. Calculate the acid discharge temperature after 1 h. The capacity of each tank is 4500 kg of acid and the overall coefficient of heat transfer in the hot tank is 1150 W/m² °C and in the colder tank 750 W/m² °C. These coefficients may be assumed constant.

Solution

The steady state conditions of the cooling system are shown in Fig. 2.7, and the steady state data calculated from the illustrated conditions are:

Water rate before failure of supply $= 0.96$ kg/s
Intermediate water temperature between the two tanks $= 40$ °C
Heat transfer area of coil in hot tank $= 6.28$ m²
Heat transfer area of coil in cold tank' $= 8.65$ m²

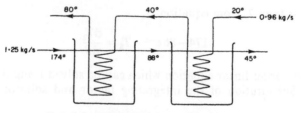

FIG. 2.7. Sulphuric acid cooling system

Water Fails

Because the holding time of each tank is 1 h, it is more convenient numerically to let θ be the time in hours rather than seconds, convert the flow rates to kg/h and allow for a factor of 3600 when the heat transfer coefficients are used. Thus, for the purpose of the analysis let

Acid flow rate be M kg/h $= 4500$

Acid heat capacity be C J/kg °C $= 1500$

Water heat capacity be C_w J/kg °C $= 4200$

Feed acid temperature be T_0 °C $= 174$

Acid temperature ex tank 1 be T_1 °C

Acid temperature ex tank 2 be T_2 °C

Capacity of each tank be V kg $= 4500$

Then the heat balances are as follows.

$$\text{Input} \qquad \text{Output} \qquad \text{Accumulation}$$

$$\text{Tank 1} \qquad MCT_0 - MCT_1 = VC\frac{dT_1}{d\theta} \qquad\qquad \text{I}$$

$$\text{Tank 2} \qquad MCT_1 - MCT_2 = VC\frac{dT_2}{d\theta} \qquad\qquad \text{II}$$

These are simultaneous first order differential equations, but they can be solved in succession. Because $M = V$ in the numerical example, equation I becomes

$$T_0 - T_1 = \frac{dT_1}{d\theta} \qquad\qquad \text{III}$$

which has the solution,

$$T_0 - T_1 = Ke^{-\theta} = 174 - T_1 \qquad\qquad \text{IV}$$

and when $\theta = 0$, $T_1 = 88\ °C$ the steady state temperature. Therefore when $\theta = 1\cdot 0$,

$$T_1 = 174 - 86\,e^{-1\cdot 0} = 142\cdot 4\ °C \qquad\qquad \text{V}$$

Similarly for tank 2, from equation II,

$$174 - 86\,e^{-\theta} - T_2 = \frac{dT_2}{d\theta} \qquad\qquad \text{VI}$$

This is a first order linear equation which can be solved using the integrating factor e^{θ}. Substitution of the integrating factor and solution of equation VI gives

$$T_2 = 174 - (86\theta + 129)\,e^{-\theta} \qquad\qquad \text{VII}$$

which gives for $\theta = 1\cdot 0$,

$$T_2 = 94\cdot 9\ °C \qquad\qquad \text{VIII}$$

Water Supply Restored

Water flow rate is now 4500 kg/h but for the purpose of the following derivation let the water rate be W kg/h, let the water supply temperature be $t_3\ °C$, let the water temperature ex tank 2 be $t_2\ °C$, and let the water temperature ex tank 1 be $t_1\ °C$. Then the heat balances are as follows.

$$\text{Input} \qquad\qquad \text{Output} \qquad\qquad \text{Accumulation}$$

$$\text{Tank 1} \quad WC_w t_2 + MCT_0 - (WC_w t_1 + MCT_1) = VC\frac{dT_1}{d\theta} \qquad \text{IX}$$

$$\text{Tank 2} \quad WC_w t_3 + MCT_1 - (WC_w t_2 + MCT_2) = VC\frac{dT_2}{d\theta} \qquad \text{X}$$

The heat transfer rate equations for the two tanks are

$$WC_w(t_1 - t_2) = U_1 A_1 \left[\frac{(T_1 - t_1) - (T_1 - t_2)}{\ln(T_1 - t_1) - \ln(T_1 - t_2)} \right] \qquad \text{XI}$$

and

$$WC_w(t_2 - t_3) = U_2 A_2 \left[\frac{(T_2 - t_2) - (T_2 - t_3)}{\ln(T_2 - t_2) - \ln(T_2 - t_3)} \right] \qquad \text{XII}$$

These equations simplify to

$$\frac{T_1 - t_1}{T_1 - t_2} = \exp\left(-\frac{U_1 A_1}{WC_w}\right) = \alpha \quad \text{(say)} \qquad \text{XIII}$$

and

$$\frac{T_2 - t_2}{T_2 - t_3} = \exp\left(-\frac{U_2 A_2}{WC_w}\right) = \beta \quad \text{(say)} \qquad \text{XIV}$$

Equations XIII and XIV can be written more simply,

$$T_1(1 - \alpha) = t_1 - \alpha t_2 \qquad \text{XV}$$

$$T_2(1 - \beta) = t_2 - \beta t_3 \qquad \text{XVI}$$

Equations IX, X, XV and XVI are the four equations which have to be solved simultaneously. The difficulty is that of elimination. Hence the following table, Table 2.2, gives the frequency of the appearance of the four variables.

TABLE 2.2. *Frequency of Appearance of Variables*

T_1	T_2	t_1	t_2
1	1	1	1
+	+	1	1
1	1		1
1			1
3 plus+	2 plus+	2	4

In Table 2.2, "1" indicates the appearance of the variable and "+" its derivative. From Table 2.2, t_1 should be eliminated first, then t_2, because both T_1 and T_2 also appear in a differentiated form. Substitution of equation XV into equation IX, and using the numerical property that $M = W = V$, gives

$$C_w t_2 + CT_0 - CT_1 - C_w(1 - \alpha)T_1 - C_w \alpha t_2 = C\frac{dT_1}{d\theta} \qquad \text{XVII}$$

Now eliminate t_2 from the system by substituting from equation XVI into equations XVII and X thus,

$$C_w(1-\alpha)(1-\beta)T_2+C_w(1-\alpha)\beta t_3+CT_0-CT_1-C_w(1-\alpha)T_1 = C\frac{dT_1}{d\theta} \quad \text{XVIII}$$

and

$$C_w t_3+CT_1-CT_2-C_w(1-\beta)T_2-C_w\beta t_3 = C\frac{dT_2}{d\theta} \quad \text{XIX}$$

Rearranging equation XIX,

$$CT_1 = C\frac{dT_2}{d\theta}+(C_w-\beta C_w+C)T_2-C_w(1-\beta)t_3 \quad \text{XX}$$

and differentiating,

$$C\frac{dT_1}{d\theta} = C\frac{d^2 T_2}{d\theta^2}+(C_w-\beta C_w+C)\frac{dT_2}{d\theta} \quad \text{XXI}$$

Using equations XX and XXI to eliminate T_1 and its differential coefficient from equation XVIII gives

$$C^2\frac{d^2 T_2}{d\theta^2}+(2C+2C_w-\alpha C_w-\beta C_w)C\frac{dT_2}{d\theta}$$
$$+[C^2+CC_w(1-\alpha\beta)+C_w^2(1-\alpha)(1-\beta)]T_2$$
$$= [C(1-\alpha\beta)+C_w(1-\alpha)(1-\beta)]C_w t_3+C^2 T_0 \quad \text{XXII}$$

Equation XXII is a second order linear differential equation with constant coefficients. The operator does not have simple factors and it is advantageous to introduce the numerical values at this stage.

$$\frac{U_1 A_1}{WC_w} = \frac{1150\times 3600\times 6\cdot 28}{4500\times 4200} = 1\cdot 376, \quad \therefore \quad \alpha = 0\cdot 2526$$

$$\frac{U_2 A_2}{WC_w} = \frac{750\times 3600\times 8\cdot 65}{4500\times 4200} = 1\cdot 236, \quad \therefore \quad \beta = 0\cdot 2906$$

$$\therefore \quad (1-\alpha)(1-\beta) = 0\cdot 530 \quad \text{and} \quad (1-\alpha\beta) = 0\cdot 927$$

Equation XXII thus becomes

$$\frac{d^2 T_2}{d\theta^2}+6\cdot 08\frac{dT_2}{d\theta}+7\cdot 75T_2 = 309 \quad \text{XXIII}$$

which has the solution

$$T_2 = A e^{-4\cdot 26\theta}+B e^{-1\cdot 82\theta}+39\cdot 9 \quad \text{XXIV}$$

The initial conditions for this solution are the final temperatures from the previous operation given by equations V and VIII. Before these can be used,

T_1 has to be determined from equations XXIV and XX. The boundary conditions give

$$A + B = 55 \cdot 0 \qquad\qquad \text{XXV}$$

and

$$-1 \cdot 27A + 1 \cdot 17B = 63 \cdot 0 \qquad\qquad \text{XXVI}$$

Solving for A and B, and substituting in equation XXIV gives

$$T_2 = 0 \cdot 6\, e^{-4 \cdot 26\theta} + 54 \cdot 4\, e^{-1 \cdot 82\theta} + 39 \cdot 9 \qquad\qquad \text{XXVII}$$

so that the acid discharge temperature after 1 h is 48·8 °C.

2.7. Conclusions

Methods have been presented in this chapter for solving the simpler types of differential equations which the chemical engineer is likely to encounter. There is no general method of solution but the methods included should solve most first order equations.

Non-linear second order equations are frequently insoluble analytically, but the procedures of Section 2.4 may be useful in providing at least an approximate yet valuable solution.

The most important type of equation dealt with is the linear differential equation with constant coefficients. The complementary function can always be found easily and the methods of undetermined coefficients and inverse operators are recommended for the determination of the particular integral.

Chapter 3

SOLUTION BY SERIES

3.1. Introduction

METHODS of solution were presented in Chapter 2 for a limited number of types of ordinary differential equation. Most first order equations were solved in principle, and higher order equations were classified into linear and non-linear differential equations. It was shown that non-linear equations can be solved in the special cases where either x or y is absent, or the equation is homogeneous; and that the solution of any linear differential equation can be resolved into the sum of a complementary function and a particular integral. The complementary function was found in all cases where the coefficients of the derivatives are constant, or the reduced equation is homogeneous, but the general linear equation with variable coefficients was not solved. Particular integrals were found by a variety of methods, some of these involving knowledge of the corresponding complementary function. The purpose of the present chapter is to solve the general linear ordinary differential equation with variable coefficients. The conditions which these coefficients must satisfy are given in the chapter, but they are invariably satisfied in problems of chemical engineering importance. As is usual in treatments of this subject, only the complementary function will be found in most cases, the particular integral can then be found by using the methods of Chapter 2.

When the differential equation has constant coefficients, the solution involves the exponential function. This function, in common with many others, can be expanded into a series of ascending powers of x. The series itself can be considered to be a solution of the differential equation, a property which may be demonstrated by differentiation and substitution into the equation. Methods given in this chapter will show that solutions in the form of infinite series can be obtained for many linear differential equations with variable coefficients. Of the series that arise, only a few can be summed allowing the solution to be expressed in closed form, but the other series can be accepted as solutions provided that they are convergent. For practical purposes, the series will be useful only if they are rapidly convergent. It is therefore necessary to consider the convergence of series, before differential equations can be solved by their use.

3.2. Infinite Series

A series of numbers

$$u_1 + u_2 + u_3 + \ldots + u_n = S_n \tag{3.1}$$

is classified by determining the behaviour of S_n as n increases to infinity.

(a) If $S_n \to S$ some finite number as $n \to \infty$, the series is termed "convergent".

(b) If $S_n \to \pm \infty$ as $n \to \infty$, the series is termed "divergent".

(c) In other cases, the series is termed "oscillatory".

When considering oscillatory series, the values of S_n may oscillate between either finite or infinite limits and a further classification can be based on this distinction.

In the application of series to differential equations, u_n will be a function of x, consequently S_n will also be a function of x. Because x may take positive or negative values, or even complex values, it is desirable to allow u_n to be complex. The properties of complex numbers will be investigated in Chapter 4, but for those readers who are not familiar with the elementary properties of complex numbers, the term "absolute value" refers to the numerical magnitude of the number and is signified by vertical lines.

$$|u_n| \equiv \text{absolute value of } u_n.$$

The series (3.1) is said to be "absolutely convergent" if the series

$$|u_1| + |u_2| + |u_3| + \dots + |u_n| \tag{3.2}$$

is convergent.

Example 1.

$$u_n = z^n$$
$$\therefore \quad S_n = z + z^2 + z^3 + \dots + z^n \qquad\qquad \text{I}$$

This series is a geometrical progression with common ratio z.

$$\therefore \quad S_n = \frac{z(1 - z^n)}{1 - z} \qquad\qquad \text{II}$$

$$= \frac{z}{1 - z} - \frac{z^{n+1}}{1 - z}$$

(a) If $|z| < 1$, $z^{n+1} \to 0$ as $n \to \infty$,

$$\therefore \quad S_n \to z/(1 - z)$$

Hence the series is convergent.

(b) If $|z| > 1$, $|z^{n+1}| \to \infty$ as $n \to \infty$, therefore the series diverges if z is real and positive, or oscillates for all other values of z.

(c) If $z = 1$, $S_n = n$ which tends to infinity as n increases indefinitely. Hence the series is divergent.

(d) If $|z| = 1$ and $z \neq 1$, the series oscillates within finite limits.

This series thus illustrates all four types of behaviour.

3.2.1. *Properties of Infinite Series*

The following are three useful properties of infinite series.

(i) If a series contains only positive real numbers or zero, then it must be either convergent or divergent; it cannot oscillate.

(ii) If a series is convergent, then $u_n \to 0$ as $n \to \infty$. This is a property of a convergent series; it is not a proof of convergence. A divergent series for which $u_n \to 0$ is illustrated in example 2, part (b) below.

(iii) If a series is absolutely convergent, then it is also convergent.

Statement (iii) can be proved as follows by referring to (i). If u_n is real, put $u_n = a_n - b_n$ where $a_n = \frac{1}{2}(|u_n| + u_n)$ and $b_n = \frac{1}{2}(|u_n| - u_n)$. Thus a_n, the series of positive terms, is separated from $-b_n$, the set of negative terms, so that neither a_n nor b_n is negative.

$$a_n \leqslant |u_n|$$

$$\therefore \quad a_1 + a_2 + a_3 + \ldots + a_n \leqslant |u_1| + |u_2| + \ldots + |u_m| \leqslant S$$

Since the series a_n only contains positive or zero terms, and the sum cannot exceed S, then statement (i) proves that $\sum a_n$ is convergent.

Similarly, $\sum b_n$ is convergent.

$$\therefore \quad \sum u_n = \sum a_n - \sum b_n$$

must tend to some finite limit between $+S$ and $-S$. If u_n is complex, a similar proof can be constructed by separating real and imaginary parts.

Thus, absolute convergence implies more than mere convergence.

3.2.2. *Comparison Test*

Only convergent series will be of importance in the solution of chemical engineering problems, but the following tests will be stated in full as they determine convergence or divergence. It is important to appreciate in these tests that if it is shown that a particular series is not convergent, this does not necessarily mean that it is divergent; it may oscillate.

The comparison test is the simplest of all and consists of two parts:

(i) If $|u_n| \leqslant v_n$ for all $n > N$ where N is some finite integer, and $\sum v_n$ is convergent, then $\sum u_n$ is absolutely convergent.

(ii) If $|u_n| \geqslant v_n$ for all $n > N$, and $\sum v_n$ is divergent to $+\infty$, then $\sum u_n$ is also divergent.

This test involves comparing the series under investigation with known convergent or divergent series, on a term-by-term basis, and it becomes more useful as the behaviour of more series becomes known. Comparison with the simple geometrical progression with a suitable common ratio as in example 1 is very useful for part (i) of the test, and the divergent series given in example 2(b) below is useful in part (ii) of the test.

Example 2.

$$u_n = n^{-p} \quad (p \text{ real})$$

$$S_n = 1 + 2^{-p} + 3^{-p} + \ldots + n^{-p} = \sum n^{-p}$$

The above series has different properties for different values of p.

(a) $p > 1$.

$$1 \qquad\qquad\qquad = 1$$

$$(\tfrac{1}{2})^p + (\tfrac{1}{3})^p \qquad\qquad < 2(\tfrac{1}{2})^p = (\tfrac{1}{2})^{p-1}$$

$$(\tfrac{1}{4})^p + (\tfrac{1}{5})^p + (\tfrac{1}{6})^p + (\tfrac{1}{7})^p \qquad < 4(\tfrac{1}{4})^p = (\tfrac{1}{4})^{p-1}$$

$$\ldots \text{etc.}$$

The rth line contains 2^{r-1} terms and begins with $(\tfrac{1}{2})^{rp}$.

$$\therefore \quad \sum n^{-p} < 1 + (\tfrac{1}{2})^{p-1} + (\tfrac{1}{4})^{p-1} + \ldots$$

The right-hand side is a geometrical progression with a finite sum to infinity. Therefore the series is convergent by Section 3.2.1(i).

(b) $p = 1$. In order to show that the series is divergent in this case, the following algebraic relationship is required.

$$\frac{1}{n-k} + \frac{1}{n+k} = \frac{2n}{n^2 - k^2} > \frac{2}{n}$$

Grouping the terms of the series as shown below, and applying the above algebraic inequality, gives

$$\tfrac{1}{2} + \tfrac{1}{3} + \tfrac{1}{4} = (\tfrac{1}{2} + \tfrac{1}{4}) + \tfrac{1}{3} > \tfrac{2}{3} + \tfrac{1}{3} = 1$$

$$\tfrac{1}{5} + \tfrac{1}{6} + \tfrac{1}{7} + \tfrac{1}{8} + \tfrac{1}{9} + \tfrac{1}{10} + \tfrac{1}{11} + \tfrac{1}{12} + \tfrac{1}{13}$$

$$= (\tfrac{1}{5} + \tfrac{1}{13}) + (\tfrac{1}{6} + \tfrac{1}{12}) + (\tfrac{1}{7} + \tfrac{1}{11}) + (\tfrac{1}{8} + \tfrac{1}{10}) + \tfrac{1}{9}$$

$$> \tfrac{2}{9} + \tfrac{2}{9} + \tfrac{2}{9} + \tfrac{2}{9} + \tfrac{1}{9} = 1$$

$$\therefore \quad (1) + (\tfrac{1}{2} + \tfrac{1}{3} + \tfrac{1}{4}) + (\tfrac{1}{5} + \ldots \tfrac{1}{13}) + (\tfrac{1}{14} + \ldots + \tfrac{1}{40}) + \ldots > 1 + 1 + 1 + 1 + \ldots$$

The rth bracket contains 3^{r-1} terms centred on 3^{1-r}. Therefore, the sum of the first $\tfrac{1}{2}(3^r - 1)$ terms is greater than r. By taking sufficient terms, the sum can be made to exceed any chosen finite value, and hence the series is divergent.

(c) $p < 1$.

$$\left(\frac{1}{n}\right)^p > \frac{1}{n} \quad \text{for } n > 1$$

The series is thus divergent by comparison with part (b) using the test of Section 3.2.2(ii).

3.2.3. Ratio Test

This also consists of two parts:

(i) If

$$\left| \frac{u_n}{u_{n+1}} \right| \geqslant k > 1$$

for all $n > N$, then the series is absolutely convergent.

(ii) If

$$\left|\frac{u_n}{u_{n+1}}\right| < 1$$

for all $n > N$, then the series is not convergent.

Regarding the first part of the test it is necessary to find a fixed finite number k to satisfy the inequality for all $n > N$. In the second part, if the series only contains positive terms, the test proves that the series is divergent, but in general the series may oscillate.

When the second part is not satisfied and $k = 1$ in the first part, thus violating the inequality, the test is not conclusive. That is, the series may be convergent, divergent or oscillatory, the test will not determine which.

Example 3.

$$S_n = 1 + \frac{1}{2} + \frac{1}{3} + \frac{1}{4} + \ldots + \frac{1}{n}$$

$$\left|\frac{u_n}{u_{n+1}}\right| = \frac{n+1}{n} = 1 + \frac{1}{n}$$

$$\therefore \quad |u_n/u_{n+1}| > 1 \quad \text{for all } n > N$$

but it is impossible to find a value of k such that part (i) of the ratio test is satisfied. The test is therefore inconclusive, but this series has already been proved divergent in Example 2(b).

3.2.4. *Alternating Series*

$$u_1 - u_2 + u_3 - u_4 + \ldots \tag{3.3}$$

is an alternating series if $u_n > 0$ for all n. If, in addition, the magnitude of successive terms decreases for all values of n and $u_n \to 0$ as $n \to \infty$, then the series is convergent.

Proof

$$S_{2m} = (u_1 - u_2) + (u_3 - u_4) + \ldots + (u_{2m-1} - u_{2m})$$

Since each term in the brackets is positive and the next term u_{2m+1} must be added, S_{2m} and S_{2m+1} must both be greater than zero.

$$S_{2m+1} = u_1 - (u_2 - u_3) - (u_4 - u_5) - \ldots - (u_{2m} - u_{2m+1})$$

Again, each term in the brackets is positive, and removing the last term to obtain S_{2m} reduces the value further. Therefore, S_{2m} and S_{2m+1} are both less than u_1.

$$\therefore \quad 0 < S_n < u_1 \quad \text{for all } n.$$

Both S_{2m} and S_{2m+1} increase as m increases but never exceed u_1, and hence they are separately convergent. Because $u_n \to 0$ as $n \to \infty$, S_{2m} and S_{2m+1} must both converge to the same limit S. Therefore the alternating series (3.3) is convergent.

3.3. POWER SERIES

Consider the power series

$$a_0 + a_1 z + a_2 z^2 + \ldots = \sum_0^\infty a_n z^n \tag{3.4}$$

If $|a_n/a_{n+1}| \to R$ as $n \to \infty$, then the ratio test can be applied as follows.

$$\left| \frac{u_n}{u_{n+1}} \right| = \left| \frac{a_n z^n}{a_{n+1} z^{n+1}} \right| = \left| \frac{a_n}{a_{n+1}} \right| \frac{1}{|z|} \to \frac{R}{|z|} \tag{3.5}$$

as $n \to \infty$. Therefore, if $|z| < R$, $|u_n/u_{n+1}|$ will tend to some limit greater than unity and will be convergent by part (i) of the ratio test. In detail, if $|z| = R - \varepsilon$ where ε is some small positive number less than R,

$$\frac{R}{|z|} = \frac{R}{R - \varepsilon}$$

$$= 1 + \frac{\varepsilon}{R - \varepsilon}$$

$$> 1 + (\varepsilon/R)$$

Taking $k = 1 + (\varepsilon/R)$ in Section 3.2.3 and substituting into equation (3.5) proves that the series (3.4) is convergent.

If $|z| > R$, a similar proof shows that the series (3.4) is divergent. Therefore, $\sum a_n z^n$ is convergent for $|z| < R$ and divergent for $|z| > R$. R is termed the "radius of convergence" of the power series and, as shown above, R is the limit of the ratio of successive coefficients. If $|a_n/a_{n+1}| \to \infty$ as $n \to \infty$, then the power series will always converge. If $|a_n/a_{n+1}| \to 0$ as $n \to \infty$, then the power series will only converge if $z = 0$, i.e. if all terms are zero.

3.3.1. *Binomial Series*

The series obtained by expanding $(1 + z)^p$ is

$$1 + pz + \frac{p(p-1)}{1.2} z^2 + \frac{p(p-1)(p-2)}{1.2.3} z^3 + \ldots = \sum_0^\infty \binom{p}{n} z^n \tag{3.6}$$

where $\binom{p}{n}$ signifies the binomial coefficient, z may be complex, but p is real.

In this case

$$a_n = \frac{p(p-1)(p-2)(\ldots)(p-n+1)}{1.2.3.\ldots.n}$$

$$\therefore \quad \left|\frac{a_n}{a_{n+1}}\right| = \left|\frac{n+1}{p-n}\right| \to 1 \quad \text{as } n \to \infty$$

Therefore, the binomial series is convergent for $|z| < 1$.

3.3.2. Exponential Series

$$e^z = 1 + z + \frac{z^2}{2!} + \frac{z^3}{3!} + \ldots = \sum_0^\infty \frac{z^n}{n!} \tag{3.7}$$

Here, $a_n = 1/(n!)$

$$\therefore \quad |a_n/a_{n+1}| = n+1 \to \infty \quad \text{as } n \to \infty.$$

Therefore, the exponential series is always convergent.

3.3.3. Logarithmic Series

$$\ln(1+z) = z - \tfrac{1}{2}z^2 + \tfrac{1}{3}z^3 - \tfrac{1}{4}z^4 + \ldots = \sum_{,1}^\infty \frac{(-1)^{n-1}z^n}{n} \tag{3.8}$$

Here, $a_n = (-1)^{n-1}/n$

$$\therefore \quad \left|\frac{a_n}{a_{n+1}}\right| = \frac{n+1}{n} \to 1 \quad \text{as } n \to \infty.$$

Therefore, the series for $\ln(1+z)$ is convergent for $|z| < 1$.

3.3.4. Trigonometric Series

$$\sin z = z - \frac{z^3}{3!} + \frac{z^5}{5!} - \ldots = \sum_0^\infty \frac{(-1)^n z^{2n+1}}{(2n+1)!} \tag{3.9}$$

$$\cos z = 1 - \frac{z^2}{2!} + \frac{z^4}{4!} - \ldots = \sum_0^\infty \frac{(-1)^n z^{2n}}{(2n)!} \tag{3.10}$$

These two series are convergent for all values of z.

3.3.5. Hyperbolic Series

$$\sinh z = z + \frac{z^3}{3!} + \frac{z^5}{5!} + \ldots = \sum_0^\infty \frac{z^{2n+1}}{(2n+1)!} \tag{3.11}$$

$$\cosh z = 1 + \frac{z^2}{2!} + \frac{z^4}{4!} + \ldots = \sum_0^\infty \frac{z^{2n}}{(2n)!} \tag{3.12}$$

These two series are convergent for all values of z.

The above series and many others can be obtained by applying Taylor's theorem, which develops a power series in ascending powers of the variable.

3.3.6. *Taylor's Theorem*

If x is a real variable, and $f(x)$ can be differentiated n times at $x = a$, then

$$f(a+h) = f(a) + hf'(a) + \tfrac{1}{2}h^2 f''(a) + \ldots + \frac{h^{n-1}}{(n-1)!} f^{(n-1)}(a) + \frac{h^n}{n!} [f^{(n)}(a) + \delta]$$

(3.13)

where dashes denote differentiation with respect to x, and $\delta \to 0$ as $h \to 0$.

Most functions of a continuous independent variable which arise in chemical engineering can be differentiated indefinitely at all points of the problem, and the conditions for the application of Taylor's theorem are always satisfied.

3.3.7. *L'Hôpital's Rule*

It is sometimes necessary to resolve an indeterminate fraction of the form $0/0$. The difficulty usually arises for some special value of a variable in a case such as

$$y = \frac{f(x)}{g(x)}$$

(3.14)

where $f(x)$ and $g(x)$ are differentiable n times, and $f(a) = g(a) = 0$. A value of y can be calculated for all values of x except $x = a$. Expanding $f(x)$ and $g(x)$ about the point $x = a$ by Taylor's theorem,

$$f(x) = f(a) + (x-a)f'(a) + \tfrac{1}{2}(x-a)^2 f''(a) + \ldots$$

$$g(x) = g(a) + (x-a)g'(a) + \tfrac{1}{2}(x-a)^2 g''(a) + \ldots$$

$$\therefore \quad y = \frac{f(a) + (x-a)f'(a) + \tfrac{1}{2}(x-a)^2 f''(a) + \ldots}{g(a) + (x-a)g'(a) + \tfrac{1}{2}(x-a)^2 g''(a) + \ldots}$$

(3.15)

If $f(a) = g(a) = 0$ and $g'(a) \neq 0$, then a factor $(x-a)$ can be taken out of both numerator and denominator of equation (3.15) giving

$$y = \frac{f'(a) + \tfrac{1}{2}(x-a)f''(a) + \ldots}{g'(a) + \tfrac{1}{2}(x-a)g''(a) + \ldots}$$

(3.16)

In the limit as $x \to a$,

$$y \to \frac{f'(a)}{g'(a)}$$

(3.17)

Example 1. Evaluate $y = (\sin x)/x$ when $x = 0$.

The value of y can be determined for any value of x no matter how small, but not for zero. The series expansion of $\sin x$, equation (3.9) can be used in this example, giving

$$y = \left(x - \frac{x^3}{3!} + \frac{x^5}{5!} - \dots\right)\Big/ x$$

$$= 1 - \frac{x^2}{3!} + \frac{x^4}{5!} - \dots$$

which is valid for all x no matter how small. In the limit as $x \to 0$, it can be seen that $y \to 1$ and this is the obvious value to take for y when $x = 0$.

This process of resolving indeterminate fractions can be repeated k times provided that $k < n$.

$$f(a) = f'(a) = \dots = f^{(k-1)}(a) = 0$$

and

$$g(a) = g'(a) = \dots = g^{(k-1)}(a) = 0.$$

The first k terms in both numerator and denominator of equation (3.15) are zero, a factor $(x-a)^k$ can be taken out of the remaining terms and the limit as $x \to a$ will give

$$y = f^{(k)}(a)/g^{(k)}(a) \tag{3.18}$$

As soon as a non-zero derivative occurs in either series, both series must be terminated and the ratio evaluated as zero, infinity, or any finite value.

Example 2. Find the limit as $x \to 0$ of

$$y = \frac{x^2 \sin x}{1 - \cos x}$$

With the above notation, $f(x) = x^2 \sin x$, $g(x) = 1 - \cos x$.

$$\therefore \quad f(0) = g(0) = 0$$

$$\therefore \quad f'(x) = 2x \sin x + x^2 \cos x \qquad\qquad f'(0) = 0$$

and

$$g'(x) = \sin x \qquad\qquad g'(0) = 0$$

$$\therefore \quad f''(x) = 2 \sin x + 4x \cos x - x^2 \sin x \quad f''(0) = 0$$

and

$$g''(x) = \cos x \qquad\qquad g''(0) = 1$$

$$\therefore \quad \text{as } x \to 0, \quad y \to \frac{f(0)}{g(0)} \to \frac{f'(0)}{g'(0)} \to \frac{f''(0)}{g''(0)} \to 0$$

When the first non-vanishing derivative occurs in the denominator, as here, the limit is zero, and when the first non-vanishing derivative occurs in the numerator, the limit is infinite. If, as in the first example, both derivatives of the same order become non-zero simultaneously, then the limit is finite.

3.3.8. *Application of Series in Chemical Engineering*

Some of the problems encounted in chemical engineering prove very difficult by standard analytical methods, and some even defy analytical solution. Consequently, methods of successive approximation, and the numerical methods of Chapter 11 are used extensively. All of these methods can only be used when they are convergent since, in practice, a fundamentally infinite process is being terminated at some finite point. Later in this chapter, series solutions will be developed for some problems, and although a series solution may be convergent and therefore satisfactory from a mathematical point of view, it may converge too slowly to be of any practical value. Unless a series is rapidly convergent requiring only about five terms to express the solution adequately, the numerical work becomes prohibitive and the solution is of little practical value. No harm can come from this situation, since the eventual solution will be valid mathematically, and in Section 3.4.6, Example 6 illustrates this situation. The danger lies in the opposite situation, when the first few terms of a series decrease rapidly giving an apparently useful solution, while the series as a whole is divergent and therefore not valid mathematically. A divergent series is of no practical value, and therefore any series obtained must always be checked for convergence.

3.4. Method of Frobenius

In Section 2.5.1, the linear differential equation with constant coefficients was solved and shown to have solutions for the complementary function of the types

$$y = A_1 e^{m_1 x} + A_2 e^{m_2 x} \tag{2.45}$$

$$y = (A + Bx) e^{mx}$$

and

$$y = (A \cos \alpha x + B \sin \alpha x) e^{\beta x} \tag{2.49}$$

By referring to Sections 3.3.2 and 3.3.4, all of these solutions can be expanded in ascending power series of x. Thus for the first case,

$$y = A(1 + m_1 x + \tfrac{1}{2} m_1^2 x^2 + \ldots) + B(1 + m_2 x + \tfrac{1}{2} m_2^2 x^2 + \ldots) \tag{3.19}$$

This series form of y can be accepted as a solution of the equation provided that the differential equation is satisfied by it and the series is convergent. it can be shown that the first condition is satisfied by differentiating the series term by term and substituting the result into the differential equation when all terms will cancel out. The series can be checked for convergence by any of the tests given in Section 3.3.

Because the linear differential equation with constant coefficients always possesses a valid series solution, it is natural to expect at least some differential equations with variable coefficients to possess series solutions. Since the majority of series cannot be summed, it is to be expected that some solutions must be left in series form.

If the general second order differential equation can be expressed in the form

$$x^2 \frac{d^2 y}{dx^2} + x F(x) \frac{dy}{dx} + G(x) y = 0 \tag{3.20}$$

where
$$F(x) = F_0 + F_1 x + F_2 x^2 + \ldots \tag{3.21}$$

and
$$G(x) = G_0 + G_1 x + G_2 x^2 + \ldots \tag{3.22}$$

with equations (3.21) and (3.22) having a radius of convergence R; then the equation can be solved completely by the method of Frobenius in the form of a power series which is also convergent for $|x| \leqslant R$.

In order to solve equation (3.20), put

$$y = \sum_0^\infty a_n x^{n+c} \quad (a_0 \neq 0) \tag{3.23}$$

where c is the arbitrary (usually fractional) power of x in the first term of the series.

$$\therefore \quad \frac{dy}{dx} = \sum_0^\infty a_n (n+c) x^{n+c-1} \tag{3.24}$$

$$\therefore \quad \frac{d^2 y}{dx^2} = \sum_0^\infty a_n (n+c)(n+c-1) x^{n+c-2} \tag{3.25}$$

Substituting these equations into equation (3.20) gives

$$\sum a_n(n+c)(n+c-1)x^{n+c} + (F_0 + F_1 x + F_2 x^2 + \ldots) \sum a_n(n+c) x^{n+c}$$
$$+ (G_0 + G_1 x + G_2 x^2 + \ldots) \sum a_n x^{n+c} = 0 \tag{3.26}$$

where the extra powers of x in equation (3.20) have been absorbed in the series for y and its derivatives.

The lowest power of x which appears in equation (3.26) is x^c, and coefficients of x^c can therefore be equated, giving

$$a_0 c(c-1) + F_0 a_0 c + G_0 a_0 = 0$$
$$\therefore \quad c^2 + (F_0 - 1)c + G_0 = 0 \tag{3.27}$$

This quadratic equation for the index of the first term (c) is called the "indicial equation".

Equating the coefficients of the next power x^{c+1},

$$a_1(1+c)c + F_0 a_1(1+c) + F_1 a_0 c + G_0 a_1 + G_1 a_0 = 0$$
$$\therefore \quad a_1[(c+1)^2 + (F_0-1)(c+1) + G_0] + a_0(F_1 c + G_1) = 0 \tag{3.28}$$

which determines a_1 in terms of a_0 and c.

The coefficient of a_1 in equation (3.28) can be obtained from the left-hand side of equation (3.27) by replacing c by $(c+1)$. In general, the coefficient of a_j in the equation obtained by equating coefficients of x^{c+j} can be obtained from equation (3.27) by replacing c by $(c+j)$. If the two roots of equation (3.27) are denoted by c_1 and c_2 and they differ by a positive integer j so that $c_2 = c_1 + j$; when the root $c = c_1$ is used, the coefficient of a_j on the first occasion that it appears will be zero by the above explanation since $(c_1 + j)$ satisfies (3.27). But this equation should determine a_j, which is impossible when a_j is not present in the equation. The method thus encounters difficulties if the roots of equation (3.27) differ by an integer.

The general solution cannot be considered further, since the methods diverge into four channels depending upon the properties of the roots of the indicial equation (3.27). The treatment of each case will now be pursued separately.

3.4.1. *Case I. Roots of Indicial Equation Different, but not by an Integer*

In this case, equation (3.28) determines a_1 in terms of a_0 and c. By equating coefficients of increasing powers of x as they appear in equation (3.26), all coefficients can be determined in terms of a_0 and c. The two different values of c which are solutions of equation (3.27) each give rise to an independent series having its first coefficient as an arbitrary constant. The sum of these two series is the complete solution of equation (3.20) for this case.

Example 1.

$$4x\frac{d^2 y}{dx^2}+6\frac{dy}{dx}+y = 0 \qquad\qquad \text{I}$$

The series substitution (3.23) and its differentiated forms are

$$y = a_0 x^c + a_1 x^{c+1} + \ldots = \sum_0^\infty a_n x^{n+c}$$

$$\frac{dy}{dx} = a_0 c x^{c-1} + a_1(c+1) x^c + \ldots = \sum_0^\infty a_n(n+c) x^{n+c-1}$$

$$\frac{d^2 y}{dx^2} = a_0 c(c-1) x^{c-2} + a_1(c+1) cx^{c-1} + \ldots = \sum_0^\infty a_n(n+c)(n+c-1) x^{n+c-2}$$

and these may be substituted into equation I to give

$$4[a_0 c(c-1) x^{c-1} + a_1(c+1) cx^c + \ldots] + 6[a_0 cx^{c-1} + a_1(c+1) x^c + \ldots]$$

$$+ a_0 x^c + a_1 x^{c+1} + \ldots = 0 \qquad\qquad \text{II}$$

Equating coefficients of like powers of x, starting with the smallest x^{c-1},

$$\therefore \quad 4a_0 c(c-1) + 6a_0 c = 0 \qquad\qquad \text{III}$$

$$\therefore \quad 4a_1(c+1) c + 6a_1(c+1) + a_0 = 0 \qquad\qquad \text{IV}$$

$$\therefore \quad 4a_2(c+2)(c+1) + 6a_2(c+2) + a_1 = 0 \qquad\qquad \text{V}$$

and for the coefficient of x^{c+r},

$$4a_{r+1}(c+r+1)(c+r) + 6a_{r+1}(c+r+1) + a_r = 0 \qquad\qquad \text{VI}$$

The indical equation III has the two roots

$$c = 0 \quad \text{or} \quad c = -\tfrac{1}{2} \qquad\qquad \text{VII}$$

When $c = 0$, the series reduces to the simple series solution with constant first term, and equations IV, V and VI become

$$6a_1 + a_0 = 0, \qquad \therefore \quad a_1 = -a_0/6$$

$$8a_2 + 12a_2 + a_1 = 0, \qquad \therefore \quad a_2 = -a_1/20$$

and

$$4r(r+1)a_{r+1} + 6(r+1)a_{r+1} + a_r = 0$$

$$\therefore \quad (2r+2)(2r+3)a_{r+1} + a_r = 0$$

$$\therefore \quad \frac{a_{r+1}}{a_r} = \frac{-1}{(2r+2)(2r+3)} \qquad\qquad \text{VIII}$$

Equation VIII must be true for all values of $r \geqslant 0$, and it can be seen that the previous answers are confirmed by using $r = 0$ and $r = 1$. Replacing r with $(r-1)$ gives

$$\frac{a_r}{a_{r-1}} = \frac{-1}{2r(2r+1)}$$

Repeating this process of reducing the value of r by unity each time gives

$$\frac{a_{r-1}}{a_{r-2}} = \frac{-1}{(2r-2)(2r-1)} \quad \text{etc.}$$

until

$$\frac{a_1}{a_0} = \frac{-1}{2(3)}$$

Multiplying the last r equations together gives

$$\frac{a_r}{a_0} = \frac{(-1)^r}{(2r+1)!} \qquad\qquad \text{IX}$$

The solution therefore takes the form

$$y = \sum_0^\infty \frac{(-1)^n a_0}{(2n+1)!} x^n$$

$$= a_0\left(1 - \frac{x}{3!} + \frac{x^2}{5!} - \frac{x^3}{7!} + \ldots\right) \qquad\qquad \text{X}$$

The alternating signs and odd factorials indicate a strong simularity between this series and the series (3.9) for $\sin z$. Thus the first solution corresponding to $c = 0$ is

$$y = a_0(\sin \sqrt{x})/\sqrt{x} \qquad\qquad \text{XI}$$

However, when $c = -\frac{1}{2}$, equations IV and VI become

$$-a_1 + 3a_1 + a_0 = 0 \quad \text{and} \quad a_{r+1}(2r+1)(2r-1) + 3a_{r+1}(2r+1) + a_r = 0$$

Rearranging,

$$\frac{a_{r+1}}{a_r} = \frac{-1}{(2r+1)(2r+2)}$$

similarly,

$$\frac{a_r}{a_{r-1}} = \frac{-1}{(2r-1)2r} \quad \text{etc.}$$

until

$$\frac{a_1}{a_0} = \frac{-1}{1 \cdot 2}$$

Multiplying these equations together gives

$$a_r = \frac{(-1)^r a_0}{(2r)!} \qquad\qquad \text{XII}$$

Substituting into the general series gives the solution

$$y_2 = a_0 x^{-\frac{1}{2}}\left[1 - \frac{x}{2!} + \frac{x^2}{4!} - \cdots\right] = \sum_0^\infty \frac{(-1)^n a_0 x^{n-\frac{1}{2}}}{(2n)!}$$

$$\therefore \quad y_2 = a_0 (\cos \sqrt{x})/\sqrt{x} \qquad\qquad \text{XIII}$$

The general solution of equation I is thus

$$y = (A \sin \sqrt{x} + B \cos \sqrt{x})/\sqrt{x} \qquad\qquad \text{XIV}$$

The coefficient a_0 which appears in both equations XI and XIII should be different in the two solutions which are independent. It is good practice to change both of these coefficients to new arbitrary constants as in equation XIV.

Equation VI, which was obtained by equating coefficients of a general power of x, is called the "recurrence relation" because it enables any coefficient a_r to be determined in terms of the earlier coefficients in the series. When only two coefficients are involved in the recurrence relation it is a simple matter to find the general term such as equation XII, whereas when more than two are involved it is virtually impossible to develop an analytical expression for the general coefficient.

3.4.2. Case II. Roots of Indicial Equation Equal

The first part of the complementary function can be found by using the method described above, but only one series containing one arbitrary constant can be determined because the indicial equation has equal roots. The second solution can be found by assuming that the two roots are different and taking the limit as the second root approaches the first through variable values of c. The complete solution for different roots can be written

$$y = Au(x, c_1) + Bu(x, c_2)$$
$$= (A + B) u(x, c_1) + B[u(x, c_2) - u(x, c_1)]$$
$$= (A + B) u(x, c_1) + B(c_2 - c_1)\left[\frac{u(x, c_2) - u(x, c_1)}{c_2 - c_1}\right]$$

Redefining the arbitrary constants A and B, in terms of α and β,

$$y = \alpha u(x, c_1) + \beta\left[\frac{u(x, c_2) - u(x, c_1)}{c_2 - c_1}\right] \tag{3.29}$$

If the limit as $c_2 \to c_1$ is taken of equation (3.29), the term in square brackets becomes the partial derivative of u with respect to c and evaluated at $c = c_1$. The complete solution for the case of equal roots of the indicial equation is thus

$$y = \alpha u(x, c_1) + \beta\left(\frac{\partial u}{\partial c}\right)_{c=c_1} \tag{3.30}$$

Example 2.

$$x\frac{d^2 y}{dx^2} + (1 - x)\frac{dy}{dx} - y = 0 \tag{I}$$

Put

$$y = \sum_0^\infty a_n x^{n+c} \tag{3.23}$$

and its differentiated forms (3.24) and (3.25) into equation I and equate coefficients of x^{c-1}, x^c, and x^{c+r} to obtain

$$a_0 c(c - 1) + a_0 c = 0 \tag{II}$$

$$a_1(c + 1)^2 = a_0(c + 1) \tag{III}$$

and

$$a_{r+1}(r + c + 1)^2 = a_r(r + c + 1) \tag{IV}$$

The indicial equation II has a double root $c = 0$ and since r is a positive integer, $(r + c + 1)$ cannot be zero for any value of r. Hence equation IV becomes the recurrence relation

$$a_{r+1} = a_r/(r + c + 1) \tag{V}$$

By using the recurrence relation r times, the following expression can be obtained for the general term.

$$a_{r+1} = \frac{a_0}{(r + c + 1)(r + c)(...)(c + 1)}$$

or

$$a_n = \frac{a_0}{(n + c)(n + c - 1)(...)(c + 1)} \tag{VI}$$

The function $u(x, c)$ is now established by putting equation VI into the general series (3.23).

$$u(x, c) = \sum_0^\infty \frac{a_0 x^{n+c}}{(n+c)(n+c-1)(...)(c+1)} \qquad \text{VII}$$

$$= a_0 x^c + \frac{a_0}{c+1} x^{c+1} + \frac{a_0}{(c+1)(c+2)} x^{c+2} + ...$$

When $c = 0$, $u(x, c)$ satisfies equation I and is the first solution,

$$y_1 = u(x, 0) = a_0\left[1 + x + \frac{x^2}{2!} + \frac{x^3}{3!} + ...\right] = a_0 e^x \qquad \text{VIII}$$

The second solution as indicated by equation (3.30) is thus

$$y_2 = \frac{\partial}{\partial c}\left[a_0 \sum_0^\infty \frac{x^{n+c}}{(c+1)(c+2)(...)(c+n)}\right]_{c=0} \qquad \text{IX}$$

The following two points of technique are useful in order to perform the required differentiation efficiently.

(i)
$$x^{n+c} = e^{(n+c)\ln x} \qquad (3.31)$$

which may be verified by taking the logarithm of both sides.

(ii) If
$$y = f_1(c) f_2(c) f_3(c)...f_r(c) \qquad (3.32)$$

then
$$\ln y = \ln f_1(c) + \ln f_2(c) + ... + \ln f_r(c)$$

Differentiating, and multiplying throughout by y,

$$\frac{dy}{dc} = y\left[\frac{f_1'(c)}{f_1(c)} + \frac{f_2'(c)}{f_2(c)} + ... + \frac{f_r'(c)}{f_r(c)}\right] \qquad (3.33)$$

where (') denotes differentiation with respect to c.

Continue the example by applying these principles to equation IX, taking

$$f_1(c) = 1/(c+1), \quad ..., \quad f_n(c) = 1/(c+n), \quad f_{n+1}(c) = e^{(n+c)\ln x}$$

$$\therefore \frac{f_1'(c)}{f_1(c)} = \frac{-1}{c+1}, \quad ..., \quad \frac{f_n'(c)}{f_n(c)} = \frac{-1}{c+n}, \quad \frac{f_{n+1}'(c)}{f_{n+1}(c)} = \ln x$$

$$\therefore y_2 = \left[a_0 \sum_0^\infty \frac{x^{n+c}}{(c+1)(c+2)(...)(c+n)}\left(\frac{-1}{c+1} - \frac{1}{c+2} - ... - \frac{1}{c+n} + \ln x\right)\right]_{c=0}$$

$$= a_0 \sum_0^\infty \frac{x^n}{n!}\ln x - a_0 \sum_1^\infty \frac{x^n}{n!}\left(1 + \frac{1}{2} + \frac{1}{3} + ... + \frac{1}{n}\right)$$

$$= a_0 e^x \ln x - a_0 \sum_1^\infty \frac{x^n}{n!}\left(1 + \frac{1}{2} + \frac{1}{3} + ... + \frac{1}{n}\right) \qquad \text{X}$$

Referring to equation VIII, it can be seen that the second solution consists of the first solution multiplied by $\ln x$, added to a new series. This relationship always exists between the two solutions when the second solution is found by using the above method.

The complete solution of equation I is thus

$$y = Au(x,0) + B(\partial u/\partial c)_{c=0}$$

$$= A\,e^{x} + B\left[e^{x}\ln x - \sum_{1}^{\infty}\frac{x^{n}}{n!}\left(1 + \frac{1}{2} + \frac{1}{3} + \ldots + \frac{1}{n}\right)\right] \qquad \text{XI}$$

3.4.3. *Case IIIa. Roots of Indicial Equation Differing by an Integer*

This case is very similar to Case II in that the second solution involves a partial differentiation with respect to c, thus generating a term $\ln x$. The function $u(x,c)$ has to be defined as before such that the recurrence relation is satisfied but not the indicial equation. The solution can be shown to take the form

$$y = Au(x, c_2) + B\frac{\partial}{\partial c}[(c - c_1)u(x, c)]_{c=c_1} \qquad (3.34)$$

The reason for the above form of the second solution is best visualized by reference to an example.

Example 3.

$$x(1 - x)\frac{d^2 y}{dx^2} + (2 - 5x)\frac{dy}{dx} - 4y = 0 \qquad \text{I}$$

With the usual series substitution (3.23) and following the same method as before, the indicial equation is

$$a_0 c(c - 1) + 2a_0 c = 0 \qquad \text{II}$$

the first recurrence relation is

$$a_1(c + 1) = a_0(c + 2) \qquad \text{III}$$

and the general recurrence relation is

$$a_{r+1} = \frac{r + c + 2}{r + c + 1}a_r \qquad \text{IV}$$

The roots of the indicial equation are

$$c_1 = -1 \quad \text{or} \quad c_2 = 0$$

As mentioned earlier in Section 3.4, when the roots of the indicial equation differ by an integer j, difficulties arise in the first recurrence relation in which a_j appears. In this example, $j = 1$, and equation III is the first equation in which a_1 occurs. When $c = -1$, equation III gives

$$a_1 0 = a_0$$

$$\therefore \quad a_1 = \infty$$

The series arising from $c = -1$ is thus of no value, but the series based on the larger root $c = 0$ is still valid. By successive use of the recurrence relation IV, the general term becomes

$$a_{r+1} = \frac{r+c+2}{c+1} a_0 \qquad \qquad \text{V}$$

and for the first solution with $c = 0$, this simplifies to

$$a_{r+1} = (r+2) a_0$$

$$\therefore \quad y_1 = a_0(1 + 2x + 3x^2 + ...)$$

$$= a_0(1-x)^{-2} \qquad \qquad \text{VI}$$

by the binomial series (3.6).

Define the function $u(x, c)$ again, which satisfies the recurrence relation IV and its solution V, but not necessarily the indicial equation II.

$$u(x, c) = \sum_0^\infty \frac{n+c+1}{c+1} a_0 x^{n+c} \qquad \qquad \text{VII}$$

If this function is substituted into the left-hand side of the differential equation I, all terms cancel out with the exception of the term identical with the indicial equation. Thus,

$$x(1-x)\frac{d^2 u}{dx^2} + (2-5x)\frac{du}{dx} - 4u = a_0 c(c+1) x^{c-1} \qquad \qquad \text{VIII}$$

If $c = 0$, the right-hand side of equation VIII vanishes thus verifying that $y_1 = u(x, 0)$ is the first solution of equation I. However, although $c = -1$ also causes the right-hand side of equation VIII to vanish it does not yield a solution because $u(x, -1) \to \infty$. An artificial way of generating a second solution is to multiply equation VIII by the factor $(c+1)$ to cancel out the denominator of $u(x, c)$ and create a factor $(c+1)^2$ on the right-hand side of equation VIII. This squared factor now provides the same features as a repeated root of the indicial equation and it can be shown that the second solution takes the form

$$y_2 = \frac{\partial}{\partial c}[(c+1) u(x, c)]_{c=-1} \qquad \qquad \text{IX}$$

But, from equation VII

$$(c+1)u(x, c) = \sum_0^\infty (n+c+1)a_0 x^{n+c} \qquad \qquad \text{X}$$

$$\therefore \quad \frac{\partial}{\partial c}(c+1)u(x, c) = \sum_0^\infty a_0 x^{n+c}[(n+c+1)\ln x+1]$$

$$\therefore \quad y_2 = a_0(\ln x)(1+2x+3x^2+...)+a_0(x^{-1}+1+x+...)$$

$$= \frac{a_0 \ln x}{(1-x)^2} + \frac{a_0}{x(1-x)} \qquad \qquad \text{XI}$$

Once again, the term multiplying $\ln x$ is the first solution. The complete solution of equation I is thus

$$y = \frac{A}{(1-x)^2} + \frac{B\ln x}{(1-x)^2} + \frac{B}{x(1-x)} \qquad \qquad \text{XII}$$

3.4.4. *Case IIIb. Roots of Indicial Equation Differing by an Integer*

There is just one further variation in the type of solution which can arise in the method of Frobenius. In the last example, equation III gave

$$a_1(c+1) = a_0(c+2)$$

and when $c = -1$, a_1 became infinite. Had the right-hand side also contained a factor $(c+1)$ this particular difficulty would not have arisen, equation III would have been automatically satisfied, and both parts of the solution would have been normal power series as in Case I. An example of this situation follows.

Example 4.
$$x\frac{d^2 y}{dx^2} + (x-1)\frac{dy}{dx} - y = 0 \qquad \qquad \text{I}$$

With the usual series substitution (3.23) and following the same method as before, the indicial equation is

$$a_0 c(c-1) - a_0 c = 0 \qquad \qquad \text{II}$$

and the general recurrence relation is

$$a_{r+1}(r+c+1)(r+c-1) + a_r(r+c-1) = 0 \qquad \qquad \text{III}$$

Since the roots of equation II differ by 2, consider the first of equations III in which a_2 occurs. Viz.

$$a_2(c+2)c + a_1 c = 0 \qquad \qquad \text{IV}$$

When $c = 0$, the coefficient of a_2 in equation IV is zero as in Section 3.4.3, but both terms in equation IV are zero when $c = 0$, and a_2 remains indeterminate instead of infinite.

Therefore, if $c = 2$, equation III can be simplified by cancellation; but if $c = 0$, the term $(r+c-1)$ can only be cancelled if $r > 1$.

$$\therefore \quad a_{r+1}(r+c+1) = -a_r \quad \text{(unless } c = 0, r = 1) \qquad\qquad \text{V}$$

The first solution is given by $c = 2$, thus

$$a_{r+1}(r+3) = -a_r$$

and by successive substitution,

$$a_{r+1} = \frac{(-1)^{r+1} 2a_0}{(r+3)!}$$

$$\therefore \quad y_1 = \sum_0^\infty \frac{(-1)^n 2a_0}{(n+2)!} x^{n+2}$$

$$= 2a_0\left(\frac{x^2}{2!} - \frac{x^3}{3!} + \frac{x^4}{4!} - \cdots\right)$$

$$= 2a_0(e^{-x} - 1 + x) \qquad\qquad \text{VI}$$

The second solution, when $c = 0$, becomes

$$y_2 = a_0 - a_0 x + a_2 x^2 - \frac{a_2 x^3}{3} + \frac{a_2 x^4}{3.4} - \cdots$$

$$= a_0(1-x) + 2a_2\left(\frac{x^2}{2!} - \frac{x^3}{3!} + \frac{x^4}{4!} - \cdots\right)$$

$$= a_0(1-x) + 2a_2(e^{-x} - 1 + x) \qquad\qquad \text{VII}$$

All coefficients after the first two can be found in terms of a_2 and there are thus two arbitrary constants in the second solution making three in all. Inspection shows that the series associated with a_2 in equation VII is identical with the first solution as given by equation VI. One solution has thus been obtained twice, and in Case IIIb when a coefficient becomes indeterminate, the series obtained by using the smaller root is in fact the complete solution. The complete solution of equation I can therefore be written

$$y = A(1-x) + Be^{-x} \qquad\qquad \text{VIII}$$

where $A = a_0 - 2a_2$ and $B = 2a_2$ in equation VII.

3.4.5. *Summary of the Method of Frobenius*

The examples given above have been specially selected so that the series solutions could all be summed. The solutions can be checked quite easily by differentiation and substitution, thus demonstrating that the series method gives correct results. It must be emphasized that the methods of Chapter 2 should always be tried before resorting to the method of

Frobenius. Too much time should not be devoted to trying to sum the series obtained because on many occasions the solutions of problems of engineering importance will not be expressible in closed analytical form.

The method has only been described for expressing the solution as a series of powers of x. Solutions in powers of $(x - x_0)$ can also be obtained by moving the origin along the x axis to x_0 and then proceeding as before. The method will be applicable if the convergence conditions are satisfied at the new origin.

The differential equation

$$x^2 \frac{d^2 y}{dx^2} + xF(x)\frac{dy}{dx} + G(x)y = 0 \tag{3.20}$$

can be solved by putting

$$y = \sum_0^\infty a_n x^{n+c} \quad (a_0 \neq 0) \tag{3.23}$$

if $F(x)$ and $G(x)$ can be expanded in a convergent series of non-negative powers of x for all $|x| < R$. The solution obtained will also be convergent for all $|x| < R$.

The indicial equation is found by equating the coefficients of the lowest power of x occurring in equation (3.20) after substituting equation (3.23). The recurrence relation is found by equating coefficients of x^{c+r} in the same equation. If the two roots of the indicial equation are $c = c_1$ and $c = c_2$ with $c_2 = c_1 + j$ and $j \geq 0$, then the following four cases arise.

Case I (j not an integer or zero). Two separate series of the form (3.23) are obtained, one for $c = c_1$ and the other for $c = c_2$. These two series added together with arbitrary constants give the complete solution.

Case II (j is zero). Define the function $u(x, c)$ which satisfies the recurrence relation but not necessarily the indicial equation. The solution then takes the form

$$y = Au(x, c_1) + B\left(\frac{\partial u}{\partial c}\right)_{c=c_1} \tag{3.30}$$

Case IIIa (j is an integer). Find the first recurrence relation in which a_j appears. When $c = c_1$, the coefficient of a_j will be zero, and if the rest of the equation is not zero the solution takes the form

$$y = Au(x, c_2) + B\frac{\partial}{\partial c}[(c - c_1)u(x, c)]_{c=c_1} \tag{3.34}$$

where $u(x, c)$ satisfies the recurrence relation but not necessarily the indicial equation.

Case IIIb (j is an integer). When $c = c_1$, the first recurrence relation in which a_j appears is identically zero leaving a_j indeterminate. The complete solution can be found in the form of equation (3.23) by putting $c = c_1$ when a_0 and a_j will be the two arbitrary constants.

3.4.6. *Illustrative Examples*

Example 5. Temperature Distribution in a Transverse Fin of Triangular Cross-section. The shape of the cooling fin is illustrated in Fig. 3.1 where the radius of the pipe (a) is 8 cm, the radius of the rim of the fin (b) is 20 cm, and the coordinate x m is measured inwards from the rim of the fin. There are two natural origins for the coordinate, but since the temperature distribution in the vicinity of the pipe axis is of no interest, the origin is

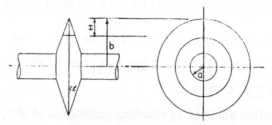

FIG. 3.1. Transverse cooling fin

taken on the rim instead. Assuming that the fin is thin, temperature variations normal to the central plane of the fin will be neglected. The thermal conductivity of the fin (k) is 380 W/m °C, and the surface heat transfer coefficient (h) is 12 W/m² °C. Denoting temperature by T °C with T_A °C representing the air temperature, the heat balance can be taken as follows.

The area available for heat conduction is

$$2\pi(b-x)(2x\tan\alpha) = 4\pi x(b-x)\tan\alpha \qquad\text{I}$$

Input
$$= -k4\pi x(b-x)\tan\alpha\frac{dT}{dx}$$

Output
$$= -k4\pi x(b-x)\tan\alpha\frac{dT}{dx} + \frac{d}{dx}\left[-k4\pi x(b-x)\tan\alpha\frac{dT}{dx}\right]\delta x$$
$$+ 2h2\pi(b-x)(\delta x\sec\alpha)(T-T_A)$$

Accumulation $= 0$

Since input − output = accumulation,

$$\frac{d}{dx}\left[k4\pi x(b-x)\tan\alpha\frac{dT}{dx}\right]\delta x - 2h2\pi(b-x)(\delta x\sec\alpha)(T-T_A) = 0$$

$$\therefore \quad k\sin\alpha\frac{d}{dx}\left[x(b-x)\frac{dT}{dx}\right] - h(b-x)(T-T_A) = 0 \qquad\text{II}$$

Putting $K = h/k\sin\alpha$ and $y = T-T_A$,

$$\therefore \quad x(b-x)\frac{d^2y}{dx^2} + (b-2x)\frac{dy}{dx} - K(b-x)y = 0 \qquad\text{III}$$

The appropriate boundary conditions are

at $\qquad\qquad\qquad x = 0, \quad y$ remains finite $\qquad\qquad$ IV

at $\qquad\qquad\qquad x = b-a, \quad T = T_B$

$\qquad\qquad\qquad \therefore \quad y = T_B - T_A \qquad\qquad\qquad$ V

where T_B is the temperature of the pipe contents.

Using the Frobenius substitutions (3.23), (3.24) and (3.25) in equation III gives

$$b \sum a_n(n+c)(n+c-1)x^{n+c-1} - \sum a_n(n+c)(n+c-1)x^{n+c}$$

$$+ b \sum a_n(n+c)x^{n+c-1} - 2 \sum a_n(n+c)x^{n+c} - Kb \sum a_n x^{n+c}$$

$$+ K \sum a_n x^{n+c+1} = 0 \qquad\qquad\qquad\qquad \text{VI}$$

The indicial equation obtained by equating coefficients of x^{c-1}, is

$$ba_0 c(c-1) + ba_0 c = 0 \qquad\qquad\qquad \text{VII}$$

$$\therefore \quad c^2 = 0$$

Equation III thus gives a Case II solution with equal roots. With the knowledge that the second solution will contain $\ln x$ which becomes infinite when $x = 0$ violating boundary condition IV, it may be concluded that the second solution has no physical significance in this example. Therefore it is only necessary to find the first solution with $c = 0$ and the remaining work can be simplified by omitting c.

Equating coefficients of the next power of x which is x^c,

$$ba_1 - Kba_0 = 0$$

$$\therefore \quad a_1 = Ka_0 \qquad\qquad\qquad\qquad \text{VIII}$$

The recurrence relation is

$$ba_{r+1}(r+1)r - a_r r(r-1) + ba_{r+1}(r+1) - 2a_r r - Kba_r + Ka_{r-1} = 0$$

$$\therefore \quad ba_{r+1}(r+1)^2 = a_r[r(r+1) + Kb] - Ka_{r-1} \qquad\qquad \text{IX}$$

This recurrence relation is different from all previous ones since it relates three of the original coefficients a_{r+1}, a_r, and a_{r-1}. It is thus difficult to determine a general formula for a_r in terms of a_0, although a solution may be possible by the methods of Chapter 9. Unless such a solution is possible, the second part of the complementary function in this example of equal roots cannot be found by the method of Section 3.4.3 even if it is required. The appropriate first solution can be found however on a term-by-term basis. Introducing the numerical values given in the problem, together with $T_A = 16$, $T_B = 100$, and $\alpha = 5°$ gives

$$K = 12/380 \times 0.0872 = 0.3621.$$

Equations VIII and IX become

$$a_1 = 0{\cdot}3621a_0$$

$$a_{r+1}(r+1)^2 = a_r[5r(r+1)+0{\cdot}3621]-1{\cdot}810a_{r-1} \qquad \text{X}$$

$$\therefore \quad a_2(2)^2 = 0{\cdot}3621a_0(10{\cdot}3621)-1{\cdot}810a_0$$

$$= 1{\cdot}9421a_0$$

$$\therefore \quad a_2 = 0{\cdot}4855a_0$$

Using equation X again

$$a_3(3)^2 = 0{\cdot}4855a_0(30{\cdot}3621)-1{\cdot}810 \times 0{\cdot}3621a_0$$

$$= 14{\cdot}085a_0$$

$$\therefore \quad a_3 = 1{\cdot}565a_0 \quad \text{etc.}$$

It can be seen that the most important terms in equation X are the first two, and the later behaviour of the series will approximate to

$$a_{r+1}(r+1)^2 = 5a_r r(r+1)$$

$$\therefore \quad a_{r+1} = \frac{5r}{r+1}a_r \qquad \text{XI}$$

By Section 3.3 this approximate series has a radius of convergence of 0·2 in agreement with the general statement of the convergence conditions related to equation (3.20). Because the fin is defined to occupy the range $0 \leqslant x \leqslant 0{\cdot}12$, the series solution is convergent throughout the fin.

Evaluating further coefficients from equation X gives

$$y = a_0(1+0{\cdot}3621x+0{\cdot}4855x^2+1{\cdot}565x^3+5{\cdot}85x^4+23{\cdot}4x^5+97{\cdot}5x^6+...) \quad \text{XII}$$

Boundary condition V becomes:

$$\text{at} \quad x = 0{\cdot}12, \quad y = 100-16 = 84$$

$$\therefore \quad 84 = a_0(1+0{\cdot}0435+0{\cdot}0070+0{\cdot}0027+0{\cdot}0012+0{\cdot}0006+0{\cdot}0003)$$

$$\therefore \quad a_0 = 84/1{\cdot}0553 = 79{\cdot}6 \qquad \text{XIII}$$

Equation XII gives the temperature distribution throughout the fin, and equation XIII shows that the rim temperature is 95·6 °C.

The rate of removal of heat by the fin can be calculated from the temperature gradient at the base. Thus

$$Q = k4\pi(b-a)a\tan\alpha(dT/dx)_{x=b-a}$$

$$= 4{\cdot}01(dT/dx)_{x=0{\cdot}12}$$

$$= 319(0{\cdot}3621+0{\cdot}1165+0{\cdot}0676+0{\cdot}0404+0{\cdot}0243+0{\cdot}0145+0{\cdot}0087+...)$$

$$= 205 \text{ W}$$

by differentiating equation XII, and using equation XI to estimate the continuation of the series to infinity.

Example 6. Tubular Gas Preheater. A supply of hot air is to be obtained by drawing cool air through a heated cylindrical pipe. The pipe is 0·1 m diameter and 1·2 m long, and is maintained at a temperature $T_w = 300\ ^\circ\text{C}$ throughout its length. The properties of the air are:

Heat capacity (C_p)	$= 1000\ \text{J/kg}\ ^\circ\text{C}$
Thermal conductivity (k)	$= 0\cdot035\ \text{W/m}\ ^\circ\text{C}$
Density (ρ)	$= 0\cdot8\ \text{kg/m}^3$
Flow rate (u)	$= 0\cdot009\ \text{m}^3/\text{s}$
(the above are all mean values)	
Inlet temperature	$= 20\ ^\circ\text{C}$

Overall heat transfer coefficient $(h) = 10x^{-\frac{1}{4}}\ \text{W/m}^2\ ^\circ\text{C}$

where x is the distance measured in metres from the pipe inlet.

Assuming that heat transfer takes place by conduction within the gas in an axial direction, mass flow of the gas in an axial direction, and by the above variable heat transfer coefficient from the walls of the tube, find the temperature of the exit gas.

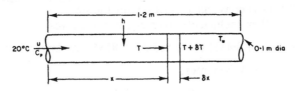

FIG. 3.2. Gas preheater

Solution

With the symbols given above and illustrated in Fig. 3.2, the following heat balance can be established.

	Input	Output
By conduction	$-kA\dfrac{dT}{dx}$	$-kA\dfrac{dT}{dx}+\dfrac{d}{dx}\left(-kA\dfrac{dT}{dx}\right)\delta x$
By mass flow	$u\rho C_p T$	$u\rho C_p\left(T+\dfrac{dT}{dx}\delta x\right)$
Wall heat transfer	$\pi dh(T_w-T)\,\delta x$	—

where A is the cross-sectional area of the pipe and d its diameter. Since in steady state the accumulation is zero,

$$-kA\frac{dT}{dx}+u\rho C_p T+\pi dh(T_w-T)\,\delta x$$

$$=-kA\frac{dT}{dx}+\frac{d}{dx}\left(-kA\frac{dT}{dx}\right)\delta x+u\rho C_p\left(T+\frac{dT}{dx}\delta x\right)$$

$$\therefore\quad \pi dh(T_w-T)=-kA\frac{d^2T}{dx^2}+u\rho C_p\frac{dT}{dx}\qquad\qquad\text{I}$$

Rearranging,

$$\frac{d^2T}{dx^2}-\frac{u\rho C_p}{kA}\frac{dT}{dx}+\frac{\pi dh}{kA}(T_w-T)=0\qquad\qquad\text{II}$$

Putting $t=T_w-T$,

$$\frac{d^2t}{dx^2}-\frac{u\rho C_p}{kA}\frac{dt}{dx}-\frac{4h}{kd}t=0\qquad\qquad\text{III}$$

Inserting the numerical values,

$$\frac{d^2t}{dx^2}-26\,200\frac{dt}{dx}-11\,430x^{-\frac12}t=0\qquad\qquad\text{IV}$$

Put $x=z^2$ to rationalize the coefficient of t,

$$\therefore\quad z\frac{d^2t}{dz^2}-(1+52\,400z^2)\frac{dt}{dz}-45\,720z^2t=0\qquad\qquad\text{V}$$

If an attempt is made to solve this equation by the method of Frobenius, the results for the first few terms are,

$$c=0\quad\text{or}\quad 2\quad\text{from the indicial equation.}$$

Taking

$$c=2$$

$$a_1=0\qquad\qquad a_3=3048a_0$$

$$a_2=13\,000a_0\qquad a_4=1\cdot14\times10^8\,a_0$$

The coefficients are increasing at an alarming rate, yet a test of equation V shows that the series must be convergent. This is a case mentioned earlier where a convergent series is of no practical value since more than 100 terms would have to be calculated to determine even the first solution. The trouble arises because two of the coefficients in equation IV are much larger than the coefficient of d^2t/dx^2. The second derivative arises from the gas conduction term and if this is neglected, equation IV becomes

$$\frac{dt}{dx} + 0.436x^{-\frac{1}{2}}t = 0 \qquad\qquad \text{VI}$$

which can be solved by separating the variables, viz.

$$t = \alpha \exp(-0.872x^{\frac{1}{2}}) \qquad\qquad \text{VII}$$

The boundary condition at $x = 0$, is $t = 300 - 20 = 280$.

$$\therefore \quad t = 280\exp(-0.872x^{\frac{1}{2}}) \qquad\qquad \text{VIII}$$

$$\therefore \quad T = 300 - 280\exp(-0.872x^{\frac{1}{2}})$$

The exit gas temperature is therefore 192 °C.

The approximation of neglecting the second differential term in equation IV can be checked by comparing the neglected term with one of the other terms. Differentiating VIII twice gives

$$\frac{dt}{dx} = -122.1x^{-\frac{1}{2}}\exp(-0.872x^{\frac{1}{2}})$$

$$\frac{d^2 t}{dx^2} = 122.1(\tfrac{1}{2}x^{-\frac{3}{2}} + 0.436x^{-1})\exp(-0.872x^{\frac{1}{2}})$$

$$\therefore \quad \frac{d^2 t}{dx^2}\Big/ 26\,200\frac{dt}{dx} = \frac{\tfrac{1}{2}x^{-\frac{3}{2}} + 0.436x^{-1}}{26\,200x^{-\frac{1}{2}}}$$

$$= \frac{1}{52\,400x} + \frac{1}{60\,090x^{\frac{1}{2}}}$$

which is small except when x is small. If $x = 10^{-3}$, the error ratio $= 2\%$.

The error has thus been made over a negligible part of the pipe and this justifies the above result.

3.5. BESSEL'S EQUATION

The differential equation

$$x^2\frac{d^2 y}{dx^2} + x\frac{dy}{dx} + (x^2 - k^2)y = 0 \qquad\qquad (3.35)$$

is known as Bessel's equation of order k, where k is a positive or zero constant. The equation arises so frequently in practical problems that it has been studied in great detail and this section will be devoted to giving an outline of the derivation of the various solutions by using the method of Frobenius. Because the equation occurs so frequently, the series solutions have been standardized and tabulated† thus eliminating the necessity of working through the series solution for each individual problem. Many other second order differential equations can be reduced to Bessel's equation (3.35) by a suitable substitution, and they too can then be solved by using the standard tables.

Applying the method of Frobenius by putting

$$y = \sum_0^\infty a_n x^{n+c} \tag{3.23}$$

and its differential forms (3.24) and (3.25) into equation (3.35),

$$\sum_0^\infty a_n(n+c)(n+c-1)x^{n+c} + \sum_0^\infty a_n(n+c)x^{n+c} + (x^2-k^2)\sum_0^\infty a_n x^{n+c} = 0$$

The indicial equation is

$$a_0 c(c-1) + a_0 c - k^2 a_0 = 0$$

which has the two roots

$$c = k \quad \text{or} \quad -k \tag{3.36}$$

The difference between the roots is thus $2k$, and the type of solution will depend upon the nature of k.

Equating coefficients of x^{c+1},

$$a_1(c+1)c + a_1(c+1) - k^2 a_1 = 0$$

$$\therefore \quad a_1(c+1+k)(c+1-k) = 0 \tag{3.37}$$

Because $c = \pm k$, and $k \geqslant 0$, then $c+1+k > 0$, but for the special case $k = \frac{1}{2}$ $c = -\frac{1}{2}$, $c+1-k = 0$. Therefore, unless

$$k = \frac{1}{2} \quad \text{and} \quad c = -\frac{1}{2}, \quad a_1 = 0 \tag{3.38}$$

The recurrence relation is obtained by equating coefficients of x^{c+r}.

$$a_r(r+c)(r+c-1) + a_r(r+c) - k^2 a_r + a_{r-2} = 0$$

$$\therefore \quad a_r(r+c)^2 - a_r k^2 = -a_{r-2} \tag{3.39}$$

$$\therefore \quad a_r = \frac{-a_{r-2}}{(r+c+k)(r+c-k)} \tag{3.40}$$

The first solution for $c = k$, will be normal in all cases and the recurrence relation becomes

$$a_r = \frac{-a_{r-2}}{r(r+2k)} \tag{3.41}$$

---

† For example, Jahnke and Emde, "Tables of Functions with Formulae and Curves", Dover Publications, New York; or Watson, "Theory of Bessel Functions", Cambridge University Press (1922).

Combining equations (3.38) and (3.40) gives

$$a_1 = a_3 = a_5 = \ldots = 0 \qquad (3.42)$$

and r can be replaced by $2m$ in the recurrence relation (3.41)

$$a_{2m} = \frac{-a_{2m-2}}{2m(2m+2k)}$$

Replacing m by $(m-1)$, $(m-2)$, etc., and making successive substitutions as before,

$$a_{2m} = \frac{(-1)^m a_0}{2^{2m} m!(m+k)(m+k-1)(\ldots)(k+1)}$$

$$= \frac{(-\frac{1}{4})^m a_0 \Gamma(k+1)}{m! \Gamma(m+k+1)}$$

where $\Gamma(k+1)$ is the gamma function or generalized factorial whose properties will be discussed in Chapter 5.

The first solution of Bessel's equation is thus

$$y_1 = 2^k \Gamma(k+1) a_0 \sum_{m=0}^{\infty} \frac{(-1)^m (\frac{1}{2}x)^{2m+k}}{m! \Gamma(m+k+1)}$$

The Bessel function of the first kind of order k is defined as

$$J_k(x) = \sum_{m=0}^{\infty} \frac{(-1)^m (\frac{1}{2}x)^{2m+k}}{m! \Gamma(m+k+1)} \qquad (3.43)$$

Defining a new arbitrary constant, the first solution of equation (3.35) is thus

$$y_1 = A J_k(x) \qquad (3.44)$$

The form of the second solution, for $c = -k$, depends on whether or not $2k$ is an integer, and the method of solution diverges into four channels corresponding with the cases of the method of Frobenius.

3.5.1. *Case I (2k is not an integer or zero)*

In this case the second solution is obtained from the first by replacing k in equation (3.43) by $-k$, giving the complete solution

$$y = A J_k(x) + B J_{-k}(x) \qquad (3.45)$$

3.5.2. *Case II (k = 0)*

Returning to the recurrence relation (3.40) and putting $k = 0$,

$$a_r = \frac{-a_{r-2}}{(r+c)^2}$$

Since $a_1 = 0$ from equation (3.38), r can be replaced by $2m$ again and the expression for the coefficient of the general term is

$$a_{2m} = \frac{(-1)^m a_0}{(2m+c)^2(2m+c-2)^2(\ldots)(c+2)^2}$$

Defining $u(x, c)$ as given in Section 3.4.2,

$$u(x, c) = \sum_{m=0}^{\infty} \frac{(-1)^m a_0 x^{2m+c}}{(2m+c)^2(2m+c-2)^2(\ldots)(c+2)^2} \tag{3.46}$$

Differentiating equation (3.46) with respect to c and putting $c = 0$ gives the second solution,

$$y_2 = \sum_{m=0}^{\infty} \frac{(-1)^m a_0(\tfrac{1}{2}x)^{2m}}{(m!)^2}\left[\ln x - 1 - \frac{1}{2} - \frac{1}{3} - \ldots - \frac{1}{m}\right]$$

$$= a_0\left[J_0(x)\ln x - \sum_{m=1}^{\infty} \frac{(-1)^m(\tfrac{1}{2}x)^{2m}}{(m!)^2}\left(1 + \frac{1}{2} + \ldots + \frac{1}{m}\right)\right]$$

The expression in brackets is the Neumann form of the second solution, but an alternative form named after Weber is more frequently used. The Weber form is obtained by adding $(\gamma - \ln 2)J_0(x)$ to the Neumann form and multiplying the result by $(2/\pi)$. Thus

$$y_2 = BY_0(x)$$

where

$$Y_0(x) = \frac{2}{\pi}[\ln(\tfrac{1}{2}x) + \gamma]J_0(x) - \frac{2}{\pi}\sum_{m=1}^{\infty} \frac{(-1)^m(\tfrac{1}{2}x)^{2m}}{(m!)^2}\left(1 + \frac{1}{2} + \ldots + \frac{1}{m}\right)$$

is the Weber form of the Bessel function of the second kind of order zero, and γ is Euler's constant which is defined by

$$\gamma = \lim_{m\to\infty}\left(1 + \frac{1}{2} + \frac{1}{3} + \ldots + \frac{1}{m} - \ln m\right) = 0.5772\ldots$$

The complete solution of equation (3.35) with $k = 0$ is thus

$$y = AJ_0(x) + BY_0(x) \tag{3.47}$$

3.5.3. Case IIIa (k is an integer)

$2k$ is an even integer and the first recurrence relation in which a_{2k} appears is, from equation (3.39),

$$a_{2k}(2k+c)^2 - a_{2k}k^2 = -a_{2k-2}$$

or

$$a_{2k}(k+c)(3k+c) = -a_{2k-2}$$

When $c = -k$, the left-hand side is zero but the right-hand side is not zero because a_{2k-2} is an even coefficient. The method for Case IIIa given in

Section 3.4.4 must therefore be followed. The eventual solution using

$$y_2 = \frac{\partial}{\partial c}\left[(c+k)\,u(x, c)\right]_{c=-k}$$

is

$$y_2 = BY_k(x) = \frac{2}{\pi}B[\ln\tfrac{1}{2}x + \gamma]\,J_k(x) - \frac{B}{\pi}\sum_{m=0}^{k-1}\frac{(k-m-1)!}{m!}(\tfrac{1}{2}x)^{2m-k}$$

$$- \frac{B}{\pi}\sum_{m=0}^{\infty}\frac{(-1)^m(\tfrac{1}{2}x)^{2m+k}}{m!(m+k)!}\left[1+\frac{1}{2}+\dots+\frac{1}{m}+1+\frac{1}{2}+\dots\frac{1}{m+k}\right]$$

where γ is again Euler's constant, and the Weber form has been found by adding $a_0(\gamma - \ln 2)\,J_k(x)$ to the Neumann form and multiplying by $(2/\pi)$. $Y_k(x)$ is the Weber form of the Bessel function of the second kind of order k.
    The complete solution of equation (3.35) is thus

$$y = AJ_k(x) + BY_k(x) \tag{3.48}$$

when k is an integer.

3.5.4. *Case IIIb (2k is an odd integer)*

The first recurrence relation in which a_{2k} appears is, from equation (3.39),

$$a_{2k}(k+c)(3k+c) = -a_{2k-2} \tag{3.49}$$

When $c = -k$, the left-hand side is zero, but the right-hand side is an odd coefficient and in all cases considered so far, the odd coefficients have been zero. This must now be investigated further. If $k = \tfrac{1}{2}$, equation (3.37) shows that a_1, the critical coefficient, is indeterminate and the solution follows Case IIIb giving two normal power series. If $k = \tfrac{3}{2}$, the critical recurrence relation (3.49) becomes

$$a_3(\tfrac{3}{2}+c)(\tfrac{9}{2}+c) = -a_1$$

and since $a_1 = 0$, a_3 is indeterminate. Continuing this argument, a_{2k} will always be indeterminate if $2k$ is odd, and the solution again consists of two normal series.

$$y = AJ_k(x) + BJ_{-k}(x) \tag{3.50}$$

The general solution of Bessel's equation (3.35) can therefore be written

$$y = AJ_k(x) + BJ_{-k}(x) \tag{3.45, 3.50}$$

if k is not an integer or zero, or

$$y = AJ_k(x) + BY_k(x) \tag{3.47, 3.48}$$

if k is an integer or zero.
    Examples of the use of the above Bessel functions occur quite frequently in chemical engineering when the problem involves partial differential

equations, but their use in ordinary differential equations is rare in chemical engineering but more frequent in other branches of engineering. Rather than present an irrelevant example at this point, the reader is referred to Example 3 in Section 8.7.5 and Section 8.7.3, where the above solutions form one stage in the solution of certain partial differential equations.

3.5.5. *Modified Bessel's Equation*

Bessel's equation (3.35) has now been solved for all real values of x, and the solution given as the sum of two linearly independent functions of x each multiplied by an arbitrary constant. Two kinds of Bessel function have been defined both of which are needed to express the solution if k is an integer or zero, but two versions of only the first kind if k is not an integer or zero.

A further type of differential equation can be solved with the aid of Bessel functions, it is

$$x^2\frac{d^2 y}{dx^2} + x\frac{dy}{dx} - (x^2 + k^2)y = 0 \tag{3.51}$$

Equation (3.51) can be obtained from Bessel's equation (3.35) by replacing x with ix where $i = \sqrt{(-1)}$. The solution of equation (3.51) can thus be written

$$y = AJ_k(ix) + BJ_{-k}(ix) \tag{3.52}$$

if k is not an integer or zero, or

$$y = AJ_k(ix) + BY_k(ix) \tag{3.53}$$

if k is an integer or zero.

In Section 2.5.2 it was found that the solution of the second order linear differential equation with constant coefficients could be expressed more conveniently in terms of real trigonometric functions than complex exponential functions when the auxiliary equation had complex roots. Here instead of using the Bessel functions of a complex variable, it is similarly convenient to define modified Bessel functions of a real variable so that the above solutions can be expressed in real form. The necessary definitions are

$$I_k(x) = [\exp(-\tfrac{1}{2}k\pi i)]\,J_k(ix) \tag{3.54}$$

$$= \sum_{m=0}^{\infty} \frac{(\tfrac{1}{2}x)^{2m+k}}{m!\,\Gamma(m+k+1)}$$

$$K_k(x) = \{\exp[\tfrac{1}{2}(k+1)\,\pi i]\}\,[J_k(ix) + iY_k(ix)] \tag{3.55}$$

$$= (-1)^{k+1}[\ln(\tfrac{1}{2}x) + \gamma]\,I_k(x) + \frac{1}{2}\sum_{m=0}^{k-1}\frac{(-1)^m(k-m-1)!}{m!}(\tfrac{1}{2}x)^{2m-k}$$

$$+ \frac{1}{2}\sum_{m=0}^{\infty}\frac{(-1)^k(\tfrac{1}{2}x)^{2m+k}}{m!(m+k)!}\left[1 + \frac{1}{2} + \dots + \frac{1}{m} + 1 + \frac{1}{2} + \dots + \frac{1}{m+k}\right]$$

The solution of equation (3.51) is thus

$$y = AI_k(x) + BI_{-k}(x) \qquad (3.56)$$

if k is not an integer or zero, or

$$y = AI_k(x) + BK_k(x) \qquad (3.57)$$

if k is an integer or zero.

$I_k(x)$ is called the modified Bessel function of the first kind of order k, and $K_k(x)$ is the modified Bessel function of the second kind of order k.

Example. Heat Loss Through Pipe Flanges. Two thin wall metal pipes of 2·5 cm external diameter and joined by flanges 1·25 cm thick and 10 cm diameter, are carrying steam at 120 °C. If the conductivity of the flange metal $k = 400$ W/m °C and the exposed surfaces of the flanges lose heat to the surroundings at $T_1 = 15$ °C according to a heat transfer coefficient $h = 12$ W/m² °C, find the rate of heat loss from the pipe, and the proportion which leaves the rim of the flange.

Solution

It is only necessary to consider one flange with one exposed circular face and an exposed rim. Take a radial coordinate r measured in metres from the axis of the pipe and consider the heat balance over an element of width δr, as shown in Fig. 3.3.

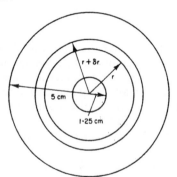

FIG. 3.3. Pipe flange

Input $\qquad = -k(2\pi r 0\cdot0125)\dfrac{dT}{dr} = -0\cdot025\pi kr\dfrac{dT}{dr}$

Output $\qquad = -0\cdot025\pi kr\dfrac{dT}{dr} + \dfrac{d}{dr}\left[-0\cdot025\pi kr\dfrac{dT}{dr}\right] + 2\pi rh\delta r(T - T_1)$

Accumulation $= 0$

Simplification of the heat balance gives

$$r\frac{d^2 T}{dr^2} + \frac{dT}{dr} - \frac{80h}{k}r(T - T_1) = 0 \qquad\qquad \text{I}$$

Putting $y = T - T_1$ and $x = r\sqrt{(80h/k)}$ gives the standard form of the equation,

$$x^2\frac{d^2 y}{dx^2} + x\frac{dy}{dx} - x^2 y = 0 \qquad\qquad \text{II}$$

Comparison with equation (3.51) shows that equation II is a modified Bessel equation of zero order. According to equation (3.57) the solution is thus

$$y = AI_0(x) + BK_0(x) \qquad\qquad \text{III}$$

The boundary conditions are

at $\qquad\qquad\qquad r = 0.0125, \quad T = 120 \qquad\qquad$ IV

and at $\qquad\qquad r = 0.05, \quad -k\dfrac{dT}{dr} = h(T - T_1) \qquad\qquad$ V

where equation V states that the heat conducted through the metal to the rim must equal the heat lost from the rim surface to the surrounding air.

Changing the variables, and introducing the numerical values, equations IV and V become

at $\qquad\qquad\qquad x = 0.0195, \quad y = 105 \qquad\qquad\qquad$ VI

and at $\qquad\qquad x = 0.078, \quad \dfrac{dy}{dx} = -0.0195y \qquad\qquad$ VII

Differentiating equation III, and using the properties (3.73), (3.74), (3.86), and (3.87), listed at the end of this chapter,

$$\frac{dy}{dx} = AI_1(x) - BK_1(x) \qquad\qquad \text{VIII}$$

Substituting equations VIII and III into equation VII gives

$$AI_1(0.078) - BK_1(0.078) = -0.0195[AI_0(0.078) + BK_0(0.078)] \qquad \text{IX}$$

and substituting equation III into equation VI gives

$$AI_0(0.0195) + BK_0(0.0195) = 105 \qquad\qquad \text{X}$$

Finding the values of the various Bessel functions from tables† and solving equations IX and X gives

$$A = 103.0, \quad B = 0.477$$

---

† "British Association Mathematical Tables", Vol. VI. Cambridge University Press (1937).

Putting these values into equation III and reverting to the original variables gives the temperature distribution

$$T = 15 + 103 \cdot 0 I_0(1 \cdot 55r) + 0 \cdot 477 K_0(1 \cdot 55r) \qquad\qquad \text{XI}$$

The heat conducted from the pipe by the flange is given by

$$Q_1 = -2\pi(0 \cdot 0125)^2 k \left(\frac{\mathrm{d}T}{\mathrm{d}r}\right)_{r=0 \cdot 0125}$$

$$= 0 \cdot 609[BK_1(0 \cdot 0195) - AI_1(0 \cdot 0195)]$$

$$= 14 \cdot 3 \text{ W}$$

and the heat lost through the rim is given by

$$Q_2 = -2\pi(0 \cdot 05)(0 \cdot 0125) k \left(\frac{\mathrm{d}T}{\mathrm{d}r}\right)_{r=0 \cdot 05}$$

$$= 2 \cdot 435[BK_1(0 \cdot 078) - AI_1(0 \cdot 078)]$$

$$= 4 \cdot 96 \text{ W}$$

Therefore, the pipe loses 14·3 W through each flange, and 35% of this is lost through the rim.

The above problem is similar to the problem of a flat circular cooling fin, which has been solved by Gardner† who presents solutions in the form of Bessel functions for the efficiencies of a variety of shapes of extended cooling surfaces.

3.6. Properties of Bessel Functions

Just as trigonometric functions have well-known properties and relationships, so do Bessel functions, although these are not so widely known. Only a few of the more useful properties will be listed here, and the reader is referred to any of the specialized books on Bessel functions for a more comprehensive list. The formulae will be given in sections on limiting behaviour, differential properties, and integral properties.

3.6.1. *Behaviour Near the Origin*

As $x \to 0$, the Bessel function tends to the same limit as its leading term. $J_k(x)$ and $I_k(x)$ start at x^k and are therefore zero when x is zero for $k > 0$. Therefore,

when $k = 0,$ $J_0(0) = I_0(0) = 1$ (3.58)

when $k > 0,$ $J_k(0) = I_k(0) = 0$ (3.59)

when non-integer $k > 0,$ $J_{-k}(0) = \pm I_{-k}(0) = \pm\infty$ (3.60)

† Gardner, K. A. *Trans. A.S.M.E.* **67**, 621 (1945).

The ambiguous sign is resolved in any particular case by reference to equations (3.43) or (3.54) where the sign is determined by the sign of $\Gamma(k+1)$ given in Section 5.3. In all physical cases, if $x = 0$ is a point in the system, no physical property can be infinite and the part of the solution containing $J_{-k}(x)$ or $I_{-k}(x)$ must be eliminated anyway. The above physical argument also applies to $Y_k(x)$ and $K_k(x)$ which are both infinite at the origin for all values of k. If k is not zero, the infinity is due to a negative power of x; if k is zero, the infinity is due to the term $\ln x$. Therefore, for all values of k,

$$- Y_k(0) = K_k(0) = \infty \qquad (3.61)$$

If the origin is a point in the calculation field, then $J_k(x)$ and $I_k(x)$ are the only physically permissible solutions.

3.6.2. Asymptotic Behaviour for Large x

If $x \gg k$, then the term k^2 in Bessel's equation (3.35) can be neglected in comparison with x^2 in the last term. Equation (3.35) becomes

$$x^2 \frac{d^2 y}{dx^2} + x \frac{dy}{dx} + x^2 y = 0 \qquad (3.62)$$

A substitution suggested by the first two terms is

$$z = x^{\frac{1}{2}} y \qquad (3.63)$$

$$\therefore \quad \frac{dz}{dx} = x^{\frac{1}{2}} \frac{dy}{dx} + \tfrac{1}{2} x^{-\frac{1}{2}} y$$

$$\therefore \quad \frac{d^2 z}{dx^2} = x^{\frac{1}{2}} \frac{d^2 y}{dx^2} + x^{-\frac{1}{2}} \frac{dy}{dx} - \tfrac{1}{4} x^{-\frac{3}{2}} y$$

$$\therefore \quad x^{\frac{1}{2}} \frac{d^2 z}{dx^2} = x^2 \frac{d^2 y}{dx^2} + x \frac{dy}{dx} - \tfrac{1}{4} y \qquad (3.64)$$

Substituting (3.63) and (3.64) into equation (3.62) gives

$$x \frac{d^2 z}{dx^2} + (\tfrac{1}{4} x^{-1} + x) z = 0$$

Again, $\tfrac{1}{4} x^{-1}$ is small compared with x if $x \gg 1$.

$$\therefore \quad \frac{d^2 z}{dx^2} + z = 0 \qquad (3.65)$$

This is the equation of simple harmonic motion which has the solution

$$z = A \cos x + B \sin x$$

Substituting back into equation (3.63),

$$y = A x^{-\frac{1}{2}} \cos x + B x^{-\frac{1}{2}} \sin x \qquad (3.66)$$

With the above simple treatment, it has been shown that for $x \gg 1$ and $x \gg k$, the Bessel functions behave as damped oscillations of period 2π. The evaluation of A and B to simulate the behaviour of the defined Bessel functions is an involved matching procedure, the result of which is

$$J_k(x) \simeq \sqrt{\left(\frac{2}{\pi x}\right)} \cos\left(x - \frac{\pi}{4} - \frac{k\pi}{2}\right) \tag{3.67}$$

$$Y_k(x) \simeq \sqrt{\left(\frac{2}{\pi x}\right)} \sin\left(x - \frac{\pi}{4} - \frac{k\pi}{2}\right) \tag{3.68}$$

A similar procedure may be adopted with equation (3.51) for the modified Bessel functions. The result is

$$I_k(x) \simeq \frac{1}{\sqrt{(2\pi x)}} e^x \tag{3.69}$$

$$K_k(x) \simeq \sqrt{\left(\frac{\pi}{2x}\right)} e^{-x} \tag{3.70}$$

Thus, $I_k(x)$ is the only Bessel function of the four which diverges to infinity as x increases; the other three decay to zero.

3.6.3. Differential Properties

By differentiating the series (3.43) term by term it can be shown that

$$x\frac{d}{dx}J_k(\alpha x) = kJ_k(\alpha x) - \alpha x J_{k+1}(\alpha x)$$

$$= \alpha x J_{k-1}(\alpha x) - kJ_k(\alpha x) \tag{3.71}$$

The corresponding formulae for the other Bessel functions are:

$$x\frac{d}{dx}Y_k(\alpha x) = kY_k(\alpha x) - \alpha x Y_{k+1}(\alpha x)$$

$$= \alpha x Y_{k-1}(\alpha x) - kY_k(\alpha x) \tag{3.72}$$

$$x\frac{d}{dx}I_k(\alpha x) = kI_k(\alpha x) + \alpha x I_{k+1}(\alpha x)$$

$$= \alpha x I_{k-1}(\alpha x) - kI_k(\alpha x) \tag{3.73}$$

$$x\frac{d}{dx}K_k(\alpha x) = kK_k(\alpha x) - \alpha x K_{k+1}(\alpha x)$$

$$= -\alpha x K_{k-1}(\alpha x) - kK_k(\alpha x) \tag{3.74}$$

3.6.4. *Integral Properties*

$$\int \alpha x^k J_{k-1}(\alpha x) \, dx = x^k J_k(\alpha x) \tag{3.75}$$

$$\int \alpha x^k Y_{k-1}(\alpha x) \, dx = x^k Y_k(\alpha x) \tag{3.76}$$

$$\int \alpha x^k I_{k-1}(\alpha x) \, dx = x^k I_k(\alpha x) \tag{3.77}$$

$$\int \alpha x^k K_{k-1}(\alpha x) \, dx = - x^k K_k(\alpha x) \tag{3.78}$$

The following integrals are needed for demonstrating the orthogonality of the Bessel functions (see Section 8.6). Since only $J_k(x)$ and $I_k(x)$ are finite at the origin and therefore the only functions of practical importance at the origin, the orthogonality integrals need only be quoted for these two functions.

$$\int_0^x J_k(\alpha x) J_k(\beta x) x \, dx = \frac{x}{\alpha^2 - \beta^2} [\alpha J_k(\beta x) J_{k+1}(\alpha x) - \beta J_k(\alpha x) J_{k+1}(\beta x)] \tag{3.79}$$

$$\int_0^x [J_k(\alpha x)]^2 x \, dx = \tfrac{1}{2} x^2 [J_k{}^2(\alpha x) - J_{k-1}(\alpha x) J_{k+1}(\alpha x)] \tag{3.80}$$

$$\int_0^x I_k(\alpha x) I_k(\beta x) x \, dx = \frac{x}{\alpha^2 - \beta^2} [\alpha I_k(\beta x) I_{k+1}(\alpha x) - \beta I_k(\alpha x) I_{k+1}(\beta x)] \tag{3.81}$$

$$\int_0^x [I_k(\alpha x)]^2 x \, dx = \tfrac{1}{2} x^2 [I_k{}^2(\alpha x) - I_{k-1}(\alpha x) I_{k+1}(\alpha x)] \tag{3.82}$$

3.6.5. *Bessel Functions of Negative Order*

Tables of Bessel functions are only given for zero and positive order, the negative order Bessel functions have to be evaluated using recursion formulae as follows. By rearranging equation (3.71)

$$\alpha x J_{k-1}(\alpha x) = 2k J_k(\alpha x) - \alpha x J_{k+1}(\alpha x) \tag{3.83}$$

Similar formulae can be obtained for the other Bessel functions by rearranging equations (3.72), (3.73), and (3.74). If the Bessel functions of order k and $k+1$ are known, the Bessel function of order $k-1$ can be calculated.

If k is also an integer, substitution into equation (3.43) and using the properties of the gamma function, Section 5.3, it can be shown that

$$J_{-k}(\alpha x) = (-1)^k J_k(\alpha x) \tag{3.84}$$

Similarly,

$$Y_{-k}(\alpha x) = (-1)^k Y_k(\alpha x) \qquad (3.85)$$

$$I_{-k}(\alpha x) = I_k(\alpha x) \qquad (3.86)$$

$$K_{-k}(\alpha x) = K_k(\alpha x) \qquad (3.87)$$

which are only true for integer values of k.

3.6.6. *Bessel Functions of Half Integer Order*

Bessel's equations (3.35) can be solved analytically for the case $k = \frac{1}{2}$ by substituting

$$z = x^{\frac{1}{2}} y \qquad (3.63)$$

as in Section 3.6.2. This gives

$$\frac{d^2 z}{dx^2} + z = 0$$

which has the solution

$$z = A \sin x + B \cos x$$

$$y = (A \sin x + B \cos x)/x^{\frac{1}{2}}$$

Writing out the first few terms of the series for $J_{\frac{1}{2}}(x)$ from equation (3.43) it can be verified that

$$J_{\frac{1}{2}}(x) = \sqrt{\left(\frac{2}{\pi x}\right)} \sin x \qquad (3.88)$$

and

$$J_{-\frac{1}{2}}(x) = \sqrt{\left(\frac{2}{\pi x}\right)} \cos x \qquad (3.89)$$

Similarly, it can be shown that

$$I_{\frac{1}{2}}(x) = \sqrt{\left(\frac{2}{\pi x}\right)} \sinh x \qquad (3.90)$$

and

$$I_{-\frac{1}{2}}(x) = \sqrt{\left(\frac{2}{\pi x}\right)} \cosh x \qquad (3.91)$$

Equation (3.83) and the similar equation established from equation (3.73) allow the J or I functions of half integer order to be expressed in terms of trigonometric or hyperbolic functions.

Chapter 4

COMPLEX ALGEBRA

4.1. Introduction

In the following paragraphs, the algebra of complex numbers will be described and the principles and properties of a complex variable introduced.

The concepts of a complex number and a complex variable are utilized in many branches of applied mathematics and the simplest of these, the complex roots of a quadratic equation, have already been introduced in Chapter 2. It will be seen later that the basis of operational calculus is the integration of functions of complex variables. Consequently an understanding of the fundamental properties of complex quantities is considered essential for the appreciation of many mathematical methods employed in the solution of chemical engineering problems. However, complex numbers have little direct application in chemical engineering and therefore most of the illustrative examples are necessarily of a pure mathematical nature.

4.2. The Complex Number

A complex number is an expression of the type $x+iy$, where both x and y are real numbers and the symbol

$$i = \sqrt{-1} \tag{4.1}$$

i is therefore not a real number and is defined as the fundamental "imaginary" number. The product iy will also be imaginary and every complex number will have a real part and an imaginary part. In the number $x+iy$, the real part is x and the imaginary part is y.

The algebraic operations of addition, subtraction, multiplication, and division can be performed on complex numbers by treating i as an unknown constant, and replacing (i^2) by (-1) on every occasion that it appears. Thus, if two complex numbers are represented by z_1 and z_2, where

$$z_1 = x_1+iy_1 \quad \text{and} \quad z_2 = x_2+iy_2 \tag{4.2}$$

then the addition of these two complex numbers is as follows.

$$z_1+z_2 = x_1+iy_1+x_2+iy_2$$
$$= (x_1+x_2)+i(y_1+y_2) \tag{4.3}$$

Subtraction gives

$$z_1-z_2 = x_1+iy_1-x_2-iy_2$$
$$= (x_1-x_2)+i(y_1-y_2) \tag{4.4}$$

Multiplication gives

$$z_1 z_2 = (x_1 + iy_1)(x_2 + iy_2)$$
$$= (x_1 x_2 - y_1 y_2) + i(x_1 y_2 + x_2 y_1) \tag{4.5}$$

Division gives

$$\frac{z_1}{z_2} = \frac{x_1 + iy_1}{x_2 + iy_2} = \frac{x_1 + iy_1}{x_2 + iy_2} \frac{x_2 - iy_2}{x_2 - iy_2}$$

$$= \left(\frac{x_1 x_2 + y_1 y_2}{x_2^2 + y_2^2} \right) + i \left(\frac{y_1 x_2 - y_2 x_1}{x_2^2 + y_2^2} \right) \tag{4.6}$$

The first three operations are quite straightforward but in the division operation the complex number was eliminated from the denominator by multiplying both numerator and denominator by a specially chosen complex number. This number, which is obtained by reversing the sign of the imaginary part of the denominator, is called the "complex conjugate" of the denominator, and Section 4.6 is devoted to the further properties of conjugate numbers.

The two complex numbers z_1 and z_2 will be equal if their difference is zero, and by inspection of the subtraction operation (4.4) it can be seen that $z_1 - z_2 = 0$ only if $x_1 - x_2 = 0$ and $y_1 - y_2 = 0$. That is, z_1 and z_2 are only equal if their real and imaginary parts are separately equal.

In each of the algebraic operations presented in equations (4.3) to (4.6) the mathematical procedure was identical to that which would be followed if the same operation was being performed on a real number. The resulting number was in general complex containing both a real and an imaginary part; although either part could be zero.

4.3. THE ARGAND DIAGRAM

Early in the nineteenth century Argand suggested that a complex number could be represented by a line in a plane in much the same way as a vector is represented (cf. Chapter 7). The value of the complex number could be expressed in terms of two axes of reference, and he suggested that one axis be called the real axis and the other axis arranged perpendicular to the first be called the imaginary axis. Then a complex number $z = x + iy$ could be represented by a line in the plane having projections x on the real axis and y on the imaginary axis. Such a line is illustrated in Fig. 4.1. Figure 4.1 is called the Argand diagram and the plane in which the line is drawn is called the Argand plane of z. The line OP represents the complex number $4 + 3i$ whose real part has a value of 4 and imaginary part 3. On the other hand, the line OQ represents the complex number $4i - 3$ with real part -3 and imaginary part 4. The lengths of OP and OQ are equal but the complex numbers they represent are unequal because the real parts and the imaginary parts of each are different. This emphasizes the fact that any equation involving complex numbers is equivalent to two real equations obtained by separately equating the real and imaginary parts.

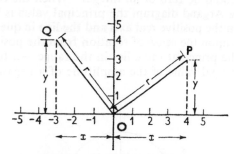

FIG. 4.1. The Argand diagram

4.3.1. *Modulus and Argument*

For each of the complex numbers represented by the lines OP and OQ, the lengths of the lines are 5. That is,

$$\sqrt{(4^2+3^2)} = \sqrt{[(-3)^2+4^2]} = 5 \qquad (4.7)$$

The value of the length of the line representing the complex number is called the "modulus" or "absolute value" of the number and is usually written in the alternative forms,

$$r = |z| = \text{mod } z = \sqrt{(x^2+y^2)} \qquad (4.8)$$

The inclination of the line representing the complex number to the positive real axis is called the "amplitude", "argument", or "phase" of z, and is usually written,

$$\theta = \text{amp } z = \arg z = \tan^{-1}(y/x) \qquad (4.9)$$

Solving equations (4.8) and (4.9) for x and y gives

$$x = r\cos\theta \quad \text{and} \quad y = r\sin\theta \qquad (4.10)$$

From the above it can be seen that a complex number can be expressed in terms of cartesian coordinates as $x+iy$, or with the aid of equations (4.10) in polar coordinates as

$$z = r\cos\theta + ir\sin\theta = r(\cos\theta + i\sin\theta) = r\,\text{cis}\,\theta \qquad (4.11)$$

where cis θ is a convenient notation which should be read "$\cos\theta + i\sin\theta$".

4.4. PRINCIPAL VALUES

When a complex number is expressed in polar coordinates, the principal value of θ is always implied unless otherwise stated. That is in Fig. 4.1,

$$\theta_P = \tan^{-1}(\tfrac{3}{4}) \quad = \quad 37° = 0.645 \text{ radians}$$

$$\theta_Q = \tan^{-1}(-\tfrac{4}{3}) = 143° = 2.495 \text{ radians}$$

If the angle is not restricted to its principal value $\text{amp}(4+3i)$ would be equal to $(0\cdot645+2n\pi)$ radians, and $\text{amp}(4i-3)$ would be equal to $(2\cdot495+2n\pi)$ radians, where n could be zero or an integer. When the complex number is represented on the Argand diagram the principal value is the smaller of the two angles between the positive real axis and the line in question. The sign of the angle depends upon the sense of rotation from the positive real axis, and this implies that the principal value lies in the range $-\pi$ to $+\pi$ as shown in Fig. 4.2. The principal value of the amplitude of a negative real number is

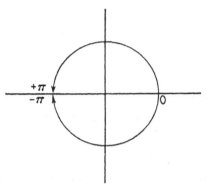

Fig. 4.2. Principal value of the argument

conventionally taken as $+\pi$. It is sometimes more convenient to choose a different range of values for θ but the range is chosen here for consistency with the definitions to be used in the theory of complex variables at the end of this chapter.

4.5. Algebraic Operations on the Argand Diagram

In Section 4.2 the addition, subtraction, multiplication, and division of two complex numbers were carried out. If the Argand diagram describes all the properties of complex numbers it should be possible to carry out the above algebraic operations on the diagram. Thus consider Fig. 4.3(a) in which the complex numbers represented by the lines OP and OQ in Fig. 4.1 are redrawn. If z_3 is the sum of z_1 represented by OP and z_2 represented by OQ, then

$$z_3 = z_1 + z_2 = (x_1 + x_2) + i(y_1 + y_2)$$

or
$$x_3 = x_1 + x_2 \quad \text{and} \quad y_3 = y_1 + y_2 \tag{4.12}$$

Equations (4.12) give the coordinates of z_3 on the Argand diagram as shown in Fig. 4.3(a) by the line OR. Using the same numerical values as before,

$$x_3 = 4 - 3 = 1 \quad \text{and} \quad y_3 = 3 + 4 = 7$$

$$\therefore \quad z_3 = x_3 + iy_3 = 1 + 7i \tag{4.13}$$

It is easily shown that the point R can be located geometrically by completing the parallelogram OPQR as in Fig. 4.3(a).

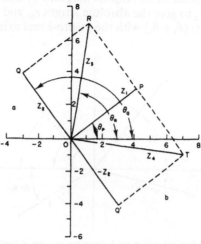

FIG. 4.3. Addition and subtraction on the Argand diagram

In the same way, the subtraction of two complex numbers can be expressed in the form of the addition of z_1 to minus z_2 where

$$-z_2 = -(4i-3) = 3-4i \tag{4.14}$$

By completing the parallelogram OPTQ′ in Fig. 4.3(b) to locate the diagonal OT, the difference between the complex numbers z_1 and z_2 (i.e. $z_4 = z_1 - z_2$) is obtained. The coordinates of z_4 can be verified algebraically as before, thus,

$$z_4 = z_1 + (-z_2) = (4+3) + i(3-4) = 7-i.$$

From the elementary geometrical property of triangles that the sum of the absolute lengths of any two sides of a triangle must exceed the third side, there follows the important result,

$$|z_1 + z_2| \leqslant |z_1| + |z_2| \tag{4.15}$$

In order to illustrate multiplication and division on the Argand diagram, it is first necessary to show how the multiplication and division of two complex numbers are expressed in terms of polar coordinates. Thus

$$z_5 = z_1 z_2 = r_1 r_2 (\cos \theta_1 + i \sin \theta_1)(\cos \theta_2 + i \sin \theta_2)$$

$$= r_1 r_2 [(\cos \theta_1 \cos \theta_2 - \sin \theta_1 \sin \theta_2) + i(\sin \theta_1 \cos \theta_2 + \cos \theta_1 \sin \theta_2)]$$

$$= r_1 r_2 [\cos(\theta_1 + \theta_2) + i \sin(\theta_1 + \theta_2)] \tag{4.16}$$

or $z_5 = r_5(\cos \theta_5 + i \sin \theta_5)$

where $$r_5 = r_1 r_2 \quad \text{and} \quad \theta_5 = \theta_1 + \theta_2 \tag{4.17}$$

Therefore, to multiply two complex numbers, it is necessary to multiply the moduli and add the arguments. Hence the multiplication of two complex numbers can be expressed on an Argand diagram by multiplication of the two real numbers r_1 and r_2 to give the absolute length r_5, and drawing the line OU of length r_5 at an angle $(\theta_1 + \theta_2)$ with the positive real axis as illustrated in Fig. 4.4.

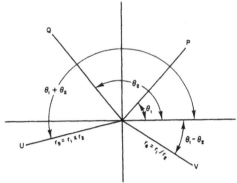

FIG. 4.4. Multiplication and division on the Argand diagram

The division of z_1 by z_2 can be expressed in terms of polar coordinates as follows.

$$\frac{z_1}{z_2} = \frac{r_1(\cos\theta_1 + i\sin\theta_1)}{r_2(\cos\theta_2 + i\sin\theta_2)}$$

$$= \frac{r_1(\cos\theta_1 + i\sin\theta_1)(\cos\theta_2 - i\sin\theta_2)}{r_2(\cos^2\theta_2 + \sin^2\theta_2)}$$

$$= \frac{r_1}{r_2}[\cos(\theta_1 - \theta_2) + i\sin(\theta_1 - \theta_2)] \tag{4.18}$$

where the denominator was again made real by multiplying numerator and denominator by the complex conjugate of the denominator.

Therefore, to divide two complex numbers it is necessary to divide the moduli and subtract the arguments. Hence the division of two complex numbers can be expressed on the Argand diagram by dividing the real number r_1 by the real number r_2, to give an absolute length of r_6, then drawing the line OV at an angle $(\theta_1 - \theta_2)$ with the positive real axis. This is illustrated in Fig. 4.4.

In the multiplication and division operations θ_1 and θ_2 were the principal values of the arguments of the numbers. However $(\theta_1 + \theta_2)$ and $(\theta_1 - \theta_2)$ need not be the principal values of the arguments of z_5 and z_6. Thus consider the complex numbers $z_1 = (3i - 4)$ and $z_2 = (i - 1)$ with arguments $\theta_1 = 143°$ and $\theta_2 = 135°$. By equation (4.5) $z_1 z_2 = 1 - 7i$ and $\theta_5 = \theta_1 + \theta_2 = 278°$. The principal value of the argument lying between $-180°$ and $+180°$ is $-82°$ and this value may be used in subsequent calculations unless there is some overriding consideration (e.g. see Section 4.12.3).

It has been shown that it is possible to represent simple addition, subtraction, multiplication and division of two complex numbers on the Argand diagram. Therefore it is obvious that by a series of geometric operations it is possible to describe any process involving complex numbers on an Argand diagram provided that they consist of a combination of the above four operations.

4.6. CONJUGATE NUMBERS

Two complex numbers such as $x+iy$ and $x-iy$ of which the real and imaginary parts are of equal magnitude, but in which the imaginary parts are of opposite sign are said to be conjugate numbers. On the Argand diagram they can be considered to be mirror images of each other in the real axis. Conjugate numbers have already been used in the definition of the division process. Usually, the conjugate of a complex number z is written as $\bar{z}$, and throughout this chapter conjugates of complex numbers will be written accordingly.

If the imaginary part is zero, the real number is its own conjugate. The sum and the product of a complex number with its conjugate are always real. Thus,

$$(x+iy)+(x-iy) = 2x \tag{4.19}$$

and
$$(x+iy)(x-iy) = x^2-(iy)^2 = x^2+y^2 \tag{4.20}$$

i.e. the final result is a real number since $i^2 = -1$. The division of a true complex by its conjugate will not produce a real number. These relations between sums and products can be extended to more than one complex number and its conjugate. Thus if w and z be two complex numbers whose conjugates are $\bar{w}$ and $\bar{z}$ respectively, let

$$w = r_1(\cos \theta + i \sin \theta) \tag{4.21}$$

then
$$\bar{w} = r_1(\cos \theta - i \sin \theta) \tag{4.22}$$

also let
$$z = r_2(\cos \phi + i \sin \phi) \tag{4.23}$$

then
$$\bar{z} = r_2(\cos \phi - i \sin \phi) \tag{4.24}$$

$$\therefore \quad w+z = r_1(\cos \theta + i \sin \theta) + r_2(\cos \phi + i \sin \phi)$$
$$= (r_1 \cos \theta + r_2 \cos \phi) + i(r_1 \sin \theta + r_2 \sin \phi) \tag{4.25}$$

$$\therefore \quad \overline{w+z} = (r_1 \cos \theta + r_2 \cos \phi) - i(r_1 \sin \theta + r_2 \sin \phi) \tag{4.26}$$

But
$$\bar{w}+\bar{z} = r_1(\cos \theta - i \sin \theta) + r_2(\cos \phi - i \sin \phi)$$
$$= (r_1 \cos \theta + r_2 \cos \phi) - i(r_1 \sin \theta + r_2 \sin \phi) \tag{4.27}$$

From equations (4.26) and (4.27)

$$\overline{w+z} = \bar{w}+\bar{z} \tag{4.28}$$

That is the conjugate of the sum of the two complex numbers is equal to the sum of the conjugates of the complex numbers. In fact this principle can be extended to any number of complex numbers.

In a similar manner it can be shown that if the complex number p is the product of $w\bar{z}$. That is

$$p = w\bar{z} \tag{4.29}$$

$$\bar{p} = \bar{w}z \tag{4.30}$$

Also if the complex number q is related to w and z by

$$q = w/z \tag{4.31}$$

then
$$\bar{q} = \bar{w}/\bar{z} \tag{4.32}$$

It is left to the reader to verify equations (4.29) to (4.32).

4.7. De Moivre's Theorem

This important theorem can be enunciated as follows: For all rational values of n, $\cos n\theta + i \sin n\theta$ is one of the values of $(\cos\theta + i\sin\theta)^n$. The proof of De Moivre's theorem depends on the value of n. That is, on whether n is a positive or negative integer, or a real fraction. Irrational numbers such as π are not included, but this is an academic restriction. The verification of this theorem when n is a positive integer follows from Section 4.5 where the product of two complex numbers $z_1 z_2$ was shown to be

$$r_1 r_2 [\cos(\theta_1 + \theta_2) + i\sin(\theta_1 + \theta_2)]$$

Hence, taking $z_1 = z_2 = z$, equation (4.16) gives

$$z^2 = r^2(\cos\theta + i\sin\theta)^2 = r^2(\cos 2\theta + i\sin 2\theta)$$

A further application of equation (4.16) to the complex numbers z and z^2 gives

$$z^3 = r^3(\cos\theta + i\sin\theta)^3 = r^3(\cos 3\theta + i\sin 3\theta)$$

The above operations can be extended to z^n giving

$$(\cos\theta + i\sin\theta)^n = \cos n\theta + i\sin n\theta \tag{4.33}$$

As stated above, equation (4.33) is true for all rational values of n. In particular, one of the two square roots of a complex number can be found by putting $n = \frac{1}{2}$. Thus

$$\sqrt{z} = (\sqrt{r})(\cos\tfrac{1}{2}\theta + i\sin\tfrac{1}{2}\theta) \tag{4.34}$$

4.8. Trigonometrical-exponential Identities

The convergence tests discussed in Chapter 3 are applicable to a series of complex numbers and in particular the exponential series

$$e^z = 1 + z + \frac{z^2}{2!} + \dots + \frac{z^n}{n!} + \dots \tag{3.7}$$

is convergent for all complex values of z. If $z = iy$, equation (3.7) becomes

$$e^{iy} = 1 + iy - \frac{y_2}{2!} - \frac{iy^3}{3!} + \frac{y^4}{4!} + \cdots \tag{4.35}$$

$$= \left(1 - \frac{y^2}{2!} + \frac{y^4}{4!} - \cdots\right) + i\left(y - \frac{y^3}{3!} + \frac{y^5}{5!} - \cdots\right) \tag{4.36}$$

Each of the series in equation (4.36) can be recognized by comparing with equations (3.9) and (3.10) in Chapter 3. Thus

$$e^{iy} = \cos y + i \sin y \tag{4.37}$$

or more generally

$$e^z = e^{x+iy} = e^x(\cos y + i \sin y) \tag{4.38}$$

Similarly it can be shown that

$$e^{-iy} = \cos y - i \sin y \tag{4.39}$$

Addition of equations (4.37) and (4.39) gives

$$e^{iy} + e^{-iy} = 2 \cos y$$

or

$$\cos y = \frac{e^{iy} + e^{-iy}}{2} \tag{4.40}$$

subtraction of equation (4.39) from equation (4.37) gives

$$e^{iy} - e^{-iy} = 2i \sin y$$

or

$$\sin y = \frac{e^{iy} - e^{-iy}}{2i} \tag{4.41}$$

The identities developed above between real trigonometrical functions and imaginary exponential functions can be extended to include hyperbolic functions. Thus, using equation (4.40) to find the cosine of a pure imaginary quantity,

$$\cos ix = \frac{e^{i^2x} + e^{-i^2x}}{2} = \frac{e^{-x} + e^x}{2}$$

which defines the hyperbolic cosine function,

$$\cos ix = \frac{e^x + e^{-x}}{2} = \cosh x \tag{4.42}$$

Similarly,

$$\sin ix = \frac{e^{i^2x} - e^{-i^2x}}{2i} = i\frac{e^x - e^{-x}}{2} = i \sinh x \tag{4.43}$$

which defines the hyperbolic sine function.

The above relationships can be extended to define a hyperbolic tangent function as,

$$\tan ix = \frac{\sin ix}{\cos ix} = \frac{i \sinh x}{\cosh x} = i \tanh x$$

$$\therefore \quad \tan ix = i \tanh x \tag{4.44}$$

All properties of hyperbolic functions can be obtained from the corresponding trigonometrical identities by using equations (4.42), (4.43), and (4.44).

4.9. THE COMPLEX VARIABLE

In the paragraphs above, the complex number z has been expressed in terms of the fixed real quantities x and y by the relation $z = x + iy$. If however the quantities x and y are variables, then so is z. In fact the quantity z becomes a "complex variable", and since z is related to the variables x and y, the way in which z varies with x and y can be expressed by a curve in a plane having coordinates x and y. The plane is called the "z plane".

If the complex variable z is related functionally to another complex variable w so that

$$w = f(z) \tag{4.45}$$

and the complex variable w can be expressed by the real quantities u and v so that $w = u + iv$, then when z varies in the z plane, w will follow a curve corresponding to equation (4.45) in the "w plane" whose coordinates are u and v. Since $z = x + iy$ and $w = u + iv$, u and v must be real functions of both x and y. This will become clear in the following example.

Example. If $z = x + iy$, and $w = u + iv$, where

$$w = z^2 \qquad\qquad\qquad \text{I}$$

express u and v in terms of x and y.

Substitute x, y, u, and v for w and z in equation I.

$$\therefore \quad u + iv = (x + iy)^2 = x^2 + 2ixy - y^2 \qquad\qquad \text{II}$$

Equating real and imaginary parts gives

$$u = x^2 - y^2 \quad \text{and} \quad v = 2xy \qquad\qquad \text{III}$$

which determine u and v as functions of x and y.

Because the complex variable w has a functional relationship to z expressed by equation (4.45), derivatives of w with respect to z will exist provided that $f(z)$ is continuous in the region where the derivatives are required. Thus a complex variable $w = f(z)$ is said to be continuous at a point p if $f(p)$ has a finite value at $z = p$, and if the limit of $f(z)$ as z approaches p is $f(p)$. That is,

$$\lim_{z \to p} f(z) = f(p) \tag{4.46}$$

If $f(z) = z^{-1}$ and the point p is the origin, the value of $f(z)$ at the origin depends upon the way in which the origin is approached. For instance, if z approaches zero along the positive real axis,

$$\lim_{x \to 0+} z^{-1} = \infty \qquad (4.47)$$

whereas if z approaches zero along the negative real axis,

$$\lim_{x \to 0-} z^{-1} = -\infty \qquad (4.48)$$

The limit of z^{-1} as z approaches zero does not therefore exist, and the function z^{-1} has a "singularity" at $z = 0$.

On the other hand, if $f(z) = z^4$, the function is continuous everywhere in the z plane but becomes infinite as z becomes infinite. Furthermore, z^4 has only one value at any given value of z and therefore it is said to be a "single-valued" function. However, if w is related to z by $w = z^{\frac{1}{2}}$, we can have more than one value for each value of z and w is then said to be a "many valued" function.

Difficulties can arise therefore in a number of ways when treating complex variables, but many of these can be resolved by confining the function to regions in the complex plane in which the functional relationships apply. This fact will be better understood later in the chapter.

4.10. Derivatives of a Complex Variable

Consider the complex variable

$$w = f(z) \qquad (4.45)$$

to be a continuous function, and let $w = u + iv$ and $z = x + iy$ as before. Then the partial derivative of w with respect to x can be obtained in two ways, thus

$$\frac{\partial w}{\partial x} = \frac{\partial u}{\partial x} + i \frac{\partial v}{\partial x} \qquad (4.49)$$

and

$$\frac{\partial w}{\partial x} = \frac{df}{dz} \frac{\partial z}{\partial x} = \frac{df}{dz} \qquad (4.50)$$

since $\partial z / \partial x = 1$. Equating these two expressions

$$\frac{df}{dz} = \frac{\partial u}{\partial x} + i \frac{\partial v}{\partial x} \qquad (4.51)$$

Similarly, differentiating w with respect to y gives

$$\frac{\partial w}{\partial y} = \frac{\partial u}{\partial y} + i \frac{\partial v}{\partial y} \qquad (4.52)$$

and

$$\frac{\partial w}{\partial y} = \frac{df}{dz} \frac{\partial z}{\partial y} = i \frac{df}{dz} \qquad (4.53)$$

since $\partial z / \partial y = i$.

$$\therefore \quad i \frac{df}{dz} = \frac{\partial u}{\partial y} + i \frac{\partial v}{\partial y} \qquad (4.54)$$

Eliminating df/dz from equations (4.51) and (4.54) gives

$$i\frac{\partial u}{\partial x} - \frac{\partial v}{\partial x} = i\frac{df}{dz} = \frac{\partial u}{\partial y} + i\frac{\partial v}{\partial y} \qquad (4.55)$$

Since u, v, x, and y are all real, the real and imaginary parts of equation (4.55) must be equated.

$$\therefore \quad \frac{\partial u}{\partial x} = \frac{\partial v}{\partial y} \quad \text{and} \quad \frac{\partial u}{\partial y} = -\frac{\partial v}{\partial x} \qquad (4.56)$$

If equations (4.56) are satisfied, the derivative dw/dz becomes a unique single-valued function which can be used in the mathematical solution of engineering problems. Hence the derivative of a complex variable has certain restrictions which derivatives of real variables do not have and these are described by equations (4.56). They are known as the "Cauchy–Riemann" conditions and they must be satisfied for the derivative of a complex number to have any meaning.

4.11. ANALYTIC FUNCTIONS

A function $w = f(z)$ of the complex variable $z = x + iy$ is called an analytic or regular function within a region R, if at all points z_0 in the region it satisfies the following conditions.
  (i) It is single valued in the region R.
  (ii) It has a unique finite value.
  (iii) It has a unique finite derivative at z_0 which satisfies the Cauchy–Riemann conditions.
Only functions which satisfy these conditions can be utilized in pure and applied mathematics, and therefore the whole of the following treatment will deal with analytic functions. However, when a mathematical analysis of an engineering problem involving complex functions is being undertaken, care should be exercised to confirm that the function satisfies the criteria stated above. The following examples illustrate the application of these rules.

Example 1. If $w = z^3$, show that the function satisfies the Cauchy–Riemann conditions and state the region wherein the function is analytic.

$$w = z^3 = (x+iy)^3 = x^3 + 3ix^2 y - 3xy^2 - iy^3 \qquad \text{I}$$

But $w = u + iv$, and equating real and imaginary parts,

$$u = x^3 - 3xy^2 \quad \text{and} \quad v = 3x^2 y - y^3 \qquad \text{II}$$

$$\therefore \quad \frac{\partial u}{\partial x} = 3x^2 - 3y^2, \qquad \frac{\partial v}{\partial y} = 3x^2 - 3y^2 \qquad \text{III}$$

and $\qquad \dfrac{\partial u}{\partial y} = -6xy, \qquad \dfrac{\partial v}{\partial x} = 6xy \qquad \text{IV}$

By inspection of equations III and IV it can be seen that the Cauchy–Riemann conditions (4.56) are satisfied. Furthermore, for all finite values of z, w is finite; hence the function $w = z^3$ is analytic in any region of finite size.

Example 2. If $w = z^{-1}$, show that the function satisfies the Cauchy–Riemann conditions and state the region wherein the function is analytic.

$$w = \frac{1}{z} = \frac{1}{x+iy} = \frac{x-iy}{x^2+y^2} \qquad \text{I}$$

$$\therefore \quad u = \frac{x}{x^2+y^2} \quad \text{and} \quad v = \frac{-y}{x^2+y^2} \qquad \text{II}$$

From equations II,

$$\frac{\partial u}{\partial x} = \frac{\partial v}{\partial y} = \frac{y^2-x^2}{(x^2+y^2)^2} \qquad \text{III}$$

and

$$\frac{\partial u}{\partial y} = -\frac{\partial v}{\partial x} = \frac{-2xy}{(x^2+y^2)^2} \qquad \text{IV}$$

Equations III and IV show that the Cauchy–Riemann conditions are satisfied at a general point but consider the behaviour at the origin.

$\partial u/\partial x$ is obtained by keeping y constant, and to reach the origin, y must be constant at zero, hence equation II becomes

$$u = \frac{1}{x}$$

$$\therefore \quad \frac{\partial u}{\partial x} = \frac{-1}{x^2} \qquad \text{V}$$

which tends to negative infinity as x tends to zero through either positive or negative values.

Similarly, $\partial v/\partial y$ at the origin must be evaluated when $x = 0$. Equation II simplifies to

$$v = \frac{-1}{y}$$

$$\therefore \quad \frac{\partial v}{\partial y} = \frac{1}{y^2} \qquad \text{VI}$$

which tends to positive infinity as y tends to zero through either positive or negative values. Therefore, from equations V and VI, $\partial u/\partial x$ and $\partial v/\partial y$ diverge to opposite extremes at the origin, and one half of the Cauchy–Riemann conditions is not satisfied. It can be shown that the other half of the condition is satisfied, i.e.

$$\frac{\partial u}{\partial y} = -\frac{\partial v}{\partial x} = 0 \qquad\qquad \text{VII}$$

The Cauchy–Riemann conditions are not fully satisfied at the origin, w becomes infinite at the origin, and the function $w = z^{-1}$ is therefore analytic everywhere in the z plane with the exception of the one point $z = 0$.

4.12. SINGULARITIES

It has been shown in the examples above that the function $w = z^3$ is analytic everywhere except at $z = \infty$ whilst the function $w = z^{-1}$ is analytic everywhere except at $z = 0$. In fact no function except a constant is analytic throughout the complex plane, and every function of a complex variable has one or more points in the z plane where it ceases to be analytic. These points are called "singularities", and they may be classified as follows.

(a) Poles or unessential singularities.
(b) Essential singularities.
(c) Branch points.

The singularities (a) and (b) are characteristic of single-valued functions whereas (c) arises with multivalued functions. Each will be discussed below.

4.12.1. *Poles or Unessential Singularities*

A pole is a point in the complex plane at which the value of a function becomes infinite. Thus $w = z^{-1}$ is infinite at $z = 0$, and therefore the function $w = z^{-1}$ is said to have a pole at the origin. However, in the neighbourhood of the origin, this function is finite and consequently analytic but at the pole it ceases to be analytic. The pole in $w = z^{-1}$ is said to be "first order" whilst the function $w = z^{-2}$ possesses a second order pole. The order of a pole is determined in the following manner.

If $w = f(z)$ becomes infinite at the point $z = a$, then define

$$g(z) = (x-a)^n f(z) \qquad\qquad (4.57)$$

where n is an integer. If it is possible to find a finite value of n which makes $g(z)$ analytic at $z = a$, then the pole of $f(z)$ has been "removed" in forming $g(z)$. The order of the pole is defined as the minimum integer value of n for which $g(z)$ is analytic at $z = a$.

The function

$$w = \frac{1}{z^p(z-a)^q} \qquad\qquad (4.58)$$

where p and q are both positive and finite, is a function containing multiple poles. That is, it contains a pole of order p at the origin, and a pole of order q at $z = a$. If either p or q is not an integer, the order of the pole is the next higher integer than p or q respectively.

4.12.2. *Essential Singularities*

Certain functions of complex variables have an infinite number of terms which all approach infinity as the complex variable approaches a specific value. These could be thought of as poles of infinite order, but as the singularity cannot be removed by multiplying the function by a finite factor, they cannot be poles. This type of singularity is called an essential singularity and is portrayed by functions which can be expanded in a descending power series of the variable. The classical example is $e^{1/z}$ which has an essential singularity at $z = 0$. Similarly the series

$$f(z) = \sum_{n=0}^{n=\infty} b_n(z-a)^{-n} \qquad (4.59)$$

has an essential singularity at $z = a$.

Essential singularities can be distinguished from poles by the fact that they cannot be removed by multiplying by a factor of finite value. Thus consider the function

$$w = e^{1/z} \qquad (4.60)$$

Using equation (3.7) to expand the exponential function,

$$w = 1 + \frac{1}{z} + \frac{1}{2!z^2} + \dots + \frac{1}{n!z^n} + \dots \qquad (4.61)$$

Since w is infinite at the origin, try to remove the singularity by multiplying by z^p where p is some finite integer.

$$\therefore \quad z^p w = z^p + z^{p-1} + \frac{z^{p-2}}{2!} + \dots \frac{z^{p-n}}{n!} + \dots \qquad (4.62)$$

The series in equation (4.62) continues indefinitely and eventually n will be greater than p. Equation (4.62) thus consists of a finite number of positive powers of z, followed by an infinite number of negative powers of z. Since all terms of equation (4.62) are positive, then $z^p w \to \infty$ as $z \to 0$. It is therefore impossible to find a finite value of p which will remove the singularity in $e^{1/z}$ at the origin. The singularity is thus "essential".

Finally, it can be shown that the behaviour of a function in the neighbourhood of a pole and in the neighbourhood of an essential singularity is quite different. At a pole, the modulus of the function increases symmetrically about the pole to give an infinite pinnacle at the pole; whereas at an essential singularity, the function behaves very peculiarly. For example, Mclachlan[†] has shown that the function $e^{1/z}$ does not have the characteristics of a pole as z goes to zero. Thus for a fixed value of the argument of z in the range $0 < \theta < \frac{1}{2}\pi$ the modulus traces out a spiral of ever-increasing radius in the w plane; whilst at $\theta = 0$ the function oscillates. This behaviour is most unlike that of a function near a pole.

† Mclachlan, N. W. "Complex Variable Theory and Transform Calculus", p. 16. Cambridge University Press (1955).

4.12.3. *Branch Points*

The singularities described above arise from the non-analytic behaviour of single-valued functions. However multi-valued functions such as $w = z^{1/n}$ (where n is a positive integer) frequently arise in the solution of engineering problems. Therefore, this type of function must be accommodated in the theory of complex variables. Thus consider the function

$$w = z^{\frac{1}{2}} \tag{4.63}$$

and let z vary in such a way that it follows a circular path whose centre is the origin. Converting to polar coordinates using equations (4.11) and (4.37).

$$z = re^{i\theta} \tag{4.64}$$

where r is a constant and θ a variable. Substituting equation (4.64) into equation (4.63) gives

$$w = \sqrt{(r)}e^{\frac{1}{2}i\theta} \tag{4.65}$$

Hence, when the variable z makes one traverse of the circumference of the circle in the z plane, θ passes through 2π radians, but w with a constant modulus of $\sqrt{(r)}$ will only pass through π radians by equation (4.65). If z makes a second circuit of the circle in the z plane with θ changing from 2π to 4π, the function w will complete the circle as shown dotted in the w plane of Fig. 4.5. Consequently for any value of z represented by a point on the circumference of the circle in the z plane, there will be two corresponding values of w represented by points in the w plane. One value will be on the solid line semicircle where $0 < \frac{1}{2}\theta < \pi$, and the other will be on the dotted line semicircle where $\pi < \frac{1}{2}\theta < 2\pi$. Hence, by the simple expedient of restricting the values of θ to the range $0 \leqslant \theta < 2\pi$, the function w becomes single valued, and the Cauchy–Riemann conditions can be expressed as follows.

$$w = u + iv = \sqrt{(r)}e^{\frac{1}{2}i\theta}$$

$$\therefore \quad u = \sqrt{(r)}\cos\tfrac{1}{2}\theta \quad \text{and} \quad v = \sqrt{(r)}\sin\tfrac{1}{2}\theta$$

$$\therefore \quad \frac{\partial u}{\partial r} = \frac{1}{2\sqrt{r}}\cos\tfrac{1}{2}\theta \qquad \frac{\partial v}{\partial r} = \frac{1}{2\sqrt{r}}\sin\tfrac{1}{2}\theta$$

and
$$\frac{\partial u}{\partial \theta} = -\tfrac{1}{2}\sqrt{(r)}\sin\tfrac{1}{2}\theta \qquad \frac{\partial v}{\partial \theta} = \tfrac{1}{2}\sqrt{(r)}\cos\tfrac{1}{2}\theta$$

which gives

$$\frac{\partial u}{\partial r} = \frac{1}{r}\frac{\partial v}{\partial \theta} \quad \text{and} \quad \frac{\partial v}{\partial r} = -\frac{1}{r}\frac{\partial u}{\partial \theta} \tag{4.66}$$

Equations (4.66) are the Cauchy–Riemann conditions in polar coordinates. The final step to equations (4.66) can only be made in this example if θ is restricted to the range $0 \leqslant \theta \leqslant 2\pi$.

The restriction which makes a multi-valued function single valued within a given region enables these functions to be employed mathematically. The region where the function is single valued is called the "branch", and the particular value of z at which the function becomes infinite or zero is called the "branch point". In the example above, the origin is the branch point.

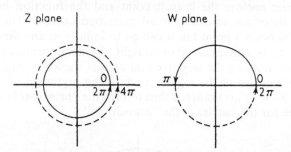

FIG. 4.5. Branch point of $w = z^{\frac{1}{2}}$ on the z and w complex planes

The function $w = z^{\frac{1}{2}}$ was multi-valued because the contour followed by z enclosed the branch point at the origin. However, this is a particular case, since the complex variable z may travel along any open curve in the z plane or any closed curve which may or may not enclose the branch point. Thus let the function $w = z^{\frac{1}{2}}$ follow the contour shown in Fig. 4.6 which does not enclose the branch point. On this occasion the angle θ, which is the argument of z, changes from θ_1 to θ_2 and back again to θ_1, and does not change by 2π radians in traversing the contour. Hence the function is single valued around the contour in Fig. 4.6 without any action being taken to restrict the values of θ. This shows that unless the contour of z encloses the branch point, the angle θ can only have one value at each point on the contour and this value cannot be altered by 2π by traversing the contour.

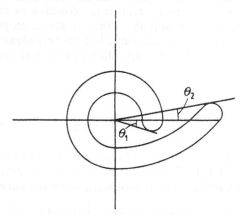

FIG. 4.6. Contour of $w = z^{\frac{1}{2}}$ not enclosing branch point

It has now been demonstrated that a function is only multi-valued around closed contours which enclose the branch point, so it is only necessary to eliminate such contours and the function will become single valued. The simplest way of doing this is to erect a barrier from the branch point to infinity and not allow any curve to cross the barrier. With this convention, no curve can ever enclose the branch point and the function becomes single valued and therefore analytic for all permitted curves. The barrier must start from the branch point but it can go to infinity in any direction in the z plane, and may be either curved or straight. In most normal applications the barrier is drawn along the negative real axis as shown in Fig. 4.7 so that it confines the function to the region in which the argument of z is within the range $-\pi < \theta < \pi$. This branch is then termed the "principal branch", and the correct name for the barrier is the "branch cut".

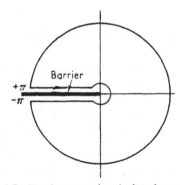

Fig. 4.7. Barrier erected to isolate branch point

The above discussion refers to a single branch point at the origin. If the z plane contains only one branch point which is not at the origin, it is a simple matter to transfer the origin to the branch point. When the z plane contains more than one branch point, it is necessary to draw a branch cut from each branch point to infinity. There is much scope for choosing a suitable set of branch cuts in such cases but no new ideas are involved and this complication does not occur very often in practical applications.

4.13. Integration of Functions of Complex Variables, and Cauchy's Theorem

The successive values of a complex variable z can be represented by a curve in the complex plane, and the function $w = f(z)$ will have a particular value at each point on this curve. Therefore the integral of $f(z)$ with respect to z is the sum of the products $f_M(z)\,\delta z$ along the curve in the complex plane, where $f_M(z)$ is the mean value of $f(z)$ in the length δz of the curve. That is,

$$\lim_{\delta z \to 0} \sum f_M(z)\,\delta z = \int_C f(z)\,dz \qquad (4.67)$$

where the suffix C under the integral sign specifies the curve in the z plane along which the integration is performed. However, if $w = u + iv = f(z)$ and $z = x + iy$ are substituted into equation (4.67),

$$\int_C f(z)\, dz = \int_C (u + iv)(dx + i\, dy)$$

$$= \int_C (u\, dx - v\, dy) + i \int_C (v\, dx + u\, dy) \qquad (4.68)$$

Equation (4.68) expresses the integration of the complex function $w = f(z)$ with respect to z. If both w and z were real, the integral would become

$$\int_C u\, dx$$

since the values of v and y would be zero, which shows that the integration of real functions is a special case of a contour integration along the real axis.

4.13.1. Cauchy's Theorem

If any function is analytic within and upon a closed contour, the integral taken around the contour is zero.

The above theorem can be proved using a very concise procedure by referring to Stokes' theorem, which is

$$\int_C (P\, dx + Q\, dy) = \iint_A \left(\frac{\partial Q}{\partial x} - \frac{\partial P}{\partial y} \right) dx\, dy \qquad (4.69)$$

where P and Q are functions of x and y, and the surface integral on the right-hand side is taken over the area A enclosed by the closed contour C. Stokes' theorem will be proved in Section 7.7.4 to which the reader is referred.

If KLMN represents a closed curve in Fig. 4.8 and there are no singularities of $f(z)$ within or upon the contour, the value of the integral of $f(z)$ around the contour is expressed by equation (4.68); and since the curve is closed, each

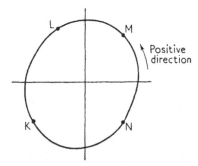

FIG. 4.8. Contour integration of $f(z)$ around closed curve

integral on the right-hand side of equation (4.68) can be restated as a surface integral using Stokes' theorem (4.69), thus

$$\int_C (u\,dx - v\,dy) = -\iint_A \left(\frac{\partial v}{\partial x} + \frac{\partial u}{\partial y}\right) dx\,dy \qquad (4.70)$$

and

$$\int_C (v\,dx + u\,dy) = \iint_A \left(\frac{\partial u}{\partial x} - \frac{\partial v}{\partial y}\right) dx\,dy \qquad (4.71)$$

But for an analytic function, by the Cauchy–Riemann conditions (4.56) each integral on the right-hand side is zero.

$$\therefore \quad \int_C f(z)\,dz = 0 \qquad (4.72)$$

which proves Cauchy's theorem.

In Fig. 4.8, let the region of the complex plane within and upon the contour KLMN contain no singularities. Therefore a function of z, $f(z)$ will be analytic within and upon the contour and Cauchy's theorem (4.72) is valid. If now the integration of $f(z)$ is started at the point K on the contour and terminated at the point M, then the value of the integral along the curve KNM will be

$$\int_{KNM} f(z)\,dz \qquad (4.73)$$

This will be positive by the conventional order of the suffices of the integral corresponding with the description of the path of integration. If the integration is continued along MLK following the positive direction of the arrow in Fig. 4.8, the value of the second integral is

$$\int_{MLK} f(z)\,dz \qquad (4.74)$$

However, the value of the integral around the closed contour KNMLK is zero by Cauchy's theorem and therefore

$$\int_{KNM} f(z)\,dz + \int_{MLK} f(z)\,dz = 0 \qquad (4.75)$$

or reversing the order of the suffices of the second integral and therefore its sign,

$$\int_{KNM} f(z)\,dz = \int_{KLM} f(z)\,dz \qquad (4.76)$$

That is, the value of an integral of $f(z)$ between two points in the complex plane is independent of the path of integration, provided that the function is analytic everywhere within the region containing all of the paths.

Cauchy's theorem and its corollary that the value of the contour integral is independent of the path are of great value for the integration of complex functions of complex variables. The following examples indicate extensions of these ideas.

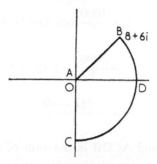

FIG. 4.9. Two equivalent contours

Example 1. Show that the value of $\int z^2\,dz$ between $z = 0$ and $z = 8 + 6i$ is the same whether the integration is carried out along the path AB or around the path ACDB in Fig. 4.9. The path AB is given by the equation

$$y = \tfrac{3}{4}x \qquad\qquad\qquad \text{I}$$

$$\therefore\quad z^2 = (x + \tfrac{3}{4}ix)^2 = \frac{7 + 24i}{16}x^2 \qquad\qquad \text{II}$$

and

$$dz = dx + \tfrac{3}{4}i\,dx \qquad\qquad\qquad \text{III}$$

$$\therefore\quad \int_0^{8+6i} z^2\,dz = \int_0^8 \frac{7+24i}{16}\,\frac{4+3i}{4}\,x^2\,dx \qquad \text{IV}$$

$$= \frac{-44 + 117i}{64}\left[\frac{x^3}{3}\right]_0^8$$

$$= \frac{-352 + 936i}{3} \qquad\qquad\qquad \text{V}$$

Consider the integration along the curve ACDB in parts.
(a) Along AC, $x = 0$ and thus $z = iy$.

$$\therefore\quad \int_0^{-10i} z^2\,dz = -i\int_0^{-10} y^2\,dy = \left[-\frac{1}{3}iy^3\right]_0^{-10} = \frac{1000i}{3} \qquad \text{VI}$$

(b) The arc CDB is of constant radius 10, therefore $z = 10\,e^{i\theta}$.

$$\therefore\quad \int_{-10i}^{8+6i} z^2\,dz = \int_{-\frac{1}{2}\pi}^{\alpha} 100\,e^{2i\theta}\,10i\,e^{i\theta}\,d\theta \qquad \text{VII}$$

where
$$\alpha = \tan^{-1}\tfrac{3}{4} = 36° 52'$$

$$\therefore \int_{-10i}^{8+6i} z^2\,dz = 1000i\left[\frac{1}{3i}e^{3i\theta}\right]_{-\frac{1}{2}\pi}^{\alpha}$$

$$= \frac{1000}{3}(\cos 3\alpha + i\sin 3\alpha - i)$$

$$= \frac{1000}{3}(-0.352 + 0.936i - i)$$

$$= \frac{-352 - 64i}{3} \qquad\qquad\text{VIII}$$

The complete integral round ACDB is the sum of equations VI and VIII which gives

$$\frac{-352 + 936i}{3}$$

as before. The integral of z^2 from $z = 0$ to $z = 8 + 6i$ is thus independent of the path taken.

Example 2. Evaluate $\int_C z^{-2}\,dz$ around a circle with its centre at the origin. Let $z = re^{i\theta}$, then $dz = ire^{i\theta}\,d\theta$ since r is constant.

$$\therefore \int_C \frac{dz}{z^2} = \int_0^{2\pi} \frac{ir\,e^{i\theta}\,d\theta}{r^2 e^{2i\theta}} = \frac{i}{r}\int_0^{2\pi} e^{-i\theta}\,d\theta \qquad\qquad\text{I}$$

$$= \frac{i}{r}\left[\frac{e^{-i\theta}}{-i}\right]_0^{2\pi}$$

$$= 0$$

That is, $\int_C z^{-2}\,dz = 0$ although the function is not analytic at the origin.

Similarly, $\int_C z^{-n}\,dz$ is always zero for any integer value of n other than unity.

Example 3. Evaluate $\int_C (dz/z)$ around a circle with its centre at the origin. As in the previous example, put $z = re^{i\theta}$ with r constant.

$$\therefore \int_C \frac{dz}{z} = \int_0^{2\pi} \frac{ir\,e^{i\theta}\,d\theta}{re^{i\theta}}$$

$$= i[\theta]_0^{2\pi}$$

$$= 2\pi i$$

This result is one of the fundamentals of contour integration.

4.13.2. *Cauchy's Integral Formula*

Consider Fig. 4.10 and let a complex function $f(z)$ be analytic upon and within the solid line contour C. Let a be a point within the closed contour such that $f(z)$ is not zero and define a new function $g(z)$ by

$$g(z) = \frac{f(z)}{z-a} \qquad (4.77)$$

This function is analytic within the contour C except at the point a where a simple pole will exist. If the pole is isolated by drawing a small circle γ around a and joining γ to C as shown dotted in Fig. 4.10, the value of the integral around the complete modified contour will be zero by Cauchy's

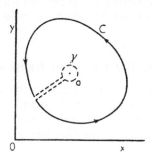

FIG. 4.10. Integration contour for $\displaystyle\int \frac{f(z)}{z-a}\, dz$

theorem. The straight dotted lines joining the outside contour C and the inner circle γ are drawn very close together and their paths will be synonymous. Since integration along them will be in opposite directions and $g(z)$ is analytic in the region containing them, the net value of the integral along the straight dotted lines will be zero.

$$\therefore \quad \int_C \frac{f(z)\,dz}{z-a} - \int_\gamma \frac{f(z)\,dz}{z-a} = 0 \qquad (4.78)$$

The negative sign before the second integral is necessary because the contour traverses the curve γ in a clockwise, negative sense. Let the value of $f(z)$ on γ be

$$f(z) = f(a) + \varepsilon \qquad (4.79)$$

where ε is a small quantity.

$$\therefore \quad \int_C \frac{f(z)\,dz}{z-a} - \int_\gamma \frac{f(a)\,dz}{z-a} - \int_\gamma \frac{\varepsilon\,dz}{z-a} = 0 \qquad (4.80)$$

Since $f(a)$ is constant, the second integral can be evaluated by putting $z = a + r e^{i\theta}$, and as θ varies by 2π radians the curve γ is described.

$$\therefore \quad \int_\gamma \frac{f(a)\,dz}{z-a} = if(a) \int_0^{2\pi} \frac{r e^{i\theta}\,d\theta}{r e^{i\theta}}$$

$$= 2\pi\, if(a) \tag{4.81}$$

As the radius of the circle γ tends to zero, ε also tends to zero and the third integral in equation (4.80) becomes insignificant.

$$\therefore \quad f(a) = \frac{1}{2\pi i} \int_C \frac{f(z)\,dz}{z-a} \tag{4.82}$$

which is "Cauchy's Integral Formula". It permits the evaluation of a function at any point within a closed contour when the value of the function on the contour is known.

4.14. THE THEORY OF RESIDUES

The following section of complex variable theory is paramount to the understanding of the Laplace transformation and its effective application to chemical engineering problems. The theory of residues is an extension of Cauchy's theorem for the case when $f(z)$ has a singularity at some point within the contour C.

For simplicity, take a coordinate system with its origin at the singularity of $f(z)$, with the contour C enclosing the origin but no other singularities of $f(z)$. If the singularity at the origin is a pole of order N, equation (4.57) shows that the function

$$g(z) = z^N f(z)$$

will be analytic at all points within the contour C. The new function $g(z)$ can therefore be expanded in an ascending power series in z. It follows that $f(z)$ is also expandable in a power series starting at z^{-N}. Thus

$$f(z) = \frac{B_N}{z^N} + \frac{B_{N-1}}{z^{N-1}} + \dots + \frac{B_1}{z} + \sum_{n=0}^{\infty} C_n z^n \tag{4.83}$$

which is known as the Laurent expansion.

Equation (4.83) can be used to obtain the integral of $f(z)$ round any contour C enclosing the singularity as follows. The infinite series of positive powers of z is analytic within and upon C and the integral of these terms will be zero by Cauchy's theorem. Referring to Example 2 in Section 4.13.1, the integral of each of the other terms on the right-hand side of equation (4.83) will also be zero except for the term (B_1/z). Example 3 of Section 4.13.1 shows that this integral will be $2\pi i B_1$, or

$$\int_C f(z)\,dz = 2\pi i B_1 \tag{4.84}$$

Hence the value of the contour integral of a complex function is $2\pi i$ multiplied by the coefficient of z^{-1} in the Laurent expansion of the complex function. This coefficient is known as the "residue" of the function at the pole.

The above analysis is valid when the singularity is not at the origin, thus if the pole is at $z = z_0$, the Laurent expansion is obtained as a series of powers of $(z - z_0)$ and the residue of the function at z_0 is the coefficient of $(z - z_0)^{-1}$ in the Laurent expansion in the neighbourhood of z_0. If the Laurent expansion is valid near an essential singularity, the coefficient B_1 is still termed the residue at the singularity.

Example 1. Evaluate $\int_C[e^z\,dz/(z-a)^3]$ around a circle centred at the origin.

If $|z| < |a|$, the function is analytic within the contour and

$$\int_C \frac{e^z\,dz}{(z-a)^3} = 0 \qquad\qquad \text{I}$$

by Cauchy's theorem.

If $|z| > |a|$, there is a pole of order 3 at $z = a$ within the contour. Therefore transfer the origin to $z = a$ by putting $\lambda = z - a$. Then

$$\int_C \frac{e^z\,dz}{(z-a)^3} = \int_C \frac{e^{\lambda+a}\,d\lambda}{\lambda^3}$$

$$= e^a \int_C\left[\frac{1}{\lambda^3} + \frac{1}{\lambda^2} + \frac{1}{2!\,\lambda} + \frac{1}{3!} + \frac{\lambda}{4!} + \dots\right]d\lambda \qquad\qquad \text{II}$$

All integrals on the right-hand side are zero except the third term which gives the residue $\frac{1}{2}e^a$

$$\therefore \quad \int_C \frac{e^{\lambda+a}\,d\lambda}{\lambda^3} = \int_C \frac{e^z\,dz}{(z-a)^3} = 2\pi i(\tfrac{1}{2}\,e^a) = \pi i\,e^a$$

4.14.1. *Evaluation of Residues without the Laurent Expansion*

With a large number of expressions encountered in applied science, the Laurent expansion is not immediately apparent to engineers. When this is the case, the following procedure is helpful. The complex function $f(z)$ can be expressed in terms of a numerator and a denominator if it has any singularities. That is,

$$f(z) = \frac{F(z)}{g(z)} \qquad\qquad (4.85)$$

and if a simple pole exists at $z = a$, then $(z - a)$ is a factor of $g(z)$ so that $g(z)$ can be written $(z - a)\,G(z)$ with $G(z)$ not zero at $z = a$. Furthermore, the Laurent expansion of $f(z)$ will exist even though it is not apparent, and therefore it can be written in the general form of equation (4.83),

$$f(z) = \frac{B_1}{z-a} + b_0 + b_1(z-a) + \ldots + b_n(z-a)^n + \ldots \qquad (4.86)$$

Equation (4.86) has been terminated at the first inverse term because there is only a simple pole at $z = a$. Multiplying both sides by $(z-a)$ and then putting $z = a$ gives

$$B_1 = (z-a)f(z)\big|_{z=a} = F(a)/G(a) \qquad (4.87)$$

which evaluates the residue at a simple pole. This is a very convenient method which will be illustrated by examples below, but certain points must be emphasized first. Thus in equation (4.85), $g(z)$ must contain one more factor $(z-a)$ than $F(z)$ for $f(z)$ to have a simple pole. Finally, if $g(z)$ has a number of factors so that

$$f(z) = \frac{F(z)}{(z-a_1)(z-a_2)(\ldots)(z-a_n)} \qquad (4.88)$$

$f(z)$ will have a residue at each pole, $z = a_1, a_2, \ldots, a_n$. Each residue can be evaluated independently by the method given above.

Example 2. Evaluate the residues of $z/(z^2 + z - 12)$

$$f(z) = \frac{z}{z^2 + z - 12} = \frac{z}{(z+4)(z-3)}$$

which has poles at $z = 3$ and $z = -4$.

Using equation (4.87), the residue at $z = 3$ is $3/7$; and the residue at $z = -4$ is $-4/(-7) = 4/7$.

The sum of the residues is

$$\tfrac{3}{7} + \tfrac{4}{7} = 1$$

Example 3. Evaluate the residues of $e^z/(z^2 + w^2)$

$$f(z) = \frac{e^z}{(z+iw)(z-iw)} \qquad \text{I}$$

which has poles at $z = \pm iw$.

The residue at the pole $z = iw$ is

$$\frac{e^z}{z+iw}\bigg|_{z=iw} = \frac{e^{iw}}{2iw} \qquad \text{II}$$

The residue at the pole $z = -iw$ is

$$\frac{e^z}{z-iw}\bigg|_{z=-iw} = \frac{-e^{-iw}}{2iw} \qquad \text{III}$$

The sum of the residues of $e^z/(z^2+w^2)$ is

$$\frac{e^{iw}-e^{-iw}}{2iw} = \frac{\sin w}{w} \qquad\qquad \text{IV}$$

The above method of determining the residue of a function appears to depend on the factorization of the denominator. However, the method is generally applicable to the evaluation of the residue at a simple pole even when the denominator cannot be factorized. If $f(z)$ is expressed as in equation (4.85) and the denominator $g(z)$ does not factorize, the same procedure may still be followed. Thus the residue of $f(z)$ at $z = a$ is

$$B_1 = \frac{(z-a)\,F(z)}{g(z)}\bigg|_{z=a} = \frac{0}{0} \qquad\qquad (4.89)$$

which is indeterminate. Therefore, by L'Hôpital's rule, Section 3.3.7,

$$\frac{d(z-a)\,F(z)/dz}{dg(z)/dz} = \frac{F(z)+(z-a)\,F'(z)}{g'(z)}$$

$$\therefore \quad B_1 = \frac{F(a)}{g'(a)} \qquad\qquad (4.90)$$

where $g'(a)$ is the differential coefficient of $g(z)$ with respect to z, evaluated at $z = a$.

Example 4. Evaluate $\int_C (e^z\, dz/\sin nz)$ around a circle with centre at the origin and radius $|z| < \pi/n$.

The residue at the pole where $z = 0$ is given by equation (4.90) as

$$B_1 = \frac{e^z}{d\sin nz/dz}\bigg|_{z=0} = \frac{e^z}{n\cos nz}\bigg|_{z=0} = \frac{1}{n}$$

$$\therefore \quad \int_C \frac{e^z\, dz}{\sin nz} = 2\pi i\left(\frac{1}{n}\right) = \frac{2\pi i}{n}$$

4.14.2. *Evaluation of Residues at Multiple Poles*

If $f(z)$ has a pole of order n at $z = a$ and no other singularity, $f(z)$ can be written in the form

$$f(z) = \frac{F(z)}{(z-a)^n} \qquad\qquad (4.91)$$

where n is a finite integer, and $F(z)$ is analytic at $z = a$. $F(z)$ can be expanded by the Taylor series to give

$$F(z) = F(a)+(z-a)\,F'(a)+\frac{(z-a)^2}{2!}F''(a)+\ldots+\frac{(z-a)^n}{n!}F^n(a)+\ldots \quad (4.92)$$

Dividing throughout by $(z-a)^n$ gives

$$f(z) = \frac{F(a)}{(z-a)^n} + \frac{F'(a)}{(z-a)^{n-1}} + \ldots + \frac{F^{n-1}(a)}{(n-1)!\,(z-a)} + \ldots \tag{4.93}$$

which is the Laurent expansion of $f(z)$ and therefore the residue at $z = a$ is the coefficient of $(z-a)^{-1}$ in equation (4.93). Therefore, the residue at a pole of order n situated at $z = a$ is given by

$$B_1 = \frac{1}{(n-1)!} \frac{d^{n-1}}{d\,z^{n-1}} [(z-a)^n f(z)]_{z=a} \tag{4.94}$$

This reduces to equation (4.87) for a simple pole when $n = 1$.

Example 5. Evaluate $\int_C [\cos 2z \, dz/(z-a)^3]$ around a circle of radius $|z| > |a|$.

$\cos 2z/(z-a)^3$ has a pole of order 3 at $z = a$, and the residue must be evaluated using equation (4.94).

$$\therefore \quad B_1 = \frac{1}{2!} \frac{d^2}{d\,z^2} \left[\frac{(z-a)^3 \cos 2z}{(z-a)^3} \right]_{z=a}$$

$$= \tfrac{1}{2}(-4 \cos 2z)|_{z=a}$$

$$= -2 \cos 2a$$

$$\therefore \quad \int_C \frac{\cos 2z \, dz}{(z-a)^3} = 2\pi i(-2\cos 2a) = -4\pi i \cos 2a$$

Chapter 5

FUNCTIONS AND DEFINITE INTEGRALS

5.1. Introduction

THE study of differential equations is one of the many branches of mathematics in which the object is to determine a functional relationship between two or more variables. The symbolism which is used to describe these functional relationships is largely a matter of convention; when a particular function is found convenient, it enters general usage and its properties are investigated. Elementary functions such as polynomials, powers, logarithms, exponentials, trigonometric and hyperbolic functions should be familiar to the reader already, and in Chapter 3 it was shown that the four kinds of Bessel function were useful for expressing the solutions of a particular class of differential equations. Other functions have been defined as the solutions of differential equations, and the Legendre polynomials which will be introduced in Chapter 8 is one such group.

Functions are also defined by integrals which cannot be evaluated in terms of the above elementary functions, and the purpose of this present chapter is to introduce some of these functions, enunciate their properties, and draw attention to their tabulated values as an aid to calculation. It will be shown that a variety of integrals can be expressed in terms of these new functions and thus be evaluated from tables.

Methods for evaluating some definite integrals by contour integration were presented in Chapter 4, and a further method by differentiating with respect to a parameter will be included in the present chapter.

5.2. The Error Function

This function which occurs in the theory of probability, distribution of residence times, conduction of heat, and diffusion of matter is defined by the integral,

$$\operatorname{erf} x = \frac{2}{\sqrt{\pi}} \int_0^x e^{-z^2} \, dz \tag{5.1}$$

The error function is illustrated in Fig. 5.1 which shows $\operatorname{erf} x$ as the area beneath the curve and between the ordinates $z = 0$ and $z = x$ when $2/\sqrt{(\pi)}\, e^{-z^2}$ is plotted against z. In equation (5.1) z is a "dummy variable" because it only enables the curve of Fig. 5.1 to be described and any variable would do this. The variable z is eliminated by the limits of integration thus leaving x as the only variable. The factor $2/\sqrt{\pi}$ is introduced for convenience so that

$$\operatorname{erf} \infty = 1 \tag{5.2}$$

The truth of equation (5.2) can be proved by the following argument. To simplify the algebra, remove the normalizing factor $2/\sqrt{\pi}$ and consider two

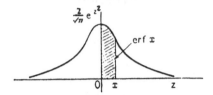

FIG. 5.1. The error function

versions of the integral, thus

$$I = \int_0^R e^{-x^2}\,dx = \int_0^R e^{-y^2}\,dy \tag{5.3}$$

Considering that x and y are two independent cartesian coordinates, equation (5.3) can be written

$$I^2 = \int_0^R \int_0^R e^{-(x^2+y^2)}\,dx\,dy \tag{5.4}$$

which can be interpreted as a volume as shown in Fig. 5.2. If the variables x and y are changed to polar coordinates r and θ using the equations

$$x = r\cos\theta, \quad y = r\sin\theta \tag{5.5}$$

then the element of area $(dx\,dy)$ has to be replaced by $(r\,dr\,d\theta)$. Also, the range of integration has to be altered from the square OABC, to the quadrant OAC which results in an error denoted by ε. Thus, equation (5.4) can be

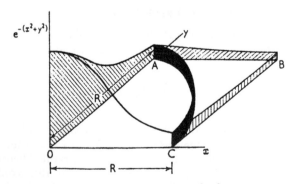

FIG. 5.2. Evaluation of erf ∞

written

$$I^2 = \int_0^R \int_0^{\frac{1}{2}\pi} e^{-r^2} r \, dr \, d\theta + \varepsilon \qquad (5.6)$$

The volume represented by ε has a base area which is less than $\frac{1}{2}R^2$ and a maximum height of e^{-R^2}. Thus

$$\varepsilon < \tfrac{1}{2}R^2 e^{-R^2} \qquad (5.7)$$

and as $R \to \infty$, $\varepsilon \to 0$.

Evaluating the double integral in equation (5.6) gives

$$I^2 = \tfrac{1}{4}\pi - \tfrac{1}{4}\pi e^{-R^2} + \varepsilon \qquad (5.8)$$

Therefore, as $R \to \infty$, $I^2 \to \tfrac{1}{4}\pi$. Comparing equations (5.1) and (5.3) gives the required result which is equation (5.2).

5.2.1. *Properties of the Error Function*

Differentiating equation (5.1) gives directly

$$\frac{d}{dx}\operatorname{erf} x = \frac{2}{\sqrt{\pi}} e^{-x^2} \qquad (5.9)$$

Using this result to integrate equation (5.1) by parts gives

$$\int \operatorname{erf} x \, dx = x \operatorname{erf} x - \int x \frac{2}{\sqrt{\pi}} e^{-x^2} dx + C$$

$$= x \operatorname{erf} x + \frac{1}{\sqrt{\pi}} e^{-x^2} + C \qquad (5.10)$$

where C is the constant of integration. Equation (5.10) is sometimes tabulated under the symbol "$\operatorname{ierf} x$" with $C = -1/\sqrt{\pi}$ so that $\operatorname{ierf} 0 = 0$.

Another related function which is sometimes tabulated is the complementary error function "$\operatorname{erfc} x$". This is defined by the equation

$$\operatorname{erfc} x = 1 - \operatorname{erf} x$$

$$= \frac{2}{\sqrt{\pi}} \int_x^\infty e^{-z^2} dz \qquad (5.11)$$

5.3. THE GAMMA FUNCTION

This is defined by the integral

$$\Gamma(n) = \int_0^\infty t^{n-1} e^{-t} dt \qquad (5.12)$$

for positive values of n. t is again a dummy variable since the value of the definite integral is independent of t. The variable t is only used to describe the function to be integrated. As an illustration of this fact, take the special case $n = 1$, thus

$$\Gamma(1) = \int_0^\infty e^{-t} dt$$

$$\therefore \quad \Gamma(1) = 1 \tag{5.13}$$

which is independent of t.

The most important property of the gamma function can be derived by integrating (5.12) by parts.

$$\Gamma(n) = \int_0^\infty t^{n-1} e^{-t} dt$$

$$= [-t^{n-1} e^{-t}]_0^\infty + (n-1) \int_0^\infty t^{n-2} e^{-t} dt$$

$$\therefore \quad \Gamma(n) = (n-1)\Gamma(n-1) \tag{5.14}$$

since the first term is zero at both limits of integration provided that $n > 1$.

Repeated applications of equation (5.14) for integer values of n gives

$$\Gamma(n) = (n-1)(n-2)(\ldots)(2)(1)\Gamma(1)$$

$$= (n-1)! \tag{5.15}$$

by using equation (5.13). The gamma function is thus a generalized factorial, and for positive integer values of n, the gamma function can be replaced by a factorial as given by equation (5.15). For non-integer values of n, the defining integral (5.12) has been evaluated numerically for $1 < n < 2$ and this table can be extended to all positive values of n by repeated use of equation (5.14).

The defining integral (5.12) is not valid for negative values of n, but the gamma function of a negative quantity can be found by using equation (5.14) to extend the definition. Thus, rearranging equation (5.14),

$$\Gamma(n-1) = \Gamma(n)/(n-1)$$

$$= \Gamma(n+1)/\{n(n-1)\} \quad \text{etc.} \tag{5.16}$$

This process can be repeated until the gamma function on the right-hand side has an argument lying in the range 1 to 2. However, if n is zero or a negative integer, the gamma function becomes infinite. An example of this can be seen by putting $n = 1$ into equation (5.14), thus

$$\Gamma(1) = 0\Gamma(0) \tag{5.17}$$

But equation (5.13) shows that $\Gamma(1) = 1$. Therefore, $\Gamma(0)$ must be infinite to satisfy equation (5.17). It is obvious that the sequence of operations (5.16) will eventually involve $\Gamma(0)$ whenever n is a negative integer, and thus the gamma function is infinite for all negative integers.

Figure 5.3 shows how $\Gamma(n)$ varies with n. For negative values of n, $\Gamma(n)$ is alternatively positive and negative between successive integer values, and for positive values of n the curve has a minimum value of 0·8856 at $n = 1·4616$.

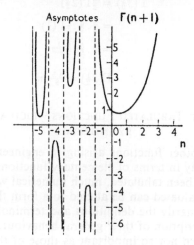

FIG. 5.3. The gamma function

$\Gamma(\tfrac{1}{2})$ can also be evaluated analytically as follows. By definition, equation (5.12),

$$\Gamma(\tfrac{1}{2}) = \int_0^\infty t^{-\frac{1}{2}} e^{-t} \, dt \qquad (5.18)$$

The negative square root can be removed by making the substitution $t = x^2$,

$$\therefore \quad dt = 2x \, dx.$$

$$\therefore \quad \Gamma(\tfrac{1}{2}) = \int_0^\infty x^{-1} e^{-x^2} 2x \, dx$$

$$= 2 \int_0^\infty e^{-x^2} \, dx \qquad (5.19)$$

Comparing equations (5.19) and (5.1) gives

$$\Gamma(\tfrac{1}{2}) = \sqrt{(\pi)} \, \mathrm{erf} \, \infty$$

and using equation (5.2) gives finally,

$$\Gamma(\tfrac{1}{2}) = \sqrt{\pi} \qquad (5.20)$$

Other half integer gamma functions can be found using equations (5.14) and (5.16), as illustrated by the following examples.

Example 1. Evaluate $\Gamma(3\frac{1}{2})$.

Successive application of equation (5.14) gives

$$\Gamma(3\tfrac{1}{2}) = \tfrac{5}{2}\Gamma(2\tfrac{1}{2})$$

$$= \tfrac{5}{2}\cdot\tfrac{3}{2}\cdot\tfrac{1}{2}\Gamma(\tfrac{1}{2})$$

$$= \tfrac{5}{2}\cdot\tfrac{3}{2}\cdot\tfrac{1}{2}\sqrt{\pi}$$

by equation (5.20).

$$\therefore \quad \Gamma(3\tfrac{1}{2}) = \tfrac{15}{8}\sqrt{\pi}$$

5.4. Other Tabulated Functions which are Defined by Integrals

There are many other functions arising in engineering which cannot be integrated analytically in terms of elementary functions. Since the values of many integrals have been tabulated, much numerical work can be avoided if the integral to be evaluated can be altered to a form that is tabulated. This section contains primarily the definitions and terminology of some of these functions and a description of their general behaviour. Their properties are neither so well known nor so important as those of the earlier functions in this chapter; nevertheless, a knowledge of the definitions is valuable particularly when using the Laplace transform.

5.4.1. *The Beta Function*

The defining integral contains two parameters instead of one, and the integral is taken between fixed finite limits, viz.

$$B(p,q) = \int_0^1 t^{p-1}(1-t)^{q-1}\,dt \tag{5.21}$$

Both p and q must be positive to ensure convergence of the integral at its limits.

An alternative form can be obtained for the defining integral by making the substitution $t = \sin^2\theta$, $1-t = \cos^2\theta$,

$$\therefore \quad dt = 2\sin\theta\cos\theta\,d\theta$$

Thus

$$B(p,q) = 2\int_0^{\frac{1}{2}\pi} \sin^{2p-1}\theta\cos^{2q-1}\theta\,d\theta \tag{5.22}$$

By substituting $t = \cos^2\theta$ in equation (5.21), an equation can be obtained which is similar to equation (5.22) but with p and q interchanged, thus showing that the beta function is symmetrical in its two arguments p and q.

A further useful substitution into equation (5.21) is

$$t = \frac{x}{1+x}, \quad 1-t = \frac{1}{1+x}, \quad dt = \frac{dx}{(1+x)^2}$$

which gives

$$B(p,q) = \int_0^\infty \frac{x^{p-1}\,dx}{(1+x)^{p+q}} \qquad (5.23)$$

For the special case $p+q = 1$, equation (5.23) becomes

$$B(1-q,q) = \int_0^\infty \frac{x^{-q}\,dx}{1+x} \qquad (5.24)$$

This integral can be evaluated using the methods developed in Chapter 4 by integrating $z^{-q}/(1-z)$ round the contour shown in Fig. 5.4 where the complex z plane has a branch cut along the negative real axis because q is not an integer. The result is

$$B(1-q,q) = \frac{\pi}{\sin q\pi} \qquad (5.25)$$

for $0<q<1$.

It can be shown that the beta and gamma functions are related by

$$\Gamma(p)\,\Gamma(q) = B(p,q)\,\Gamma(p+q) \qquad (5.26)$$

for all positive values of p and q.

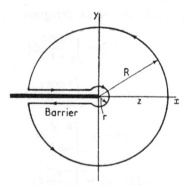

FIG. 5.4. Integration contour of beta function

In the special case $p+q = 1$, using equations (5.13) and (5.25) proves the useful formula

$$\Gamma(q)\,\Gamma(1-q) = \pi/\sin q\pi \qquad (5.27)$$

which is valid for $0<q<1$.

5.4.2. The Elliptic Integral of the First Kind

This is usually defined by

$$F(k,\phi) = \int_0^\phi \frac{d\Phi}{\sqrt{(1-k^2\sin^2\Phi)}} \qquad (5.28)$$

where k is termed the "modulus" and ϕ the "amplitude". The integral occurs in the study of periodic motions when the amplitude is not small and the motion is not therefore simple harmonic. Hence the terminology for k and ϕ. The complete elliptic integral of the first kind is a function of k only and is defined by

$$F(k, \tfrac{1}{2}\pi) = K(k) \tag{5.29}$$

5.4.3. The Elliptic Integral of the Second Kind

This is defined in a similar manner by

$$E(k, \phi) = \int_0^\phi \sqrt{(1 - k^2 \sin^2 \Phi)}\, d\Phi \tag{5.30}$$

where again k is the modulus and ϕ the amplitude. This integral measures length round the arc of an ellipse from the minor axis, where ϕ is the complement of the normal parametric variable of an ellipse and k is its eccentricity. The above connection with the geometry of the ellipse gives rise to the name "elliptic integral". The complete elliptic integral of the second kind is again defined as

$$E(k, \tfrac{1}{2}\pi) = E(k) \tag{5.31}$$

5.4.4. The Sine, Cosine, and Exponential Integrals

$$Si(x) = \int_0^x \frac{\sin t}{t}\, dt \tag{5.32}$$

$$Ci(x) = \int_\infty^x \frac{\cos t}{t}\, dt \tag{5.33}$$

$$Ei(x) = \int_{-\infty}^x \frac{e^t}{t}\, dt \tag{5.34}$$

$$-Ei(-x) = \int_x^\infty \frac{e^{-t}}{t}\, dt \tag{5.35}$$

Of this group of functions, only the sine integral remains finite at the origin. All of the other functions are infinite at the origin; also, $Ei(x)$ is infinite as $x \to \infty$, and $Ei(-x)$ is infinite as $x \to -\infty$.

Although the sine integral cannot be evaluated in terms of elementary functions, it is possible to evaluate $Si(\infty)$. The integrand of equation (5.32) is an even function, so that

$$Si(\infty) = \frac{1}{2} \int_{-\infty}^\infty \frac{\sin t}{t}\, dt \tag{5.36}$$

and integrating a suitable function round a closed contour in the complex plane, the integral can be evaluated.

An alternative and simpler method is to introduce a suitable parameter α and evaluate the integral as follows. Define

$$I(\alpha) = \int_0^\infty \frac{e^{-\alpha t} \sin t}{t} dt \qquad (5.37)$$

and differentiate equation (5.37) with respect to α thus,

$$\frac{dI}{d\alpha} = -\int_0^\infty e^{-\alpha t} \sin t \, dt \qquad (5.38)$$

Equation (5.38) can be integrated by parts twice, or a complex method can be adopted.

$$\frac{dI}{d\alpha} = -\operatorname{Im} \int_0^\infty e^{-\alpha t} e^{it} dt$$

$$= \operatorname{Im} \left[\frac{1}{\alpha - i} e^{-(\alpha - i)t} \right]_0^\infty$$

$$= -\operatorname{Im} \left[\frac{\alpha + i}{\alpha^2 + 1} \right]$$

$$= -\operatorname{Im} \left[\frac{\alpha}{\alpha^2 + 1} + i \frac{1}{\alpha^2 + 1} \right]$$

$$\therefore \quad \frac{dI}{d\alpha} = -\frac{1}{\alpha^2 + 1} \qquad (5.39)$$

On integration, this yields

$$I = -\int \frac{d\alpha}{\alpha^2 + 1}$$

$$\therefore \quad I = A - \tan^{-1} \alpha \qquad (5.40)$$

where A is an arbitrary constant which can be evaluated from a known result at any value of α. When $\alpha = \infty$, equation (5.37) shows $I(\alpha) = 0$.

$$\therefore \quad 0 = A - \tfrac{1}{2}\pi$$

Substituting this value of A into equation (5.40) gives

$$I(\alpha) = \tfrac{1}{2}\pi - \tan^{-1} \alpha \qquad (5.41)$$

The infinite sine integral (5.36) is given by equation (5.37) with $\alpha = 0$, therefore from equation (5.41).

$$I(0) = \tfrac{1}{2}\pi$$

and

$$Si(\infty) = \int_0^\infty \frac{\sin t}{t} dt = \tfrac{1}{2}\pi \qquad (5.42)$$

5.5. EVALUATION OF DEFINITE INTEGRALS

The method used above to evaluate $Si(\infty)$ is of wider application, particularly in some of the transform methods given in Chapter 8. It is often unnecessary to introduce a parameter such as α above, because a suitable parameter is already present within the integral. Consider the following example of an integral of this type.

Example 1. Evaluate

$$I(\alpha) = \int_0^\infty e^{-x^2} \frac{\sin 2\alpha x}{x} dx \qquad \qquad \text{I}$$

This is a much more difficult example; however, following the same method and differentiating equation I with respect to α,

$$\therefore \quad \frac{dI}{d\alpha} = \int_0^\infty 2 e^{-x^2} \cos 2\alpha x \, dx \qquad \qquad \text{II}$$

The primary effect of this differentiation has been to remove the term x from the denominator of the integrand. A further differentiation will introduce a factor x into the numerator and facilitate an integration by parts, thus,

$$\frac{d^2 I}{d\alpha^2} = -\int_0^\infty 4x e^{-x^2} \sin 2\alpha x \, dx \qquad \qquad \text{III}$$

$$= \left[2 e^{-x^2} \sin 2\alpha x - \int 4\alpha e^{-x^2} \cos 2\alpha x \, dx \right]_0^\infty \qquad \text{IV}$$

On the right-hand side of equation IV, the first term vanishes at both limits, and the second term is a constant multiple of the right-hand side of equation II. Therefore,

$$\frac{d^2 I}{d\alpha^2} = -2\alpha \frac{dI}{d\alpha} \qquad \qquad \text{V}$$

The dependent variable does not occur explicitly in equation V and the method of Section 2.4.1 can be used. Thus, put

$$p = \frac{dI}{d\alpha} \qquad \qquad \text{VI}$$

separate the variables, and obtain the solution for p in terms of α, viz.

$$p = A e^{-\alpha^2} \qquad \qquad \text{VII}$$

where A is the constant of integration which can be evaluated for the special case $\alpha = 0$, when equation II gives

$$p = \frac{dI}{d\alpha} = \sqrt{\pi} \operatorname{erf} \infty = \sqrt{\pi}$$

$$\therefore \quad p = \frac{dI}{d\alpha} = \sqrt{(\pi)} e^{-\alpha^2} \qquad \qquad \text{VIII}$$

Integrating again with respect to α and using equation (5.1),

$$I = \tfrac{1}{2}\pi \operatorname{erf}\alpha + B \qquad\qquad \text{IX}$$

where B, the constant of integration, can be evaluated by using equation I when $\alpha = 0$. Thus,

if $\qquad\qquad \alpha = 0, \quad I = 0; \quad \therefore \quad B = 0.$

The evaluation of the integral in equation I is thus

$$I(\alpha) = \int_0^\infty e^{-x^2}\frac{\sin 2\alpha x}{x}\,dx = \tfrac{1}{2}\pi \operatorname{erf}\alpha \qquad\qquad \text{X}$$

The above method consists essentially of finding a differential equation V which the given integral satisfies, solving the differential equation and hence obtaining an alternative form for the solution. This process will inevitably introduce arbitrary constants of integration and these must be evaluated by referring to the original integral which is the desired particular solution of the differential equation.

In the previous section, $Si(\infty)$ was shown to be a solution of equation (5.39). It is not always possible to find a suitable differential equation; for example, the gamma function defined by equation (5.12) does not satisfy any differential equation with rational coefficients. Nevertheless, the above method and the method of contour integration are two advanced techniques which will successfully evaluate many definite integrals.

Tables of the functions discussed in this chapter are readily available in:

Dwight, H. B. "Mathematical Tables" (2nd ed.). Dover (1958).

Jahnke, E. and Emde, F. "Tables of Functions" (4th ed.). Dover (1945).

"Standard Mathematical Tables" (11th ed.). Chemical Rubber Publishing Company (1957).

An index of published tables has been compiled by:

Fletcher, A., Miller, J. C. P., Rosenhead, L. and Comrie. "An Index of Mathematical Tables." Scientific Computing Services, London (1962).

Chapter 6

THE LAPLACE TRANSFORMATION

6.1. INTRODUCTION

THE Laplace transformation is the name given to a particular "operational method" of solving differential equations. Essentially, an operational method is a technique whereby an ordinary differential equation is converted into an equivalent algebraic form which can be solved by the laws of elementary algebra. The method will also convert a partial differential equation into an equivalent easily solvable ordinary differential equation. There are many such operational methods but the Laplace transformation is a particularly useful mathematical tool for solving engineering problems because the boundary conditions are introduced into the equation prior to its solution.

The application of operational methods for the solution of engineering problems was first made by Heaviside (1850–1925), and although his methods were largely intuitive they were successful. This stimulated the interest of other mathematicians who developed the theoretical basis of Heaviside's methods and extended them into what is known as the Laplace transformation. The essential features of the method and its application to the solution of chemical engineering problems involving differential equations are presented in the following sections, although the methods of Chapter 2 are normally used for solving these problems. The most useful applications of the Laplace transformation, to partial differential equations and difference-differential equations, will be given in Chapters 8 and 9 respectively.

6.2. THE LAPLACE TRANSFORM

If $f(t)$ is a continuous function of an independent variable t for all values of t greater than zero, then the integral with respect to t of the product of $f(t)$ with e^{-st} between the limits 0 and ∞ is defined as the Laplace transform of $f(t)$. That is,

$$\int_0^\infty e^{-st} f(t)\, dt = \mathscr{L}[f(t)] \tag{6.1}$$

For the present, the only conditions that will be placed on the integral in equation (6.1) are that the parameter s must be large enough to make the integral convergent at the upper limit and t must be positive.

The Laplace transform of a function can be denoted by placing $\mathscr{L}$ before the function as shown in equation (6.1). To be exact, this signifies that the operation of the Laplace transformation has been carried out on the function

$f(t)$ and the modern practice is to write $\bar{f}(s)$ for the transform of the function to imply the connection between $f(t)$ and its transform $\bar{f}(s)$. Thus

$$\mathcal{L}[f(t)] = \bar{f}(s) \tag{6.2}$$

To familiarize the reader with the method, the Laplace transforms of some simple functions will now be derived. Thus if $f(t) = 1$, then

$$\mathcal{L}[1] = \int_0^\infty 1\, e^{-st}\, dt = \left[\frac{e^{-st}}{-s}\right]_0^\infty = \frac{1}{s} \tag{6.3}$$

or $\bar{f}(s) = 1/s$.

From equation (6.3) it can be seen that the Laplace transform of a constant K is

$$\mathcal{L}[K] = K\mathcal{L}[1] = K/s \tag{6.4}$$

since the constant can be inside or outside the integral in equation (6.3). This simple result follows from the fact that the Laplace transformation is a linear operation.

To illustrate the procedure further, let the Laplace transform of t^n be required. Thus

$$\mathcal{L}[t^n] = \int_0^\infty t^n e^{-st}\, dt \tag{6.5}$$

The integration variable in equation (6.5) can be changed by putting

$$z = st$$

$$\therefore \quad dz = s\, dt$$

$$\therefore \quad \mathcal{L}[t^n] = s^{-n-1} \int_0^\infty z^n e^{-z}\, dz \tag{6.6}$$

Comparing equation (6.6) with the definition of the gamma function, equation (5.12), gives

$$\mathcal{L}[t^n] = \Gamma(n+1)/s^{n+1} \tag{6.7}$$

If n is a positive integer, the property of the gamma function given by equation (5.15) shows that

$$\mathcal{L}[t^n] = n!/s^{n+1} \tag{6.8}$$

Finally, let $f(t) = e^{at}$; and the Laplace transform of this function is

$$\bar{f}(s) = \int_0^\infty e^{at} e^{-st}\, dt = \int_0^\infty e^{-(s-a)t}\, dt = \frac{1}{s-a} \tag{6.9}$$

In the same way the Laplace transforms of other simple functions can be obtained. Tables of transforms are shown in the appendix at the end of this book and the reader is encouraged to verify some of the examples given.

6.2.1. *Laplace Transforms of Derivatives*

The method given in Section 6.2 for determining the Laplace transform of a function can be applied to the differential coefficient of any continuous function. Thus let the derivative of a function be $df(t)/dt$, then the Laplace transform of this derivative is

$$\int_0^\infty \frac{df(t)}{dt} e^{-st} dt = [f(t) e^{-st}]_0^\infty + s \int_0^\infty f(t) e^{-st} dt \qquad (6.10)$$

and since the Laplace transform of $f(t)$ is $\bar{f}(s)$, equation (6.10) becomes

$$\mathscr{L}\left[\frac{df(t)}{dt}\right] = s\bar{f}(s) - f(0) \qquad (6.11)$$

Similarly, the second derivative of a function $d^2f(t)/dt^2$ can be converted to its Laplace transform by the same procedure to give

$$\mathscr{L}\left[\frac{d^2f(t)}{dt^2}\right] = s^2\bar{f}(s) - sf(0) - f'(0) \qquad (6.12)$$

where $f(0)$ and $f'(0)$ are the values of the function and its derivative when the independent variable is zero.

Finally the integration processes described above can be extended to finding the Laplace transform of an nth order derivative of a function. The result is

$$\mathscr{L}\left[\frac{d^nf(t)}{dt^n}\right] = s^n\bar{f}(s) - [s^{n-1}f(0) + s^{n-2}f'(0)... + sf^{(n-2)}(0)$$

$$+ f^{(n-1)}(0)] \qquad (6.13)$$

where $f'(0), f^{(2)}(0), ..., f^{(n-1)}(0)$ are the values of the first, second, up to the $(n-1)$th derivative of the function when the independent variable is zero.

The Laplace transforms of the different order derivatives presented above are only valid for continuous functions. If the function undergoes a step-change in the range in which the independent variable is being considered, each of the equations (6.11) to (6.13) have to be modified. This will be considered later, in Section 6.5.

Equations (6.11) to (6.13) introduce the fundamental operational properties of the Laplace transformation. These are the properties that introduce the boundary conditions $f(0)$, $f'(0)$, etc. into the problem before solution of the equation. At the same time, the differential equation is reduced to an algebraic equation in terms of the operator s. However, it should be noted that the operation described is only applicable to "initial value" problems. That is, the value of the function and its derivatives must be known when the independent variable is zero. This becomes apparent by inspection of the transforms of the derivatives.

Finally, equation (6.13) can be utilized to find the Laplace transform of a function without carrying out the integration procedure, and to illustrate the method, the Laplace transform of $\sin \alpha t$ will be evaluated.

Let
$$f(t) = \sin \alpha t$$

Then
$$f'(t) = \alpha \cos \alpha t$$

and
$$f''(t) = -\alpha^2 \sin \alpha t$$

Also
$$f(t)|_{t=0} = \sin \alpha t|_{t=0} = 0$$

and
$$f'(t)|_{t=0} = \alpha \cos \alpha t|_{t=0} = \alpha$$

Since
$$\mathscr{L}(f''(t)) = s^2 \bar{f}(s) - sf(0) - f'(0) \qquad (6.12)$$

$$\therefore \quad -\mathscr{L}[\alpha^2 \sin \alpha t] = s^2 \mathscr{L}[\sin \alpha t] - 0 - \alpha$$

or
$$\mathscr{L}[\sin \alpha t] = \frac{\alpha}{s^2 + \alpha^2} \qquad (6.14)$$

which is transform number 15 in the table in the Appendix.

6.2.2. The Shifting Theorem

This theorem is most valuable in obtaining the transform of an exponential type of function and also in carrying out the inverse transformation which will be described later. The shifting theorem may be enunciated as follows:

If the Laplace transform of $f(t)$ is $\bar{f}(s)$, then the Laplace transform of $e^{-\alpha t} f(t)$ is $\bar{f}(s+\alpha)$, where α is a finite constant. That is, the multiplication of $f(t)$ by $e^{-\alpha t}$ in effect shifts the origin of s to $(s+\alpha)$.

This theorem can be proved by carrying out the integration procedure on the general function $f(t)$, but it will be more illuminating to verify it by applying the theorem to the following example.

Example. Determine the Laplace transform of $e^{-\alpha t} \cos \beta t$. Transform number 16 shows that if $f(t) = \cos \beta t$, then $\bar{f}(s) = s/(s^2 + \beta^2)$. Therefore, by the shifting theorem

$$\mathscr{L}[e^{-\alpha t} \cos \beta t] = \frac{s+\alpha}{(s+\alpha)^2 + \beta^2} \qquad \text{I}$$

which is the desired result. However, this will be verified by integration. Thus

$$\int_0^\infty e^{-\alpha t} \cos \beta t \, e^{-st} \, dt = \mathscr{R} \int_0^\infty e^{-\alpha t} e^{i\beta t} e^{-st} \, dt \qquad \text{II}$$

$$= \mathscr{R} \int_0^\infty e^{-(\alpha+s-i\beta)t} \, dt$$

$$= \mathscr{R} \left[\frac{e^{-(\alpha+s-i\beta)t}}{-(\alpha+s-i\beta)} \right]_0^\infty$$

$$= \mathscr{R}[1/(\alpha+s-i\beta)]$$

Using the division rule for complex numbers, and discarding the imaginary part confirms the result given in equation I.

6.3. The Inverse Transformation

In order that the operational method of the Laplace transformation can be exploited for the solution of engineering problems, it must be convenient to convert the transform back to a function of the independent variable. That is, it is necessary to perform the operation

$$\mathscr{L}^{-1}[\bar{f}(s)] = f(t) \tag{6.15}$$

in order to complete the solution of the problem. On occasion this can be done by inspection, by looking up the function corresponding to the transform in a table of Laplace transforms, such as the one reproduced at the end of this book. Some manipulation of the constants is often necessary as illustrated by the following example.

Example 1. Find the inverse transform of $s/(s^2 - 2s + 5)$. Completing the square of the denominator in an attempt to find factors gives

$$\bar{f}(s) = \frac{s}{(s-1)^2 + 4} \tag{I}$$

Applying the shifting theorem,

$$\bar{f}(s+1) = \frac{s+1}{s^2 + 2^2} \tag{II}$$

But the right-hand side of equation II is the sum of transforms number 15 and 16 at the end of the book.

$$\therefore \quad e^{-t}f(t) = \cos 2t + \tfrac{1}{2}\sin 2t$$

$$\therefore \quad f(t) = e^{t}(\cos 2t + \tfrac{1}{2}\sin 2t) \tag{III}$$

is the inverse transform of $\bar{f}(s) = s/(s^2 - 2s + 5)$.

6.3.1. *Inverting the Transform Using Partial Fractions*

The Laplace transform of a function is an expression involving the parameter s which may be very complex. When the expression is of the form

$$\bar{f}(s) = \frac{\theta(s)}{\phi(s)} \tag{6.16}$$

where $\theta(s)$ and $\phi(s)$ are polynomials in s, and $\phi(s)$ is of higher degree than $\theta(s)$, $\bar{f}(s)$ can be expanded into its partial fractions. If the complex roots of $\phi(s)$ are also resolved, then all terms in the resolution must be of the type $1/(s-a)^n$ with n an integer and a a real or complex constant. The inversion of

$\bar{f}(s)$ is then simply a case of repeated application of transform number 10 in the tables, viz.

If
$$\bar{f}(s) = \frac{1}{(s-a)^n}$$

then
$$f(t) = \frac{1}{(n-1)!} t^{n-1} e^{at}$$

For use with partial fractions, this transform pair is well worth remembering. Alternatively, transform number 3 can be memorized and the shifting theorem applied to it. The following example illustrates this technique.

Example 2. Determine $f(t)$ if $\bar{f}(s) = 1/(s+a)(s+b)^2$. Expanding by partial fractions gives

$$\bar{f}(s) = \frac{1}{(s+a)(s+b)^2} = \frac{A}{s+a} + \frac{B}{s+b} + \frac{C}{(s+b)^2} \qquad \text{I}$$

$$\therefore \quad A(s+b)^2 + B(s+a)(s+b) + C(s+a) = 1 \qquad \text{II}$$

Evaluating C by putting $s = -b$ gives

$$C = 1/(a-b) \qquad \text{III}$$

Evaluating A by putting $s = -a$ gives

$$A = 1/(a-b)^2 \qquad \text{IV}$$

Evaluating B by equating coefficients of s^2 in equation II gives

$$B = -1/(a-b)^2 \qquad \text{V}$$

$$\therefore \quad (a-b)^2 \bar{f}(s) = \frac{1}{s+a} - \frac{1}{s+b} + \frac{a-b}{(s+b)^2} \qquad \text{VI}$$

Inverting the right-hand side of equation VI term by term using transform number 10 (or special cases 8 and 9) gives

$$(a-b)^2 f(t) = e^{-at} - e^{-bt} + (a-b)t\,e^{-bt} \qquad \text{VII}$$

$$\therefore \quad f(t) = \mathscr{L}^{-1}\left[\frac{1}{(s+a)(s+b)^2}\right] = \frac{e^{-at} + [(a-b)t-1]e^{-bt}}{(a-b)^2} \qquad \text{VIII}$$

The expansion of the transform by the use of partial fractions is quickly and conveniently accomplished when the denominator contains only simple roots; but when the denominator contains a repeated root of high order, resolution of the coefficients is a lengthy, time-consuming process. However, the evaluation of coefficients by the theory of residues is much more straightforward and is recommended under such circumstances. This method is discussed in Section 6.8 where it is shown to be the most powerful method of obtaining the inverse transform.

6.3.2. *Application to the Solution of Ordinary Differential Equations*

The primary application of the Laplace transformation is the provision of a powerful operational method for removing derivatives from an equation. Therefore, before proceeding further with the properties of transformation, the operational characteristics will be applied to the solution of ordinary differential equations in order to illustrate the technique. The method depends on the fact that an ordinary differential equation can be converted into an equivalent algebraic equation called the "subsidiary equation" in which the Laplace operator s is the independent variable. The procedure is illustrated by the following example.

Example 3.   Solve $(d^2 y/dt^2) + 4y = 3$, where $y(0) = y'(0) = 1$.   Using equation (6.12) to transform the derivative gives

$$\mathscr{L}\left[\frac{d^2 y}{dt^2}\right] = s^2 \bar{y}(s) - s - 1 \qquad\qquad \text{I}$$

and transforming the constant gives

$$\mathscr{L}[3] = 3/s \qquad\qquad \text{II}$$

Substituting these transforms into the differential equation gives the subsidiary equation

$$s^2 \bar{y}(s) - s - 1 + 4\bar{y}(s) = 3/s \qquad\qquad \text{III}$$

Solving the algebraic equation III for $\bar{y}(s)$ gives

$$\bar{y}(s) = \frac{s}{s^2 + 4} + \frac{1}{s^2 + 4} + \frac{3}{s(s^2 + 4)} \qquad\qquad \text{IV}$$

Inverting equation IV term by term, using transforms number 16, 15, and 19 respectively from the table at the back of the book gives

$$y = \cos 2t + \tfrac{1}{2} \sin 2t + \tfrac{3}{4}(1 - \cos 2t) \qquad\qquad \text{V}$$

which will simplify to

$$y = \tfrac{1}{2} \sin 2t + \tfrac{1}{4} \cos 2t + \tfrac{3}{4} \qquad\qquad \text{VI}$$

6.4. PROPERTIES OF THE LAPLACE TRANSFORMATION

Before further application of the Laplace transformation to the solution of engineering problems is made, some of its most important properties will be discussed, so that this operational method can be applied to the solution of more complicated problems.

6.4.1. *Differentiation of the Transform with Respect to the Operator s*

If $f(t)$ is a continuous function of the independent variable t, and its Laplace transform is $\bar{f}(s)$, then differentiation of the transform with respect to s corresponds to multiplying the function by $-t$.

This property can be proved as follows.

$$\bar{f}(s) = \int_0^\infty f(t)\,e^{-st}\,dt \tag{6.17}$$

by definition, and differentiating equation (6.17) with respect to s gives

$$\frac{d}{ds}[\bar{f}(s)] = \int_0^\infty -tf(t)\,e^{-st}\,dt \tag{6.18}$$

But the integral in equation (6.18) is the Laplace transform of $-tf(t)$ by the definition (6.1). Hence the property is proved.

6.4.2. *The Laplace Transform of the Integral of a Function*

If $f(t)$ is an integrable function of the independent variable t and its Laplace transform is $\bar{f}(s)$, then division of $\bar{f}(s)$ by s corresponds to integration of $f(t)$ between the limits 0 and t.

The above property can be verified most easily by showing that the converse is true. Thus

$$\mathscr{L}\left[\int_0^t f(t)\,dt\right] = \int_0^\infty \left[\int_0^t f(t)\,dt\right] e^{-st}\,dt \tag{6.19}$$

The right-hand side can be integrated by parts by letting

$$u = \int_0^t f(t)\,dt$$

and then

$$du = f(t)\,dt$$

$$dv = e^{-st}\,dt$$

so that

$$v = -e^{-st}/s$$

Substitution of these parts into equation (6.19) gives

$$\mathscr{L}\left[\int_0^t f(t)\,dt\right] = -\left[\frac{e^{-st}}{s}\int_0^t f(t)\,dt\right]_0^\infty + \frac{1}{s}\int_0^\infty f(t)\,e^{-st}\,dt \tag{6.20}$$

The first term is zero at the upper limit because of the exponential term, and zero at the lower limit because of the range of integration. Therefore

$$\mathscr{L}\left[\int_0^t f(t)\,dt\right] = \frac{1}{s}\bar{f}(s) \tag{6.21}$$

The results of Sections 6.4.1 and 6.4.2 are of great value in the solution of engineering problems. They find extensive use in the solution of differential equations with variable coefficients and the method of solution will be illustrated by the next example.

Example. Using the data given below obtain a preliminary estimate of the diameter of the tubes to be installed in a fixed bed catalytic reactor which is to be used for the synthesis of vinyl chloride from acetylene and hydrogen chloride. The tubes are to contain the mercuric chloride catalyst deposited on 2·5 mm particles of carbon and the heat of the reaction is to be employed to generate steam at 120 °C for the remainder of the process. To do this, the temperature of the inside surface of the tubes should be constant at 149 °C.

Effective thermal conductivity of bed (k_E) = 25·4 kJ/m °K h

Heat of reaction at bed temperature $(-\Delta H)$ = $1·07 \times 10^5$ kJ/kg mol†

Bulk density of bed (ρ) = 290 kg/m³

The rate of reaction is a function of temperature, concentration and the various adsorption coefficients,‡ but for the preliminary estimate assume that the rate of reaction can be expressed as

$$r = r_0(1 + AT) \text{ kg moles/h kg catalyst}$$

where $r_0 = 0·12$, $A = 0·043$, and T is the temperature in degrees Kelvin above a datum of 366 °K.

The maximum allowable catalyst temperature to ensure a satisfactory life is 525 °K (that is, $T = 159$).

Solution

Consider Fig. 6.1, which illustrates the following additional symbols describing the process.

G = flow rate in kg/h m² of reactor section

R = tube radius in m

x = radial coordinate

z = axial coordinate from inlet

C_p = heat capacity of the gas in kJ/kg °K

Taking a heat balance over the element illustrated in Fig. 6.1 gives

$$\text{Input} = -2\pi x k_E \frac{\partial T}{\partial x} \delta z + 2\pi x \delta x G C_p T - 2\pi x \delta x \delta z \rho \Delta H r \qquad \text{I}$$

$$\text{Output} = -2\pi x k_E \frac{\partial T}{\partial x} \delta z + \frac{\partial}{\partial x}\left[-2\pi x k_E \frac{\partial T}{\partial x} \delta z\right] \delta x$$

$$+ 2\pi x \delta x G C_p \left(T + \frac{\partial T}{\partial z} \delta z\right) \qquad \text{II}$$

Accumulation = 0, because the process is at steady state.

† Calculated from heats of formation, etc.
‡ Wesselhoft, R. D., Woods, J. M. and Smith, J. M. *A.I.Ch.E.J.* **5**, 361 (1959).

But,   Input − Output = Accumulation   (1.25)

and dividing throughout by $2\pi k_E x \delta x \delta z$ gives

$$\frac{\partial^2 T}{\partial x^2} + \frac{1}{x}\frac{\partial T}{\partial x} - \frac{GC_p}{k_E}\frac{\partial T}{\partial z} - \frac{\rho \Delta Hr}{k_E} = 0 \qquad\qquad \text{III}$$

Equation III is a well-known partial differential equation expressing the relationship between the temperature and the dimensions of the reactor. If, for the preliminary estimate, it is assumed that the temperature of the bed reaches a maximum at all radii, at the same distance z from the reactor

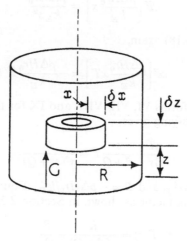

FIG. 6.1. Fixed bed reactor

entrance, then at this radial section of the bed $\partial T/\partial z$ will be zero even though the actual temperatures will vary with x. The temperature will be a maximum at the tube axis and this must not exceed 525 °K. Hence at the tube axis $\partial T/\partial x$ will be zero also, but will have a finite value at all radii greater than zero. Hence the partial differential equation III becomes an ordinary differential equation at this particular value of z. That is,

$$\frac{d^2 T}{dx^2} + \frac{1}{x}\frac{dT}{dx} - \frac{\rho \Delta Hr_0}{k_E}(1 + AT) = 0 \qquad\qquad \text{IV}$$

Equation IV can be easily converted to a Bessel equation of zero order and solved by the method given in Chapter 3, but on this occasion it will be solved using the Laplace transformation. The equation can be rewritten

$$x\frac{d^2 T}{dx^2} + \frac{dT}{dx} - \frac{\rho \Delta Hr_0}{k_E}(1 + AT)x = 0 \qquad\qquad \text{V}$$

Using equations (6.12) and (6.18) with x as the independent variable

$$\mathscr{L}\left[x\frac{d^2 T}{dx^2}\right] = -\frac{d}{ds}[s^2 \bar{T}(s) - sT(0) - T'(0)]$$

$$= -s^2 \bar{T}'(s) - 2s\bar{T}(s) + T(0) \qquad \text{VI}$$

and using equation (6.11)

$$\mathscr{L}\left[\frac{dT}{dx}\right] = s\bar{T}(s) - T(0) \qquad \text{VII}$$

Also,

$$\mathscr{L}\left[\frac{\rho\Delta Hr_0}{k_E}x\right] = \frac{\rho\Delta Hr_0}{k_E s^2} \qquad \text{VIII}$$

and using equation (6.18) again,

$$\mathscr{L}\left[\frac{\rho\Delta Hr_0}{k_E}xT\right] = -\frac{\rho\Delta Hr_0}{k_E}\bar{T}'(s) \qquad \text{IX}$$

Substituting the transforms VI, VII, VIII, and IX for the equivalent terms in equation V and simplifying gives

$$\frac{d\bar{T}}{ds} + \frac{s}{s^2 + Q}\bar{T} = \frac{P}{s^2(s^2 + Q)} \qquad \text{X}$$

where $Q = -\rho\Delta Hr_0 A/k_E$ and $P = -\rho\Delta Hr_0/k_E$. Equation X can be solved by using an integrating factor as shown in Section 2.3.4 giving the result

$$\bar{T}(s) = \frac{K}{\sqrt{(s^2 + Q)}} - \frac{P}{Qs} \qquad \text{XI}$$

This can be inverted using the tables, numbers 55 and 1.

$$\therefore \quad T = KJ_0(x\sqrt{Q}) - \frac{1}{A} \qquad \text{XII}$$

where K is the arbitrary constant of integration which can be evaluated as follows.
When $x = 0$, $J_0(x\sqrt{Q}) = 1$, and $T = 159$

$$\therefore \quad 159 = K - 1/A \qquad \text{XIII}$$

but

$$A = 0 \cdot 043$$

$$\therefore \quad 1/A = 23 \cdot 3 \quad \text{and} \quad K = 182 \cdot 3$$

Now

$$Q = \frac{290 \times 1 \cdot 07 \times 10^5 \times 0 \cdot 12 \times 0 \cdot 043}{25 \cdot 4} = 6300$$

$$\therefore \quad \sqrt{Q} = 79 \cdot 4$$

and equation XII becomes

$$T = 182 \cdot 3 J_0(79 \cdot 4x) - 23 \cdot 3 \qquad \text{XIV}$$

At the boundary, $x = R$ and $T = 149 - 93 = 56$

$$\therefore \quad 56 = 182 \cdot 3 J_0(79 \cdot 4R) - 23 \cdot 3$$

$$\therefore \quad J_0(79 \cdot 4R) = 0 \cdot 435$$

Use of a table of Bessel functions gives

$$79 \cdot 4R = 1 \cdot 64$$

$$\therefore \quad R = 2 \cdot 07 \text{ cm}$$

For safety, the diameter of the tubes that would be installed in the reactor should be less than 4·14 cm, as for this diameter the cooling rate would be just sufficient to prevent the temperature of the catalyst at the centre of the tube exceeding 525 °K. Hence tubes of diameter 4·0 cm would be recommended.

6.5. THE STEP FUNCTIONS

It is possible for chemical engineering processes and equipment to undergo sudden changes in their operating conditions; for example, the steam supply could suddenly change or a vacuum pump or compressor could fail. It is desirable to be able to predict what effects these sudden changes will have on the process and quality of the product; and this can be accomplished by means of the step functions which will be considered below.

6.5.1. *The Unit Step Function*

This is illustrated in Fig. 6.2, which shows that the value of the function is zero for values of the independent variable less than b, and unity for values of

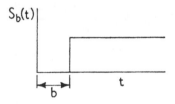

FIG. 6.2. The unit step function

the independent variable greater than b. The Laplace transform of the unit step function can be obtained in the usual way by integration. Thus

$$\mathscr{L}[S_b(t)] = \int_0^b 0 \cdot e^{-st} \, dt + \int_b^\infty 1 \cdot e^{-st} \, dt = e^{-bs}/s \qquad (6.22)$$

If the value of b is zero, the transform would reduce to $1/s$ which is the transform of unity. Thus multiplication of the transform of unity by e^{-bs} indicates that the function is zero over the interval 0 to b and unity for values of t greater than b. This concept can be extended to any other functions which are zero for specific intervals or which undergo sudden changes as illustrated by the unit impulse function below.

6.5.2. *The Unit Impulse Function*

The unit step function was characterized by having zero value for all values of the independent variable less than b, and a value of unity for values of t between b and infinity. If the value of this function is suddenly reduced to zero when the value of the independent variable is $(b+c)$ the function will appear as an impulse when plotted as shown in Fig. 6.3. Furthermore, if the

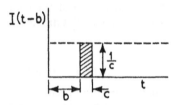

FIG. 6.3. The unit impulse function

height of the impulse between b and $(b+c)$ is $1/c$, the area beneath the curve will be unity. As $c \to 0$, the area will remain constant at unity, and the limit defines the unit impulse function as

$$I(t-b) = \lim_{c \to 0} [S_b(t) - S_{b+c}(t)]/c \qquad (6.23)$$

Taking the Laplace transform of both sides, using equation (6.22) gives

$$\mathscr{L}[I(t-b)] = \lim_{c \to 0} \left[\frac{e^{-bs} - e^{-(b+c)s}}{cs} \right] \qquad (6.24)$$

By L'Hôpital's rule, Section 3.3.7, this limit becomes

$$\mathscr{L}[I(t-b)] = e^{-bs} \qquad (6.25)$$

Frequently, the unit impulse function is called the "Dirac delta function" and is written $\delta(t-b)$. It expresses the rate of change of the unit step function.

6.5.3. *The Staircase Function*

This is shown in Fig. 6.4 where it can be seen that the function is formed by the successive addition of unit step functions at 0, b, $2b$, $3b$, ..., etc. Hence the Laplace transform of the staircase function is obtained by adding together the transforms of each step function. Thus

$$\mathcal{L}[S(b,t)] = \frac{1}{s} + \frac{e^{-bs}}{s} + \frac{e^{-2bs}}{s} + \ldots + \frac{e^{-nbs}}{s} + \ldots$$

$$= (1 + e^{-bs} + e^{-2bs} + \ldots e^{-nbs} + \ldots)/s \qquad (6.26)$$

The infinite series in equation (6.26) is a geometrical progression which can be summed to give the result

$$\mathcal{L}[S(b,t)] = \frac{1}{s(1-e^{-bs})} \qquad (6.27)$$

Use will be made of these functions later in the text but the use of the unit step functions will be illustrated by the following example.

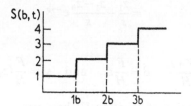

FIG. 6.4. The staircase function

Example. A feed solution containing z_0 mass fraction of a particular volatile component is fed to a single effect evaporator at a rate of F kg/h. At steady-state operating conditions V kg/h of vapour and L kg/h of liquid are produced. If the concentration of the feed solution suddenly changes to a higher constant concentration z_1 and this persists for τ h before returning to z_0, develop equations to predict the changes in concentration of the volatile component in the product streams during and after this disturbance. It may be assumed that the vapour and liquid rates remain constant and that the liquid hold-up in the evaporator is H kg. The volatility of the vaporizable component may be expressed by the relationship

$$y = mx \qquad \text{I}$$

where x is the mass fraction of the volatile component in the liquid and y is the equilibrium mass fraction in the vapour.

Solution

The steady-state mass balance is

$$Fz_0 = Vy_0 + Lx_0 \qquad \text{II}$$

The unsteady-state mass balance during and after the transient is

$$Fz - Vy - Lx = H\frac{dx}{dt} \qquad \text{III}$$

Eliminating y by substituting equation I into equation III gives

$$\frac{dx}{dt} = \frac{F}{H}z - \frac{mV+L}{H}x \qquad\qquad \text{IV}$$

The behaviour of z with time consists of a step up from z_0 to z_1 at $t = 0$ and a step down from z_1 to z_0 at $t = \tau$. This can be described by the equation

$$z = z_0 + (z_1 - z_0)\left[S_0(t) - S_\tau(t)\right] \qquad\qquad \text{V}$$

Using equation (6.22)

$$\mathcal{L}(z) = \frac{z_0}{s} + \frac{z_1 - z_0}{s}(1 - e^{-\tau s})$$

$$= \frac{z_1}{s} - \frac{z_1 - z_0}{s}e^{-\tau s} \qquad\qquad \text{VI}$$

Putting

$$\alpha = \frac{mV+L}{H}, \quad \beta = \frac{F}{H}z_1, \quad \text{and} \quad \gamma = \frac{F}{H}(z_1 - z_0)$$

the Laplace transformation of equation IV becomes

$$s\bar{x} - x_0 = \frac{\beta}{s} - \frac{\gamma}{s}e^{-\tau s} - \alpha\bar{x} \qquad\qquad \text{VII}$$

Solving algebraically gives

$$\bar{x} = \frac{x_0}{s+\alpha} + \frac{\beta}{s(s+\alpha)} - \frac{\gamma}{\alpha}\left(\frac{1}{s} - \frac{1}{s+\alpha}\right)e^{-\tau s}$$

This can be inverted using transforms number 8, 12, and 61 from the table and the shifting theorem. The result is

$$x = x_0 e^{-\alpha t} + \frac{\beta}{\alpha}(1 - e^{-\alpha t}) - \frac{\gamma}{\alpha}\left[1 - e^{-\alpha(t-\tau)}\right]S_\tau(t) \qquad\qquad \text{VIII}$$

which checks to give $x = x_0$ at $t = 0$. When t becomes very large, equation VIII becomes

$$x = 0 + \frac{\beta}{\alpha} - \frac{\gamma}{\alpha}$$

$$= \frac{F}{mV+L}z_0 \qquad\qquad \text{IX}$$

Eliminating y_0 between equations I and II gives

$$Fz_0 = (mV+L)x_0$$

Hence equation IX shows that x again assumes the value x_0 when t becomes large. The response indicated by equation VIII is illustrated in Fig. 6.5.

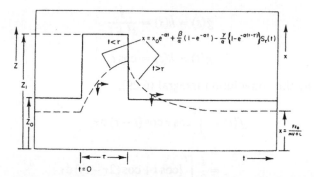

FIG. 6.5. Response to a feed change in a single effect evaporator

6.6. CONVOLUTION

It frequently happens in solving problems by the Laplace transformation that the final transform of the equation is a product of two easily identifiable transforms, but these are difficult to resolve in the form of a summation. When this arises, the inverse transformation can be accomplished by the method of "convolution".

Space does not permit the derivation of the convolution integral and therefore only the mechanics of its use will be explained. Any reader interested in the derivation is advised to consult Thomson† or Churchill.‡ Thus if the transform of the equation is $\bar{f}(s)$ and this is composed of the factors $\bar{g}(s)$ and $\bar{h}(s)$ whose inverse transforms are recognizable from Laplace transform tables, the inverse transform of $\bar{f}(s)$ is obtained as follows.

Let the inverse transform of $\bar{g}(s)$ be $g(t)$, and the inverse transform of $\bar{h}(s)$ be $h(t)$. Then since

$$\bar{f}(s) = \bar{g}(s)\,\bar{h}(s) \tag{6.28}$$

$$f(t) = \mathscr{L}^{-1}[\bar{g}(s)\,\bar{h}(s)] = \int_0^t g(\tau)\,h(t-\tau)\,\mathrm{d}\tau = \int_0^t h(\tau)\,g(t-\tau)\,\mathrm{d}\tau \tag{6.29}$$

Equation (6.29) gives the convolution integral of $f(t)$ and implies its symmetry. The convolution integral is frequently given the shorthand form

$$f(t) = g(t)\,h^*(t) \tag{6.30}$$

The use of the convolution integral to find an inverse transform will be illustrated in the next example.

† Thomson, W. T. "Laplace Transformation". Longmans Green & Co., London
‡ Churchill, R. V. "Operational Mathematics". McGraw-Hill Book Co., New York

Example. Find the inverse transform of $s^2/(s^2+1)^2$.
Transform number 16 in the tables shows that if

$$\bar{g}(s) = \bar{h}(s) = \frac{s}{s^2+1}$$

then
$$g(t) = h(t) = \cos t$$

Therefore, by the convolution integral (6.29),

$$f(t) = \int_0^t \cos \tau \cos (t-\tau)\,d\tau \qquad\qquad \text{I}$$

$$= \frac{1}{2}\int_0^t [\cos t + \cos (2\tau - t)]\,d\tau \qquad\qquad \text{II}$$

$$= \tfrac{1}{2}[\tau \cos t + \tfrac{1}{2}\sin (2\tau - t)]_0^t$$

$$= \tfrac{1}{2}(t \cos t + \tfrac{1}{2}\sin t + \tfrac{1}{2}\sin t) \qquad\qquad \text{III}$$

$$= \tfrac{1}{2}(t \cos t + \sin t) \qquad\qquad \text{IV}$$

The answer given by equation IV checks with transform number 23 in
the tables.

6.7. INVERSION BY ELEMENTARY INTEGRATION

It was shown in Section 6.4.2 that dividing the transform of a function
$f(t)$ by s was equivalent to integrating the function between the limits 0 and t.
Thus the power of s in the denominator of $\bar{f}(s)$ can be increased in an integral
manner by successive applications of equation (6.21) as illustrated in the next
example. When this process is combined with the shifting theorem (Section
6.2.2), a powerful method of inversion results.

Example. Find $f(t)$ if $\bar{f}(s) = 1/s^2(s+a)$.
Transform number 8 shows that if $\bar{f_1}(s) = 1/(s+a)$, then $f_1(t) = e^{-at}$.
Application of equation (6.21) gives

$$\bar{f_2}(s) = \frac{1}{s}\bar{f_1}(s) = \frac{1}{s(s+a)} \qquad\qquad \text{I}$$

$$\therefore \ f_2(t) = \int_0^t e^{-at}\,dt$$

$$= \frac{1-e^{-at}}{a} \qquad\qquad \text{II}$$

A further application of equation (6.21) to this result gives

$$\bar{f_3}(s) = \frac{1}{s}\bar{f_2}(s) = \frac{1}{s^2(s+a)} \qquad\qquad \text{III}$$

$$\therefore \quad f_3(t) = \int_0^t (1-\mathrm{e}^{-at})\frac{\mathrm{d}t}{a}$$

$$= \frac{1}{a}\left[t + \frac{\mathrm{e}^{-at}}{a}\right]_0^t$$

$$= \frac{1}{a^2}(at + \mathrm{e}^{-at} - 1) \qquad\qquad \text{IV}$$

Therefore, if $f(s) = 1/s^2(s+a)$, then

$$f(t) = (\mathrm{e}^{-at} + at - 1)/a^2 \qquad\qquad \text{V}$$

6.8. INVERSION OF THE LAPLACE TRANSFORM BY CONTOUR INTEGRATION

In the above theory of the Laplace transformation, the various methods given for the inversion of the transform to regenerate a function of the independent variable depended on the resolution of the transform into partial fractions, or modification of the transforms appearing in the tables by using convolution or the shifting theorem. These methods are usually adequate for solving problems involving ordinary differential equations, but often fail when applied to complex transforms resulting from the solution of partial differential equations. These complex transforms can be inverted by means of a contour integral, and the following section will explain this method in preparation for its application to the solution of partial differential equations in Chapter 8.

The basis of the inversion of a transform by contour integration is the Mellin–Fourier theorem, which may be formally stated by the following expressions without placing any restrictions on the functions. Thus if

$$\left.\begin{aligned}\bar{f}(s) &= \int_0^\infty \mathrm{e}^{-st} f(t)\,\mathrm{d}t \\[2mm] f(t) &= \frac{1}{2\pi i}\lim_{\beta\to\infty}\int_{\alpha-i\beta}^{\alpha+i\beta} \mathrm{e}^{st}\bar{f}(s)\,\mathrm{d}s\end{aligned}\right\} \qquad (6.31)$$

then

where α is a constant greater than the real part of any singularity in the transform $\bar{f}(s)$.

The Mellin–Fourier theorem is frequently referred to simply as the inversion theorem and it can be deduced directly from Fourier's integral theorem which is comprehensively dealt with in most university texts on mathematics. Thus

Fourier's integral theorem can be expressed

$$f(t) = \frac{1}{2\pi} \int_{-\infty}^{\infty} e^{i\lambda t} \int_{-\infty}^{\infty} e^{-i\lambda t} f(t) \, dt \, d\lambda \tag{6.32}$$

where $f(t)$ is any function of t, periodic or otherwise.

In applied mathematics, t cannot be less than zero and therefore let $f(t) = 0$ for $t < 0$, in which case equation (6.32) becomes

$$f(t) = \frac{1}{2\pi} \int_{-\infty}^{\infty} e^{i\lambda t} \int_{0}^{\infty} e^{-i\lambda t} f(t) \, dt \, d\lambda \tag{6.33}$$

Put $i\lambda = s - \alpha$, where α is chosen to ensure convergence of the first integral

$$\therefore \quad f(t) = \frac{1}{2\pi i} \int_{\alpha-i\infty}^{\alpha+i\infty} e^{st} e^{-\alpha t} \int_{0}^{\infty} e^{-st} e^{\alpha t} f(t) \, dt \, ds \tag{6.34}$$

$$\therefore \quad e^{\alpha t} f(t) = \frac{1}{2\pi i} \int_{\alpha-i\infty}^{\alpha+i\infty} e^{st} \int_{0}^{\infty} e^{-st} [e^{\alpha t} f(t)] \, dt \, ds \tag{6.35}$$

which is equivalent to equation (6.31) applied to $e^{\alpha t} f(t)$.

Therefore, the Mellin–Fourier integral theorem provides a method of obtaining the inverse transform directly by integration in the complex plane along a line parallel to the imaginary axis, and at distance α along the positive real axis from the origin. The value of this integral may be obtained as follows. Consider the contour shown in Fig. 6.6 where the arc of the circle of radius R encloses all the singularities. The integral around this contour is the sum of the integrals along the straight path AB from $-i\beta$ to $+i\beta$, and around the arc BCA. The integration path from A to B as $\beta \to \infty$ is known as the "first Bromwich path". Expressed symbolically,

$$\int_{ABCA} e^{st} \bar{f}(s) \, ds = \int_{\alpha-i\beta}^{\alpha+i\beta} e^{st} \bar{f}(s) \, ds + \int_{BCA} e^{st} \bar{f}(s) \, ds \tag{6.36}$$

If equation (6.36) is to be of any value, the last integral must vanish as $R \to \infty$. The condition for this will now be established.

On the arc BDCEA, $|s| = R$ and it will be assumed that $|\bar{f}(s)| < \varepsilon$ for all points on the arc.

$$\therefore \quad \left| \int_{DCE} e^{st} \bar{f}(s) \, ds \right| < \varepsilon \left| \int_{DCE} e^{st} \, ds \right| \tag{6.37}$$

Because e^{st} is analytic everywhere except at infinity, the integral on the right-hand side of equation (6.37) will have the same value for all paths between $s = -iR$ and $s = iR$.

$$\therefore \quad \left| \int_{DCE} e^{st} \bar{f}(s) \, ds \right| < \varepsilon \left| \left[\frac{e^{st}}{t} \right]_{-iR}^{iR} \right| = \frac{2\varepsilon}{t} \left| \sin Rt \right| \tag{6.38}$$

Along the arc BD, $|\bar{f}(s)| < \varepsilon$, $|e^{st}| \leqslant e^{\alpha t}$, and it is certainly true that the length of the integration path is less than 2α.

$$\therefore \quad \left| \int_{BD} e^{st} \bar{f}(s) \, ds \right| < \varepsilon \, e^{\alpha t} \, 2\alpha \tag{6.39}$$

Similarly, the integral along the arc EA can be shown to be less than $2\alpha\varepsilon \, e^{\alpha t}$. Adding these results together,

$$\left| \int_{BCA} e^{st} \bar{f}(s) \, ds \right| < \varepsilon \left(4\alpha \, e^{\alpha t} + \frac{2}{t} \left| \sin Rt \right| \right) \tag{6.40}$$

As $R \to \infty$, $\beta \to \infty$, but $\sin Rt \leqslant 1$. Therefore the term in the brackets in equation (6.40) remains finite as $R \to \infty$. If $\varepsilon \to 0$ as $R \to \infty$, then

$$\int_{BCA} e^{st} \bar{f}(s) \, ds \to 0 \quad \text{as } R \to \infty \tag{6.41}$$

Therefore, provided $|\bar{f}(s)| \to 0$ as $|s| \to \infty$ in the left-hand side half-plane, the last integral in equation (6.36) can be neglected and this gives

$$f(t) = \frac{1}{2\pi i} \int_{\alpha - i\infty}^{\alpha + i\infty} e^{st} \bar{f}(s) \, ds = \frac{1}{2\pi i} \int_{ABCA} e^{st} \bar{f}(s) \, ds \tag{6.42}$$

The evaluation of the integrals in equation (6.42) depends on the type of singularity of $\bar{f}(s)$.

6.8.1. Inversion when the Singularities are Poles

In this case, the last integral in equation (6.42) can be evaluated by the theory of residues given in Section 4.14. Thus

$$f(t) = \frac{1}{2\pi i} \int_{Br_1} e^{st} \bar{f}(s) \, ds = \frac{1}{2\pi i} \int_{ABCA} e^{st} \bar{f}(s) \, ds$$

$$= \sum [\text{residues of } e^{st} \bar{f}(s)] \tag{6.43}$$

Example 1. Find the inverse transform of $(2s+1)/[s(s^2+1)]$ by contour integration.

The given function has three simple poles, at $s = 0$, i, and $-i$. Using equation (4.87) to evaluate each residue separately gives:

Residue at $s = 0$,

$$\rho_0 = \lim_{s \to 0} \left[\frac{(2s+1)s}{s(s^2+1)} \right] e^{st} = 1 \qquad\qquad \text{I}$$

Residue at $s = i$,

$$\rho_i = \lim_{s \to i} \left[\frac{(2s+1)(s-i)}{s(s-i)(s+i)} \right] e^{st} = \frac{1+2i}{-2} e^{it} \qquad\qquad \text{II}$$

Residue at $s = -i$,

$$\rho_{-i} = \lim_{s \to -i} \left[\frac{(2s+1)(s+i)}{s(s-i)(s+i)} \right] e^{st} = \frac{1-2i}{-2} e^{-it} \qquad \text{III}$$

$$\therefore \quad f(t) = \rho_0 + \rho_i + \rho_{-i} = 1 - \frac{e^{it} + e^{-it}}{2} - i(e^{it} - e^{-it}) \qquad \text{IV}$$

$$\therefore \quad f(t) = 1 - \cos t + 2 \sin t \qquad \text{V}$$

6.8.2. *Inversion when the Singularity is a Branch Point*

In Section 4.12.3 it was shown that the path of a contour integral cannot cross a "barrier" erected from a branch point, and the integration path must consequently be modified to exclude this point. If the Laplace transform $f(s)$ contains such a branch point at the origin and the branch cut follows the negative real axis, then the integration path of Fig. 6.6 must be modified as shown in Fig. 6.7. On the assumption that this transform does not contain

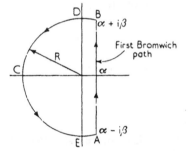

FIG. 6.6. Contour integration of equation (6.31)

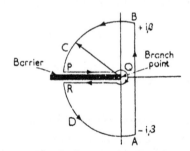

FIG. 6.7. Contour integration along second Bromwich path

other singularities, the value of the integral around the closed contour ABCPQRDA is zero by Cauchy's theorem. Furthermore, as $\beta \to \infty$, the value of the integral around the arcs BCP and RDA is zero. The value of the integral along the first Bromwich path is thus equivalent to the value along the path RQP which is called the "second Bromwich path", or

$$f(t) = \frac{1}{2\pi i} \int_{RQP} e^{st} f(s) \, ds = \frac{1}{2\pi i} \int_{Br_2} e^{st} f(s) \, ds \qquad (6.44)$$

Example 2. Evaluate the inverse transform of $1/s\sqrt{(s+1)}$.

In Section 6.4.2 it was shown that division of a transform by s corresponded to integrating the function between the limits 0 and t. Therefore let

$$\frac{df(t)}{dt} = \mathscr{L}^{-1} \left[\frac{1}{\sqrt{(s+1)}} \right] \qquad \text{I}$$

The inversion of equation I can be done directly from the tables, but here it will be inverted by contour integration. Since the function has no poles but a branch point at $s = -1$, the origin will be moved to the branch point by putting $\lambda = s+1$, and equation (6.44) will be applied. Thus

$$\frac{df(t)}{dt} = \frac{1}{2\pi i} \int_{Br_2} \frac{e^{(\lambda-1)t} d\lambda}{\sqrt{\lambda}} \qquad \text{II}$$

Br_2 for this function is illustrated in Fig. 6.7 and the value of the integral along the arcs is zero. The integral along RQP will now be evaluated in three parts as follows.

(i) Integration along RQ. The line RQ is defined by $\lambda = x e^{-i\pi}$ with x varying from ∞ to 0.

$$\therefore \quad \lambda = x e^{-i\pi} = x(\cos \pi - i \sin \pi) = -x \qquad \text{III}$$

$$\therefore \quad d\lambda = -dx$$

$$\therefore \quad \sqrt{\lambda} = x^{\frac{1}{2}} e^{-\frac{1}{2}\pi i} = -i\sqrt{x} \qquad \text{IV}$$

$$\therefore \quad \int_{RQ} \frac{e^{(\lambda-1)t}}{\sqrt{\lambda}} d\lambda = -i e^{-t} \int_{\infty}^{0} \frac{e^{-xt}}{\sqrt{x}} dx = i e^{-t} \int_{0}^{\infty} x^{-\frac{1}{2}} e^{-xt} dx \qquad \text{V}$$

(ii) Integration around the circle Q of radius δ

Let $\lambda = \delta e^{i\theta}$ where θ varies from $-\pi$ to π.

$$\therefore \quad d\lambda = i\delta e^{i\theta} d\theta \qquad \text{VI}$$

$$\therefore \quad \sqrt{\lambda} = \delta^{\frac{1}{2}} e^{\frac{1}{2}i\theta} \qquad \text{VII}$$

$$\therefore \quad \int_{Q} \frac{e^{(\lambda-1)t}}{\sqrt{\lambda}} d\lambda = \lim_{\delta \to 0} i e^{-t} \delta^{\frac{1}{2}} \int_{-\pi}^{\pi} e^{\delta(\cos\theta + i\sin\theta)t + \frac{1}{2}i\theta} d\theta = 0 \qquad \text{VIII}$$

(iii) Integration along QP. By the same procedure as in part (i), the value of this integral is

$$\int_{QP} \frac{e^{(\lambda-1)t}}{\sqrt{\lambda}} d\lambda = i e^{-t} \int_{0}^{\infty} x^{-\frac{1}{2}} e^{-xt} dx \qquad \text{IX}$$

Adding together equations V, VIII and IX and substituting into equation II gives

$$\frac{df(t)}{dt} = \frac{2i e^{-t}}{2\pi i} \int_{0}^{\infty} x^{-\frac{1}{2}} e^{-xt} dx \qquad \text{X}$$

The integral in equation X can be evaluated by putting $xt = \alpha^2$.

$$\therefore \quad \frac{df(t)}{dt} = \frac{2 e^{-t}}{\pi \sqrt{t}} \int_{0}^{\infty} e^{-\alpha^2} d\alpha \qquad \text{XI}$$

$$= e^{-t}/\sqrt{(\pi t)}$$

by equation (5.2). Integrating gives

$$f(t) = \frac{1}{\sqrt{\pi}} \int_0^t t^{-\frac{1}{2}} e^{-t} \, dt \qquad\qquad \text{XII}$$

$$\therefore \quad f(t) = \mathscr{L}^{-1}\left[\frac{1}{s\sqrt{(s+1)}}\right] = \text{erf}\sqrt{t} \qquad\qquad \text{XIII}$$

The last few stages of this calculation involve the properties of the error function from Chapter 5.

6.8.3. *Inversion when there are Both Poles and a Branch Point*

When there are pole singularities in addition to the branch point, equation (6.44) is still applicable but the contour Br_2 must be modified to exclude all of the singularities. This is done by drawing a circle around each singularity and connecting it to the main contour with a pair of parallel coincident straight lines. Figure 6.8 shows the case of a branch point at the origin and two poles which are self-conjugate. Br_2 is the curve GFEDC and the Mellin–Fourier integral is evaluated as follows.

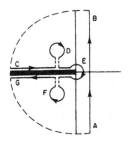

FIG. 6.8. Modification of Br_2 for poles and a branch point

By Cauchy's theorem, the integral of $e^{st} f(s)$ round the closed curve ABCDEFGA is zero. The parts of this contour which have been drawn with dotted lines make no contribution to the integral; the sections BC and GA because of equation (6.41), and the pairs of parallel lines because they are coincident and are traversed in opposite directions. Thus,

$$\int_{Br_1} e^{st} f(s) \, ds = \int_{GFEDC} e^{st} f(s) \, ds$$

The circles enclosing the poles at D and F are both described in the positive (anticlockwise) direction and make a contribution $2\pi i$ (residue at the pole) towards the integral along Br_2.

Hence, the transform can be inverted by adding $2\pi i$ times the sum of the residues at all of the poles of $e^{st} f(s)$, to the piecewise integration along GEC as for an ordinary branch cut without poles. That is,

$$f(t) = \frac{1}{2\pi i} \int_{Br_2} e^{st} \bar{f}(s) \, ds = \frac{1}{2\pi i} \int_{GEC} e^{st} \bar{f}(s) \, ds$$

$$+ \sum [\text{residues of } e^{st} \bar{f}(s)] \qquad (6.45)$$

Equation (6.45) summarizes the method of inversion of the Laplace transform by contour integration. To use it, if $\bar{f}(s)$ does not contain a branch point, the integral along GEC will be zero and equation (6.45) simplifies to equation (6.43), but in general the full equation (6.45) is needed. The integral along GEC can always be performed in a piecewise manner as in Example 2 above, and the residues of $\bar{f}(s) e^{st}$ can always be found by the methods given in Section 4.14. Thus any transform can be inverted provided the conditions for equation (6.41) are satisfied. That is, $|\bar{f}(s)| \to 0$ as $|s| \to \infty$. Any transform arising in connection with a physically reasonable problem will satisfy this condition so that the solution can be found in principle by the method of residues.

6.9. APPLICATION OF THE LAPLACE TRANSFORM TO AUTOMATIC CONTROL THEORY

The study or design of automatic control systems for chemical processes requires setting up differential equations that will predict the dynamic behaviour of a piece of equipment, and in particular the response of the equipment to sudden changes. The example in Section 6.5 illustrates how these problems can be solved by means of the Laplace transformation. However, let it be assumed that a simple automatic control system is described by the linear differential equation

$$A \frac{d^2 y}{dt^2} + B \frac{dy}{dt} + Cy = f(t) \qquad (6.46)$$

The Laplace transform of this equation is

$$\bar{y}(s) = \frac{\bar{f}(s)}{As^2 + Bs + C} + \frac{(As + B) y(0) + Ay'(0)}{As^2 + Bs + C} \qquad (6.47)$$

where $y(0)$ and $y'(0)$ are the values of the measured variable, i.e. temperature, pressure, at a time $t = 0$, $\bar{y}(s)$ is called the "response transform". The first transform on the right of equation (6.47) is the transform of the steady-state solution after the change, and the last term on the right is the transform of the transient solution. That is, the first term on the right is characteristic of the control system and the last term is characteristic of the initial disturbance. Furthermore, the denominator of the terms on the right-hand side of equation (6.47) determines the way in which the system reaches its new equilibrium position and therefore $(As^2 + Bs + C)^{-1}$ is the important group in automatic control theory and is called the "transfer function".

The transfer function is defined as the response of a control system to a disturbance of unit impulse from the steady-state conditions. Therefore referring to equation (6.47) it will be appreciated that the origin can be chosen so that $y(0)$ and $y'(0)$ are both zero when $t = 0$. The transfer function is now characterized by

$$\frac{\bar{y}(s)}{\bar{f}(s)} = \frac{1}{As^2 + Bs + C} \tag{6.48}$$

which defines it as the ratio of the Laplace transform of the response variable to the Laplace transform of the disturbing variable. In addition, since this ratio is equal to the reciprocal of the transform of the differential equation describing the system, it can be evaluated only for equations that can be transformed. However, this need not be a serious restriction because in many cases an approximate subsidiary equation can be found without undue simplification; on the other occasions it may be necessary to obtain the transfer function experimentally because of the complexity of the control system.

The transfer function describing a part of or the complete control system is usually inserted into a "block diagram". Such diagrams are used extensively in automatic control theory to depict the action of the variables involved. They are similar to energy flow diagrams but the arrowed lines to and from the different blocks indicate input and output signals. A typical block diagram of an automatic control system is illustrated in Fig. 6.9(b). The corresponding process diagram, depicting an electrical heating coil installed in a stirred tank through which a liquid to be heated is passed, is shown in Fig. 6.9(a).

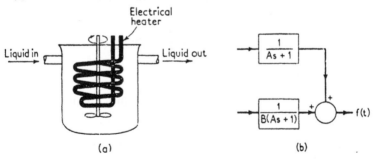

FIG. 6.9. Block diagram of electrical heating process

In Fig. 6.9(b) the circle represents an algebraic operation and the two plus signs imply that the output signal $f(t)$ is obtained by summing the outputs from the two transfer functions. Each transfer function is enclosed in a rectangle, and the transfer function together with its block is called a dynamic function because it indicates a rate of change. In the particular example it indicates a rate of heat input.

The only rules to be remembered in the construction of block diagrams are:

(a) Only one line may enter, and only one line may leave a block representing a dynamic function.

(b) Only two lines may enter and one line leave a circle representing an algebraic operation.

(c) If a control system is described by one or more linear differential equations, the dynamic functions and algebraic operations can be rearranged without altering the controllability of the process.

For further details of these points the reader is advised to consult "Automatic Process Control" by Eckman,† as the present text is only concerned with the mathematical background. However, it is possible to rearrange or combine the transfer functions of a series of controlled elements in such a way that the whole process may be represented by blocks whose transfer functions have poles or zeros that are easily identified. This is very important in the analysis of systems containing a large number of variables because the roots of the transformed equation in the denominator of the transfer function may not be easily obtained if its degree is high.

6.9.1. *Control Systems*

Chemical processes are controlled by either the "open-loop" or "closed-loop" control system, and in all but the most simple cases, closed-loop control is employed. Thus, in the open-loop system the composition of a stream may be controlled as shown in Fig. 6.10(a) by proportioning the flow of A by

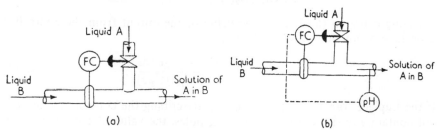

Fig. 6.10. Open and closed loops

means of the flow meter in the line through which the liquid B passes, and the flow controller is only able to take a preset action according to the signal it receives from this meter. Any other changes in the process conditions, such as the concentration of A or B prior to mixing, will have no effect on the controller, and will lead to deviations from the chosen position. If, however, a detecting device such as a pH meter is placed in the product line as shown in Fig. 6.10(b) so that the concentration of A in B can be measured, and a signal representing the measured concentration changes fed back to the controller, the possible deviation can be corrected. That is "feedback" is an essential feature of a closed-loop system.

† Eckman, D. P. "Automatic Process Control." Wiley & Sons, New York (1958.)

A block diagram of a closed-loop system is illustrated in Fig. 6.11 in which the transfer functions of the different elements depicting the controlling

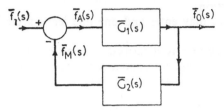

FIG. 6.11. Typical closed loop block diagram

means have been grouped together in the combined transfer function $\bar{G}_1(s)$, whilst the transfer function of the feedback element (i.e. the measuring element) is shown in the block $\bar{G}_2(s)$. Thus, making use of the rules for dynamic and algebraic functions, the subsidiary equations for different parts of this control process can be written as

$$\bar{f}_M(s) = \bar{G}_2(s)\bar{f}_0(s) \tag{6.49}$$

$$\bar{f}_A(s) = \bar{f}_1(s) - \bar{f}_M(s) = \bar{f}_1(s) - \bar{G}_2(s)\bar{f}_0(s) \tag{6.50}$$

$$\bar{f}_0(s) = \bar{G}_1(s)\bar{f}_A(s) = \bar{G}_1(s)[\bar{f}_1(s) - \bar{G}_2(s)\bar{f}_0(s)] \tag{6.51}$$

$$\therefore \ \bar{f}_0(s) = \frac{\bar{G}_1(s)\bar{f}_1(s)}{1 + \bar{G}_1(s)\bar{G}_2(s)} \tag{6.52}$$

Inverting equation (6.52) gives the value of the output from the controlled process as a function of time. That is,

$$\bar{f}_0(t) = \frac{1}{2\pi i} \int_{\mathrm{Br}_1} \frac{\bar{G}_1(s)\bar{f}_1(s)e^{st}\,ds}{1 + \bar{G}_1(s)\bar{G}_2(s)} \tag{6.53}$$

If the Laplace transform of the function describing the control system does not contain any singularities other than poles, the value of the integral can be obtained from equation (6.43) as

$$\bar{f}_0(t) = \sum \left[\text{residues of } \frac{\bar{G}_1(s)\bar{f}_1(s)e^{st}}{1 + \bar{G}_1(s)\bar{G}_2(s)} \right] \tag{6.54}$$

and the behaviour of the control system, after being subjected to a disturbance, can be predicted.

6.9.2. Control System Stability

To be useful, a control system must be stable and some indication of this may be obtained by inspection of the residues in equation (6.54). Thus if any of the poles of the numerator or zeros of the denominator have positive real parts, then $f_0(t)$ will contain a term $e^{\alpha t}$ which will increase to infinity as the time t increases and hence lead to instability. Consequently, all of the relevant poles and zeros must have negative real parts so that any disturbance will decay exponentially with time and the control system will be stable. This implies that all of the poles of the integrand of equation (6.53) must be in the left half of the s plane for the system to be stable.

If, in addition, there are poles on the imaginary axis, the control system will fluctuate indefinitely, whilst a pole at the origin, associated with other poles in the negative half plane, indicates that the system will fluctuate about a new position after a disturbance. This means that the control system will experience an "offset".

6.9.3. Stability Criteria

The transfer functions describing a control system may be very complicated, containing numerous poles and zeros, with the result that the procedure of evaluating the residues is very lengthy and time-consuming. Consequently some "short-cut methods, indicating whether or not a control system will be stable, have been devised. These will not reveal the extent of the instability but for obvious reasons this is not necessary. Neither will these methods indicate the sensitivity of a control system to a disturbance; for a control system could be mathematically stable but functionally useless because the time required to attain steady state could be very large.

Two methods extensively used to test the stability of a control system are those of Routh and Hurwitz, and Nyquist. Essentially the Routh–Hurwitz criteria of stability have their basis in matrix theory which will be treated in Chapter 12; here it will suffice to explain the mechanics of the method so that the section will be complete in itself. Thus let the denominator be expressed in the form of a polynomial, viz.

$$1 + \bar{G}_1(s)\,\bar{G}_2(s) = p_0 s^n + p_1 s^{n-1} + \ldots + p_n = 0 \qquad (6.55)$$

The stability of the control system will depend upon the roots of the subsidiary equation (6.55) and it can be assessed as follows. The system will be unstable if

(a) the coefficients in equation (6.55) vary in sign,

(b) some coefficients are zero.

Both of these conditions can be assessed by inspection. However they must be further tested by "Routh's Rules" for the assessment of the number of roots of an equation that have positive real parts. Routh proposed that the coefficients of terms in an equation be arranged into test functions R_i as follows. Thus in relation to equation (6.55)

$$R_0 = p_0$$

$$R_1 = p_1$$

$$R_2 = p_2 - \frac{p_0 p_3}{p_1}$$

$$R_3 = \frac{p_1 p_2 p_3 - p_0 p_3{}^2 - p_1{}^2 p_4 + p_0 p_1 p_5}{p_1 p_2 - p_0 p_3} \quad \text{etc.}$$

Routh's test functions can be conveniently expressed in terms of "test determinants" which are more rational expressions for the test functions. They are

$$\Delta_0 = p_0, \quad \Delta_1 = p_1, \quad \Delta_2 = \begin{vmatrix} p_1 & p_0 \\ p_3 & p_2 \end{vmatrix}$$

$$\Delta_3 = \begin{vmatrix} p_1 & p_0 & 0 \\ p_3 & p_2 & p_1 \\ p_5 & p_4 & p_3 \end{vmatrix}, \quad \Delta_4 = \begin{vmatrix} p_1 & p_0 & 0 & 0 \\ p_3 & p_2 & p_1 & p_0 \\ p_5 & p_4 & p_3 & p_2 \\ p_7 & p_6 & p_5 & p_4 \end{vmatrix}$$

In these determinants it will be noticed that the order of the determinant is the same as the suffix of Δ with the exception that Δ_0 is of first order. In addition, the elements of each determinant are arranged in ascending order down the diagonals, and in descending order horizontally from left to right. The determinants are related to the test functions as follows.

$$R_0 = \Delta_0$$
$$R_1 = \Delta_1$$
$$R_2 = \Delta_2/\Delta_1$$
$$R_3 = \Delta_3/\Delta_2$$
$$\vdots$$
$$R_n = \Delta_n/\Delta_{n-1}$$

In fact,
$$R_n = \Delta_n/\Delta_{n-1} = p_n$$

The necessary and sufficient condition for the control system to be stable is that all of the test functions, or all of the test determinants, shall be positive.

Example 1. The denominator of the complete transfer function of a control system can be represented by

$$5s^4 + 3(\alpha - 1)s^3 + 2\alpha s^2 + (\alpha - 2)s - (\alpha - 3) \qquad \text{I}$$

Discuss the stability of the system.

Inspection of equation I shows that if α is less than 2, then p_3 is negative; and if α is greater than 3, then p_4 is negative. Hence the system may be stable if α has values between 2 and 3. This will be confirmed using Routh's test functions which are:

$$R_0 = 5$$
$$R_1 = 3(\alpha - 1)$$
$$R_2 = 2\alpha - \frac{5(\alpha - 2)}{3(\alpha - 1)}$$
$$R_3 = \frac{3(\alpha - 1)(2\alpha)(\alpha - 2) - 5(\alpha - 2)^2 + 9(\alpha - 1)^2(\alpha - 3)}{6\alpha(\alpha - 1) - 5(\alpha - 2)}$$

and
$$R_4 = p_4 = 3 - \alpha$$

These can be simplified to give

$$R_0 = 5 \qquad\qquad\qquad\qquad \text{II}$$

$$R_1 = 3(\alpha - 1) \qquad\qquad\qquad \text{III}$$

$$R_2 = \frac{6\alpha^2 - 11\alpha + 10}{3(\alpha - 1)} \qquad\qquad \text{IV}$$

$$R_3 = \frac{15\alpha^3 - 68\alpha^2 + 95\alpha - 47}{6\alpha^2 - 11\alpha + 10} \qquad \text{V}$$

$$R_4 = 3 - \alpha \qquad\qquad\qquad\qquad \text{VI}$$

The quadratic in the numerator of R_2 and the denominator of R_3 has complex roots and is therefore positive for all real values of α. Hence, provided $1 < \alpha < 3$, R_0, R_1, R_2, and R_4 must be positive. It is now necessary to determine if there are any values of α within this range that make the numerator of R_3 positive. Denote this numerator by $f(\alpha)$.

$$\therefore \quad f(\alpha) = 15\alpha^3 - 68\alpha^2 + 95\alpha - 47 \qquad \text{VII}$$

To form an idea of the shape of this function it can be differentiated, viz.

$$\frac{df}{d\alpha} = 45\alpha^2 - 136\alpha + 95 \qquad\qquad \text{VIII}$$

Equation VIII has zeros at $\alpha = 1\cdot10$ and $\alpha = 1\cdot93$. The first corresponds to a maximum and the second to a minimum. Substitution into equation VII shows that
if
$$\alpha = 1\cdot10, \quad f(\alpha) = -4\cdot8 \qquad\qquad \text{IX}$$

Since this is a maximum of $f(\alpha)$, $f(\alpha)$ must be negative for all $\alpha < 1\cdot93$. The available range of values for α is now restricted to the original range $2 < \alpha < 3$ approximately and choosing a few values in this range gives

$$\alpha = 2, \quad f(\alpha) = -9\cdot0$$

$$\alpha = 2\cdot5, \quad f(\alpha) = -0\cdot125$$

$$\alpha = 3, \quad f(\alpha) = 31\cdot0$$

Therefore $\alpha = 2\cdot5$ is an approximate root of equation VII with $f(\alpha) = 0$. This value can be improved by the method to be given in Chapter 11, thus proving that the system described by equation I is stable for

$$2\cdot503 < \alpha < 3\cdot0 \qquad\qquad\qquad \text{X}$$

6.9.4. *The Nyquist Diagram*

The Nyquist criteria of stability makes use of the complex character of the Laplace parameter s, and the fact that the denominator of an unstable control loop transfer function will have roots with positive real parts situated in the right half of the s plane. Hence if the contour of integration of equation (6.54) is drawn in this part of the s plane as shown in Fig. 6.12, it will enclose all of the zeros and poles of $1 + \bar{G}_1(s)\,\bar{G}_2(s)$ with positive real parts. Furthermore, it can be shown that within any closed contour

$$\begin{bmatrix} \text{Number} \\ \text{of zeros} \end{bmatrix} - \begin{bmatrix} \text{Number} \\ \text{of poles} \end{bmatrix} = \frac{1}{2\pi} \begin{bmatrix} \text{Change in arg of} \\ \text{function around contour} \end{bmatrix} \quad (6.56)$$

Nyquist stated that $[\arg \bar{f}(s)/2\pi]_c$ was equal to the number of times the contour encircled the origin of $\bar{f}(s)$. If this is N, the number of zeros is Z, and the number of poles is P, then equation (6.56) becomes

$$Z - P = N \quad (6.57)$$

In these equations, an nth order zero or pole is considered to be n zeros or poles of first order.

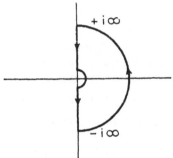

FIG. 6.12. Nyquist contour

Nyquist used equation (6.57) to restate the stability requirement of equation (6.54). A control system is stable if there are no zeros of the function $[1 + \bar{G}_1(s)\,\bar{G}_2(s)]$ in the positive half plane, and this means that if the sum of the number of encirclements of the origin of this function and the number of poles enclosed by the contour is zero then the control system is stable. The Nyquist diagram consists of plotting the "transfer locus" of $[\bar{G}_1(s)\,\bar{G}_2(s)]$ as s traverses the contour of Fig. 6.12. The origin of the function $[1 + \bar{G}_1(s)\,\bar{G}_2(s)]$ is $(-1, 0)$ and provided $|\bar{G}_1(s)\bar{G}_2(s)| \to 0$ as $|s| \to \infty$ the infinite semicircle can be ignored. The poles of $\bar{G}_1(s)$ and $\bar{G}_2(s)$ have to be located in any case thus determining P, and the number of times the point $(-1, 0)$ is enclosed by the transfer locus determines N and the Nyquist criteria can be applied. Usually it is only necessary to plot the transfer locus of $s = i\omega$ for $0 < \omega < \infty$ because the locus between 0 and $-\infty$ is just its reflection in the real axis.

The Nyquist criteria can be applied by plotting the function $[\bar{G}_1(s)\,\bar{G}_2(s)]$ from arbitrarily chosen values of ω between 0 and ∞, or by plotting directly experimental phase and magnitude data. Since it is not essential to factorize the open-loop transfer function in order to obtain the number of poles, the stability of the system can be assessed by using the value of N obtained from the transfer locus and the value of P obtained by inspection. Unless $Z = P + N = 0$, then the control system is unstable. These principles will now be illustrated by an example.

Example 2. Estimate the stability of an automatic control system whose combined transfer function is

$$\bar{G}_1(s)\,\bar{G}_2(s) = 9/(10s+1)^3 \qquad \text{I}$$

Put

$$s = i\omega$$

then

$$\frac{9}{(10i\omega+1)^3} = \frac{9}{(10i\omega+1)^3}\frac{(1-10i\omega)^3}{(1-10i\omega)^3} \qquad \text{II}$$

$$= \frac{9(1-300\omega^2)}{(1+100\omega^2)^3} - i\left[\frac{9(30\omega-1000\omega^3)}{(1+100\omega^2)^3}\right] \qquad \text{III}$$

In equation III the function has been separated into its real and imaginary parts. These are the coordinates of the transfer function in the Nyquist diagram and Table 6.1 shows the evaluation of them.

The final two columns of Table 6.1 are plotted in Fig. 6.13 to form the Nyquist diagram for the transfer function $9/(10s+1)^3$. This function does

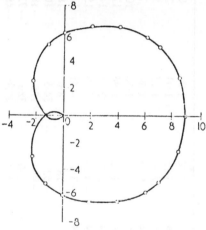

FIG. 6.13. Stability diagram for the transfer function $\dfrac{9}{(10s+1)^3}$

not have any poles in the right half plane as can be seen by inspection, however the transfer locus encircles the point $(-1,0)$ twice.

$$\therefore \quad N = 2 \qquad \text{IV}$$

Therefore, by equation (6.57)

$$Z = N+P = 2$$

and the control system is unstable.

In the example given, the Laplace transform could have been inverted and the stability checked as a function of time just as rapidly as Table 6.1 and Fig. 6.13 were prepared. However, it must be remembered that with an actual control system, the transfer function would be very complicated, and it would then be advantageous to prepare the transfer locus and use the Nyquist criteria to test stability.

Table 6.1. *Transfer Locus of* $9/(10s+1)^3$

ω	ω^2	ω^3	$9(1-300\omega^2)$	$(1+100\omega^2)^3$	$9(30\omega-1000\omega^3)$	$\dfrac{9(1-300\omega^2)}{(1+100\omega^2)^3}$	$\dfrac{9(30\omega-1000\omega^3)}{(1+100\omega^2)^3}$
0	0	0	9·00	1·00	0·00	9·00	0·00
0·01	1×10^{-4}	1×10^{-6}	8·73	1·03	2·69	8·48	2·61
0·02	4×10^{-4}	8×10^{-6}	7·92	1·12	5·33	7·08	4·76
0·03	9×10^{-4}	27×10^{-6}	6·57	1·30	7·85	5·05	6·03
0·04	16×10^{-4}	64×10^{-6}	4·68	1·56	10·23	3·00	6·56
0·05	25×10^{-4}	$1·25\times10^{-4}$	2·25	1·96	12·37	1·15	6·30
0·06	36×10^{-4}	$2·16\times10^{-4}$	-0·72	2·51	14·25	-0·28	5·68
0·07	49×10^{-4}	$3·43\times10^{-4}$	-4·23	3·30	15·9	-1·28	4·81
0·08	64×10^{-4}	$5·12\times10^{-4}$	-8·3	4·41	17·0	-1·88	3·85
0·10	0·0100	0·0010	-18	8·0	18·0	-2·25	2·25
0·15	0·0225	0·0034	-52	34·5	10·1	-1·50	0·29
0·20	0·0400	0·0080	-99	125	-18·0	-0·79	-0·15
0·25	0·0625	0·0156	-160	380	-72·9	-0·42	-0·19
0·4	0·16	0·064	-423	4920	-468	-0·09	-0·09
0·8	0·64	0·512	-1719	275000	-4392	-0·01	-0·01

Chapter 7

VECTOR ANALYSIS

7.1. Introduction

In Chapter 4 it was shown that a complex number consisted of two independent parts, a real part and an imaginary part. The one symbol z was used to represent a combination of two symbols x and y so that manipulations with a single symbol were much quicker than the corresponding elementary operations on the separate variables. It is natural to enquire whether this principle can be extended so that more than two variables can be combined into a single symbol thus shortening the work in multi-variable systems. The Argand diagram is a plane two-dimensional representation of the two parts of a complex number which lie along two orthogonal axes, and a natural extension is to represent three parts of a "number" along three orthogonal axes in three-dimensional space. This new type of "number" is a vector.

Further extension of this idea cannot be visualized, but any number of variables can be grouped into a single symbol in two ways; the first, by means of matrices, will be discussed in Chapter 12, and the second using tensors will be described briefly here. Tensors are introduced to indicate a general type of number of which vectors are a special case. The two variables which are combined in a complex number are distinguished as a real part and an imaginary part, and the principal difference between tensors and matrices is the labelling and ordering of the many distinct parts. The ordinary operations of arithmetic have to be restricted in the study of complex numbers by defining principal values, giving a special rule for division, etc. With vectors, the restrictions are more severe as will be shown in the following sections of this chapter.

The symbolism and operations of vector analysis constitute a powerful shorthand method for describing complex problems and expressing equations in general terms. For example, in Section 1.8, the steady-state temperature distribution in a cylindrical conductor was shown to satisfy the equation

$$\frac{d^2 T}{dr^2} + \frac{1}{r}\frac{dT}{dr} = 0 \tag{7.1}$$

It is a simple matter to show that the corresponding temperature distribution in a flat conductor with insulated edges satisfies the equation

$$\frac{d^2 T}{dx^2} = 0 \tag{7.2}$$

187

Both of the above equations relate to the one process, heat transfer, but to systems having different geometrical shapes. The utility of vector analysis is that the following equation in vector form will represent all steady-state heat conduction processes with constant thermal conductivity in any number of dimensions and for any shape of solid. Thus

$$\nabla^2 T = 0 \qquad (7.3)$$

The rules of vector analysis are explained in the earlier sections of this chapter, so that the second half of the chapter can be devoted to the use and interpretation of Hamilton's operator (∇) which is the fundamental symbol of this compact notation.

Vector equations are usually used to describe general processes such as the above, but when a particular example is considered, the vector form (7.3) is expanded into the appropriate particular form (7.1) or (7.2) and the completion of the problem is outside the domain of vector analysis. The procedure is illustrated in the last part of this chapter with reference to the Navier–Stokes equations which describe fluid flow, since it is in this field of chemical engineering that vector analysis is usually used.

7.2. Tensors

The notation of tensors can be introduced by referring to complex numbers. The complex number z can be written in the alternative ways

$$z = x + iy = z_1 + iz_2 \qquad (7.4)$$

where z_m is referred to as a tensor of first rank because one suffix m is needed to specify it. It is understood that the subscript m takes either of two values, 1 for the real part and 2 for the imaginary part. A more general tensor z_m can have more than two parts if m is allowed to have more than two values. The resulting parts z_1, z_2, z_3, ... etc are all independent of one another just as z_1 and z_2 are independent in equation (7.4). By reference to the Argand diagram this can be interpreted as extending the number of dimensions of the system. If m is restricted to three values, the system is three-dimensional and can be illustrated diagrammatically, whereas if m takes values greater than three, systems of higher dimensions can be described but not visualized pictorially.

The notion of a tensor can be further generalized by using more than one subscript, thus z_{mn} is a tensor of second rank in which both m and n can take integer values within the dimensions of the system. For instance, in three dimensions the tensor z_{mn} represents the nine quantities $z_{11}, z_{12}, z_{13}, z_{21}, z_{22}, z_{23}, z_{31}, z_{32}, z_{33}$, whilst in four dimensions, z_{mn} represents sixteen quantities. For a clearer distinction between the first and second subscript, the one index is sometimes raised to the position of a superscript, thus z_m^n. The symbolism for the general tensor is now obvious, it consists of a main symbol such as z with any number of associated indices which may be subscripts or superscripts. Each index is allowed to take any integer value up to the chosen dimensions of the system. The number of indices associated with the tensor is called the "rank" of the tensor.

7.2.1. *Tensors of Zero Rank, Scalars*

If the tensor has no index, then it only consists of one quantity independent of the number of dimensions of the system. The value of this quantity is independent of the complexity of the system, possesses magnitude only and is called a "scalar". No direction can be associated with a scalar since direction can only be specified in a system of known dimensionality. All physical properties are examples of scalars. For example, density, mass, specific heat, coefficient of viscosity, diffusion coefficient, thermal conductivity, etc. Other examples are the scalar point functions such as temperature, concentration, and pressure which are all signified by a number which may vary with position but not depend upon direction. Energy and time are also scalars.

7.2.2. *Tensors of First Rank, Vectors*

When the tensor has a single subscript it consists of as many elements as the number of dimensions of the system. For practical purposes this number is three and the tensor has three elements which are normally called components. The tensor of first rank is alternatively named a "vector". Vectors have both magnitude and direction and can be represented in three dimensions by a straight line in space. The most common examples of vectors are force, velocity, momentum, angular velocity, weight and area.

The above classification of scalars and vectors should help to clear some common misconceptions about the nature of the properties listed. For instance, mass is a scalar and depends only upon the substance involved and not upon its environment, whereas weight is a force resulting from the action of gravity upon the mass. Weight is thus a vector since it acts in a direction chosen by the gravitational field in which the scalar mass is situated. Newton's second law of motion is a vector equation relating the two vectors force and acceleration by means of a proportionality constant, mass.

Pressure is a scalar quantity because it acts equally in all directions and thus has no special direction. Scalar pressure becomes vector force when a surface is defined to sustain the pressure; the direction of the force vector is associated with the orientation of the surface and thus surface area is a vector quantity.

The results of various products between vectors and scalars will be better understood later in this chapter when multiplication of vectors has been defined especially the reasons why momentum is a vector, whereas energy is a scalar.

7.2.3. *Tensors of Second Rank*

The one tensor of second rank which occurs frequently in engineering is the stress tensor. This of course has a double subscript and has both a magnitude and two directions associated with it. In three dimensions the stress tensor consists of nine quantities which can be arranged in a matrix form thus,

$$T_{mn} = \begin{bmatrix} T_{11} & T_{12} & T_{13} \\ T_{21} & T_{22} & T_{23} \\ T_{31} & T_{32} & T_{33} \end{bmatrix} \tag{7.5}$$

The physical interpretation is as follows. Take three cartesian coordinate axes, x, y, z, and define an elementary rectangular box of dimensions δx, δy, δz. There are three typical faces to the box each having a normal parallel to one of the coordinate axes. Each face of this box has a force acting upon it, and this force can be resolved into three components parallel to the three axes. This is illustrated in Fig. 7.1 for two of the faces where the normal

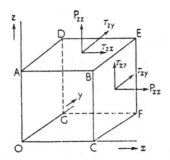

FIG. 7.1. The stress tensor

components have been labelled with p and the shear components with τ. To distinguish between these various forces it is necessary to specify both the direction of the plane and the line of action. The nine elements of the stress tensor (7.5) can be associated with these individual forces by using the first subscript to denote the plane and the second subscript to denote the direction of the force. Thus

$$T_{mn} = \begin{bmatrix} p_{xx} & \tau_{xy} & \tau_{xz} \\ \tau_{yx} & p_{yy} & \tau_{yz} \\ \tau_{zx} & \tau_{zy} & p_{zz} \end{bmatrix} \tag{7.6}$$

where the subscripts 1, 2, 3 have been replaced by x, y, z to restrict the tensor to three dimensions. τ_{xy} is read as "the shear force on the x facing plane acting in the y direction". It can be seen that one component of the force on a plane acts in the normal direction whilst the other two are shear forces along the surface. The forces illustrated in the diagram are the forces acting on the element, and the reaction forces exerted by the element on neighbouring elements are obviously all reversed in direction. For this reason, the components of stress acting on a plane facing in a positive axis direction (e.g. ABED) all act in the positive direction of the corresponding axis; whereas the components acting on a plane facing in a negative direction (e.g. OABC) all act in the negative direction of the corresponding axis. It is to be noted that the repeated suffix gives the normal forces, and the unlike suffices denote the shear forces. Also, the normal forces include the pressure, but with reversed sign because the force due to pressure acts inwards towards the element.

Tensor calculus, which is a specialized field of mathematics, has little direct application to chemical engineering at present since chemical engineers confine their attention to three dimensions. Consequently, only the above special kinds of tensor (scalar and vector) which have physical interpretations will be considered in this chapter. If the reader is interested in the manipulation of tensors he is referred to Craig or Solkolnikoff.†

In the subsequent discussion, all quantities will be referred to as either vectors or scalars and all further remarks will apply to three dimensions. The subscripts will be labelled by the coordinate variables rather than by numbers for extra clarity. Bold face type will be used for vectors, and ordinary type for scalars.

7.3. ADDITION AND SUBTRACTION OF VECTORS

Any vector can be represented geometrically by a straight line with an arrow, the length of the line representing the magnitude of the vector and the direction of the line as indicated by the arrow representing the direction of the vector. The addition of two vectors is defined geometrically by the well-known "triangle of forces". The start of the line representing the second vector is superimposed on the end of the line representing the first vector. The sum of the two vectors is represented by the line joining the start of the first vector to the end of the second vector as illustrated in Fig. 7.2.

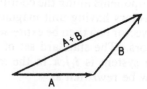

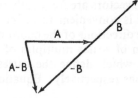

FIG. 7.2. Addition of vectors　　　　FIG. 7.3. Subtraction of vectors

The product of two scalars is obtained according to the rules of ordinary arithmetic and to multiply a vector by a scalar, the magnitude of the vector is multiplied by the scalar and the direction remains unchanged. Multiplication of a vector by a negative number also involves reversing the direction of the vector. Thus to subtract one vector from another it is only necessary to reverse the direction of the vector to be subtracted and add the other to it as illustrated in Fig. 7.3.

† Craig, H. V. "Vector and Tensor Analysis." McGraw-Hill (1943). Sokolnikoff, I. S. "Tensor Analysis." Wiley & Sons (1951).

7.3.1. *Components*

The process of addition described above can be reversed, allowing a vector to be resolved into equivalent constituent parts called "components". If all components of a vector are added together using the above rule, the original vector is returned. A vector can be resolved into components in an infinite variety of ways but the resolution into components in three specified non-coplanar directions is unique. This fundamental property is utilized by defining a coordinate system referred to three mutually orthogonal axes. Define the three cartesian coordinate axes x, y, z as shown in Fig. 7.4, and

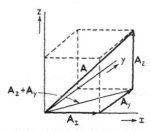

FIG. 7.4. Resolving into components

represent the vector A by a line starting from the origin. Complete the rectangular box which has edges parallel to the axes, and the vector A as diagonal, and denote the edges of the box by the vectors A_x, A_y, A_z. The subscripts denote the directions of the vectors and it is obvious that all parallel edges have the same length and are denoted by the same vector. Applying the rule of addition twice it can be seen that

$$A = A_x + A_y + A_z \qquad (7.7)$$

Since vectors are frequently resolved into components along the coordinate axes, it is convenient to define a set of unit vectors having unit magnitudes and the directions of the coordinate axes. Any vector can then be expressed as the sum of scalar multiples of these unit vectors. The standard set of unit vectors which define the cartesian coordinate system is i, j, k in the x, y, z directions respectively. Equation (7.7) can now be rewritten as

$$A = A_x i + A_y j + A_z k \qquad (7.8)$$

where A_x, A_y, A_z are the scalar lengths of the edges of the box, and the unit vectors i, j, k define the directions. Two applications of the theorem of Pythagoras in Fig. 7.4 shows that the magnitude of A, which is denoted by $|A|$ or A by analogy with complex numbers, is given by

$$|A| = \sqrt{(A_x{}^2 + A_y{}^2 + A_z{}^2)} \qquad (7.9)$$

7.3.2. *Position Vectors*

The position of a point in space can be specified relative to an origin by defining the vector joining the origin to the point. Such a vector is called a "position vector". The above definition of position is independent of any coordinate system and only requires the definition of an origin. An analogy which may be helpful in grasping this point is the reply to a stranger who asks for directions to a particular destination. If the reply given is "travel 500 metres in a NNE direction", then a complete coordinate system is implicitly involved; whereas if the information is given in the form of an arm pointing out the direction with the instruction "travel 500 metres over there", the arm is a vector indicating direction and the spoken instruction gives the magnitude. The first reply presupposes that both parties can locate the exact northerly direction, while the second reply is much more direct and useful.

A vector defining the position of a point is usually given the symbol r which can be resolved into components in the same way as any other vector once a coordinate system has been defined. Thus

$$r = x+y+z \tag{7.10}$$

or

$$r = xi+yj+zk \tag{7.11}$$

The point can therefore be specified by its coordinates either by equation (7.10) in a vector form or by equation (7.11) in a scalar form. The latter is more usual and it is implied that x is measured in a direction defined by i and and similarly for y and z. Thus when a point is specified by (x, y, z) it is vitally important that the order of the coordinates should not be disturbed.

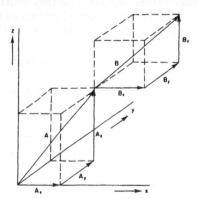

FIG. 7.5. Addition of vectors by components

Returning to the addition of two vectors A and B, and referring to Fig. 7.5, it can be seen that the sum of the vectors can be obtained by summing the separate components. Thus if

$$A = A_x i+A_y j+A_z k \tag{7.12}$$

and

$$B = B_x i+B_y j+B_z k \tag{7.13}$$

then

$$A+B = (A_x+B_x)i+(A_y+B_y)j+(A_z+B_z)k \tag{7.14}$$

also

$$A-B = (A_x-B_x)i+(A_y-B_y)j+(A_z-B_z)k \tag{7.15}$$

For two vectors to be equal, their difference must be zero, and from equation (7.15), A can only equal B if

$$A_x - B_x = 0, \quad A_y - B_y = 0, \quad \text{and} \quad A_z - B_z = 0 \quad (7.16)$$

Thus a single vector equation represents three simultaneous scalar equations.

7.3.3. *Properties of Addition*

Both the associative law and the commutative law of algebra for the addition of vectors are valid. That is,

$$A + (B + C) = (A + B) + C \qquad (7.17)$$

and
$$A + B = B + A \qquad (7.18)$$

The distributive law is also valid for the multiplication of the sum of two vectors by a scalar. Thus

$$m(A + B) = mA + mB \qquad (7.19)$$

There is no divergence therefore from the normal rules of algebra when addition and subtraction of vectors is considered.

7.3.4. *Geometrical Applications*

In general, two vectors are sufficient to define a plane, and if a linear relationship exists between three vectors then they must be coplanar. This is obvious from the rule of addition because when two of the vectors have been added together, they must have the same direction as the third vector.

Example 1. If A and B are two position vectors, find the equation of the straight line passing through the end points of A and B.

Referring to Fig. 7.6 where the origin is O and C is the position vector of any other point on the line through A and B, the vector joining A to B is

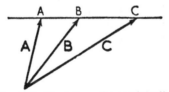

FIG. 7.6. Equation of a straight line

$B - A$. Similarly, the vector joining B to C is $C - B$. If ABC is to be a straight line, then these two vectors have the same direction and can only differ in magnitude. Therefore

$$C - B = m(B - A) \qquad \text{I}$$

where m is a variable scalar and C is the variable vector. Equation I can be rearranged to give

$$C = (m + 1)B - mA \qquad \text{II}$$

or
$$mA - (m + 1)B + C = 0 \qquad \text{III}$$

Equation II is the standard equation of a straight line with m the independent variable and C the dependent variable. Equation III is in the form of a linear relationship between three vectors, the general form of which is

$$pA + qB + rC = 0 \qquad\qquad \text{IV}$$

Comparison of equations III and IV shows that equation III is the special case in which $p + q + r = 0$. Thus, the general result that if three position vectors are related by a linear equation then they are coplanar. If the sum of the three coefficients is also zero, then the end points of the vectors are colinear.

Example 2. Prove that the medians of a triangle are concurrent.

Figure 7.7 illustrates the problem with reference to the triangle ABC, where L, M, N are the mid-points of the sides. It is assumed that BM and AL intersect at P and that CN does not necessarily pass through P.

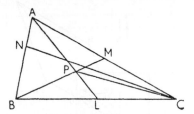

FIG. 7.7. Medians of a triangle

Taking C as origin, and denoting the position vectors of A and B by A and B, then the equations of the lines BM and AL can be written as follows using the results of Example 1. Thus the equation of the line through B and M is

$$\alpha = (s+1)B - s(\tfrac{1}{2}A) \qquad\qquad \text{I}$$

and as the parameter s varies, the vector α determines the various points along the line BM. Similarly, the equation of AL is

$$\alpha' = (t+1)A - t(\tfrac{1}{2}B) \qquad\qquad \text{II}$$

Since P is the point of intersection of AL and BM, the position vector of P is given by $\alpha = \alpha'$, or

$$(s+1)B - \tfrac{1}{2}sA = (t+1)A - \tfrac{1}{2}tB \qquad\qquad \text{III}$$

$$\therefore \quad (t+1+\tfrac{1}{2}s)A = (s+1+\tfrac{1}{2}t)B \qquad\qquad \text{IV}$$

Equation IV relates two vectors which have different directions and the equality can only be satisfied by both vectors having zero magnitude.

$$\therefore \quad t+1+\tfrac{1}{2}s = 0 \quad \text{and} \quad s+1+\tfrac{1}{2}t = 0 \qquad\qquad \text{V}$$

The solution of equations V is

$$s = t = -\tfrac{2}{3}$$

and the vector CP becomes $\tfrac{1}{3}(B+A)$ from equation I.

The vector CN is the sum of the vectors along CB and BN; and thus the position vector of N is given by

$$B + \tfrac{1}{2}(A - B) = \tfrac{1}{2}(A + B)$$

Therefore the vectors CN and CP have the same direction $(A + B)$ and thus CN must pass through P.

Example 3. *Generalized Vector Method for Stagewise Processes.* It has been shown by Lemlich and Leonard† that many of the existing difference point constructions for determining the number of theoretical steps in a stagewise process are variations of a more general vector method. They have also shown that the comprehensive vector method suggests many hitherto untried constructions and this example from their work derives the standard rectangular diagram from the general vector diagram.

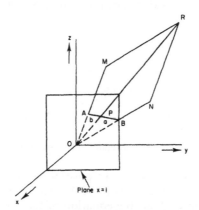

FIG. 7.8. Representation of a stagewise process

In any stagewise process there is more than one property to be conserved and for the purpose of this example it will be assumed that the three properties, enthalpy (H), total mass flow (M), and mass flow of one component (C) are conserved. Instead of considering three separate scalar balances, one vector balance can be taken by using a set of cartesian coordinates in the following manner. Using x to measure M, y to measure H, and z to measure C, then any process stream can be represented by a vector thus,

$$OM = M_1 i + H_1 j + C_1 k \qquad\qquad \text{I}$$

where OM is the vector illustrated in Fig. 7.8. Similarly, a second stream can be represented by

$$ON = M_2 i + H_2 j + C_2 k \qquad\qquad \text{II}$$

and using vector addition,

$$OR = OM + ON = (M_1 + M_2) i + (H_1 + H_2) j + (C_1 + C_2) k \qquad\qquad \text{III}$$

† Lemlich, R. and Leonard, R. A. *A.I.Ch.E.J.* 8, 214 (1962).

Thus OR which represents the sum of the two streams must be a constant vector for the three properties to be conserved within the system. Hence, to perform a calculation, when either of the streams OM or ON is determined, the other is obtained by subtraction from the constant OR. It will now be shown that the intersection of these vectors with the plane $x = 1$, as illustrated in Fig. 7.8, gives the usual enthalpy–concentration diagram for the Ponchon–Savarit method.

The constant line OR will cross the plane $x = 1$ at point P which will be a fixed point. The variable vectors OM and ON will also cross the plane at A and B respectively, and since OMRN is a parallelogram, APB will be a straight line. Let

$$\frac{AP}{PB} = \frac{b}{a} \qquad\qquad\qquad \text{IV}$$

as illustrated in Fig. 7.8. Then by the triangle rule of addition,

$$aOA = aOP + aPA = aOP + \frac{ab}{a+b} BA \qquad\qquad \text{V}$$

and

$$bOB = bOP + bPB = bOP + \frac{ab}{a+b} AB \qquad\qquad \text{VI}$$

Adding the extreme parts of equations V and VI gives

$$aOA + bOB = (a+b) OP \qquad\qquad\qquad \text{VII}$$

Because equation IV only defines a ratio between a and b, the magnitude of a can now be chosen so that

$$aOA = OM \qquad\qquad\qquad \text{VIII}$$

Since OB is in the same direction as ON and OP is in the same direction as OR, it can be proved in a similar manner to that used in Example 2 that equations III and VII are now identical, i.e.

$$bOB = ON \qquad\qquad\qquad \text{IX}$$

and

$$(a+b) OP = OR \qquad\qquad\qquad \text{X}$$

Using similar triangles and equations I and VIII, the x component of OM can be written in two equivalent ways.

$$a = M_1 \qquad\qquad\qquad \text{XI}$$

Similarly,

$$b = M_2 \qquad\qquad\qquad \text{XII}$$

and substituting into equation IV gives

$$\frac{AP}{PB} = \frac{M_2}{M_1} \qquad\qquad\qquad \text{XIII}$$

the normal lever law.

Using equations I, II, III, VIII, IX, X, XI, and XII, the coordinates of points A, B, and P can be determined as

$$\text{point A is } \left(1, \ \frac{H_1}{M_1}, \ \frac{C_1}{M_1}\right)$$

$$\text{point B is } \left(1, \ \frac{H_2}{M_2}, \ \frac{C_2}{M_2}\right)$$

$$\text{point C is } \left(1, \ \frac{H_1+H_2}{M_1+M_2}, \ \frac{C_1+C_2}{M_1+M_2}\right)$$

Comparing these with the original interpretation of the coordinate axes, the appropriate two-dimensional coordinates are the enthalpy and the mass fraction of the important component.

Thus the normal constructions in the enthalpy–concentration diagram are valid as a special case of a more general vector method. By taking the intersections of the vectors with different planes, different two-dimensional diagrams can be obtained. If the chosen plane is symmetrically inclined to all axes, then the familiar triangular diagram is obtained. The work of Lemlich and Leonard is extended to these cases and also to four dimensions. Suggested extension to higher orders is difficult to visualize and matrix methods may be more appropriate. The example is, however, sufficient indication of the generality of the vectorial approach which gives a clearer insight into the connection between related problems.

7.4. MULTIPLICATION OF VECTORS

In multiplication, the first serious divergence from normal algebra arises because it is necessary to define two distinct kinds of multiplication. Two physical situations in which the interplay of two vectors gives rise to the result are, firstly, the work done by a force during a displacement when both the force and the displacement are vectors; and, secondly, the moment of a force about an origin, when both the force and the position vector of its point of application must be specified. These two interactions between vectors are fundamentally different as will be shown in the present section.

7.4.1. *Scalar or Dot Product*

The scalar product of two vectors is defined as the product of the magnitude of the one vector with the magnitude of the component of the other vector resolved along it. This product is signified by placing a dot between the vectors to be multiplied together. Symbolically,

$$A.B = |A||B| \cos \theta = B.A \tag{7.20}$$

where θ is the angle between the vectors as shown in Fig. 7.9. The result of the operation is a scalar quantity, hence the name for the product.

One physical interpretation of this product is the calculation giving the work done by a force during a displacement. The fundamental definition of the work done is the magnitude of the force multiplied by the distance moved by its point of application in the direction of the force. Alternatively, the work done is the displacement multiplied by the component of the force causing the displacement. That these two alternatives are equivalent is obvious from the definition, equation (7.20). Work and hence energy are thus scalar quantities which arise from the multiplication of two vectors.

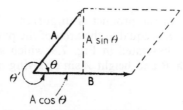

FIG. 7.9. Scalar and vector products

When a vector is multiplied by itself in this fashion, the scalar result can be written in the equivalent forms

$$A \cdot A = A^2 = A^2 \tag{7.21}$$

The inverse process to multiplication, division, is not uniquely defined for the following reason. If an equation such as

$$A \cdot B = 0 \tag{7.22}$$

is satisfied, then any of the following three conclusions can be drawn.
(a) The vector A is zero.
(b) The vector B is zero.
(c) $\theta = 90°$, and A and B are mutually perpendicular.
The third possibility arises in equation (7.20) because $\cos \theta$ may be zero. It is therefore possible for the scalar product of two vectors to be zero when both vectors are finite. Division thus has no meaning, for if equation (7.22) could be divided by a non-zero A, the inevitable conclusion would be that B is zero which is not necessarily true.

7.4.2. Vector or Cross-product

Referring to Fig. 7.9 again in which θ is the angle between the two vectors A and B, then the vector product of A and B is defined by

$$A \wedge B = |A||B| \sin \theta \, n \tag{7.23}$$

where the symbol ($\wedge$) is used to denote a vector product, n is a unit vector along the normal to the plane containing A and B and its positive direction is determined as follows. The right-hand screw rule is applied to the sense of rotation from A to B in measuring the angle θ, thus defining the positive direction of n. Referring to Fig. 7.9, this rule indicates that n acts into the plane of the paper, thus $A \wedge B$ is also a vector into the plane of the paper. This definition is unique, because if θ were described in the opposite sense as indicated by θ', the positive direction of n would be reversed, the sign of sin θ would also be reversed, and the vector $A \wedge B$ would still act into the plane of the paper. However, the vector $B \wedge A$ has the same magnitude as $A \wedge B$ but the direction of n is reversed due to describing θ by moving from B to A. Hence

$$A \wedge B = - B \wedge A \qquad (7.24)$$

and the order of terms in the product is important. The magnitude of the vector product of A and B is equal to the area of the parallelogram formed by A and B. This is also illustrated in Fig. 7.9 which shows the base of the parallelogram of length B and height $A \sin \theta$, giving an area of $AB \sin \theta$ or $|A||B| \sin \theta$.

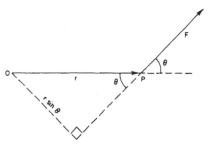

FIG. 7.10. Moments of a force

The moment of a force is defined by a vector product in the following manner. If there is a force F acting at a point P with position vector r relative to an origin O as illustrated in Fig. 7.10, then the moment of F about O is defined by

$$L = r \wedge F \qquad (7.25)$$

In order to define the positive normal n, the vectors r and F must have a common origin. If P is taken as this origin and the right-hand screw rule applied to equation (7.25) then n is seen to act out of the plane of the paper. Using the fundamental definition of L as the product of the force and the perpendicular distance of the origin from the line of action of the force gives

$$|L| = |F| r \sin \theta$$

which checks with equation (7.25) and the definition (7.23). The direction of the vector L which is outwards from the plane of the paper also corresponds with the fundamental definition of a moment.

Again, vector division is impossible, since if

$$A \wedge B = 0 \qquad (7.26)$$

then either
(a) the vector A is zero,
(b) the vector B is zero,
or   (c) θ is zero, and the vectors A and B are parallel.

7.4.3. *Properties of Multiplication*

Although the associative, distributive, and commutative laws are valid for the addition and subtraction of vectors, these rules need to be considered more carefully with respect to products. It has already been shown in equation (7.20) that the commutative law is valid for the scalar product, whereas equation (7.24) shows that the commutative law is not valid for the vector product. That is,

$$A.B = B.A \tag{7.20}$$

but
$$A \wedge B = - B \wedge A \tag{7.24}$$

The order of terms in a vector product can only be altered if the sign is reversed.

The distributive law is unreservedly valid for all vector products. That is,

$$A.(B+C) = A.B + A.C \tag{7.27}$$

$$A \wedge (B+C) = A \wedge B + A \wedge C \tag{7.28}$$

The associative law is simplified by the convention that two vectors without an operating sign between them have no meaning, thus reducing the need for brackets. For example, the following terms can only be evaluated in the one particular order indicated by brackets on the right-hand side of each equation.

$$A.BC.D = (A.B)(C.D) \tag{7.29}$$

$$A.BC = (A.B)C \tag{7.30}$$

$$A \wedge B.C = (A \wedge B).C \tag{7.31}$$

In equations (7.29) and (7.30) each dot product yields a scalar which can be treated as any other scalar, so that equation (7.29) is a product of two scalars and in equation (7.30) the magnitude of the vector C is increased by the scalar multiplier $(A.B)$. In equation (7.31) the vector product must be taken first because a vector product cannot be formed between the vector A and the scalar $(B.C)$. Brackets are usually essential when more than one vector product is involved, and then the associative law is not valid because

$$A \wedge (B \wedge C) \neq (A \wedge B) \wedge C$$

as will be shown in Section 7.4.6.

7.4.4. *Unit Vector Relationships*

It is frequently useful to resolve vectors into components along the axial directions in terms of the unit vectors i, j and k. All operations are then performed on the unit vectors and the results of products between them are all standard. Since by definition, $i, j,$ and k are mutually orthogonal, the angle θ in the definitions of the two products must be either 0 or $\pi/2$ and because

$$\sin 0 = 0 \quad \text{and} \quad \cos \pi/2 = 0$$

the following results are true.

$$i.j = j.k = k.i = 0 \qquad (7.32)$$

and
$$i \wedge i = j \wedge j = k \wedge k = 0 \qquad (7.33)$$

Because $i, j,$ and k have unit magnitude.

$$i.i = j.j = k.k = 1 \qquad (7.34)$$

By the definition of the vector product in Section 7.4.2 $i \wedge j$ is a vector normal to the plane of i and j and hence must lie in the k direction. A convention is now needed which determines whether $i \wedge j$ is in the positive or negative k direction. If $i \wedge j = k$, the set of axes is said to be "right handed", since k is defined by the right-hand screw rule from the vector product of i and j; whereas if $i \wedge j = -k$, the set of axes is left handed. It is normal to use a right-handed set of axes and it can easily be verified that

$$i \wedge j = k, \quad j \wedge k = i, \quad k \wedge i = j \qquad (7.35)$$

and
$$j \wedge i = -k, \quad k \wedge j = -i, \quad i \wedge k = -j \qquad (7.36)$$

The use of these relationships is illustrated by the straightforward application to the simple products of the two vectors

$$A = A_x i + A_y j + A_z k \qquad (7.12)$$

and
$$B = B_x i + B_y j + B_z k \qquad (7.13)$$

Thus
$$A.B = (A_x i + A_y j + A_z k).(B_x i + B_y j + B_z k)$$
$$= A_x B_x + A_y B_y + A_z B_z \qquad (7.37)$$

and
$$A \wedge B = (A_x i + A_y j + A_z k) \wedge (B_x i + B_y j + B_z k)$$
$$= A_x(B_y k - B_z j) + A_y(B_z i - B_x k) + A_z(B_x j - B_y i)$$

using equations (7.32) to (7.36). Rearrangement of the last line gives

$$A \wedge B = i(A_y B_z - A_z B_y) + j(A_z B_x - A_x B_z) + k(A_x B_y - A_y B_x) \qquad (7.38)$$

Equation (7.38) can be written in a more useful and compact form by expressing the right-hand side as a determinant, thus

$$A \wedge B = \begin{vmatrix} i & j & k \\ A_x & A_y & A_z \\ B_x & B_y & B_z \end{vmatrix} \qquad (7.39)$$

7.4.5. *Scalar Triple Product*

Because $A \wedge B$ is a vector, further products can be taken with a third vector. The scalar product of $A \wedge B$ with C is best considered geometrically as shown in Fig. 7.11, where θ is the usual angle between the two vectors $(A \wedge B)$ and C. By definition,

$$A \wedge B . C$$

is a scalar whose magnitude is the magnitude of $(A \wedge B)$ multiplied by the component of the vector C resolved along $(A \wedge B)$. But the magnitude of $(A \wedge B)$ is the area of the parallelogram formed on A and B, and since the vector $(A \wedge B)$ must be perpendicular to the plane of A and B, then the

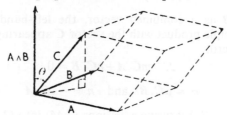

FIG. 7.11. Scalar triple product

resolved part of C along $(A \wedge B)$ must equal the height of the parallelepiped as shown in Fig. 7.11. The magnitude of $A \wedge B . C$ is thus the volume of the parallelepiped with edges parallel to A, B, and C. The result is positive or negative according as θ is acute or obtuse. Since the volume of a parallelepiped can be expressed as the product of any base with the corresponding height, and the order of vectors in a dot product is irrelevant, the scalar triple product can be written in six equivalent ways, thus

$$A \wedge B . C = A . B \wedge C = B \wedge C . A = B . C \wedge A = C \wedge A . B = C . A \wedge B$$

$$= [A, B, C] \tag{7.40}$$

where the three vectors remain in the cyclic order A, B, C and the positions of the dot and cross are arbitrary. The other six ways of writing this triple product have the same numerical value as the expressions in (7.40) but have the opposite sign due to reversing the order of terms in the vector product. Thus

$$[A, B, C] = -[A, C, B] \tag{7.41}$$

There are seven ways in which the parallelepiped can have zero volume. Any of the three vectors may be zero, any pair of vectors may be equal or parallel, or the three vectors may lie in the same plane.

7.4.6. *Vector Triple Product*

The vector product of $A \wedge B$ with a third vector C can be taken, but in this case the order of multiplication and position of brackets is of vital importance.

Referring to Fig. 7.11 again, it can be seen that the vector $A \wedge B$ is perpendicular to the plane of A and B. When the further vector product with C is taken, the resulting vector must be perpendicular to $A \wedge B$ and hence in the plane of A and B. The resulting vector triple product can therefore be resolved into components along A and B, thus

$$(A \wedge B) \wedge C = mA + nB \tag{7.42}$$

where m and n are scalar constants to be determined. Multiplying throughout equation (7.42) by C using a scalar product gives

$$C.(A \wedge B) \wedge C = mC.A + nC.B \tag{7.43}$$

Considering $A \wedge B$ as a combined vector, the left-hand side of equation (7.43) is a scalar triple product with the vector C appearing twice. Hence the left-hand side is zero.

$$\therefore \quad mC.A + nC.B = 0$$

or $$m = \alpha C.B \quad \text{and} \quad n = -\alpha C.A \tag{7.44}$$

which is equivalent. Substituting equations (7.44) into (7.42) gives

$$(A \wedge B) \wedge C = \alpha[(C.B)A - (C.A)B] \tag{7.45}$$

Since each term in equation (7.45) contains the three vectors, α must be just a number. Also, because equation (7.45) is valid for any vectors A, B, and C, it must be valid for a particular set of values for A, B, and C. Thus letting $A = i$, $B = C = j$, and using equations (7.35) and (7.36) gives

$$(A \wedge B) \wedge C = (i \wedge j) \wedge j$$

$$= k \wedge j$$

$$= -i$$

But using equations (7.32) and (7.34),

$$C.B = j.j = 1$$

$$C.A = j.i = 0$$

Using these results in equation (7.45) gives

$$-i = \alpha i$$

$$\therefore \quad \alpha = -1$$

and $$(A \wedge B) \wedge C = (A.C)B - (B.C)A \tag{7.46}$$

Similarly it can be shown that

$$A \wedge (B \wedge C) = (A.C)B - (A.B)C \tag{7.47}$$

In order to remember equations (7.46) and (7.47) the following rule is helpful.

The vectors appearing outside the brackets on the right-hand side are the vectors appearing inside the brackets on the left-hand side. The positive term is the dot product of the extreme vectors multiplied by the central vector.

7.5. DIFFERENTIATION OF VECTORS

If a vector r is a function of a scalar variable t (say time), then when t varies by an increment δt, r will vary by an increment δr. Just as in ordinary calculus, δr is a variable associated with r but it need not have either the same magnitude or direction as r. If δr tends to zero as δt tends to zero, then

$$\lim_{\delta t \to 0} \frac{\delta r}{\delta t} = \frac{dr}{dt} \tag{7.48}$$

defines the first derivative of r with respect to t.

If r is interpreted as a position vector resolved into its components, then

$$r = xi + yj + zk \tag{7.11}$$

and

$$\delta r = \delta xi + \delta yj + \delta zk$$

$$\therefore \quad \frac{\delta r}{\delta t} = \frac{\delta x}{\delta t}i + \frac{\delta y}{\delta t}j + \frac{\delta z}{\delta t}k$$

and taking the limit as $\delta t \to 0$,

$$\frac{dr}{dt} = \frac{dx}{dt}i + \frac{dy}{dt}j + \frac{dz}{dt}k \tag{7.49}$$

Differentiation of a vector with respect to a scalar is thus similar to ordinary differentiation and the rules as applied to products of vectors are unchanged, viz.

$$\frac{d}{dt}(A \cdot B) = A \cdot \frac{dB}{dt} + \frac{dA}{dt} \cdot B \tag{7.50}$$

$$\frac{d}{dt}(A \wedge B) = A \wedge \frac{dB}{dt} + \frac{dA}{dt} \wedge B \tag{7.51}$$

As t varies, the end point of the position vector r will trace out a curve in space. Taking s as a variable measuring length along this curve, the differentiation process can be performed with respect to s thus,

$$\frac{dr}{ds} = \frac{dx}{ds}i + \frac{dy}{ds}j + \frac{dz}{ds}k \tag{7.52}$$

Applying equation (7.9),

$$\left|\frac{dr}{ds}\right| = \sqrt{\left[\left(\frac{dx}{ds}\right)^2 + \left(\frac{dy}{ds}\right)^2 + \left(\frac{dz}{ds}\right)^2\right]}$$

$$= \frac{\sqrt{[(dx)^2 + (dy)^2 + (dz)^2]}}{ds}$$

But
$$(ds)^2 = (dx)^2 + (dy)^2 + (dz)^2$$

Therefore, dr/ds is a unit vector in the direction of the tangent to the curve. A further useful result can be obtained as follows.

$$\frac{d}{dt}(A^2) = \frac{d}{dt}(A.A) = 2A.\frac{dA}{dt}$$

If A is a vector of constant magnitude but variable direction, then A^2 will be a constant.

$$\therefore \quad A.\frac{dA}{dt} = 0 \tag{7.53}$$

and the vector dA/dt is perpendicular to A. An example of this is uniform motion in a circle when the acceleration is perpendicular to the instantaneous velocity.

It has been shown above that dr/ds is a unit vector and hence differentiating again with respect to s and using equation (7.53), then d^2r/ds^2 must be perpendicular to the tangent dr/ds. The direction of d^2r/ds^2 is the normal to the curve, and the two vectors defined as the tangent and normal define what is called the "osculating plane" of the curve.

7.5.1. *Partial Differentiation of Vectors*

Temperature is a scalar quantity which can depend in general upon three coordinates defining position and a fourth independent variable time. An increment δT can arise from an increment in any of the four independent variables, or from a combination of such increments. The increment δT arising from a single space increment, say δx, can be expressed as a ratio to δx and in the limit as $\delta x \to 0$, the ratio becomes a derivative. This particular derivative is denoted by $\partial T/\partial x$ which implies that the other independent variables remain constant during the limiting process, and in this respect, $\partial T/\partial x$ is a "partial derivative". $\partial T/\partial x$ is the temperature gradient in the x direction and is a vector quantity; similarly, $\partial T/\partial y$ and $\partial T/\partial z$ are also vector gradients. As will be shown in the next section, these three vectors can be compounded into a single generalized vector gradient. The other partial derivative with respect to time, $\partial T/\partial t$, is a scalar rate of change.

A dependent variable such as temperature, having the properties described above, is called a "scalar point function" and the system of variables is frequently called a "scalar field". There are many scalar point functions and amongst those of interest to chemical engineers are temperature, concentration, and pressure. In chemical engineering, there are other dependent variables which are vectorial in nature, and vary with position. Examples are velocity, heat flow rate, and mass transfer rate, which are called "vector point functions" and they constitute "vector fields". A familiar illustration of a two-dimensional vector field is the disposition of iron filings showing the field strength of a magnet. The differentiation of vector point functions will be discussed in the next section.

7.6. Hamilton's Operator ∇

Returning to the consideration of temperature as a function of the three space variables, it was shown in Section 7.5.1 that the three partial derivatives of the temperature were vector gradients. If these three vector components are added together, there results a single vector gradient

$$i\frac{\partial T}{\partial x}+j\frac{\partial T}{\partial y}+k\frac{\partial T}{\partial z}=\nabla T \tag{7.54}$$

which defines the operator ∇ for determining the complete vector gradient of a scalar point function. The operator ∇ is pronounced "del" or "nabla", and the vector ∇T is often written "grad T" for obvious reasons.

∇ can operate upon any scalar quantity as above and yield a vector gradient. Written out in an abstract manner,

$$\nabla = i\frac{\partial}{\partial x}+j\frac{\partial}{\partial y}+k\frac{\partial}{\partial z} \tag{7.55}$$

The nature of ∇T can be further illustrated by taking a scalar product of ∇T with an infinitesimal increment $\mathbf{dr}$ in the position vector $\mathbf{r}$. Thus

$$\mathbf{dr}.\nabla T = (i\,dx+j\,dy+k\,dz).\left(i\frac{\partial T}{\partial x}+j\frac{\partial T}{\partial y}+k\frac{\partial T}{\partial z}\right) \tag{7.56}$$

and using the properties of products of unit vectors (7.32) and (7.34), this simplifies to

$$\mathbf{dr}.\nabla T = \frac{\partial T}{\partial x}dx+\frac{\partial T}{\partial y}dy+\frac{\partial T}{\partial z}dz \tag{7.57}$$

Comparing equation (7.57) with equation (8.9) from the next chapter, it can be seen that

$$\mathbf{dr}.\nabla T = dT \tag{7.58}$$

If vector division were possible (which it is not) then equation (7.58) could be rearranged to show that ∇T is the ratio of dT to $\mathbf{dr}$, which is the generalized first derivative of a scalar variable with respect to an independent vector variable. The expression $dT/\mathbf{dr}$ does not exist to represent this quantity, but ∇T is the vector equivalent of the generalized gradient.

The physical interpretation of ∇T can be found by taking a variable position vector $\mathbf{r}$ to describe an isothermal surface.

$$T(x, y, z) = C$$

With this definition of $\mathbf{r}$, for any increment $\mathbf{dr}$, the corresponding increment dT will be zero because $\mathbf{dr}$ lies on the isothermal surface. But equation (7.58) shows that

$$\mathbf{dr}.\nabla T = dT = 0$$

and thus ∇T must be perpendicular to $d\mathbf{r}$. Since the vector $d\mathbf{r}$ can lie in any direction in the tangent plane to the isothermal at $\mathbf{r}$, and ∇T can only have one value at a point $\mathbf{r}$, then ∇T must be perpendicular to the tangent plane at $\mathbf{r}$. This direction is the line of most rapid change of T. Thus ∇T is a vector in the direction of the most rapid change of T, and its magnitude is equal to this rate of change.

7.6.1. *Divergence of a Vector, $\nabla.A$*

Because the operator ∇ is of vector form, a scalar product can be obtained between ∇ and any vector A, with the interpretation that ∇ is to be treated as a vector as far as the product with A is concerned, and as a differential operator upon everything which follows it. Thus, if A is resolved into components, then

$$\nabla.A = \left(i\frac{\partial}{\partial x}+j\frac{\partial}{\partial y}+k\frac{\partial}{\partial z}\right).(A_x i+A_y j+A_z k)$$

Since i, j, and k are constant both in magnitude and direction, the differential operators only take effect upon A_x, A_y, and A_z. Using the unit vector relationships (7.32) and (7.34), therefore

$$\nabla.A = \frac{\partial A_x}{\partial x}+\frac{\partial A_y}{\partial x}+\frac{\partial A_z}{\partial z} \tag{7.59}$$

Equation (7.59) has been obtained by abstract manipulations, but the operation ($\nabla.$) on a vector has an important physical interpretation.

The continuity equation expressing the law of conservation of material within a flowing fluid will be derived in Chapter 8 as

$$\frac{\partial}{\partial x}(\rho u)+\frac{\partial}{\partial y}(\rho v)+\frac{\partial}{\partial z}(\rho w)+\frac{\partial \rho}{\partial t} = 0 \tag{7.60}$$

where ρ is the density, and

$$u = ui+vj+wk \tag{7.61}$$

is the velocity. Multiplying equation (7.61) by ρ and applying the operator ($\nabla.$) to the result using equation (7.59) gives the first three terms of equation (7.60). Thus

$$\nabla.\rho u+\frac{\partial \rho}{\partial t} = 0 \tag{7.62}$$

is the vector form of the continuity equation. The physical interpretation of equation (7.62) is as follows. $\partial \rho/\partial t$ is the rate of change of mass contained in unit volume, or the rate of accumulation of mass per unit volume. Using the general conservation law (1.25), then $\nabla.\rho u$ must equal the output less the input, which is the net rate of mass flow from unit volume. In general, $\nabla.A$ is the net flux of A per unit volume at the point considered, counting vectors into the volume as negative, and vectors out of the volume as positive.

If a tubular element of volume is taken such that the vectors of the field A lie in the curved surface of the tubular element, then the only flux of A within the element must enter and leave at the ends of the tube. If $\nabla.A$ is positive, then the flux leaving the one end must exceed the flux entering at the other end. In this respect the tubular element is divergent in the direction of flow. For this reason, the operation $\nabla.$ is frequently called the "divergence" and is written

$$\nabla.A = \operatorname{div} A \tag{7.63}$$

7.6.2. Curl or Rotation of a Vector, $\nabla \wedge A$

A further differential function can be obtained using Hamilton's operator by taking a vector product between the operator and another vector. Expanding into components and using equation (7.39) to express the vector product gives

$$\nabla \wedge A = \begin{vmatrix} i & j & k \\ \dfrac{\partial}{\partial x} & \dfrac{\partial}{\partial y} & \dfrac{\partial}{\partial z} \\ A_x & A_y & A_z \end{vmatrix} = \operatorname{curl} A = \operatorname{rot} A \tag{7.64}$$

The physical interpretation of the operator $(\nabla \wedge)$ can be explained by referring to Fig. 7.12. For simplicity, it is assumed that the z component of velocity (w) is zero and that the flow is two-dimensional. This implies that all partial derivatives with respect to z are zero, and equation (7.64) simplifies to

$$\nabla \wedge u = \begin{vmatrix} i & j & k \\ \dfrac{\partial}{\partial x} & \dfrac{\partial}{\partial y} & 0 \\ u & v & 0 \end{vmatrix} = k\left(\frac{\partial v}{\partial x} - \frac{\partial u}{\partial y}\right) \tag{7.65}$$

in Fig. 7.12, an element of fluid is defined from an origin O by taking increments δx and δy along the axes. Using Taylor's theorem, the y component of velocity at A and the x component of the velocity at B will be as shown.

The side OA of the fluid element is rotating in addition to its lateral and longitudinal motion, and its angular velocity is $\partial v/\partial x$ as may be seen from the diagram. This rotation is in an anti-clockwise direction and by the right-hand screw rule acts in the positive k direction. Thus the angular velocity of OA is $k(\partial v/\partial x)$. Similarly, the angular velocity of OB is $-k(\partial u/\partial y)$. The angular velocity of the fluid element is the average of these two values, and denoting the angular velocity by ωk, then

$$\omega k = \frac{1}{2}\left(\frac{\partial v}{\partial x} - \frac{\partial u}{\partial y}\right)k \tag{7.66}$$

Comparing equations (7.65) and (7.66) shows that

$$\nabla \wedge u = 2\omega k = \zeta \tag{7.67}$$

which defines ζ, the "vorticity" of the fluid element. The vorticity is thus twice the angular velocity of the fluid element and this explains why $(\nabla \wedge)$ is called the "rotation" or "curl" operation.

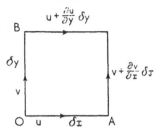

FIG. 7.12. Curl of a velocity vector

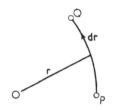

FIG. 7.13. Line integrals

7.7. INTEGRATION OF VECTORS AND SCALARS

In ordinary integration with respect to a single independent variable, the value of an integral depends upon the limits of integration only, because only one path of integration is possible. When alternative paths are available in two or more dimensions, the value of an integral depends upon both the limits and the path. It was shown in Section 4.13.1 that for an analytic function of a complex variable, the path of integration was irrelevant. These properties will now be generalized for three dimensions by using vectors.

7.7.1. *Line Integrals*

An arbitrary path of integration can be specified by defining a variable position vector r such that its end point sweeps out the curve between P and Q as shown in Fig. 7.13. A vector A can be integrated between two fixed points P and Q along the curve r by summing the components of A resolved along the curve. This integral can be written

$$\int_P^Q A \cdot dr = \int_P^Q (A_x \, dx + A_y \, dy + A_z \, dz) \tag{7.68}$$

by the definition of scalar products and resolution into components.

If the integral in equation (7.68) is to depend on P and Q but not upon the path r, then the integrand must be the differential of some function ϕ, i.e.

$$A \cdot dr = d\phi \tag{7.69}$$

But using equation (7.58),

$$dr \cdot \nabla\phi = d\phi \tag{7.70}$$

$$\therefore \quad (A - \nabla\phi) \cdot dr = 0 \tag{7.71}$$

by subtraction of equations (7.69) and (7.70). Since the result is to be independent of the path of integration, dr can have any direction at a point, and $(A - \nabla\phi)$ cannot be perpendicular to all directions. Hence,

$$A - \nabla\phi = 0 \tag{7.72}$$

This means that if a vector field A can be expressed as the gradient of a scalar field ϕ, then the line integral of the vector A between any two points P and Q is independent of the path taken.

If ϕ is a single-valued function, then the order of differentiation of ϕ with respect to any two variables is immaterial. Thus

$$\frac{\partial}{\partial y}(A_x) = \frac{\partial}{\partial y}\left(\frac{\partial \phi}{\partial x}\right) = \frac{\partial}{\partial x}\left(\frac{\partial \phi}{\partial y}\right) = \frac{\partial}{\partial x}(A_y) \tag{7.73}$$

But the k component of $\nabla \wedge A$ from equation (7.64) is

$$k\left(\frac{\partial A_y}{\partial x} - \frac{\partial A_x}{\partial y}\right)$$

which is zero by equation (7.73). Similarly, all three components of $\nabla \wedge A$ are zero if equation (7.72) is satisfied.

Hence the general result that the line integral of equation (7.68) is independent of the path of integration if $\nabla \wedge A = 0$. Also, if this is true, then A can be represented as the gradient of some scalar point function ϕ.

An example of the use of line integrals is the following. If the vector field A is a force field so that a particle at a point r experiences a force A, then the work done in moving the particle a distance δr from r is defined as the displacement times the component of force opposing the displacement. Therefore, the incremental work done (δW) is given by

$$\delta W = A . \delta r \tag{7.74}$$

Referring to Fig. 7.13, the total work done in moving the particle from P to Q is the sum of the increments along the path. As the size of the increments tends to zero, the sum becomes an integral, or

$$W = \int_P^Q A . dr \tag{7.75}$$

When this work done is independent of the path of the particle from P to Q, the force field A is said to be "conservative", and the results just enunciated are valid. Such a force field can be represented by the gradient of a scalar function. When a scalar point function is used to represent a vector field, it is called a "potential" function. Thus, the gravitational force field is represented by a gravitational potential function (potential energy), the electrostatic force field by an electric potential, and a magnetic force field by a magnetic potential. A further useful potential function will be introduced in Section 7.10.2.

7.7.2. Vector Area and Surface Integrals

An element of surface has its area defined as a vector by reference to its boundary. The magnitude of the area is defined as the maximum projected area of the element, and the direction of the vector is normal to this plane of projection. The direction of the positive normal is determined by the right-hand screw rule applied to the sense of describing the boundary of the element as illustrated in Fig. 7.14.

FIG. 7.14. Vector area

When two adjacent elements of area are added together, their separate vector areas are defined so that the directions of tracing the common part of the boundary are opposite, so that the combined area is traced out by ignoring the common boundary. Since areas are thus defined as vectors, the vector rule of addition must be applied. Hence it can be seen that the vector area of any closed figure must be zero because the projected area on any plane is doubly described, one clockwise and the other anticlockwise leading to complete cancellation.

When discussing line integrals above, the infinitesimal dr was a vector and the type of product with the vector A had to be specified. To integrate a vector over a surface, it is also necessary to specify the type of product, and although the dot product is used again, the physical interpretation is different. If there is a vector field denoted by A and the element of surface is denoted by $dS = n\,dS$ where n is the unit vector normal to the surface element, then the surface integral is usually defined as

$$\int A \cdot dS = \int A \cdot n\,dS \tag{7.76}$$

When A is a force field, the above integral gives the total force acting on the surface, so that if the surface is closed, the integral gives the resultant force on the body due to the integrated effect over the surface. If the vector A is replaced by the velocity vector u, then the surface integral defined by equation (7.76) gives the net volumetric flow across the surface.

It is possible to integrate a cross-product between A and dS, but this has very little practical value and is usually ignored.

7.7.3. Volume Integrals

The element of volume $dx\,dy\,dz$ is usually denoted by $d\sigma$ so that its shape is not restricted in any way. Although both the elements of length (dr) and surface (dS) are vectors, the element of volume ($d\sigma$) is a scalar quantity, and thus there is no cause to specify any type of multiplication.

There are relationships between these three types of integration and the following subsections will be devoted to them. If the three integrals are considered in terms of cartesian coordinates, it can be seen that a surface integration involves two linear integrations, whilst a volume integration involves three linear integrations. It seems obvious therefore that any relationship between the different types of integration will involve a number of linear differentiations to match the orders. If these general ideas can be assimilated, the following relationships may appear to be more logical.

7.7.4. Stokes' Theorem

Consider any continuous surface S having elements dS defined in the usual way and denote by C the curve describing the boundary of the surface S. The positive direction of the curve C is related to the positive direction of the normal vector n as given in the definition of vector area in Section 7.7.2. Stokes' theorem states that if there is a vector field A then the line integral of A taken round C is equal to the surface integral of $\nabla \wedge A$ taken over S. Symbolically,

$$\int_C A \cdot dr = \int_S \nabla \wedge A \cdot dS = \int_S n \cdot \nabla \wedge A \, dS \qquad (7.77)$$

It can now be seen that the result of Section 7.7.1 is a special case of the above theorem. That is, if $\nabla \wedge A = 0$, then the line integral of A around any closed curve is zero. Equation (7.77) is much more general than this previous result, and is frequently used in vector analysis. Similar results are available for other line and surface integrals but they are of rarer occurrence. The formulae are

$$\int U \, dr = \int n \wedge \nabla U \, dS \qquad (7.78)$$

$$\int A \cdot dr = \int n \wedge \nabla \cdot A \, dS \qquad (7.79)$$

$$-\int A \wedge dr = \int (n \wedge \nabla) \wedge A \, dS \qquad (7.80)$$

where all line integrals are taken along the curve C which defines the boundary of the surface S over which the surface integrals are taken. The negative sign is necessary in equation (7.80) because the vector A has been moved from the first to the last term of a vector product so that it follows the operator ∇.

Stokes' theorem can be applied to a two-dimensional system by putting

$$A = Pi + Qj$$

where P and Q are functions of x and y only and the elementary length

$$dr = dxi + dyj$$

The normal to the xy plane is in the z direction hence

$$n \, dS = dx \, dy \, k$$

and using equation (7.64) gives

$$\nabla \wedge A = k\left(\frac{\partial Q}{\partial x} - \frac{\partial P}{\partial y}\right)$$

Putting these various forms into equation (7.77) and simplifying gives

$$\int_C (P\,\mathrm{d}x + Q\,\mathrm{d}y) = \iint_S \left(\frac{\partial Q}{\partial x} - \frac{\partial P}{\partial y}\right)\mathrm{d}x\,\mathrm{d}y \qquad (7.81)$$

which was used in Section 4.13 to prove Cauchy's theorem.

7.7.5. *Gauss' Divergence Theorem*

When the continuity equation was considered in Section 7.6.1 it was shown that $\nabla \cdot \rho u$ is the net rate of mass flow from unit volume. It was also stated in Section 7.7.2 that the surface integral of the velocity vector u gives the net volumetric flow across the surface. If these two ideas are now combined by applying them to a closed surface S enclosing a volume σ, the following result is obtained by equating the two expressions for the mass flow rate.

$$\int_S \rho u \cdot \mathrm{d}S = \int_\sigma \nabla \cdot \rho u \, \mathrm{d}\sigma \qquad (7.82)$$

Equation (7.82) is valid for any vector A giving

$$\int_S A \cdot \mathrm{d}S = \int_\sigma \nabla \cdot A \, \mathrm{d}\sigma \qquad (7.83)$$

which is the "divergence theorem".

Again, this equation can be extended to other products giving the set of formulae

$$\int n U \, \mathrm{d}S = \int \nabla U \, \mathrm{d}\sigma \qquad (7.84)$$

$$\int n \cdot A \, \mathrm{d}S = \int \nabla \cdot A \, \mathrm{d}\sigma \qquad (7.85)$$

$$\int n \wedge A \, \mathrm{d}S = \int \nabla \wedge A \, \mathrm{d}\sigma \qquad (7.86)$$

where the surface integrals are taken over the surface S enclosing the volume σ over which the volume integrals are taken.

7.7.6. *Green's Theorem*

If U and V are two continuous differentiable point functions, then an application of Gauss' divergence theorem (7.83) to the function $U \nabla V$ gives

$$\int_S U \nabla V \cdot \mathrm{d}S = \int_\sigma \nabla \cdot (U \nabla V) \, \mathrm{d}\sigma \qquad (7.87)$$

The right-hand side can be expanded as the derivative of a product, thus

$$\nabla.(U\nabla V) = (\nabla U).(\nabla V) + U\nabla^2 V \tag{7.88}$$

Substituting equation (7.88) into equation (7.87) and rearranging gives

$$\int \nabla U.\nabla V\,d\sigma = \int U\nabla V.d\mathbf{S} - \int U\nabla^2 V\,d\sigma \tag{7.89}$$

Interchanging U and V throughout the above argument gives

$$\int \nabla U.\nabla V\,d\sigma = \int V\nabla U.d\mathbf{S} - \int V\nabla^2 U\,d\sigma \tag{7.90}$$

Equating equations (7.89) and (7.90) and rearranging the integrals gives

$$\int (U\nabla V - V\nabla U).d\mathbf{S} = \int (U\nabla^2 V - V\nabla^2 U)\,d\sigma \tag{7.91}$$

Equations (7.89, 7.90, 7.91) are known as various forms of "Green's theorem". As will be appreciated in the next chapter, the differential equation

$$\nabla^2 V = \nabla.\nabla V = 0 \tag{7.92}$$

is of frequent occurrence, and equation (7.91) can be very useful for dealing with such systems. It is of even more use in the theory of electricity and magnetism because both the electric and magnetic potentials satisfy equation (7.92).

7.8. STANDARD IDENTITIES

Any of the following equations can be verified quite easily by using the formulae given in the earlier parts of this chapter, but it is useful to have the following list available for reference purposes. This first group of equations arises from differentiating various products.

$$\nabla.UA = U\nabla.A + A.\nabla U \tag{7.93}$$

$$\nabla\wedge UA = U\nabla\wedge A - A\wedge\nabla U \tag{7.94}$$

$$\nabla.A\wedge B = B.\nabla\wedge A - A.\nabla\wedge B \tag{7.95}$$

$$\nabla\wedge(A\wedge B) = B.\nabla A + A\nabla.B - A.\nabla B - B\nabla.A \tag{7.96}$$

In the derivation of the next useful formula, the dual nature of ∇ causes some difficulty in the notation, but if it is understood that when no symbol follows ∇, the symbol immediately preceding it is to be differentiated, then the intermediate steps can be followed.

$$A\wedge(\nabla\wedge B) = A.B\nabla - A.\nabla B \tag{7.97}$$

$$B\wedge(\nabla\wedge A) = B.A\nabla - B.\nabla A \tag{7.98}$$

and $$\nabla A.B = B.A\nabla + A.B\nabla \tag{7.99}$$

Eliminating the two expressions on the right-hand side of equation (7.99) and rearranging gives the required formula; the notation now being normal.

$$\nabla A \cdot B = A \cdot \nabla B + B \cdot \nabla A + A \wedge (\nabla \wedge B) + B \wedge (\nabla \wedge A) \qquad (7.100)$$

If now $B = A$, equation (7.100) gives

$$\tfrac{1}{2}\nabla A^2 = A \cdot \nabla A + A \wedge (\nabla \wedge A) \qquad (7.101)$$

The next group of formulae is intended to simplify expressions involving second derivatives.

$$\nabla \wedge (\nabla \wedge A) = \nabla \nabla \cdot A - \nabla^2 A \qquad (7.102)$$

$$\nabla \wedge \nabla \cdot A = \nabla \cdot \nabla \wedge A = 0 \qquad (7.103)$$

$$\nabla \wedge \nabla U = 0 \qquad (7.104)$$

These last two equations are only valid when the order of differentiation is not important in the second mixed derivative.

7.9. Curvilinear Coordinate Systems

So far in this chapter, vector equations have been developed which are independent of the coordinate system used. Position vectors only require an origin to be defined as stated in Section 7.3.2 and thus the vector equations are very general. On occasions, a vector relationship has been illustrated by resolving the vectors into a cartesian coordinate system by using the unit vectors i, j, and k. This particular set consists of unit vectors which remain constant in both magnitude and direction, but in many problems the cartesian set of coordinates is not the most convenient. Both cylindrical polar coordinates and spherical polar coordinates are frequently used in chemical engineering problems and it is therefore necessary to be able to express a vector relationship in terms of other coordinate systems. Books devoted to vector analysis† invariably derive the necessary equations for a general coordinate system and the reader is referred to these, but all of the necessary equations will be developed here for spherical polar coordinates, and quoted for cylindrical polar coordinates. One important point which must be accepted here is that in any coordinate system other than cartesian, the unit vectors are variable in direction. For example, when a radial coordinate is used, the value of the variable increases in opposite directions at the opposite ends of a diameter, and comparing the two positions the unit radial vector is also reversed. Hence extra terms arise as a result of differentiating the unit vectors.

7.9.1. *Spherical Polar Coordinates*

These can be defined relative to the cartesian coordinate system as illustrated in Fig. 7.15. The radial distance of the point P from the origin O is denoted by r, the polar angle of OP relative to the z axis is denoted by θ, and the

† Weatherburn, C. E. "Advanced Vector Analysis." Bell & Sons, London (1951).

azimuthal angle of the plane containing OP and the z axis relative to the xz plane is denoted by ϕ. Care must be taken to ensure than (r, θ, ϕ) is a right-handed set of coordinates as further illustrated in Fig. 7.16. The volume element is defined by taking increments δr, $\delta \theta$, and $\delta \phi$ in the coordinates as

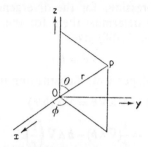

FIG. 7.15. Spherical polar coordinates

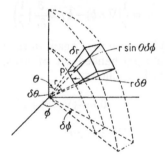

FIG. 7.16. Volume element in spherical polar coordinates

before, but now, the edges of the element are in general curved and their lengths are not equal to the simple increments. If a set of unit vectors a, b, c are defined at the point P, then the incremental position vector δr describing a general displacement from P can be written

$$\delta r = \delta r a + r \delta \theta b + r \sin \theta \delta \phi c \qquad (7.105)$$

It is now possible to expand Hamilton's operator (∇) in terms of the new coordinate system. Equation (7.58) gives the general result that

$$dr . \nabla \equiv d \qquad (7.106)$$

Equation (7.105) gives an expression for dr in the limit as $\delta r \to 0$, and the expression for (∇) which satisfies the identity (7.106) is

$$\nabla = a\frac{\partial}{\partial r} + \frac{b}{r}\frac{\partial}{\partial \theta} + \frac{c}{r \sin \theta}\frac{\partial}{\partial \phi} \qquad (7.107)$$

The various differential operations, grad, div, and curl, can now be expressed in terms of spherical polar coordinates by using this identity. The gradient of a scalar point function U is simply obtained by placing U after each term in equation (7.107). Thus

$$\nabla U = a\frac{\partial U}{\partial r} + \frac{b}{r}\frac{\partial U}{\partial \theta} + \frac{c}{r\sin\theta}\frac{\partial U}{\partial \phi} \qquad (7.108)$$

Before developing expressions for the divergence and curl of a general vector, it is necessary to determine them for the unit vectors a, b, and c. Putting $U = \theta$ in equation (7.108) gives

$$\nabla \theta = b/r \qquad (7.109)$$

because r, θ and ϕ are independent. Substituting into equation (7.104) gives

$$0 = \nabla \wedge \nabla \theta = \nabla \wedge (b/r)$$

$$= \frac{1}{r}(\nabla \wedge b) - b \wedge \nabla\left(\frac{1}{r}\right)$$

$$= \frac{1}{r}(\nabla \wedge b) - b \wedge a\frac{\partial}{\partial r}\left(\frac{1}{r}\right) \qquad (7.110)$$

where equation (7.108) has been used again in the last line. Provided (r, θ, ϕ) and hence (a, b, c) is a right-handed set of axes, equation (7.110) can be simplified and rearranged as follows.

$$0 = \frac{1}{r}(\nabla \wedge b) + c\left(-\frac{1}{r^2}\right)$$

$$\therefore \quad \nabla \wedge b = c/r \qquad (7.111)$$

Similarly, it can be shown that

$$\nabla \wedge a = 0 \qquad (7.112)$$

$$\nabla \wedge c = (a\cot\theta - b)/r \qquad (7.113)$$

To find the divergence of the unit vectors, the above expressions are used in conjunction with equation (7.95). Thus

$$\nabla . a = \nabla . (b \wedge c)$$

$$= c . \nabla \wedge b - b . \nabla \wedge c$$

$$= \frac{1}{r} + \frac{1}{r}$$

$$\therefore \quad \nabla . a = 2/r \qquad (7.114)$$

Similarly, $\nabla . b = (\cot\theta)/r \qquad (7.115)$

$$\nabla . c = 0` \qquad (7.116)$$

Assuming that the vector A can be resolved into components in terms of a, b, and c as

$$A = A_r a + A_\theta b + A_\phi c \qquad (7.117)$$

and using the above seven equations, therefore

$$\nabla . A = \frac{2A_r}{r} + \frac{A_\theta \cot \theta}{r} + \frac{\partial A_r}{\partial r} + \frac{1}{r}\frac{\partial A_\theta}{\partial \theta} + \frac{1}{r \sin \theta}\frac{\partial A_\phi}{\partial \phi}$$

$$\therefore \quad \nabla . A = \frac{1}{r^2}\frac{\partial}{\partial r}(r^2 A_r) + \frac{1}{r \sin \theta}\frac{\partial}{\partial \theta}(A_\theta \sin \theta) + \frac{1}{r \sin \theta}\frac{\partial A_\phi}{\partial \phi} \qquad (7.118)$$

and

$$\nabla \wedge A = \frac{c}{r}A_\theta + \frac{a \cot \theta - b}{r}A_\phi - \frac{c}{r}\frac{\partial A_r}{\partial \theta} + \frac{b}{r \sin \theta}\frac{\partial A_r}{\partial \phi} + c\frac{\partial A_\theta}{\partial r} - \frac{a}{r \sin \theta}\frac{\partial A_\theta}{\partial \phi}$$

$$- b\frac{\partial A_\phi}{\partial r} + \frac{a}{r}\frac{\partial A_\phi}{\partial \theta}$$

$$\therefore \quad \nabla \wedge A = \frac{a}{r \sin \theta}\left[\frac{\partial}{\partial \theta}(A_\phi \sin \theta) - \frac{\partial A_\theta}{\partial \phi}\right]$$

$$+ \frac{b}{r \sin \theta}\left[\frac{\partial A_r}{\partial \phi} - \sin \theta\frac{\partial}{\partial r}(rA_\phi)\right] + \frac{c}{r}\left[\frac{\partial}{\partial r}(rA_\theta) - \frac{\partial A_r}{\partial \theta}\right]$$

$$(7.119)$$

By applying equation (7.118) to the expansion of ∇U as given by equation (7.108), the following expression for $\nabla^2 U$ can be obtained.

$$\nabla^2 U = \frac{1}{r^2}\frac{\partial}{\partial r}\left(r^2\frac{\partial U}{\partial r}\right) + \frac{1}{r^2 \sin \theta}\frac{\partial}{\partial \theta}\left(\sin \theta\frac{\partial U}{\partial \theta}\right) + \frac{1}{r^2 \sin^2 \theta}\frac{\partial^2 U}{\partial \phi^2} \qquad (7.120)$$

The expansion of $\nabla^2 A$ is more complicated than equation (7.120) and is not usually given as a standard formula. It can be obtained however by expanding equation (7.102) by two applications of equation (7.119) and one application of each of equations (7.118) and (7.108).

7.9.2. *Cylindrical Polar Coordinates*

Following the same procedure as above, the following formulae can be derived for cylindrical polar coordinates (ω, ϕ, z) which are defined as illustrated in Fig. 7.17. z is one of the cartesian coordinates, and ϕ is the same azimuthal angle as for spherical polar coordinates.

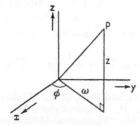

FIG. 7.17. Cylindrical polar coordinates

$$\mathbf{dr} = \delta\omega\mathbf{a} + \omega\delta\phi\mathbf{b} + \delta z\mathbf{c} \tag{7.121}$$

$$\nabla U = \mathbf{a}\frac{\partial U}{\partial \omega} + \frac{\mathbf{b}}{\omega}\frac{\partial U}{\partial \phi} + \mathbf{c}\frac{\partial U}{\partial z} \tag{7.122}$$

$$\nabla . A = \frac{1}{\omega}\frac{\partial}{\partial \omega}(\omega A_\omega) + \frac{1}{\omega}\frac{\partial A_\phi}{\partial \phi} + \frac{\partial A_z}{\partial z} \tag{7.123}$$

$$\nabla \wedge A = \mathbf{a}\left[\frac{1}{\omega}\frac{\partial A_z}{\partial \phi} - \frac{\partial A_\phi}{\partial z}\right] + \mathbf{b}\left[\frac{\partial A_\omega}{\partial z} - \frac{\partial A_z}{\partial \omega}\right] + \frac{\mathbf{c}}{\omega}\left[\frac{\partial(\omega A_\phi)}{\partial \omega} - \frac{\partial A_\omega}{\partial \phi}\right] \tag{7.124}$$

$$\nabla^2 U = \frac{1}{\omega}\frac{\partial}{\partial \omega}\left(\omega\frac{\partial U}{\partial \omega}\right) + \frac{1}{\omega^2}\frac{\partial^2 U}{\partial \phi^2} + \frac{\partial^2 U}{\partial z^2} \tag{7.125}$$

Again, $\nabla^2 A$ must be evaluated by using equations (7.102, 7.122, 7.123, 7.124).

7.10. THE EQUATIONS OF FLUID FLOW

One of the most important applications of vector analysis in chemical engineering is to the study of fluid flow. As stated previously, velocity is a vector quantity, and in a moving fluid the velocity is a vector point function. Ignoring the heating effect due to the friction in a moving fluid and assuming that no mass transfer is taking place, there are nine scalar variables related by the fluid flow equations. There are three coordinates and time which are the four independent variables; and in addition to the three velocity components there are pressure and density making five dependent variables. It will be assumed that the fluid is Newtonian so that the coefficient of viscosity remains constant. Five equations are needed because there are five dependent variables, and one of these, the continuity equation (7.62), has already been mentioned. A second equation is the equation of state relating the density to the pressure which leaves three further equations to be supplied. Instead of treating the three velocity components separately, the use of a single vector velocity variable will reduce the requirement to one vector equation and this is how vector analysis effects a simplification. This equation is obtained by applying Newton's second law of motion to a fluid element. That is, the external forces, the pressure forces, and the viscous forces are combined and equated to the product of the mass of the element and its acceleration. The details of the derivation are given in many books devoted to fluid flow[†] and in any textbook devoted to theoretical hydrodynamics.

For an incompressible fluid, the equation of motion can be written

$$\frac{\partial u}{\partial t} + u . \nabla u = -\frac{1}{\rho}\nabla p + \nu\nabla^2 u + F \tag{7.126}$$

---

† For example, Knudsen, J. G. and Katz, D. L. "Fluid Dynamics and Heat Transfer." McGraw-Hill (1958). Lamb, H. "Hydrodynamics." Cambridge University Press (1895). Milne-Thompson, L. M. "Theoretical Hydrodynamics" (2nd ed.). Macmillan (1949).

which is known as the "Navier–Stokes equation". The advantage of the above vector form is that the five terms can be easily identified as: time dependent, inertial, pressure, viscous, and external force terms respectively. The continuity equation for such an incompressible fluid is the following simplified version of equation (7.62).

$$\nabla \cdot \boldsymbol{u} = 0 \qquad (7.127)$$

F representing the external forces is usually due to gravity and gives rise to the hydrostatic pressure in the system, and if this is the case, F can be expressed as the gradient of a scalar point function and combined with the pressure term. It will therefore be ignored in the remainder of this work.

There is no general method of solution for the system of equations (7.126) and (7.127) and most attempts have centred on either of two simplifying assumptions. These theories will now be considered separately.

7.10.1. *Stokes' Approximation*

If it is assumed that the fluid velocity is everywhere small, the inertia term in equation (7.126) can be omitted. Without this term, the system of equations is linear and can therefore be solved. To find out under what conditions the approximation is justified, it is convenient to express the equations in dimensionless form by taking the ratio of all lengths to some characteristic length (L) and the ratio of all velocities to some characteristic velocity (U). Denoting the dimensionless variables by ($'$), equation (7.126) becomes

$$\frac{L}{U}\frac{\partial \boldsymbol{u}'}{\partial t} + \boldsymbol{u}' \cdot \nabla \boldsymbol{u}' = -\frac{1}{\rho U^2}\nabla p + \frac{\nu}{LU}\nabla^2 \boldsymbol{u}' \qquad (7.128)$$

In this dimensionless equation, three dimensionless groups have arisen; Ut/L, which is the dimensionless time, $2p/\rho U^2$, the dimensionless pressure coefficient, and LU/ν, the Reynolds number. It is conventional to introduce a factor of 2 into the pressure coefficient and sometimes into the Reynolds number, although in the latter case a convenient characteristic length is often defined to include this factor.

The approximation of neglecting the inertia term in comparison with the viscous term can now be interpreted as assuming that the Reynolds number is small compared with unity. Returning to the dimensional equation (7.126) and making the approximation gives

$$\frac{\partial \boldsymbol{u}}{\partial t} = -\frac{1}{\rho}\nabla p + \nu \nabla^2 \boldsymbol{u} \qquad (7.129)$$

Without vectors, the system of equations (7.127) and (7.129) would contain four dependent variables, and a lengthy piece of algebra would be required to find a single differential equation involving only one dependent variable.

However, a simple vector method is available. Operating on equation (7.129) with $(\nabla \cdot)$ gives

$$\frac{\partial}{\partial t}(\nabla \cdot u) = -\frac{1}{\rho}\nabla^2 p + \nu\nabla^2 \nabla \cdot u \tag{7.130}$$

But the continuity equation (7.127) shows that $\nabla \cdot u = 0$.

$$\therefore \quad \nabla^2 p = 0 \tag{7.131}$$

and the pressure distribution can be found in general terms. Unfortunately, the boundary conditions usually govern the velocity and not the pressure, and thus equation (7.131) is not really useful. Hence, operating upon equation (7.129) with $(\nabla \wedge)$ gives

$$\frac{\partial}{\partial t}(\nabla \wedge u) = -\frac{1}{\rho}\nabla \wedge \nabla p + \nu\nabla^2 \nabla \wedge u \tag{7.132}$$

Now equation (7.67) defines the vorticity of the fluid by

$$\nabla \wedge u = \zeta \tag{7.67}$$

and equation (7.104) shows that the pressure term is identically zero. Hence

$$\frac{\partial \zeta}{\partial t} = \nu\nabla^2 \zeta \tag{7.133}$$

which is analogous to the heat conduction or mass transfer equation, except that equation (7.133) is a vector equation whereas both of the others are scalar equations.

Equation (7.133) will now be put into a simpler form for the special case of fluid flow with an axis of symmetry by introducing the stream function ψ so that the continuity equation (7.127) is automatically satisfied. Taking spherical polar coordinates and assuming that the problem has an axis of symmetry (the z axis) then from any point P, curves Q_1 and Q_2 can be drawn to a point on the z axis as illustrated in Fig. 7.18. By rotating the plane of the

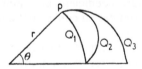

FIG. 7.18. Stokes stream function, axi-symmetrical flow

diagram about the z axis, the curves Q_1 and Q_2 become surfaces. Now the total volume of fluid crossing the surface Q_1 from right to left must equal the volume crossing Q_2 in the same direction. Also, because there is no flow across the axis of symmetry, there will be an equal volumetric flow across any other surface such as Q_3. This volumetric flow is denoted by $2\pi\psi$ where ψ is Stokes' stream function. This definition shows that $\psi = 0$ at all points on the z axis, and elsewhere it is a scalar point function.

Taking a small displacement from P in each of the coordinate directions, it can be shown that

$$u = \frac{-1}{r^2 \sin \theta} \frac{\partial \psi}{\partial \theta}, \qquad v = \frac{1}{r \sin \theta} \frac{\partial \psi}{\partial r} \qquad (7.134)$$

where u and v are respectively the r and θ components of the velocity $\boldsymbol{u}$. Putting equations (7.134) into equation (7.118) shows that the continuity equation is satisfied. The vorticity can be expressed in terms of the stream function by substituting equations (7.134) into equation (7.119). Thus

$$\boldsymbol{\zeta} = \nabla \wedge \boldsymbol{u} = \frac{c}{r}\left[\frac{\partial}{\partial r}\left(\frac{1}{\sin \theta}\frac{\partial \psi}{\partial r}\right) - \frac{\partial}{\partial \theta}\left(\frac{-1}{r^2 \sin \theta}\frac{\partial \psi}{\partial \theta}\right)\right]$$

$$= \frac{c}{r \sin \theta}\left[\frac{\partial^2 \psi}{\partial r^2} + \frac{\sin \theta}{r^2}\frac{\partial}{\partial \theta}\left(\frac{1}{\sin \theta}\frac{\partial \psi}{\partial \theta}\right)\right] \qquad (7.135)$$

$$\therefore \quad \zeta r \sin \theta = c E^2 \psi \qquad (7.136)$$

where E^2 is the differential operator in the brackets of equation (7.135). Equation (7.136) shows that the vorticity has a component in the ϕ direction only and the scalar symbol ζ can be used to represent this component.

Applying equation (7.119) to the vorticity gives

$$\nabla \wedge \boldsymbol{\zeta} = \frac{a}{r \sin \theta}\frac{\partial}{\partial \theta}(\zeta \sin \theta) - \frac{b}{r}\frac{\partial}{\partial r}(\zeta r) \qquad (7.137)$$

Using equation (7.119) again gives

$$\nabla \wedge (\nabla \wedge \boldsymbol{\zeta}) = \frac{c}{r}\left\{ -\frac{\partial^2}{\partial r^2}(\zeta r) - \frac{\partial}{\partial \theta}\left[\frac{1}{r \sin \theta}\frac{\partial}{\partial \theta}(\zeta \sin \theta)\right]\right\} \qquad (7.138)$$

Combining equations (7.127, 7.102, 7.132, 7.138) and assuming that the flow is steady (i.e. $\partial \zeta / \partial t = 0$) gives

$$\frac{\partial^2}{\partial r^2}(\zeta r) + \frac{1}{r}\frac{\partial}{\partial \theta}\left[\frac{1}{\sin \theta}\frac{\partial}{\partial \theta}(\zeta \sin \theta)\right] = 0$$

which can be rearranged to give

$$E^2(\zeta r \sin \theta) = 0$$

or
$$E^4 \psi = 0 \qquad (7.139)$$

which is the fourth order linear scalar partial differential equation describing the steady axi-symmetrical flow of an incompressible fluid at low Reynolds numbers.

Example 1. *Viscous Flow Round a Sphere.* For low Reynolds numbers (< 1) it has just been shown that the fluid flow round a sphere with its centre at the origin is given by a solution of the equation

$$E^4 \psi = 0$$

I

where the operator E^2 is defined by equation (7.135) in spherical polar coordinates as

$$E^2\psi = \frac{\partial^2\psi}{\partial r^2} + \frac{\sin\theta}{r^2}\frac{\partial}{\partial\theta}\left(\frac{1}{\sin\theta}\frac{\partial\psi}{\partial\theta}\right)$$ II

If the radius of the sphere is a, and the fluid flows with velocity U_0 along the axis of symmetry $\theta = 0$ towards the origin as shown in Fig. 7.19, the boundary conditions for large values of r are

$$\text{as } r\to\infty, \quad u\to - U_0\cos\theta, \quad v\to U_0\sin\theta$$

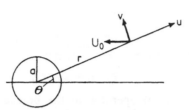

FIG. 7.19. Flow round a sphere

In terms of the stream function, these become the single condition

as $$r\to\infty, \quad \psi\to\tfrac{1}{2}U_0 r^2\sin^2\theta$$ III

The boundary conditions at the surface of the sphere of radius a which impose conditions of zero slip are

at $$r = a, \quad u = 0, \quad v = 0$$

or at $$r = a, \quad \frac{\partial\psi}{\partial r} = 0, \quad \frac{\partial\psi}{\partial\theta} = 0$$ IV

Equation III suggests trying the particular solution

$$\psi = f(r)\sin^2\theta$$ V

Substituting into equation II gives

$$E^2\psi = \frac{d^2 f}{dr^2}\sin^2\theta + \frac{\sin\theta}{r^2}\frac{\partial}{\partial\theta}\left(\frac{f}{\sin\theta}\frac{\partial\sin^2\theta}{\partial\theta}\right)$$

$$= \frac{d^2 f}{dr^2}\sin^2\theta - \frac{2f}{r^2}\sin^2\theta$$

$$\therefore \quad E^2\psi = \left(\frac{d^2 f}{dr^2} - \frac{2f}{r^2}\right)\sin^2\theta$$ VI

Denoting $(d^2 f/dr^2 - 2f/r^2)$ by $F(r)$, and applying the operator E^2 to equation VI, the steps between equations V and VI will be repeated identically with f

replaced by F. Therefore

$$E^4 \psi = \left(\frac{d^2 F}{dr^2} - \frac{2F}{r^2} \right) \sin^2 \theta \qquad \text{VII}$$

Substituting for F in terms of f and equating to zero as given by equation I,

$$\frac{d^2}{dr^2} \left(\frac{d^2 f}{dr^2} - \frac{2f}{r^2} \right) - \frac{2}{r^2} \left(\frac{d^2 f}{dr^2} - \frac{2f}{r^2} \right) = 0$$

$$\therefore \quad \frac{d^4 f}{dr^4} - \frac{4}{r^2} \frac{d^2 f}{dr^2} + \frac{8}{r^3} \frac{df}{dr} - \frac{8}{r^4} f = 0 \qquad \text{VIII}$$

Equation VIII is linear and homogeneous and has the solution

$$f = Ar^4 + Br^2 + Cr + D/r \qquad \text{IX}$$

Substituting into equation V, therefore

$$\psi = (Ar^4 + Br^2 + Cr + D/r) \sin^2 \theta \qquad \text{X}$$

is a solution of equation I. To satisfy boundary equation III

$$A = 0$$

and

$$B = \tfrac{1}{2} U_0$$

$$\therefore \quad \psi = (\tfrac{1}{2} U_0 r^2 + Cr + D/r) \sin^2 \theta \qquad \text{XI}$$

Using boundary conditions IV gives

$$\tfrac{1}{2} U_0 a^2 + Ca + D/a = 0 \qquad \text{XII}$$

and

$$U_0 a + C - D/a^2 = 0 \qquad \text{XIII}$$

Solving equations XII and XIII for C and D and substituting into equation XI gives the final solution

$$\psi = \tfrac{1}{2} U_0 r^2 \left(1 - \frac{3}{2} \frac{a}{r} + \frac{1}{2} \frac{a^3}{r^3} \right) \sin^2 \theta \qquad \text{XIV}$$

the vorticity distribution can be obtained by substituting equation XIV into equation (7.136). Thus

$$\zeta = \frac{3 U_0 a}{2 r^2} \sin \theta \qquad \text{XV}$$

All other properties of the flow can be obtained by using equation XIV in the appropriate equations to determine the velocity components, the pressure distribution, and eventually the total drag force:

$$F = 6 \pi \mu a U_0 \qquad \text{XVI}$$

7.10.2. *The Ideal Fluid Approximation*

It is assumed in this case that the viscosity of the fluid is small, i.e. that the Reynolds number is large. Thus, neglecting the viscous term in equation (7.126) gives

$$\frac{\partial u}{\partial t} + u \cdot \nabla u = -\frac{1}{\rho} \nabla p \qquad (7.140)$$

Using equation (7.101) to alter the form of the middle term and rearranging gives

$$\nabla \left(\tfrac{1}{2} u^2 + \frac{p}{\rho} \right) = u \wedge (\nabla \wedge u) - \frac{\partial u}{\partial t} \qquad (7.141)$$

If the flow is steady and the vorticity is zero, then the right-hand side of equation (7.141) vanishes and the equation reduces to Bernoulli's theorem.

$$p + \tfrac{1}{2} \rho u^2 = \text{const.} \qquad (7.142)$$

From the above analysis it can be seen that Bernoulli's theorem is only strictly true when the laminar flow is steady, incompressible, inviscid, and irrotational.

Applying the operator $(\nabla \wedge)$ to equation (7.141) and using the property (7.104) to simplify the left-hand side gives

$$0 = \nabla \wedge (u \wedge \zeta) - \frac{\partial \zeta}{\partial t}$$

and expanding the vector triple product and remembering that

$$\nabla \cdot \zeta = \nabla \cdot \nabla \wedge u = 0$$

gives
$$\frac{\partial \zeta}{\partial t} + u \cdot \nabla \zeta = 0 \qquad (7.143)$$

Equation (7.143) shows that the substantive derivative (see Section 8.2.3) of the vorticity is zero. This means that the vorticity of any fluid element remains constant, and if the fluid motion has been started from rest, then the vorticity will be zero everywhere for all time. Thus, for a fluid motion started from rest, the vorticity is zero and the flow is irrotational.

$$\therefore \quad \nabla \wedge u = 0 \qquad (7.144)$$

However, it was shown in Section 7.7.1 that if the curl of a vector is zero, then the vector itself can be expressed as the gradient of a scalar point function called a potential function. Thus, for an inviscid irrotational fluid, the velocity can be expressed as the gradient of the "velocity potential" ϕ. Thus

$$u = -\nabla \phi \qquad (7.145)$$

Combining equation (7.145) with the continuity equation (7.127) shows that the velocity potential satisfies Laplace's equation

$$\nabla^2 \phi = 0 \qquad (7.146)$$

It is possible to express ideal fluid flow in terms of the stream function ψ, and in the axi-symmetrical case equation (7.136) becomes

$$E^2 \psi = 0 \qquad (7.147)$$

because the flow is irrotational. Although the stream function representation is valid for both Stokes' flow and ideal fluid flow, the velocity potential can only be used in ideal fluid flow. In viscous flow, equation (7.144) is not valid, the system is not conservative, and energy is dissipated by the action of viscosity. Therefore, for comparison of the two solutions, both must be expressed in terms of the stream function.

Example 2. *Ideal Fluid Flow Round a Sphere.* Using the same symbols and Fig. 7.19 from the last example, the ideal fluid flow will be given by a solution of the equation

$$\nabla^2 \phi = 0 \qquad \text{I}$$

where the velocity u is given by

$$u = -\nabla\phi \qquad \text{II}$$

as derived above.

It is convenient to use spherical polar coordinates for the present problem and there is no dependence on the azimuthal angle and hence no ambiguity over the meaning of the symbol ϕ. Using equation (7.120), equation I can be written

$$\frac{1}{r^2}\frac{\partial}{\partial r}\left(r^2\frac{\partial\phi}{\partial r}\right) + \frac{1}{r^2\sin\theta}\frac{\partial}{\partial\theta}\left(\sin\theta\frac{\partial\phi}{\partial\theta}\right) = 0 \qquad \text{III}$$

and the boundary conditions can be written

at
$$r = a, \quad u = 0 \qquad \text{IV}$$

as
$$r \to \infty, \quad u \to -U_0\cos\theta, \quad v \to U_0\sin\theta \qquad \text{V}$$

where u and v are the r and θ components of the velocity u respectively.

Equation IV prohibits flow through the surface of the sphere, and equation V states that the flow at a large distance from the obstacle, in any direction, is undisturbed parallel flow. In order to solve equation III for these boundary conditions, equations V must be expressed in terms of ϕ by substituting into equation II. Thus

$$-\frac{\partial\phi}{\partial r} = -U_0\cos\theta$$

$$-\frac{1}{r}\frac{\partial\phi}{\partial\theta} = U_0\sin\theta$$

$$\therefore \quad \phi = U_0 r\cos\theta$$

$$\therefore \quad \text{as } r\to\infty, \quad \phi \to U_0 r\cos\theta \qquad \text{VI}$$

Boundary condition VI suggests trying a particular solution of the form

$$\phi = f(r)\cos\theta \qquad \text{VII}$$

in the partial differential equation III. Thus

$$\cos\theta \frac{\partial}{\partial r}\left(r^2\frac{\partial f}{\partial r}\right) + \frac{f}{\sin\theta}\frac{\partial}{\partial\theta}\left(\sin\theta\frac{\partial\cos\theta}{\partial\theta}\right) = 0 \qquad \text{VIII}$$

Performing the differentiations with respect to θ and replacing the partial derivatives with respect to r with ordinary derivatives,

$$\cos\theta \frac{d}{dr}\left(r^2\frac{df}{dr}\right) - \frac{f}{\sin\theta}(2\sin\theta\cos\theta) = 0$$

$$\therefore \quad r^2\frac{d^2f}{dr^2} + 2r\frac{df}{dr} - 2f = 0 \qquad \text{IX}$$

Equation IX is independent of θ, and since it is both linear and homogeneous it can be solved by the methods of Sections 2.4 and 2.5. The result is

$$f = Ar + B/r^2 \qquad \text{X}$$

Substituting into equation VII gives the solution

$$\phi = (Ar + B/r^2)\cos\theta$$

To satisfy condition VI, $A = U_0$,

$$\therefore \quad \phi = (U_0 r + B/r^2)\cos\theta \qquad \text{XI}$$

But $u = -\partial\phi/\partial r$ from equation II, therefore boundary condition IV becomes

at $\qquad r = a, \quad u = -(U_0 - 2B/r^3)\cos\theta = 0$

$$\therefore \quad B = \tfrac{1}{2}U_0 a^3 \qquad \text{XII}$$

Combining equations XI and XII gives

$$\phi = U_0\left(r + \frac{a^3}{2r^2}\right)\cos\theta \qquad \text{XIII}$$

which satisfies the boundary conditions IV and VI and the differential equation III and is thus the solution. The stream function form of equation XIII is

$$\psi = \tfrac{1}{2}U_0 r^2\left(1 - \frac{a^3}{r^3}\right)\sin^2\theta \qquad \text{XIV}$$

which can be checked by differentiating equations XIII and XIV and comparing the two expressions for each velocity component.

It is worth noting that the θ component of velocity does not vanish at the surface of the sphere. That is,

$$v = -\frac{1}{r}\frac{\partial \phi}{\partial \theta} = \frac{3}{2}U_0 \sin \theta \qquad \text{XV}$$

which means that there is slipping at the surface of the sphere. This is not physically reasonable and is caused by neglecting the viscous term in the equation of motion.

Comparing equation XIV in Example 1 with equation XIV in Example 2, it can be seen that the two solutions are quite different, and yet they are frequently confused because both are symmetrical about the equatorial plane of the sphere. Viscous fluid flow is described by a fourth order differential equation (7.139) and four boundary conditions can be satisfied, two of which are at the surface of the sphere, and thus both velocity components can be equated to zero at the surface. It is also true that at very low Reynolds numbers the inertial effects really are negligible and the agreement between the solution given in Example 1 and experimental results is good.

However, ideal fluid flow is described by a second order differential equation (7.146) and only two boundary conditions can be satisfied, one at large distances and one at the surface, thus only one of the velocity components at the surface of the sphere can be specified as zero. This led to a slip velocity along the surface given by equation XV in Example 2. Clearly there should not be any velocity discontinuity at the surface, and any small viscosity of the fluid would cause this velocity difference to be accommodated over a small but finite distance.

7.10.3. *Boundary Layer Theory*

Because of the inadequacy of ideal fluid flow to describe fluid motion at high Reynolds numbers, Prandtl developed the boundary layer theory by assuming that ideal fluid flow existed everywhere except in a thin layer of fluid near any solid boundary. Within this thin layer viscous effects are not negligible, velocity gradients normal to the boundary are quite large, but velocity gradients parallel to the boundary are relatively small.

The above properties of boundary layers can be derived from the Navier–Stokes equation (7.136) by making two fundamental assumptions:

(1) The thickness of the boundary layer at any point on a surface is small compared with the length of the surface to that point measured along the surface in the direction of flow.

(2) Viscous effects are confined to the boundary layer and ideal fluid flow exists outside it.

Provided that the boundary layer is thin compared with the local radius of curvature of the surface it is permissible to assume that the surface is flat so that cartesian coordinates can be used. Taking the case of two dimensional flow with the x axis along the surface in the direction of flow and the y axis normal to the surface, the steady-state Navier–Stokes equation (7.126) becomes the two equations

$$u\frac{\partial u}{\partial x}+v\frac{\partial u}{\partial y}=-\frac{1}{\rho}\frac{\partial p}{\partial x}+\nu\left(\frac{\partial^2 u}{\partial x^2}+\frac{\partial^2 u}{\partial y^2}\right) \qquad (7.148)$$

and

$$u\frac{\partial v}{\partial x}+v\frac{\partial v}{\partial y}=-\frac{1}{\rho}\frac{\partial p}{\partial y}+\nu\left(\frac{\partial^2 v}{\partial x^2}+\frac{\partial^2 v}{\partial y^2}\right) \qquad (7.149)$$

whilst the continuity equation (7.127) becomes

$$\frac{\partial u}{\partial x}+\frac{\partial v}{\partial y}=0 \qquad (7.150)$$

These three equations can be simplified within the boundary layer by choosing units for velocity and distance so that x takes values of order unity (i.e. $O(1)$) and the main stream velocity parallel to the surface is also $O(1)$. Denoting the boundary layer thickness by δ, which is small compared with unity, it is clear that within the boundary layer, y takes values between 0 and δ. Thus

$$x = O(1)$$

$$u = O(1)$$

and

$$y = O(\delta).$$

For the two terms in equation (7.150) to be the same order of magnitude,

$$v = O(\delta).$$

From these orders of magnitude it is clear that $\partial^2 u/\partial x^2$ can be neglected in equation (7.148) with respect to $\partial^2 u/\partial y^2$, but that all other terms must be of the same order of magnitude as each other. Hence equation (7.148) becomes

$$u\frac{\partial u}{\partial x}+v\frac{\partial u}{\partial y}=-\frac{1}{\rho}\frac{\partial p}{\partial x}+\nu\frac{\partial^2 u}{\partial y^2} \qquad (7.151)$$

where each term is of order unity. Therefore

$$v = O(\delta^2) \qquad (7.152)$$

In equation (7.149), $\partial^2 v/\partial x^2 \leqslant \partial^2 v/\partial y^2$ and is neglected. The remaining terms are all of order δ, thus the term

$$\frac{1}{\rho}\frac{\partial p}{\partial y} = O(\delta).$$

Hence pressure gradients across the layer are small compared with pressure gradients along the surface. Since the layer is also thin, the pressure change across the layer will be negligible.

Equation (7.152) makes more sense in the form

$$\delta = O(\nu^{\frac{1}{2}}) \qquad (7.153)$$

or taking a typical length of unity along the surface and the main stream velocity as unity, equation (7.153) may be written

$$\delta = O(\mathrm{Re}^{-\frac{1}{2}}) \qquad (7.154)$$

where Re is the Reynolds number.

The following properties of a boundary layer have thus been demonstrated.

(1) The normal component of velocity is small.

(2) The change in pressure normal to the surface across the layer is negligible.

(3) For a given geometry, the thickness of the boundary layer is inversely proportional to the square root of the Reynolds number.

(4) The Navier–Stokes equation simplifies to

$$u\frac{\partial u}{\partial x} + v\frac{\partial u}{\partial y} = -\frac{1}{\rho}\frac{\partial p}{\partial x} + v\frac{\partial^2 u}{\partial y^2} \qquad (7.151)$$

which must be solved in conjunction with the continuity equation

$$\frac{\partial u}{\partial x} + \frac{\partial v}{\partial y} = 0 \qquad (7.150)$$

The first three properties above are generally true for all boundary layers, the fourth property is specific to two-dimensional problems with surfaces of small curvature.

In a similar manner to that described in deriving equation (7.134), a two-dimensional stream function can be defined such that

$$u = -\frac{\partial \psi}{\partial y} \quad \text{and} \quad v = \frac{\partial \psi}{\partial x} \qquad (7.155)$$

so that the continuity equation (7.150) is automatically satisfied. Substituting into equation (7.151) gives

$$\frac{\partial \psi}{\partial y}\frac{\partial^2 \psi}{\partial x\,\partial y} - \frac{\partial \psi}{\partial x}\frac{\partial^2 \psi}{\partial y^2} = -\frac{1}{\rho}\frac{\partial p}{\partial x} - v\frac{\partial^3 \psi}{\partial y^3} \qquad (7.156)$$

which is a third order partial differential equation whereas Stokes' approximation led to a fourth order partial differential equation. The boundary layer thickness (δ) is also a parameter which provides sufficient flexibility so that four boundary conditions can still be satisfied. Two of these are the conditions of zero slip at the surface, and the other two are that the velocity component (u) parallel to the surface at the limit of the boundary layer equals the same velocity component in the ideal fluid flow region, and that the velocity gradient ($\partial u/\partial y$) at the edge of the layer is zero.

Although equation (7.156) is third order, it only contains first derivatives with respect to x and is thus an "initial value" problem, whereas Stokes' approximation leads to a "boundary value" problem. It will be seen later, in Chapter 11, that initial value problems are easier to solve numerically.

It can also be shown that equation (7.151) in terms of velocity components is of initial value type by following the method of Pohlhausen.† The procedure is to integrate the equation, on average only across the boundary layer, at each point on the surface. This can be achieved by integrating equation (7.151) with respect to y between the limits 0 and δ. Thus

$$\int_0^\delta \left(u\frac{\partial u}{\partial x}+v\frac{\partial u}{\partial y}\right)dy = -\frac{\delta}{\rho}\frac{\partial p}{\partial x}-v\left(\frac{\partial u}{\partial y}\right)_{y=0} \qquad (7.157)$$

Integrating the second term by parts gives

$$\int_0^\delta v\frac{\partial u}{\partial y}\,dy = [uv]_0^\delta - \int_0^\delta u\frac{\partial v}{\partial y}\,dy \qquad (7.158)$$

But the continuity equation (7.150) gives

$$\frac{\partial v}{\partial y} = -\frac{\partial u}{\partial x}$$

$$\therefore \quad (v)_{y=\delta} = -\int_0^\delta \frac{\partial u}{\partial x}\,dy$$

Hence equation (7.158) can be written

$$\int_0^\delta v\frac{\partial u}{\partial y}\,dy = -U\int_0^\delta \frac{\partial u}{\partial x}\,dy + \int_0^\delta u\frac{\partial u}{\partial x}\,dy \qquad (7.159)$$

where U is the value of u at $y = \delta$. Thus equation (7.157) becomes

$$\int_0^\delta \frac{\partial u^2}{\partial x}\,dy - U\int_0^\delta \frac{\partial u}{\partial x}\,dy = -\frac{\delta}{\rho}\frac{\partial p}{\partial x}-v\left(\frac{\partial u}{\partial y}\right)_{y=0}$$

or

$$\frac{\partial}{\partial x}\int_0^\delta \rho u^2\,dy - U\frac{\partial}{\partial x}\int_0^\delta \rho u\,dy = -\delta\frac{\partial p}{\partial x}-\tau_0 \qquad (7.160)$$

where τ_0 is the shear stress at the surface. Although δ is a function of x, the order of integration with respect to y and differentiation with respect to x can be reversed because both integrands have the value ρU^2 at the variable upper limit of integration and the extra terms cancel out. Equation (7.160) is known as the "momentum integral equation".

Because there is a negligible pressure drop across the boundary layer, $\partial p/\partial x$ can be either measured experimentally at the surface or calculated from the ideal fluid flow pattern outside the layer. The velocity (U) at the outer edge of the boundary layer can be calculated from Bernoulli's theorem, equation (7.142), once the pressure is known as a function of x. Pohlhausen's

† Pohlhausen, *Zeitschr. f. Angew. Math. u. Mech.* 1, 256 (1921).

method now requires an assumption to be made about the velocity distribution, and the most useful one is the quartic polynomial

$$\frac{u}{U} = 2\eta - 2\eta^3 + \eta^4 + \tfrac{1}{6}\Lambda\eta(1-\eta)^3 \tag{7.161}$$

where

$$\eta = y/\delta \tag{7.162}$$

and

$$\Lambda = -\frac{\delta^2}{\mu U}\frac{\partial p}{\partial x} \tag{7.163}$$

The polynomial (7.161) satisfies the boundary conditions already stated but in addition it satisfies the extra two conditions

$$\frac{\partial^2 u}{\partial y^2} = 0 \quad \text{at} \quad y = \delta \tag{7.164}$$

and

$$\frac{\partial p}{\partial x} = \mu\frac{\partial^2 u}{\partial y^2} \quad \text{at} \quad y = 0 \tag{7.165}$$

This latter condition is the degenerate form of the original boundary layer equation (7.151) at the surface.

Substitution of equation (7.161) into equation (7.160) gives a first order differential equation relating δ and functions of δ to x in terms of $\partial p/\partial x$. The integrals which arise are clearly the same functions of Λ and these have been tabulated by Howarth† so that in any given boundary layer problem, once the pressure distribution is known, the boundary layer thickness (δ) can be calculated numerically for progressively increasing values of x. The reader is referred to specialized texts for the further development of these methods.

Example 3. *Centrifugal Atomizer.* Drops are formed at a centrifugal atomizer by feeding a slurry down a vertical pipe onto the centre of a rotating disc. The slurry flows out radially towards the periphery of the disc as a thin film. It is discharged from the edge of the disc in the form of a sheet of slurry which subsequently breaks up into drops 25% larger than the thickness of this slurry sheet.

Estimate the size of the drops formed when 36 m³/h of slurry are fed onto a disc of 0·5 m diameter rotating at 2860 r.p.m. The kinematic viscosity of the slurry is 5×10^{-4} m²/s, and the feed pipe is 0·08 m diameter, terminating 0·01 m from the surface of the disc.

With no external forces, the steady-state Navier–Stokes equation (7.126) becomes

$$\boldsymbol{u}.\nabla\boldsymbol{u} = -\frac{1}{\rho}\nabla p + \nu\nabla^2\boldsymbol{u} \qquad\qquad \text{I}$$

† Howarth, *A. R. C., Rep. & Mem.* No. 1632, p. 14 (1935).

Because the film of liquid on the disc is thin and always in contact with the atmospheric pressure, all pressure gradients will be small and equation I can be further simplified to

$$\boldsymbol{u}.\nabla\boldsymbol{u} = \nu\nabla^2\boldsymbol{u} \qquad\qquad \text{II}$$

Taking cylindrical polar coordinates (r, ϕ, z) and denoting the velocity components by (u, v, w) respectively, the expansion of equation II can be made as described in Section 7.9.2. As noted earlier, the evaluation of $\nabla^2\boldsymbol{u}$ requires some care. Using equations (7.101, 7.102, 7.127), equation II can be written

$$\tfrac{1}{2}\nabla u^2 - \boldsymbol{u}\wedge(\nabla\wedge\boldsymbol{u}) = -\nu\nabla\wedge(\nabla\wedge\boldsymbol{u}) \qquad\qquad \text{III}$$

Assuming axial symmetry so that all derivatives with respect to ϕ are zero, and using equations (7.122, 7.124) the equation of motion becomes

$$u\frac{\partial u}{\partial r} + w\frac{\partial u}{\partial z} - \frac{v^2}{r} = \nu\left(\frac{\partial^2 u}{\partial r^2} + \frac{1}{r}\frac{\partial u}{\partial r} + \frac{\partial^2 u}{\partial z^2} - \frac{u}{r^2}\right) \qquad\qquad \text{IV}$$

$$u\frac{\partial v}{\partial r} + w\frac{\partial v}{\partial z} + \frac{uv}{r} = \nu\left(\frac{\partial^2 v}{\partial r^2} + \frac{1}{r}\frac{\partial v}{\partial r} + \frac{\partial^2 v}{\partial z^2} - \frac{v}{r^2}\right) \qquad\qquad \text{V}$$

$$u\frac{\partial w}{\partial r} + w\frac{\partial w}{\partial z} = \nu\left(\frac{\partial^2 w}{\partial r^2} + \frac{1}{r}\frac{\partial w}{\partial r} + \frac{\partial^2 w}{\partial z^2}\right) \qquad\qquad \text{VI}$$

and the continuity equation (7.127) becomes

$$\frac{\partial u}{\partial r} + \frac{u}{r} + \frac{\partial w}{\partial z} = 0 \qquad\qquad \text{VII}$$

Using the standard properties of boundary layers that the velocity normal to the surface (w) is small and that derivatives normal to the surface are much larger than derivatives parallel to the surface leads to simplification of equations IV and V to the form

$$u\frac{\partial u}{\partial r} + w\frac{\partial u}{\partial z} - \frac{v^2}{r} = \nu\frac{\partial^2 u}{\partial z^2} \qquad\qquad \text{VIII}$$

$$u\frac{\partial v}{\partial r} + w\frac{\partial v}{\partial z} + \frac{uv}{r} = \nu\frac{\partial^2 v}{\partial z^2} \qquad\qquad \text{IX}$$

Equation VI consists entirely of small terms, none of which are larger than the pressure gradient term which has already been ignored. The set of equations VII, VIII, and IX provide three equations for the three unknown velocity components.

These equations, VIII and IX, can be solved as before by converting each of them to a momentum integral equation. Thus

$$\int_0^\delta w \frac{\partial u}{\partial z} dz = [wu]_0^\delta - \int_0^\delta u \frac{\partial w}{\partial z} dz$$

$$= wu\Big|_\delta + \int_0^\delta \left(u \frac{\partial u}{\partial r} + \frac{u^2}{r} \right) dz$$

and

$$w\Big|_\delta = -\int_0^\delta \left(\frac{\partial u}{\partial r} + \frac{u}{r} \right) dz$$

Hence, equation VIII becomes

$$\int_0^\delta \left(u \frac{\partial u}{\partial r} + w \frac{\partial u}{\partial z} - \frac{v^2}{r} \right) dz = \int_0^\delta \left[\frac{\partial}{\partial r} (u^2) + \frac{u^2}{r} - \frac{v^2}{r} \right] dz - U \int_0^\delta \left[\frac{\partial u}{\partial r} + \frac{u}{r} \right] dz$$

$$= \nu \left[\frac{\partial u}{\partial z} \right]_0^\delta$$

or

$$\frac{\partial}{\partial r} \int_0^\delta u^2 \, dz + \frac{1}{r} \int_0^\delta (u^2 - v^2) \, dz - U \frac{\partial}{\partial r} \int_0^\delta u \, dz - \frac{U}{r} \int_0^\delta u \, dz = -\nu \frac{\partial u}{\partial z} \Big|_0 \qquad \text{X}$$

Equation X is the radial momentum integral equation which corresponds to equation (7.160). The extra terms in equation X are due to curvature as may be seen by taking a large value for the radial coordinate (r) when the extra terms would become negligible.

A similar procedure applied to equation IX gives the angular momentum integral equation

$$\frac{\partial}{\partial r} \int_0^\delta uv \, dz + \frac{2}{r} \int_0^\delta uv \, dz - v \frac{\partial}{\partial r} \int_0^\delta u \, dz - \frac{v}{r} \int_0^\delta u \, dz = -\nu \frac{\partial v}{\partial z} \Big|_0 \qquad \text{XI}$$

The next step is to assume reasonable velocity profiles for u and v within the boundary layer. Although the quartic polynomial (7.161) is more accurate, a simpler quadratic form will be chosen for this example. Thus

$$u = U(2\eta - \eta^2) \qquad \text{XII}$$

$$v = V(2\eta - \eta^2) + \omega r(1 - 2\eta + \eta^2) \qquad \text{XIII}$$

where

$$\eta = z/\delta \qquad \text{XIV}$$

The above profiles are the simplest polynomials which satisfy the boundary conditions of no slip at the surface of the disc ($\eta = 0$, $u = 0$, $v = \omega r$), and zero velocity gradients at the outer edge of the boundary layer ($\eta = 1$, $\partial u/\partial z = \partial v/\partial z = 0$). The various integrals required can now be evaluated.

$$\int_0^\delta u\,dz = U\delta \int_0^1 (2\eta - \eta^2)\,d\eta = \tfrac{2}{3}U\delta$$

Similarly,

$$\int_0^\delta u^2\,dz = \tfrac{8}{15}U^2\,\delta$$

$$\int_0^\delta uv\,dz = \tfrac{8}{15}UV\delta + \tfrac{2}{15}U\omega r\delta$$

$$\int_0^\delta v^2\,dz = \tfrac{8}{15}V^2\,\delta + \tfrac{4}{15}V\omega r\delta + \tfrac{1}{5}\omega^2 r^2\,\delta$$

These values can be substituted into equations X and XI to give, after differentiating all product terms,

$$3U\delta\frac{dU}{dr} - U^2\frac{d\delta}{dr} - \frac{U^2\delta}{r} - \frac{4V^2\delta}{r} - 2V\omega\delta - \frac{3\omega^2 r\delta}{2} = -\frac{15\nu U}{\delta} \qquad \text{XV}$$

and

$$4U\delta\frac{dV}{dr} + (\omega r - V)\left(U\frac{d\delta}{dr} + \delta\frac{dU}{dr}\right) + 3U\omega\delta + \frac{3UV\delta}{r} = \frac{15\nu}{\delta}(\omega r - V) \quad \text{XVI}$$

The mathematical model is not yet complete because nothing has been included to distinguish between the boundary layer thickness (δ) and the physical film thickness (h). Near the axis of the disc the film will be much thicker than the boundary layer, but the boundary layer thickness will grow with increasing radius until it occupies the entire film. Further out on the disc, the film will continue to thin and the boundary layer will be constrained to thin with it. Only the slurry within the boundary layer will be influenced by the rotation of the disc, and in the region of the axis there will be a pool of liquid flowing radially, without rotation, above the boundary layer. The critical radius (R) which separates these two régimes of flow will be where the total flow within the boundary layer first equals the feed rate. These two regions must be considered separately.

Inner Region ($r < R$)

At the base of the feed pipe, liquid is forced out radially onto the disc at a velocity (U_0) which can be calculated by dividing the volumetric flow rate (Q) by the area of the gap ($2\pi r_0 h_0$). Neglecting the gravitational acceleration due to the thinning film, the upper layer of fluid will continue to flow radially with constant velocity (U_0) until the boundary layer has grown to the full thickness of the film. Thus, in this region, U and V are known and either of the momentum integral equations can be used to determine δ as a function of r. However, the steepest velocity gradients are likely to be in the angular direction, so the angular momentum integral equation XVI will be solved.

Putting $U = U_0$ and $V = 0$ gives

$$rU_0 \frac{d\delta}{dr} + 3U_0\, \delta = \frac{15\nu r}{\delta}$$

which can be rearranged to the form

$$r\delta \frac{d\delta}{dr} + 3\delta^2 = \frac{15\nu}{U_0} r \qquad\qquad \text{XVII}$$

The form of equation XVII suggests the substitution

$$f = \delta^2 \qquad\qquad \text{XVIII}$$

and hence

$$\frac{df}{dr} = 2\delta \frac{d\delta}{dr}$$

$$\therefore \quad r\frac{df}{dr} + 6f = \frac{30\nu}{U_0} r \qquad\qquad \text{XIX}$$

Using the integrating factor r^5 gives

$$\frac{d}{dr}(r^6 f) = \frac{30\nu}{U_0} r^6$$

and hence

$$r^6 f = \frac{30\nu}{7U_0} r^7 + K$$

$$\therefore \quad f = \delta^2 = \left(\frac{30\nu}{7U_0}\right) r + Kr^{-6} \qquad\qquad \text{XX}$$

where K is an arbitrary constant. The factor r^{-6} will cause the last term to be infinite on the axis, therefore K will be zero. Hence

$$\delta = \sqrt{\left(\frac{30\nu r}{7U_0}\right)} \qquad\qquad \text{XXI}$$

The volumetric flow within the boundary layer (q) is given by

$$q = 2\pi r \int_0^\delta u\, dz$$

$$= \tfrac{4}{3}\pi r U \delta \qquad\qquad \text{XXII}$$

The limit of the inner region is thus given by combining equations XXI, XXII and the definition of U_0. Thus

$$Q = \tfrac{4}{3}\pi R U_0 \sqrt{\left(\frac{30\nu R}{7U_0}\right)} = 2\pi r_0 h_0 U_0$$

$$\therefore \quad Q^2 = \frac{16\pi^2 R^2 Q \times 30\nu R}{9 \times 2\pi r_0 h_0 \times 7}$$

or
$$R^3 = \frac{21 Q r_0 h_0}{80\pi\nu} \qquad\qquad \text{XXIII}$$

Outer Region $(r > R)$

Throughout this region, the thickness of the boundary layer (δ) is equal to the physical thickness of the film and equation XXII with $q = Q$ must be satisfied. That is,

$$Q = \tfrac{4}{3}\pi r U \delta \qquad\qquad \text{XXIV}$$

Since the boundary layer now extends to the free surface, both U and V, the velocity components at the free surface, will be determined by solutions of the momentum integral equations XV and XVI. Using equation XXIV to eliminate δ from both equations gives

$$\frac{9Q}{4\pi r}\frac{dU}{dr} - \frac{3QU^2}{4\pi}\left(\frac{-1}{r^2 U} - \frac{1}{rU^2}\frac{dU}{dr}\right) - \frac{3QU}{4\pi r^2} - \frac{3QV^2}{\pi r^2 U} - \frac{3QV\omega}{2\pi rU} - \frac{9Q\omega^2}{8\pi U}$$

$$= -\frac{20\pi\nu r U^2}{Q}$$

and
$$\frac{3Q}{\pi r}\frac{dV}{dr} + (\omega r - V)\left(\frac{-3Q}{4\pi r^2}\right) + \frac{9Q\omega}{4\pi r} + \frac{9QV}{4\pi r^2} = \frac{20\pi rUv}{Q}(\omega r - V)$$

which simplify to

$$\frac{dU}{dr} = \frac{V^2}{rU} + \frac{V\omega}{2U} + \frac{3\omega^2 r}{8U} - \frac{20\pi^2 \nu r^2 U^2}{3Q^2} \qquad\qquad \text{XXV}$$

and
$$\frac{dV}{dr} = \frac{20\pi^2 r^2 Uv}{3Q^2}(\omega r - V) - \frac{V}{r} - \frac{\omega}{2} \qquad\qquad \text{XXVI}$$

Equations XXV and XXVI are clearly too complicated to be integrated analytically and a numerical method must be used.

A digital computer program was written to use a standard Runge–Kutta subroutine (see Section 11.2.1) to integrate the pair of first order equations XXV and XXVI. In standard S.I. units the data used to obtain Table 7.1 was $h_0 = 0.01$ m, $r_0 = 0.04$ m, $Q = 0.01$ m³/s, $\nu = 0.0005$ m²/s, $\omega = 300$ rads/s. The step length used in the numerical integration was half of the tabular interval, i.e. $\delta r = 0.005$ m.

The thickness of the film leaving the edge of the disc is seen to be 0·763 mm and hence the predicted drop size formed under the given conditions is 0·955 mm diameter.

TABLE 7.1. *Flow on Disc Atomizer*

r	U	V	δ = h
0·08744	3·979	0	0·00686
0·09	5·465	0·447	0·00485
0·10	8·769	4·476	0·00272
0·11	10·619	10·759	0·00204
0·12	11·937	17·923	0·00167
0·13	12·893	24·927	0·00142
0·14	13·474	31·132	0·00126
0·15	13·732	36·388	0·00116
0·16	13·776	40·880	0·00108
0·17	13·703	44·881	0·00102
0·18	13·577	48·597	0·00098
0·19	13·428	52·148	0·00094
0·20	13·271	55·967	0·00090
0·21	13·112	58·975	0·00087
0·22	12·954	62·299	0·00084
0·23	12·800	65·581	0·00081
0·24	12·649	68·829	0·00079
0·25	12·503	72·048	0·000763

7.11. TRANSPORT OF HEAT, MASS, AND MOMENTUM

The partial differential equation which governs heat transfer will be derived from first principles in the next chapter, but here it will be derived by a vector method for comparison. The rate of flow of heat per unit area at any point is proportional to the temperature gradient at that point; the constant of proportionality being the thermal conductivity. Using the generalized gradient from Section 7.6 gives

$$\frac{\partial Q}{\partial t} = -k\nabla T \tag{7.166}$$

The definition of the divergence operator in Section 7.6.1 shows that $\nabla . Q$ is the net flow of heat from unit volume. However, the total heat content of unit volume is $\rho C_p T$ where C_p is the specific heat, and the conservation law states that heat can only leave at the expense of the remaining contents. Thus

$$-\frac{\partial}{\partial t}(\rho C_p T) = \frac{\partial}{\partial t}(\nabla . Q) \tag{7.167}$$

Since space derivatives and time derivatives are independent, they are commutable. Hence, operating upon equation (7.166) with $(\nabla .)$, assuming that the physical properties remain constant, and substituting into equation (7.167) gives

$$\frac{\partial T}{\partial t} = \frac{k}{C_p \rho} \nabla^2 T$$

or
$$\frac{\partial T}{\partial t} = \alpha \nabla^2 T \qquad (7.168)$$

where α is the "thermal diffusivity".

In the transport of mass by molecular diffusion the equation analogous to (7.166) is

$$\frac{\partial N}{\partial t} = - D \nabla c \qquad (7.169)$$

where N is the molar flux density, D is the diffusitivity, and c is the molar concentration. Following the same argument as for heat transfer, it can be shown that

$$\frac{\partial c}{\partial t} = D \nabla^2 c \qquad (7.170)$$

Equations (7.168) and (7.170) have been derived for systems without bulk motion, if a bulk flow is included in the derivation the time derivatives have to be replaced by the substantive time derivatives (see Section 8.2.3) and collecting the equations (7.126, 7.168, 7.170) together, the equations governing the transport processes are

$$\frac{\partial T}{\partial t} + u \cdot \nabla T = \alpha \nabla^2 T \qquad (7.171)$$

$$\frac{\partial c}{\partial t} + u \cdot \nabla c = D \nabla^2 c \qquad (7.172)$$

$$\frac{\partial u}{\partial t} + u \cdot \nabla u = \nu \nabla^2 u - \frac{1}{\rho} \nabla p + F \qquad (7.173)$$

The equations are seen to be very similar in vector form, yet equation (7.173) is quite different, having two extra terms, and being the only vector equation. The analogy between heat and mass transfer however is complete if α and D have the same value. These points are discussed more fully in the literature.†

† Klinkenberg, A. and Mooy, H. H. *Chem. Engng. Prog.*, **44**, 17 (1948). Garner, F. H. Jenson, V. G. and Keey, R. B. *Trans. Instn. Chem. Engrs.*, **37**, 191 (1959).

Chapter 8

PARTIAL DIFFERENTIATION AND PARTIAL DIFFERENTIAL EQUATIONS

8.1. INTRODUCTION

AFTER showing how conservation laws and rate equations could be used to derive algebraic equations in Chapter 1, the method was extended to systems described by an independent variable, either a distance or time. Thus ordinary differential equations were derived by considering the equilibrium of an infinitesimal element and using Taylor's theorem to introduce the derivative. In this chapter, problems requiring the specification of more than one independent variable will be considered. An example of such a problem is the change of the temperature distribution within a system as the steady-state condition is approached, when the temperature is a function both of position and time. The differentiation process can be performed relative to an incremental change in the space variable giving a temperature gradient, or with respect to an increment of the time variable giving a rate of temperature rise. These two first derivatives of the temperature have to be distinguished from each other, and the first part of this chapter clarifies the meaning of a derivative in a multi-variable system.

The formulation of partial differential equations follows the same rules as given for ordinary differential equations in Chapter 1 and in difficult problems, such as fluid flow, it is often advisable to formulate the problem in cartesian coordinates or find the vector form of the differential equation directly. The vector equation can then be expanded into a coordinate system which better suits the shape of the boundaries of the problem. For example, it is easier to specify a spherical surface by $r = a$ in spherical polar coordinates than by $x^2 + y^2 + z^2 = a^2$ in cartesian coordinates. Thus it can be seen that in problems involving more than one independent variable, the boundary conditions play an important part in determining the co-ordinate system for the problem. Special care must also be taken in formulating the boundary conditions.

One method of solution has already been used in Section 7.10 to solve problems of flow round spheres. This consisted of generalizing the functional form of the boundary conditions in the hope of finding a particular solution of the differential equation. The first method to be considered in this chapter is also to look for a particular solution of the equation, then to use the properties of linearity and symmetry to modify this particular solution to fit the problem. These two methods, though relatively simple, are not generally applicable and the more fundamental methods of separation of variables and operators, which form the major part of this chapter, are probably the most useful mathematical methods in chemical engineering.

8.2. Interpretation of Partial Derivatives

Figure 8.1 shows curves of the system

$$u = f(x, y) \tag{8.1}$$

plotted with u as a parameter. At each point in the xy plane, a value of u can be calculated from equation (8.1) and each curve is obtained by connecting points with equal values of u. Thus Fig. 8.1 can be interpreted

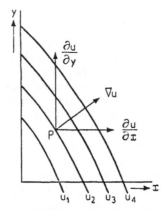

Fig. 8.1. First partial derivatives of $u = f(x, y)$

as a contour map where u represents height measured perpendicularly to the xy plane. If x is allowed to vary whilst y remains constant then in general u will also vary and the derivative of u with respect to x will be the rate of change of u relative to x, or the gradient in the chosen direction. This derivative is written

$$\left(\frac{\partial u}{\partial x}\right)_y \tag{8.2}$$

where y is placed outside the brackets to imply that y must be held constant while the derivative is being evaluated.

Similarly, the gradient along a line parallel to the y axis (x held constant) is written

$$\left(\frac{\partial u}{\partial y}\right)_x \tag{8.3}$$

In Chapter 7 it was shown that ∇u is a vector along the line of greatest slope and has a numerical value equal to that slope. This vector is also illustrated in Fig. 8.1. In the present case, z is not a relevant variable and equation (7.54) gives

$$\nabla u = i\,\frac{\partial u}{\partial x} + j\,\frac{\partial u}{\partial y} = \operatorname{grad} u \tag{8.4}$$

the gradient in the x direction can be reclaimed from equation (8.4) by taking the scalar product (Section 7.4.1) with the unit vector i in the x direction. Thus

$$i\,.\,\nabla u = \partial u/\partial x \tag{8.5}$$

Using the appropriate unit vector the gradient can be found in any direction.

If δr is an incremental displacement from P so that

$$\delta r = i\delta x + j\delta y \tag{8.6}$$

then u will change by $(\partial u/\partial x)\,\delta x$ due to the change in x, and by $(\partial u/\partial y)\,\delta y$ due to the change in y provided δx and δy are small.

$$\therefore \quad \delta u = \left(\frac{\partial u}{\partial x}\right)\delta x + \left(\frac{\partial u}{\partial y}\right)\delta y \tag{8.7}$$

and in the limit as $\delta x \to dx$ and $\delta y \to dy$, $\delta u \to du$ where

$$du = \left(\frac{\partial u}{\partial x}\right)dx + \left(\frac{\partial u}{\partial y}\right)dy \tag{8.8}$$

is called the "total differential" of u.

In general, if u is a function of many variables $x_1, x_2, x_3, \ldots, x_n$, then

$$du = \left(\frac{\partial u}{\partial x_1}\right)dx_1 + \left(\frac{\partial u}{\partial x_2}\right)dx_2 + \ldots + \left(\frac{\partial u}{\partial x_n}\right)dx_n \tag{8.9}$$

which is the fundamental equation of partial differentiation.

It will be noted that while a partial differential coefficient such as $(\partial u/\partial x_1)$ is being evaluated, all other independent variables, $x_2, x_3, \ldots, x_n$ must be held constant. These other variables should be placed outside the brackets enclosing the derivative but in most cases it is obvious which variables must be held constant and they are omitted for brevity.

8.2.1. *Relationships between Partial Derivatives*

Returning to functions of one independent variable

$$y = f(x) \tag{8.10}$$

then dy, the infinitesimal change in y, is related to dx by

$$dy = \left(\frac{df}{dx}\right)dx \tag{8.11}$$

Sometimes, y is not given as an explicit function of x such as equation (8.10) but as an implicit function thus

$$g(x, y) = 0 \tag{8.12}$$

Considering g as a third variable temporarily, equation (8.9) can be used to give

$$\mathrm{d}g = \left(\frac{\partial g}{\partial x}\right)\mathrm{d}x + \left(\frac{\partial g}{\partial y}\right)\mathrm{d}y = 0 \qquad (8.13)$$

where $\mathrm{d}g$ must equal zero to satisfy equation (8.12). By rearranging equation (8.13) an expression for $\mathrm{d}y$ can be obtained and compared with equation (8.11) to give

$$\frac{\mathrm{d}f}{\mathrm{d}x} = -\frac{(\partial g/\partial x)}{(\partial g/\partial y)} = \frac{\mathrm{d}y}{\mathrm{d}x} \qquad (8.14)$$

One important fact concerning partial derivatives is evident from equation (8.14); the symbol "∂g" cannot be cancelled out because at least a sign mistake would be caused. Thus the two parts of the ratio defining a partial derivative can never be separated and considered alone. This is in marked contrast to ordinary derivatives where the infinitesimals $\mathrm{d}x, \mathrm{d}y$ can be treated separately. To demonstrate the truth of equation (8.14), consider the following example.

Example 1. If $x^2 + y^2 = a^2$, find $\mathrm{d}y/\mathrm{d}x$.
The above equation can be solved for y giving

$$y = (a^2 - x^2)^{\frac{1}{2}} \qquad\qquad \text{I}$$

Differentiating in the usual way,

$$\frac{\mathrm{d}y}{\mathrm{d}x} = \frac{\frac{1}{2}(-2x)}{(a^2 - x^2)^{\frac{1}{2}}} = \frac{-x}{(a^2 - x^2)^{\frac{1}{2}}} \qquad\qquad \text{II}$$

Alternatively, the original equation can be written

$$g(x, y) = x^2 + y^2 - a^2 = 0 \qquad\qquad \text{III}$$

Differentiating equation III partially with respect to x keeping y constant gives

$$\frac{\partial g}{\partial x} = 2x \qquad\qquad \text{IV}$$

Similarly, $\qquad\qquad \dfrac{\partial g}{\partial y} = 2y \qquad\qquad$ V

Substituting the values given in equations IV and V into equation (8.14) gives

$$\frac{\mathrm{d}y}{\mathrm{d}x} = -\frac{2x}{2y}$$

and eliminating y by using equation I,

$$\frac{dy}{dx} = \frac{-x}{(a^2 - x^2)^{\frac{1}{2}}} \qquad\qquad \text{VI}$$

Thus, equations II and VI are identical, illustrating the validity of equation (8.14).

The above ideas can be readily extended to functions of two independent variables by generalizing equation (8.12) to

$$z = g(x, y) \qquad\qquad (8.15)$$

and hence

$$dz = \left(\frac{\partial g}{\partial x}\right) dx + \left(\frac{\partial g}{\partial y}\right) dy \qquad\qquad (8.16)$$

Equation (8.15) can also be put into the implicit form

$$F(x, y, z) = 0 \qquad\qquad (8.17)$$

and thus

$$dF = \left(\frac{\partial F}{\partial x}\right) dx + \left(\frac{\partial F}{\partial y}\right) dy + \left(\frac{\partial F}{\partial z}\right) dz = 0 \qquad\qquad (8.18)$$

Rearranging equation (8.18) gives

$$dz = -\frac{(\partial F/\partial x)}{(\partial F/\partial z)} dx - \frac{(\partial F/\partial y)}{(\partial F/\partial z)} dy \qquad\qquad (8.19)$$

and comparing equations (8.16) and (8.19) gives

$$\frac{\partial g}{\partial x} = -\frac{\partial F/\partial x}{\partial F/\partial z}, \quad \frac{\partial g}{\partial y} = -\frac{\partial F/\partial y}{\partial F/\partial z} \qquad\qquad (8.20)$$

Considering the manner in which a partial derivative is obtained, it is clear that

$$\left(\frac{\partial g}{\partial x}\right)\left(\frac{\partial x}{\partial g}\right) = 1 \qquad\qquad (8.21)$$

and eliminating $(\partial F/\partial z)$ from the two equations (8.20) gives

$$\left(\frac{\partial g}{\partial x}\right)\left(\frac{\partial y}{\partial g}\right) = \left(\frac{\partial F}{\partial x}\right)\left(\frac{\partial y}{\partial F}\right) \qquad\qquad (8.22)$$

$\partial y/\partial x$ can be obtained by holding z constant in equation (8.18), thus $dz = 0$ and

$$\frac{\partial y}{\partial x} = -\frac{(\partial F/\partial x)}{(\partial F/\partial y)} \qquad\qquad (8.23)$$

Equating (8.22) and (8.23) and rearranging gives

$$\left(\frac{\partial g}{\partial x}\right)\left(\frac{\partial y}{\partial g}\right)\left(\frac{\partial x}{\partial y}\right) = -1 \qquad\qquad (8.24)$$

or using the symmetrical version of equation (8.15), viz. (8.17), equation (8.24) can be written

$$\left(\frac{\partial z}{\partial x}\right)\left(\frac{\partial y}{\partial z}\right)\left(\frac{\partial x}{\partial y}\right) = -1 \qquad (8.25)$$

giving further indication that partial derivative signs cannot be cancelled out.

8.2.2. *Changing the Independent Variables*

Referring to equation (8.15) where x and y are the two independent variables, it may be desired to change to new independent variables u and v where the old and new variables are related by

$$\left. \begin{array}{l} x = \phi(u, v) \\ y = \psi(u, v) \end{array} \right\} \qquad (8.26)$$

Applying the general formula (8.9) to equations (8.26) gives

$$\left. \begin{array}{l} dx = \dfrac{\partial \phi}{\partial u} du + \dfrac{\partial \phi}{\partial v} dv \\[2mm] dy = \dfrac{\partial \psi}{\partial u} du + \dfrac{\partial \psi}{\partial v} dv \end{array} \right\} \qquad (8.27)$$

which can be substituted into equation (8.16), thus

$$dz = \left(\frac{\partial g}{\partial x}\right)\left(\frac{\partial \phi}{\partial u} du + \frac{\partial \phi}{\partial v} dv\right) + \left(\frac{\partial g}{\partial y}\right)\left(\frac{\partial \psi}{\partial u} du + \frac{\partial \psi}{\partial v} dv\right) \qquad (8.28)$$

The partial derivative of z with respect to u is obtained by keeping v constant, and hence $dv = 0$ in equation (8.28).

$$\frac{\partial z}{\partial u} = \frac{\partial g}{\partial x}\frac{\partial \phi}{\partial u} + \frac{\partial g}{\partial y}\frac{\partial \psi}{\partial u}$$

or using the alternative nomenclature to clarify the equation,

$$\frac{\partial z}{\partial u} = \frac{\partial z}{\partial x}\frac{\partial x}{\partial u} + \frac{\partial z}{\partial y}\frac{\partial y}{\partial u} \qquad (8.29)$$

Similarly,

$$\frac{\partial z}{\partial v} = \frac{\partial z}{\partial x}\frac{\partial x}{\partial v} + \frac{\partial z}{\partial y}\frac{\partial y}{\partial v} \qquad (8.30)$$

Equations (8.29) and (8.30) can be extended to any number of independent variables. Thus if

$$z = f(x_1, x_2, x_3, \ldots, x_n)$$

$$x_1 = g_1(u_1, u_2, u_3, \ldots, u_m)$$

$$x_2 = g_2(u_1, u_2, u_3, \ldots, u_m) \quad \text{etc.}$$

then
$$\frac{\partial z}{\partial u_1} = \frac{\partial z}{\partial x_1}\frac{\partial x_1}{\partial u_1} + \frac{\partial z}{\partial x_2}\frac{\partial x_2}{\partial u_1} + \dots + \frac{\partial z}{\partial x_n}\frac{\partial x_n}{\partial u_1} \tag{8.31}$$

The number of variables m need not be the same as the original number of variables n, but $m = n$ is the usual case.

8.2.3. *Independent Variables not Truly Independent*

On occasions there are restricting relationships between the basic independent variables. For instance, the composition of the vapour leaving a binary liquid is a function of temperature, pressure, and liquid composition, which are three independent variables. However, a restriction that the liquid is boiling is frequently applied and this gives a further equation relating temperature to pressure and liquid composition. This case can be stated symbolically thus

$$y = f(x, T, P) \tag{8.32}$$

$$T = g(x, P) \tag{8.33}$$

Applying equation (8.9) to both of these gives

$$dy = \frac{\partial f}{\partial x}dx + \frac{\partial f}{\partial T}dT + \frac{\partial f}{\partial P}dP \tag{8.34}$$

and
$$dT = \frac{\partial g}{\partial x}dx + \frac{\partial g}{\partial P}dP \tag{8.35}$$

Substituting for dT from equation (8.35) into (8.34) and rearranging gives

$$dy = \left(\frac{\partial f}{\partial x} + \frac{\partial f}{\partial T}\frac{\partial g}{\partial x}\right)dx + \left(\frac{\partial f}{\partial P} + \frac{\partial f}{\partial T}\frac{\partial g}{\partial P}\right)dP \tag{8.36}$$

and here it is not advisable to replace $(\partial g/\partial x)$ with $(\partial T/\partial x)$ because the latter term is ambiguous. It could be obtained from equation (8.32) at constant y and P, or from equation (8.33) at constant P, yielding different results. If the data are collected at constant pressure, $dP = 0$, and equation (8.36) becomes

$$\left(\frac{\partial y}{\partial x}\right)_P = \frac{\partial f}{\partial x} + \frac{\partial f}{\partial T}\frac{\partial g}{\partial x} \tag{8.37}$$

or
$$\left(\frac{\partial y}{\partial x}\right)_P = \left(\frac{\partial y}{\partial x}\right)_{P,T} + \left(\frac{\partial y}{\partial T}\right)_{x,P}\left(\frac{\partial T}{\partial x}\right)_P \tag{8.38}$$

where the derivatives with one variable held constant are obtained from equation (8.33) and the others from equation (8.32).

The above system is sometimes studied at constant temperature which precludes the elimination of dT between equations (8.34) and (8.35) because

$dT = 0$. However, a similar method will show that

$$\left(\frac{\partial y}{\partial x}\right)_T = \left(\frac{\partial y}{\partial x}\right)_{P,T} + \left(\frac{\partial y}{\partial P}\right)_{x,T}\left(\frac{\partial P}{\partial x}\right)_T \tag{8.39}$$

A further case of interest in chemical engineering is when a fluid property (e.g. temperature T) is a function of three coordinate (x, y, z) and time (t). The temperature increment determined from equation (8.9) is

$$dT = \left(\frac{\partial T}{\partial x}\right)dx + \left(\frac{\partial T}{\partial y}\right)dy + \left(\frac{\partial T}{\partial z}\right)dz + \left(\frac{\partial T}{\partial t}\right)dt \tag{8.40}$$

If a path is traversed across the system so that x, y, and z become functions of time only, equation (8.40) can be divided by dt to give

$$\frac{dT}{dt} = \left(\frac{\partial T}{\partial x}\right)\frac{dx}{dt} + \left(\frac{\partial T}{\partial y}\right)\frac{dy}{dt} + \left(\frac{\partial T}{\partial z}\right)\frac{dz}{dt} + \left(\frac{\partial T}{\partial t}\right) \tag{8.41}$$

where dx/dt, etc., are the velocity components which determine the progress along the path and dT/dt is the "total time derivative" of the temperature.

A special case of equation (8.41) is when the path of a fluid element is traversed so that $dx/dt = u$, $dy/dt = v$, and $dz/dt = w$ are the components of fluid velocity. The total time derivative is then given the special symbol D/Dt named the "substantive derivative". Thus

$$\frac{DT}{Dt} = \left(\frac{\partial T}{\partial x}\right)u + \left(\frac{\partial T}{\partial y}\right)v + \left(\frac{\partial T}{\partial z}\right)w + \frac{\partial T}{\partial t} \tag{8.42}$$

or, in a vector form,

$$\frac{DT}{Dt} = \frac{\partial T}{\partial t} + \boldsymbol{u}.\nabla T \tag{8.43}$$

The distinction between DT/Dt and $\partial T/\partial t$ can be clarified by considering a heat exchanger operating at steady state. At any fixed point in the exchanger the temperature of the fluid passing the point is constant, thus $\partial T/\partial t = 0$, since by definition this derivative is taken at fixed values of the coordinate variables. However, any element of fluid undergoes a temperature rise or fall as it passes through the exchanger, and as far as the fluid element is concerned $DT/Dt \neq 0$. Thus it is important to distinguish between the equilibrium of an element of space and the equilibrium of an element of fluid.

An example of the use of equation (8.41) is the rule for differentiating a product. Thus if

$$y = uv \tag{8.44}$$

where u and v are functions of x only, application of equation (8.41) gives

$$\frac{dy}{dx} = \frac{\partial y}{\partial u}\frac{du}{dx} + \frac{\partial y}{\partial v}\frac{dv}{dx}$$

Evaluating the partial derivatives by differentiating equation (8.44),

$$\frac{dy}{dx} = v\frac{du}{dx} + u\frac{dv}{dx} \qquad (8.45)$$

8.3. FORMULATING PARTIAL DIFFERENTIAL EQUATIONS

The method whereby a physical situation can be represented by a partial differential equation was described in Section 1.13 although the details were only described for ordinary differential equations. The first difficulty in describing the model is determining the number of independent variables needed to specify the system. The normal maximum number is four; three coordinates defining position, and time. If the problem contains either a plane of symmetry or a line of symmetry, one coordinate can be dispensed with in the appropriate coordinate system. Cartesian or cylindrical polar coordinates are suitable in cases of plane symmetry, whereas either cylindrical or spherical polar coordinates are suitable for describing axial symmetry.

In Chapter 1 the space element was defined by the two planes at x and $x + \delta x$, whereas in the general three-dimensional case the space element is defined by taking an increment in each of the coordinate directions. The relevant conservation law is still applied to this new space element, but the inventory of inputs and outputs must include the behaviour at all six faces of the space element, and not just two of the faces. The space element in spherical polar coordinates has already been illustrated in Fig. 7.16 where it can be seen that each face of the element is defined by keeping one of the coordinate variables constant, and each edge of the element is defined by keeping two coordinates constant. Thus, a version of Taylor's theorem can be applied along each edge of the element, and the value of the dependent variable can be established at each corner of the space element in terms of its value at one corner and the three partial derivatives constituting the Laplace operator.

These ideas can be better understood from actual examples, and in the following subsections a few of the more common partial differential equations will be derived. The first derivation will be given in full to clarify the method, but the later derivations will include many of the short cuts anticipating the result.

8.3.1. *Unsteady-state Heat Conduction in One Dimension*

Figure 8.2 illustrates a section of a flat wall of thickness L whose height and length are both large compared with L. If the temperature distribution is uniform throughout the wall at zero time and the heat is supplied at a fixed rate per unit area to the one surface, it is required to determine the temperature as a function of position and time.

Since the original temperature is uniform, and every part of each wall surface is subjected to the same conditions, no heat will travel parallel to the surface and the temperature will be constant in any plane parallel to the surface. Thus the temperature distribution can be specified in terms of a single coordinate denoted by x in Fig. 8.2., and the discussion can be restricted to a section of unit area through the wall. Considering the thermal equilibrium of a slice of the wall between a plane at distance x from the heated surface and a parallel plane at $x + \delta x$ from the same surface gives the following balance, where T is the temperature and k is the thermal conductivity.

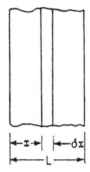

FIG. 8.2. One-dimensional heat transfer

Rate of heat input at distance x and time t is $-k(\partial T/\partial x)$.
Rate of heat input at distance x and time $t + \delta t$ is

$$-k\frac{\partial T}{\partial x} + \frac{\partial}{\partial t}\left(-k\frac{\partial T}{\partial x}\right)\delta t$$

This second rate is obtained by applying Taylor's theorem as in Chapter 1 and considering that x is a constant. Similarly, two output rates can be determined from the corresponding input rates by applying Taylor's theorem at constant t and variable x, thus:
Rate of heat output at distance $x + \delta x$ and time t is

$$-k\frac{\partial T}{\partial x} + \frac{\partial}{\partial x}\left(-k\frac{\partial T}{\partial x}\right)\delta x$$

Rate of heat output at distance $x + \delta x$ and time $t + \delta t$ is

$$-k\frac{\partial T}{\partial x} + \frac{\partial}{\partial t}\left(-k\frac{\partial T}{\partial x}\right)\delta t + \frac{\partial}{\partial x}\left[-k\frac{\partial T}{\partial x} + \frac{\partial}{\partial t}\left(-k\frac{\partial T}{\partial x}\right)\delta t\right]\delta x$$

Heat content of the element at time t is $\rho C_p T \delta x$.
Heat content of the element at time $t + \delta t$ is $\rho C_p[T + (\partial T/\partial t)\,\delta t]\,\delta x$.
Therefore accumulation of heat in time δt is $\rho C_p(\partial T/\partial t)\,\delta t\,\delta x$.
Taking average input and output rates during the time interval δt gives

$$\text{Heat input} = \left[-k\frac{\partial T}{\partial x} + \frac{1}{2}\frac{\partial}{\partial t}\left(-k\frac{\partial T}{\partial x}\right)\delta t\right]\delta t$$

$$\text{Heat output} = \left\{ -k\frac{\partial T}{\partial x} + \frac{\partial}{\partial x}\left(-k\frac{\partial T}{\partial x}\right)\delta x \right.$$

$$\left. +\frac{1}{2}\frac{\partial}{\partial t}\left[-k\frac{\partial T}{\partial x} + \frac{\partial}{\partial x}\left(-k\frac{\partial T}{\partial x}\right)\delta x\right]\delta t\right\}\delta t$$

Using the general conservation law (1.25), and cancelling some terms for simplicity,

$$-\left[\frac{\partial}{\partial x}\left(-k\frac{\partial T}{\partial x}\right)\delta x + \frac{1}{2}\frac{\partial^2}{\partial t\,\partial x}\left(-k\frac{\partial T}{\partial x}\right)\delta x\,\delta t\right]\delta t = \rho C_p\frac{\partial T}{\partial t}\delta t\,\delta x$$

Dividing throughout by $\delta x\,\delta t$ and assuming that k is constant gives

$$k\frac{\partial^2 T}{\partial x^2} + \frac{1}{2}k\frac{\partial^3 T}{\partial x^2\,\partial t}\delta t = \rho C_p\frac{\partial T}{\partial t} \tag{8.46}$$

Taking the limit of equation (8.46) as $\delta t \to 0$, the central term vanishes, and introducing the thermal diffusivity (α) as in equation (7.168) gives

$$\alpha\frac{\partial^2 T}{\partial x^2} = \frac{\partial T}{\partial t} \tag{8.47}$$

which is the required equation. It can also be obtained directly from the vector equation (7.168) by expanding (∇^2) in cartesian coordinates and deleting all y and z derivatives.

In the above derivation, it will be seen that the same equation (8.47) results if the simple input and output rates at time t are used without the added complication of using average rates. This simplification is consistent with the findings of Chapter 1, but until the reader is confident about which simplifications are justified and which are not, he is advised to allow for all possible correction terms as above. Care must be taken in each term to write down the correct number of infinitesimals (δx, δt), since these govern which terms are rejected in the step from equation (8.46) to (8.47). The boundary conditions will be derived separately in Section 8.4, and next an equation relating to unsteady-state axi-symmetrical diffusion will be derived.

8.3.2. *Mass Transfer with Axial Symmetry*

A spray column is to be used for extracting one component from a binary mixture which forms the rising continuous phase. In order to estimate the transfer coefficient it is desired to study the detailed concentration distribution around an individual droplet of the spray.

To confine attention to a single droplet, a spherical polar coordinate system is taken with its origin at the centre of the droplet. During its fall through the column, the droplet moves into contact with liquid of stronger composition so that allowance must be made for the time variation of the system. The concentration will also be a function of both the radial co-ordinate (r) and the angular coordinate (θ) but will not depend upon ϕ due to the symmetry provided that the z axis is vertical. Denoting concentration by c, diffusivity by D and the r and θ components of the continuous phase velocity by u and v respectively, a material balance can be taken over the element ABCD shown in Fig. 8.3. The element is defined in the normal

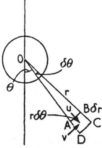

FIG. 8.3. Mass transfer with axial symmetry

manner by allowing one independent variable to vary at a time, and the diagram shows a section through what is in fact a ring-shaped element, symmetrical about the z axis.

$$\text{Area of face AB is } 2\pi r^2 \sin\theta\,\delta\theta \qquad (8.48)$$

$$\text{Area of face AD is } 2\pi r \sin\theta\,\delta r \qquad (8.49)$$

$$\text{Volume of element is } 2\pi r^2 \sin\theta\,\delta\theta\,\delta r \qquad (8.50)$$

Material is transferred across each surface of the element by two mechanisms, bulk flow and molecular diffusion. Thus the transfer rates at each surface can be written as follows.

$$\text{Input rate across AB} = \left(uc - D\frac{\partial c}{\partial r}\right)2\pi r^2 \sin\theta\,\delta\theta$$

$$\text{Input rate across AD} = \left(vc - \frac{D}{r}\frac{\partial c}{\partial \theta}\right)2\pi r \sin\theta\,\delta r$$

It should be noted that the concentration gradient in the θ direction is $(1/r)(\partial c/\partial\theta)$ as obtained from the expansion of ∇c in spherical polar co-ordinates, equation (7.108). The output rates across CD and BC are obtained from the corresponding input rates by using Taylor's theorem as in Chapter 1.

Output rate across CD

$$= \left(uc - D\frac{\partial c}{\partial r}\right)2\pi r^2 \sin\theta\,\delta\theta + \frac{\partial}{\partial r}\left[\left(uc - D\frac{\partial c}{\partial r}\right)2\pi r^2 \sin\theta\,\delta\theta\right]\delta r$$

Output rate across BC

$$= \left(vc - \frac{D}{r}\frac{\partial c}{\partial \theta}\right) 2\pi r \sin \theta \, \delta r + \frac{\partial}{\partial \theta}\left[\left(vc - \frac{D}{r}\frac{\partial c}{\partial \theta}\right) 2\pi r \sin \theta \, \delta r\right] \delta \theta$$

Accumulation rate

$$= 2\pi r^2 \sin \theta \, \delta r \, \delta \theta \frac{\partial c}{\partial t}$$

Using the general conservation law (1.25) and cancelling the obvious terms gives

$$-\frac{\partial}{\partial r}\left[\left(uc - D\frac{\partial c}{\partial r}\right) 2\pi r^2 \sin \theta \, \delta \theta\right] \delta r$$

$$-\frac{\partial}{\partial \theta}\left[\left(vc - \frac{D}{r}\frac{\partial c}{\partial \theta}\right) 2\pi r \sin \theta \, \delta r\right] \delta \theta = 2\pi r^2 \sin \theta \, \delta r \, \delta \theta \frac{\partial c}{\partial t}$$

Dividing throughout by the volume of the element (8.50).

$$\frac{\partial c}{\partial t} = -\frac{1}{r^2}\frac{\partial}{\partial r}\left(r^2 uc - r^2 D\frac{\partial c}{\partial r}\right) - \frac{1}{r\sin \theta}\frac{\partial}{\partial \theta}\left[\left(vc - \frac{D}{r}\frac{\partial c}{\partial \theta}\right)\sin \theta\right] \quad (8.51)$$

Differentiating the products and separating the velocity terms from the diffusion terms gives

$$\frac{\partial c}{\partial t} + u\frac{\partial c}{\partial r} + \frac{v}{r}\frac{\partial c}{\partial \theta} + c\left[\frac{1}{r^2}\frac{\partial}{\partial r}(r^2 u) + \frac{1}{r\sin \theta}\frac{\partial}{\partial \theta}(v\sin \theta)\right]$$

$$= \frac{D}{r^2}\frac{\partial}{\partial r}\left(r^2\frac{\partial c}{\partial r}\right) + \frac{D}{r^2\sin \theta}\frac{\partial}{\partial \theta}\left(\sin \theta\frac{\partial c}{\partial \theta}\right) \quad (8.52)$$

But the term in square brackets is the expression for $\nabla . u$ given by equation (7.118) and is thus zero because of the continuity equation.

$$\therefore \quad \frac{\partial c}{\partial t} + u\frac{\partial c}{\partial r} + \frac{v}{r}\frac{\partial c}{\partial \theta} = \frac{D}{r^2}\frac{\partial}{\partial r}\left(r^2\frac{\partial c}{\partial r}\right) + \frac{D}{r^2\sin \theta}\frac{\partial}{\partial \theta}\left(\sin \theta\frac{\partial c}{\partial \theta}\right) \quad (8.53)$$

where the right-hand side is the expansion of $\nabla^2 c$ in spherical polar coordinates as given by equation (7.120).

8.3.3. *The Continuity Equation*

This important yet simple equation expresses the fact that in a flowing fluid the mass of the fluid is conserved. Figure 8.4 shows a rectangular element of space the lengths of whose edges parallel to the three cartesian axes are δx, δy, and δz. The components of the fluid velocity are denoted by u, v, and w as shown, and the density of the fluid by ρ. The positive directions of u, v, and w must be the same as the positive directions of the axes so that fluid is considered to enter the element through faces ABCD, ABFE, and ADHE, and leave through the other three faces.

The rate at which fluid enters through face ABCD is proportional to the area of the face ($\delta y \, \delta z$), the fluid density (ρ), and the velocity component (u) perpendicular to the face. Strictly, the velocity component only has the value u at the point A, and the velocity through the face will vary slightly

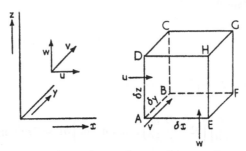

Fig. 8.4. Volume element in cartesian coordinates

from u. This is a small correction which will disappear at the end of the derivation and it will be ignored. If the reader wishes to check the above statement, all velocity components can be considered as values averaged over the relevant face, and when the limit is taken as the size of the element shrinks to zero, each average velocity tends to its point value at A. Thus

$$\text{Input rate through ABCD} = \rho u \, \delta y \, \delta z$$

$$\text{Input rate through ADHE} = \rho v \, \delta x \, \delta z$$

$$\text{Input rate through ABFE} = \rho w \, \delta x \, \delta y$$

Once again, the output rates are obtained from the corresponding input rates by applying Taylor's theorem. Thus

$$\text{Output rate through EFGH} = \rho u \, \delta y \, \delta z + \frac{\partial}{\partial x} (\rho u \, \delta y \, \delta z) \, \delta x$$

$$\text{Output rate through BCGF} = \rho v \, \delta x \, \delta z + \frac{\partial}{\partial y} (\rho v \, \delta x \, \delta z) \, \delta y$$

$$\text{Output rate through CDHG} = \rho w \, \delta x \, \delta y + \frac{\partial}{\partial z} (\rho w \, \delta x \, \delta y) \, \delta z$$

The mass of fluid contained in the element at time t is $\rho \, \delta x \, \delta y \, \delta z$. Therefore rate of increase of mass contained in element is $(\partial/\partial t)(\rho \, \delta x \, \delta y \, \delta z)$. Because

$$\text{Input—Output = Accumulation}$$

$$\therefore \quad -\frac{\partial}{\partial x}(\rho u) \, \delta x \, \delta y \, \delta z - \frac{\partial}{\partial y}(\rho v) \, \delta x \, \delta y \, \delta z - \frac{\partial}{\partial z}(\rho w) \, \delta x \, \delta y \, \delta z = \frac{\partial \rho}{\partial t} \, \delta x \, \delta y \, \delta z \tag{8.54}$$

where the obvious terms have been cancelled. Dividing by the volume of the element gives

$$\frac{\partial}{\partial x}(\rho u) + \frac{\partial}{\partial y}(\rho v) + \frac{\partial}{\partial z}(\rho w) + \frac{\partial \rho}{\partial t} = 0 \tag{8.55}$$

Taking the limit of equation (8.55) as the size of the element tends to zero leaves it unaltered, because all surplus terms have been ignored at an earlier stage in anticipation of this result.

Each term of equation (8.55) can be differentiated as a product and rearranged to give

$$\rho \frac{\partial u}{\partial x} + \rho \frac{\partial v}{\partial y} + \rho \frac{\partial w}{\partial z} + u \frac{\partial \rho}{\partial x} + v \frac{\partial \rho}{\partial y} + w \frac{\partial \rho}{\partial z} + \frac{\partial \rho}{\partial t} = 0 \tag{8.56}$$

Since ρ is a function of x, y, z, and t, its substantive derivative can be taken as in equation (8.42), thus

$$\frac{D\rho}{Dt} = u \frac{\partial \rho}{\partial x} + v \frac{\partial \rho}{\partial y} + w \frac{\partial \rho}{\partial z} + \frac{\partial \rho}{\partial t} \tag{8.57}$$

Combining equations (8.56) and (8.57) gives

$$\frac{1}{\rho} \frac{D\rho}{Dt} + \frac{\partial u}{\partial x} + \frac{\partial v}{\partial y} + \frac{\partial w}{\partial z} = 0 \tag{8.58}$$

which is the required continuity equation for a compressible fluid.

8.4. BOUNDARY CONDITIONS

In the last section the differential equation was found by taking a balance over an arbitrary element within the region of interest. The boundary conditions can always be obtained in a similar manner by considering the equilibrium of the particular element which contains the boundary. The variety of possible boundary conditions is quite small and subsections are devoted to each of four types. These results should be committed to memory, thus dispensing with the necessity for a detailed derivation on each occasion.

In the study of ordinary differential equations with only one independent variable, the boundary is defined by one particular value of the independent variable and the condition is stated in terms of the behaviour of the dependent variable at the boundary point. In the study of partial differential equations, each boundary is still defined by giving a particular value to just one of the independent variables. The boundary condition must then specify the behaviour of the dependent variable as a function of all of the other independent variables. There are three types of boundary condition arising in practice for which analytical solutions are sometimes possible and these are considered separately below. The more complicated integro-differential type of boundary condition is also derived, but problems in which they are involved are usually intractable by analytical methods. The discussion will be confined to heat transfer but the methods are generally applicable.

8.4.1. *First Type, Function Specified*

If values of the dependent variable itself are given at all points on a particular boundary, then the condition is said to be of the first type. Considering time-dependent heat transfer in one dimension as in Section 8.3.1, the temperature will be a function of both x and t, thus the boundaries will be defined as either fixed values of x or fixed values of t. Two boundary conditions of the first type for this problem are:

$$\text{(i) at} \quad t = 0, \quad T = f(x) \tag{8.59}$$

$$\text{(ii) at} \quad x = 0, \quad T = g(t) \tag{8.60}$$

Equation (8.59) gives the temperature at all values of x for one value of t, whilst equation (8.60) restricts the behaviour of the temperature at all times at one value of x.

In the case of steady heat conduction in a cylindrical conductor of finite size, it is usual to use cylindrical polar coordinates so that $z = \pm a$ defines the end faces and $r = r_0$ defines the curved surface. Hence both boundaries are defined by keeping one of the independent variables constant, and conditions of the first type take the form

$$\text{at} \quad z = a, \quad T = f(r, \theta) \tag{8.61}$$

$$\text{at} \quad r = r_0, \quad T = g(z, \theta) \tag{8.62}$$

The use of cartesian coordinates is not convenient in the above problem because the description of the boundaries is not so simple and all of the coordinates vary on the curved surface.

8.4.2. *Second Type, Derivative Specified*

The position of the boundary will still be defined as above by keeping one independent variable constant, but instead of giving the direct behaviour of the dependent variable, its derivative is restricted by a boundary condition of the second type. This situation occurs with radiant heating or cooling of a surface, electrical heating of a surface, or the special case of a thermally insulated surface. It can be seen in these cases that the heat flow rate is known but not the surface temperature. Because the heat flow rate is related to the temperature gradient, this boundary condition is of the second type, and two examples will now be considered.

Example 1. A rectangular block of metal is subject to temperature variation in all directions, but the surface at $x = 0$ is thermally insulated. Express this condition by an equation.

It is convenient to use cartesian coordinates in this problem, and the boundary condition is derived by considering the infinitesimal element residing at the boundary as shown in Fig. 8.5. Heat can only pass through the faces of the element by conduction, and the rates can be found for five of the surfaces using the same method as that used to determine the differential equation for a general volume element.

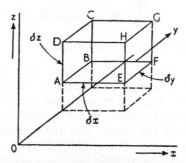

FIG. 8.5. Boundary element

$$\text{Input rate through ADHE} = -k\frac{\partial T}{\partial y}\,\delta x\,\delta z$$

$$\text{Output rate through BCGF} = -k\frac{\partial T}{\partial y}\,\delta x\,\delta z+\frac{\partial}{\partial y}\left(-k\frac{\partial T}{\partial y}\,\delta x\,\delta z\right)\delta y$$

$$\text{Input rate through ABFE} = -k\frac{\partial T}{\partial z}\,\delta x\,\delta y$$

$$\text{Output rate through DCGH} = -k\frac{\partial T}{\partial z}\,\delta x\,\delta y+\frac{\partial}{\partial z}\left(-k\frac{\partial T}{\partial z}\,\delta x\,\delta y\right)\delta z$$

$$\text{Output rate through EFGH} = -k\frac{\partial T}{\partial x}\,\delta y\,\delta z+\frac{\partial}{\partial x}\left(-k\frac{\partial T}{\partial x}\,\delta y\,\delta z\right)\delta x$$

Because the face at $x = 0$ is thermally insulated, the heat input rate through ABCD must be zero. The rate of accumulation of heat within the element is $\rho C_p\,(\partial T/\partial t)\,\delta x\,\delta y\,\delta z$ as before, and completing the balance gives (after some cancellation)

$$k\frac{\partial^2 T}{\partial y^2}\,\delta x\,\delta z\,\delta y+k\frac{\partial^2 T}{\partial z^2}\,\delta x\,\delta y\,\delta z+k\frac{\partial T}{\partial x}\,\delta y\,\delta z$$

$$+k\frac{\partial^2 T}{\partial x^2}\,\delta y\,\delta z\,\delta x = \rho C_p\frac{\partial T}{\partial t}\,\delta x\,\delta y\,\delta z \qquad \text{I}$$

Due to the third term in this equation, it is only possible to cancel out $\delta y\,\delta z$ giving

$$k\left(\frac{\partial^2 T}{\partial y^2}+\frac{\partial^2 T}{\partial z^2}+\frac{\partial^2 T}{\partial x^2}\right)\delta x+k\frac{\partial T}{\partial x} = \rho C_p\frac{\partial T}{\partial t}\,\delta x \qquad \text{II}$$

Taking the limit as the size of the element shrinks to zero, results in all of the terms of equation I tending to zero with one exception. Thus equation I becomes

$$\frac{\partial T}{\partial x} = 0 \quad \text{at} \quad x = 0 \qquad\qquad \text{III}$$

and this is the required boundary condition.

Example 2. A cylindrical furnace is lined with two uniform layers of insulating brick of different physical properties. What boundary conditions should be imposed at the junction between the layers?

Figure 8.6 illustrates the problem, where a denotes the radius of curvature of the boundary between the layers, and the element is considered to lie across the boundary as shown. The symbols T, k, C_p, and ρ will be used for the temperature, conductivity, specific heat, and density of the layers, and a subscript 1 or 2 will distinguish between the layers. Due to axial symmetry, no heat will flow across the faces of the element given by $\theta = $ constant, but allowances will be made for heat flow in the z direction.

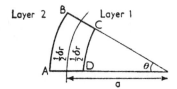

FIG. 8.6. Section through cylindrical element at boundary

One boundary condition, which is of the first type, is that the temperature must be continuous across the boundary. That is,

$$\text{at} \quad r = a, \quad T_1 = T_2 \qquad\qquad \text{I}$$

The second boundary condition is that the law of conservation of heat must be satisfied for the element shown in Fig. 8.6. The rate of flow of heat just inside the boundary of the first layer is $-k_1 a\theta\, \delta z\, (\partial T_1/\delta r)$, and using Taylor's theorem, the rate of flow of heat into the element across the face CD is

$$-k_1 r\theta\, \delta z\, \frac{\partial T_1}{\partial r} - \tfrac{1}{2}\delta r \frac{\partial}{\partial r}\left(-k_1 r\theta\, \delta z\, \frac{\partial T_1}{\partial r}\right)$$

which must be evaluated at $r = a$.

$$\therefore \quad \text{Input across CD} = -k_1 a\theta\delta z\, \frac{\partial T_1}{\partial r} + \tfrac{1}{2}k_1\, \theta\delta r\, \delta z\, \frac{\partial}{\partial r}\left(r\frac{\partial T_1}{\partial r}\right)\bigg|_a$$

Similarly,

$$\text{Output across AB} = -k_2 a\theta\delta z\, \frac{\partial T_2}{\partial r} - \tfrac{1}{2}k_2\, \theta\delta r\, \delta z\, \frac{\partial}{\partial r}\left(r\frac{\partial T_2}{\partial r}\right)\bigg|_a$$

The heat flow rates in the z direction can be written:

$$\text{Input at face } (z) = -\tfrac{1}{2}a\theta\delta r\,\frac{\partial}{\partial z}(k_1 T_1 + k_2 T_2)$$

Output at face $(z + \delta z)$

$$= -\tfrac{1}{2}a\theta\,\delta r\,\frac{\partial}{\partial z}(k_1 T_1 + k_2 T_2) + \frac{\partial}{\partial z}\left[-\tfrac{1}{2}a\theta\,\delta r\,\frac{\partial}{\partial z}(k_1 T_1 + k_2 T_2)\right]\delta z$$

Accumulation within the element

$$= \tfrac{1}{2}\rho_1 C_{p1}a\theta\,\delta r\,\delta z\,\frac{\partial T_1}{\partial t} + \tfrac{1}{2}\rho_2 C_{p2}a\theta\,\delta r\,\delta z\,\frac{\partial T_2}{\partial t}$$

Completing the heat balance and cancelling the obvious terms gives

$$-k_1 a\theta\,\delta z\,\frac{\partial T_1}{\partial r} + \tfrac{1}{2}k_1\,\theta\,\delta r\,\delta z\,\frac{\partial}{\partial r}\left(r\frac{\partial T_1}{\partial r}\right) + k_2 a\theta\,\delta z\,\frac{\partial T_2}{\partial r}$$

$$+ \tfrac{1}{2}k_2\,\theta\,\delta r\,\delta z\,\frac{\partial}{\partial r}\left(r\frac{\partial T_2}{\partial r}\right) + \tfrac{1}{2}a\theta\,\delta r\,\delta z\,\frac{\partial^2}{\partial z^2}(k_1 T_1 + k_2 T_2)$$

$$= \tfrac{1}{2}a\theta\,\delta r\,\delta z\,\frac{\partial}{\partial t}(\rho_1 C_{p1}T_1 + \rho_2 C_{p2}T_2) \qquad\qquad \text{II}$$

Dividing by $a\theta\delta z$ gives

$$-k_1\frac{\partial T_1}{\partial r} + \tfrac{1}{2}k_1\frac{\delta r}{a}\frac{\partial}{\partial r}\left(r\frac{\partial T_1}{\partial r}\right) + k_2\frac{\partial T_2}{\partial r} + \tfrac{1}{2}k_2\frac{\delta r}{a}\frac{\partial}{\partial r}\left(r\frac{\partial T_2}{\partial r}\right) + \tfrac{1}{2}\delta r\frac{\partial^2}{\partial z^2}(k_1 T_1 + k_2 T_2)$$

$$= \tfrac{1}{2}\delta r\frac{\partial}{\partial t}(\rho_1 C_{p1}T_1 + \rho_2 C_{p2}T_2) \qquad\qquad \text{III}$$

In the limit as $\delta r \to 0$, most of the terms in equation III disappear leaving

$$k_1\frac{\partial T_1}{\partial r} = k_2\frac{\partial T_2}{\partial r} \qquad\qquad \text{IV}$$

which is the second boundary condition.

These examples, particularly the second one, show that most of the effects which give rise to important terms in the differential equation are not important in the consideration of boundary conditions. The boundary condition IV does not include any terms describing the variation of temperature with movement in the z direction, the variation with time, or even the curvature of the boundary. All of these terms disappeared in the final limiting step from equation III.

Incidentally, the differential equation can also be derived from equation III by moving the element into either layer when all subscripts can be removed and a must be replaced by r. The two terms $(k\,\partial T/\partial r)$ will cancel out, followed by cancellation throughout by (δr); the limiting process will then yield the general differential equation.

$$\frac{1}{r}\frac{\partial}{\partial r}\left(r\frac{\partial T}{\partial r}\right)+\frac{\partial^2 T}{\partial z^2}=\frac{\rho C_p}{k}\frac{\partial T}{\partial t} \qquad \text{V}$$

for heat conduction in cylindrical polar coordinates with axial symmetry.

8.4.3. *Third Type, Mixed Conditions*

This type of boundary condition consists of those in which the derivative of the dependent variable is related to the boundary value of the dependent variable by a linear equation. The most usual case of this is when the surface rate of heat loss is governed by a heat transfer coefficient. Referring back to Fig. 8.5, if the surface ABCD at $x = 0$ is cooling to surroundings at a temperature (T_0) through a gas film giving a resistance which can be expressed in terms of a heat transfer coefficient (h), the boundary condition can be stated thus.

Rate at which heat is removed from the surface $= h(T-T_0)$ per unit area. Also, the rate at which heat is conducted to the surface internally $= k\,\partial T/\partial x$ per unit area. These two rates must be equal, and hence

$$k\frac{\partial T}{\partial x}=h(T-T_0) \qquad (8.63)$$

The above derivation assumes that curvature, time, and other space variations do not affect the balance, thus using the results of the previous section.

Equation (8.63) is true for any surface which faces in the negative direction of the x axis, but the sign of the left-hand side must be reversed if the surface faces in the positive direction. The signs must not be changed according to whether the surface is being heated or cooled. The only negative signs arise from the following two considerations;

(i) The heat conduction equation contains a negative sign by definition [cf. equation (1.48)].

(ii) The outward normal from the surface may be in the negative co-ordinate direction.

Equation (8.63) contains both of these negative signs and is correct as written.

8.4.4. *Integro-differential Boundary Condition*

This type of boundary condition arises most frequently in the study of mass transfer when the material crossing the boundary either enters or leaves a restricted volume and contributes to a modified driving force. Thus consider the following example.

Example 3. A solute is to be leached from a collection of porous spheres by stirring them as a suspension in a solvent. What is the correct boundary condition at the surface of one of the spheres?

The rate at which material diffuses to the surface of a porous sphere of radius a is $-4\pi a^2 D (\partial c/\partial r)|_a$, where D is an effective diffusivity and c is the concentration within the sphere. If V is the volume of solvent and C is the concentration in the bulk of the solvent, then

$$V \frac{\partial C}{\partial t} = -4N\pi a^2 D \frac{\partial c}{\partial r}\bigg|_a \qquad\qquad \text{I}$$

where N is the number of spheres. For continuity of concentration, $c = C$ at $r = a$, and equation I must be integrated to determine C. Hence the boundary condition takes the form:

$$\text{at} \quad r = a, \quad C = -\int_0^t \frac{4N\pi Da^2}{V} \frac{\partial c}{\partial r}\bigg|_a \, dt \qquad\qquad \text{II}$$

8.4.5. *Initial Value and Boundary Value Problems*

So far, the different types of boundary condition have been classified but the number of conditions required for the solution of a particular problem has not been defined. For ordinary differential equations, it was stated that the number of boundary conditions was equal to the order of the differential equation, but no corresponding rule has yet been discovered for partial differential equations. It is possible, however, to give a guiding hint which will help in the more usual problems.

Considering any one independent variable, say x, and assuming that the dependent variable is a function of x only so that all derivatives with respect to the other variables are zero, then the partial differential equation becomes an ordinary differential equation in terms of x. This simplified problem will require as many boundary conditions as its order and they must be given at fixed values of x. This number of boundary conditions is frequently the same as the number required by the corresponding partial differential equation at fixed values of x. Treating the other independent variables in the same manner gives the number of conditions needed at the boundaries of the other variables. Thus, for the equation of heat conduction in one dimension derived in Section 8.3.1,

$$\alpha \frac{\partial^2 T}{\partial x^2} = \frac{\partial T}{\partial t} \qquad\qquad (8.47)$$

this rule states that two boundary conditions are needed at fixed values of x and one at a fixed value of t. This rule cannot be relied upon in every case but is meant as a guide for common problems. For example, if it is applied to equation (8.55) it indicates that one boundary condition is required at fixed t, two at fixed r, and two at fixed θ, and yet it is not normally necessary to specify any conditions at fixed θ except that the solution should repeat itself at intervals of 2π.

The boundary conditions lead to an important classification of partial differential equations as follows. When only one condition is needed in a particular variable it is specified at one fixed value of that variable, whereas when two or more conditions are needed, they can all be specified at one value of the variable, or some can be specified at one value and the rest at another value. In the latter case, conditions are established at both ends of a range of values of an independent variable and the range is said to be "closed" by conditions at the beginning and the end of the range. Alternatively, if only one condition is stated, or all conditions are stated at one fixed value of the variable, then the behaviour of the dependent variable is restricted at the beginning of a range but no end is specified and the range is said to be "open" as far as that independent variable is concerned. If the range is closed for every independent variable, the problem is alternatively named a "boundary value" or "jury" problem. If the range of any independent variable is open, the problem is called an "initial value" or "marching" problem. This classification affects the choice of a method of solution. Thus the method of the Laplace transform, to be given in Section 8.8, can only be applied to initial value problems because the variable removed by the transformation must have an open range. When numerical methods are considered in Chapter 11 it will be seen that the calculation techniques are quite different depending on whether the problem is initial value or boundary value and that the methods are not interchangeable.

8.5. Particular Solutions of Partial Differential Equations

The general solution of a partial differential equation is not often possible, but when the boundary conditions are taken into account useful particular solutions can often be found. The methods given in this section for finding such particular solutions are the simplest to follow but the most difficult to initiate. In Section 8.7, the more powerful method of separation of variables will be described, but again the resultant expression is a collection of particular solutions which become of general use only when the conditions given in Section 8.6 are satisfied.

The other methods using various transforms are treated in Sections 8.8 and 8.9 and they will determine general solutions for particular combinations of boundary conditions classified in Section 8.4. However, a partial differential equation cannot be reduced to an ordinary differential equation by using the Laplace transform, which is the most useful of the many transform methods, unless all variables except one have open ranges. Thus the most that can be obtained in nearly all cases is a particular solution or a combination of particular solutions, and very rarely can a general solution be obtained. The particular solutions are complete and valid, and will satisfy practical requirements but an individual solution has to be obtained for each problem.

8.5.1. *Compounding the Independent Variables into One Variable*

It is always possible to make a single change of variables by replacing one variable with a new variable, but there are exceptional cases in which the elimination of one variable by substitution results in the disappearance of a second variable from the problem. The basis of the present method is to look for this possibility and hence reduce a partial differential equation in two independent variables to an ordinary differential equation.

Consider for example the one-dimensional heat conduction equation derived in Section 8.3.1:

$$\alpha \frac{\partial^2 T}{\partial x^2} = \frac{\partial T}{\partial t} \tag{8.47}$$

It is assumed that the solution of equation (8.47) can be written

$$T = f(q) \tag{8.64}$$

where

$$q = xt^n \tag{8.65}$$

n is any unknown constant and q is the new variable. Equation (8.64) can be differentiated partially with respect to both x and t as follows.

$$\frac{\partial T}{\partial x} = \frac{df}{dq} \frac{\partial q}{\partial x} = t^n \frac{df}{dq} \tag{8.66}$$

$$\therefore \quad \frac{\partial^2 T}{\partial x^2} = t^{2n} \frac{d^2 f}{dq^2} \tag{8.67}$$

and

$$\frac{\partial T}{\partial t} = \frac{df}{dq} \frac{\partial q}{\partial t} = nxt^{n-1} \frac{df}{dq} \tag{8.68}$$

Substituting equations (8.67) and (8.68) into equation (8.47),

$$\alpha t^{2n} \frac{d^2 f}{dq^2} = nxt^{n-1} \frac{df}{dq} \tag{8.69}$$

Using equation (8.65) to eliminate x from equation (8.69) gives

$$\alpha t^{2n} \frac{d^2 f}{dq^2} = nqt^{-1} \frac{df}{dq} \tag{8.70}$$

The above steps are valid for any constant value of n, and in particular they are true for

$$n = -\tfrac{1}{2} \tag{8.71}$$

With this value of n, equation (8.70) becomes independent of t giving the ordinary differential equation

$$\alpha \frac{d^2 f}{dq^2} = -\tfrac{1}{2} q \frac{df}{dq} \tag{8.72}$$

The dependent variable does not occur explicitly in equation (8.72) and hence it can be solved by the method given in Section 2.4.1. After one stage of integration,

$$\frac{df}{dq} = A\,e^{-q^2/4\alpha} \tag{8.73}$$

where A is a constant of integration. A further integration of equation (8.73) gives

$$f = B\,\mathrm{erf}(q/2\sqrt{\alpha}) + C$$

or

$$T = B\,\mathrm{erf}(x/2\sqrt{(\alpha t)}) + C \tag{8.74}$$

where B and C are arbitrary constants.

The kind of situation of which equation (8.74) is the solution can be found by putting $x = 0$ or $t = 0$ and determining the boundary conditions satisfied by equation (8.74).

If

$$x = 0, \quad T = C$$

and if

$$t = 0, \quad T = B + C$$

Thus equation (8.74) describes the temperature distribution as a function of time within a semi-infinite slab of material which is cooled on one surface. The initial temperature throughout the slab is uniform at $B + C$, and C is the temperature of the exposed face. This is the only situation described by equation (8.74); any other type of boundary condition is incompatible with the equation. The above method thus determines a particular solution which will probably not satisfy the required boundary conditions, and therefore the method is not very valuable.

8.5.2. *Superposition of Solutions*

It was shown in Section 2.5 that the complementary function of a linear ordinary differential equation could be added to the particular integral and the combination still satisfied the differential equation. Thus for any linear ordinary differential equation in which each term contains the dependent variable, solutions can be superimposed. It will now be shown that the principle of superposition can be applied to linear partial differential equations provided that every term contains the dependent variable.

Taking the general linear partial differential equation of second order in two independent variables

$$f_1\frac{\partial^2 u}{\partial x^2} + f_2\frac{\partial^2 u}{\partial x\,\partial y} + f_3\frac{\partial^2 u}{\partial y^2} + f_4\frac{\partial u}{\partial x} + f_5\frac{\partial u}{\partial y} + f_6 u = 0 \tag{8.75}$$

where f_n are functions of x and y, and assuming that $u = u_1$ and $u = u_2$ are two different particular solutions of equation (8.75), then the sum

$$u_3 = u_1 + u_2 \tag{8.76}$$

will also be a solution. Differentiating equation (8.76) gives

$$
\left.
\begin{aligned}
\frac{\partial u_3}{\partial x} &= \frac{\partial u_1}{\partial x} + \frac{\partial u_2}{\partial x} \\[2mm]
\frac{\partial^2 u_3}{\partial x^2} &= \frac{\partial^2 u_1}{\partial x^2} + \frac{\partial^2 u_2}{\partial x^2} \\[2mm]
\frac{\partial^2 u_3}{\partial x\,\partial y} &= \frac{\partial^2 u_1}{\partial x\,\partial y} + \frac{\partial^2 u_2}{\partial x\,\partial y}
\end{aligned}
\right\}
\qquad (8.77)
$$

$$\text{etc.}$$

Putting $u = u_3$ into equation (8.75) by using the set of equations (8.77) it can be seen that $u = u_3$ is also a solution when u_1 and u_2 are solutions. This is only true when the dependent variable u occurs once and once only in each term of the partial differential equation, and the above argument can obviously be extended to any number of independent variables and equations of any order. Thus the important result emerges that for a linear partial differential equation, an infinite number of particular solutions can be added together to give a further solution.

One particular solution of the heat conduction equation was found in Section 8.5.1, and using the same symbols, a more basic particular solution can be found by differentiating equation (8.47) with respect to x. Thus

$$
\alpha \frac{\partial^2}{\partial x^2}\left(\frac{\partial T}{\partial x}\right) = \frac{\partial}{\partial t}\left(\frac{\partial T}{\partial x}\right)
$$

which shows that T can be replaced by $\partial T/\partial x$ to give a further particular solution of equation (8.47). Combining equations (8.65, 8.66, 8.71, 8.73) gives

$$
\frac{\partial T}{\partial x} = A t^{-\frac{1}{2}} e^{-x^2/4\alpha t}
$$

$$
\therefore \quad T = A t^{-\frac{1}{2}} e^{-x^2/4\alpha t} \qquad (8.78)
$$

is also a solution of equation (8.47).

Considering unit cross-sectional area of a semi-infinite slab of density ρ and specific heat C_p occupying all space for positive x values, then the total heat content of the slab at any fixed time is given by

$$
H = \int_0^{\infty} \rho C_p T \, \mathrm{d}x \qquad (8.79)
$$

Using the temperature distribution given by equation (8.78),

$$
\therefore \quad H = \rho C_p A t^{-\frac{1}{2}} \int_0^{\infty} e^{-x^2/4\alpha t} \, \mathrm{d}x
$$

Because H is being evaluated at constant t, all of the present calculations are performed with t constant. Changing variables by putting $x = 2z\sqrt{(\alpha t)}$ so that $dx = 2dz\sqrt{(\alpha t)}$, hence

$$H = 2\rho C_p A \sqrt{(\alpha)} \int_0^\infty e^{-z^2} dz \qquad (8.80)$$

The above integral has been evaluated in Section 5.2.

$$\therefore \quad H = \rho C_p A \sqrt{(\alpha\pi)}$$

$$= kA\sqrt{(\pi/\alpha)} \qquad (8.81)$$

because $\qquad \alpha = k/\rho C_p \qquad (7.168)$

Equation (8.81) shows that the amount of heat in the slab remains constant for all time.

The initial temperature distribution is determined by taking the limit, as $t \to 0$, of equation (8.78). L'Hôpital's rule cannot be used because the exponent cannot be expanded by Taylor's theorem in positive powers of t. However, it can be proved that an exponential function tends to its limit more rapidly than any finite power of the variable tends to its limit, but the proof is not elementary. Equation (8.78) thus shows that when $t = 0$, $T = 0$ for all values of x except $x = 0$. At $x = 0$, the exponential factor is unity for all time so that as $t \to 0$, $T \to \infty$. Hence the heat content as given by equation (8.81) is initially concentrated at the face $x = 0$, and as time progresses it is all conducted into the slab according to equation (8.78). The following example illustrates the superposition of solutions of this type.

Example 1. Heat is supplied at a fixed rate Q W/m² to one face of a large rectangular slab of density ρ kg/m³, specific heat C_p J/kg °C, and conductivity k W/m °C, find the variation of surface temperature with time during the early stages after exposure.

The physical picture of the above situation is that heat is arriving at the surface and then being conducted into the slab according to equation (8.78) so that the temperature distribution at any time t is the accumulated effect of the heat entering the surface between zero time and time t. Thus let τ be a new time variable within the range $0 < \tau < t$ so that during a time increment $\delta\tau$, a quantity of heat $Q\,\delta\tau$ will be received by the surface. At time t, this heat will have been conducted into the slab for $(t - \tau)$ seconds and its distribution will be given by equation (8.78) as

$$\delta T = A(t - \tau)^{-\frac{1}{2}} e^{-x^2/4\alpha(t-\tau)} \qquad \text{I}$$

Using equation (8.81)

$$Q\delta\tau = kA\sqrt{(\pi/\alpha)}$$

$$\therefore \quad \delta T = \frac{Q}{k}\sqrt{\left(\frac{\alpha}{\pi}\right)}(t - \tau)^{-\frac{1}{2}} e^{-x^2/4\alpha(t-\tau)} \delta\tau \qquad \text{II}$$

Equation II gives the part of the temperature distribution at time t which is the result of the heat absorbed by the surface between time τ and $\tau + \delta\tau$. Hence the total temperature distribution can be obtained by adding together all of these temperature increments and taking the limit as $\delta\tau \to 0$. Thus

$$T = \frac{Q}{k} \sqrt{\left(\frac{\alpha}{\pi}\right)} \int_0^t (t-\tau)^{-\frac{1}{2}} e^{-x^2/4\alpha(t-\tau)} \, d\tau \qquad \text{III}$$

Equation III can be integrated in terms of the error function by making the substitution $t - \tau = 1/z^2$, but as the temperature variation at the surface is all that is required by the problem, x can be put equal to zero in equation III giving the surface temperature

$$T_s = \frac{Q}{k} \sqrt{\left(\frac{\alpha}{\pi}\right)} \int_0^t (t-\tau)^{-\frac{1}{2}} \, d\tau$$

$$= \frac{2Q}{k} \sqrt{\left(\frac{\alpha t}{\pi}\right)} \qquad \text{IV}$$

Other problems can be solved by this method, but if Q is a function of time it must remain inside the integral in equation III as $Q(\tau)$.

8.5.3. *The Method of Images*

The solutions of the conduction equation considered so far have only applied to a semi-infinite body. It is possible to extend the solutions to finite media for two types of boundary condition. Firstly, if one plane surface is insulated or the finite body has a plane of symmetry and, secondly, if the boundary condition is of the first type.

Example 2. Consider a wall of thickness $2L$ with a uniform initial temperature throughout, and let both faces be suddenly raised to the same higher temperature. Find the temperature distribution as a function of time.

Taking an x coordinate measured from one face of the slab, the initial uniform temperature as the datum and considering unit cross-sectional area, the differential equation is again

$$\alpha \frac{\partial^2 T}{\partial x^2} = \frac{\partial T}{\partial t} \qquad \text{I}$$

and the boundary conditions are

$$\text{at} \quad t = 0, \quad T = 0 \qquad \text{II}$$

$$\text{at} \quad x = 0, \quad T = T_0 \qquad \text{III}$$

$$\text{at} \quad x = L, \quad \frac{\partial T}{\partial x} = 0 \qquad \text{IV}$$

The last condition takes advantage of the symmetry of the problem so that no heat crosses the central plane. According to Section 8.5.1, the solution which satisfies conditions II and III for an infinite slab is

$$T_1 = T_0 - T_0 \operatorname{erf} [x/2 \sqrt{(\alpha t)}]$$

$$= T_0 \operatorname{erfc} [x/2 \sqrt{(\alpha t)}] \qquad\qquad \text{V}$$

This is satisfactory for small values of t, but as soon as the temperature at $x = L$ starts to rise, equation IV is violated. Since the system is symmetrical, the insulated face at $x = L$ is equivalent to another heated face at $x = 2L$ (this in fact returns to the original problem). But equation V gives the temperature distribution as a function of distance x from any heated face. The distance of a point in the body from the second heated face is $(2L - x)$, and hence the temperature distribution resulting from the heat entering the face at $x = 2L$ is given by

$$T_2 = T_0 \operatorname{erfc} [(2L - x)/2 \sqrt{(\alpha t)}] \qquad\qquad \text{VI}$$

by analogy with equation V. The combined solution becomes

$$T_3 = T_0 \operatorname{erfc} [x/2 \sqrt{(\alpha t)}] + T_0 \operatorname{erfc} [(2L - x)/2 \sqrt{(\alpha t)}] \qquad\qquad \text{VII}$$

The solution T_3 satisfies equation IV but as t increases further, the second part of equation VII violates equation III. If a heated face at $x = 2L$ causes the temperature at $x = 0$ to rise at a certain rate, then a correspondingly cooled surface at $x = -2L$ will cause a corresponding fall rate at $x = 0$. The temperature distribution caused by such a cooled face at $x = -2L$ is given by

$$T_4 = - T_0 \operatorname{erfc} [(2L + x)/2 \sqrt{(\alpha t)}] \qquad\qquad \text{VIII}$$

which can be added to equation VII to satisfy equation III again. Thus

$$T_5 = T_0 \operatorname{erfc} [x/2 \sqrt{(\alpha t)}] + T_0 \operatorname{erfc} [(2L - x)/2 \sqrt{(\alpha t)}] - T_0 \operatorname{erfc} [(2L + x)/2 \sqrt{(\alpha t)}]$$

$$\text{IX}$$

The distribution given by equation IX now violates equation IV again but a further reflection at the plane $x = L$ gives

$$T_6 = - T_0 \operatorname{erfc} [(4L - x)/2 \sqrt{(\alpha t)}] \qquad\qquad \text{X}$$

which can be added to equation IX. This process can be continued indefinitely, by treating each face at $x = 0$ and $x = L$ like a perfectly reflecting mirror so that any source to the left of $x = L$ possesses a corresponding image source to the right of $x = L$. Also, any source to the right of $x = 0$ must have a corresponding image sink to the left of $x = 0$. Thus the insulated face (or plane of symmetry) acts as a perfect reflector as far as sources and sinks are concerned, and the constant temperature face acts as a reflector which interchanges sources and sinks upon reflection. The complete solution is thus

$$T = T_0 \sum_{n=0}^{\infty} (-1)^n \{\text{erfc}\,[(x+2nL)/2\,\sqrt{(\alpha t)}] + \text{erfc}\,[(2nL+2L-x)/2\,\sqrt{(\alpha t)}]\} \qquad \text{XI}$$

This method of images was originally developed to solve linear problems in electricity and magnetism, but it can be applied to any problem involving the solution of a linear partial differential equation where superposition of solutions is permissible and the boundary conditions are of the first type, or of the second type with vanishing derivative.

It is possible to extend this image principle to cylindrical and spherical boundaries but this is rather more complicated and so specialized that the method will not be considered here.

8.6. ORTHOGONAL FUNCTIONS

Before the method of separation of variables can be used to solve partial differential equations, a fundamental relationship between the individual members of a set of functions must be investigated. The method of separation of variables normally gives the solution as the sum of an infinite set of related functions, and the method is most useful when this set of functions satisfies the following definition of orthogonality.

Two functions $\phi_n(x)$ and $\phi_m(x)$ are said to be orthogonal with respect to a weighting function $r(x)$ over the interval from a to b if

$$\int_a^b r(x)\,\phi_n(x)\,\phi_m(x)\,dx = 0 \quad \text{(for } m \neq n) \tag{8.82}$$

Thus, if there is a set of functions $\phi_n(x)$ where n is a positive integer, and the integrated product of any pair of functions with a fixed weighting function is zero, then the set of functions is orthogonal. If $m = n$ (i.e. the same function is taken twice) then the integral in equation (8.82) will not normally vanish.

Example 1. Show that the functions $\sin nx$ and $\sin mx$ are orthogonal with respect to unity in the range $0 \leqslant x \leqslant \pi$, for integer values of m and n.
Starting from the trigonometrical identity

$$2 \sin nx \sin mx = \cos (n-m)\, x - \cos (n+m)\, x \qquad \text{I}$$

and integrating over the given range, therefore

$$2 \int_0^\pi \sin nx \sin mx\, dx = \int_0^\pi \cos (n-m)\, x\, dx - \int_0^\pi \cos (n+m)\, x\, dx \qquad \text{II}$$

$$= \frac{1}{n-m} [\sin (n-m)\, x]_0^\pi - \frac{1}{n+m} [\sin (n+m)\, x]_0^\pi$$

$$= 0$$

because the sine functions vanish at both limits of integration. If $m = n$, the first integral in equation II becomes

$$\int_0^\pi dx = \pi \qquad\qquad \text{III}$$

$$\therefore \quad \int_0^\pi \sin nx \sin mx\, dx = \begin{cases} 0 & \text{if } n \neq m \\ \pi/2 & \text{if } n = m \end{cases} \qquad\qquad \text{IV}$$

which is the complete evaluation of the orthogonality integral.

8.6.1. *The Sturm–Liouville Equation*

This is the name which describes any second order linear ordinary differential equation which can be written in the form

$$\frac{d}{dx}\left[p(x)\frac{dy}{dx}\right] + [q(x) + \lambda r(x)]y = 0 \qquad\qquad (8.83)$$

where λ is a constant, and p, q, and r are functions of x. It should be noted that the first and second differential terms of a second order equation can always be combined into the form given in equation (8.83) by using an integrating factor as described in Section 2.3.4.

For any particular set of functions p, q, and r, the solution of equation (8.83) for y will depend upon λ. Hence, if λ can take a set of discrete values λ_n, a corresponding set of solutions $y = \phi_n(x)$ will be obtained. Taking any two values of λ (λ_n and λ_m) and the corresponding solutions [$y = \phi_n(x)$ and $y = \phi_m(x)$], then

$$\frac{d}{dx}\left[p(x)\frac{d\phi_n}{dx}\right] + [q(x) + \lambda_n r(x)]\phi_n = 0$$

and

$$\frac{d}{dx}\left[p(x)\frac{d\phi_m}{dx}\right] + [q(x) + \lambda_m r(x)]\phi_m = 0$$

Multiplying the first of these equations by ϕ_m, the second by ϕ_n, and subtracting the second from the first gives

$$\phi_m \frac{d}{dx}\left[p(x)\frac{d\phi_n}{dx}\right] - \phi_n \frac{d}{dx}\left[p(x)\frac{d\phi_m}{dx}\right] + (\lambda_n - \lambda_m)r(x)\phi_n\phi_m = 0 \qquad (8.84)$$

Integrating equation (8.84) with respect to x from a to b gives

$$(\lambda_n - \lambda_m)\int_a^b r(x)\phi_n\phi_m\, dx = \int_a^b \phi_n \frac{d}{dx}\left[p(x)\frac{d\phi_m}{dx}\right]dx - \int_a^b \phi_m \frac{d}{dx}\left[p(x)\frac{d\phi_n}{dx}\right]dx$$

Evaluating each of the integrals on the right-hand side by parts gives

$$(\lambda_n - \lambda_m) \int_a^b r(x)\phi_n\phi_m \, dx = \left[\phi_n p(x)\frac{d\phi_m}{dx}\right]_a^b - \int_a^b p(x)\frac{d\phi_m}{dx}\frac{d\phi_n}{dx}dx$$

$$- \left[\phi_m p(x)\frac{d\phi_n}{dx}\right]_a^b$$

$$+ \int_a^b p(x)\frac{d\phi_n}{dx}\frac{d\phi_m}{dx}dx$$

Simplifying by cancelling the two identical integrals and combining the boundary values gives

$$(\lambda_n - \lambda_m) \int_a^b r(x)\phi_n\phi_m \, dx = \left[p(x)\left(\phi_n\frac{d\phi_m}{dx} - \phi_m\frac{d\phi_n}{dx}\right)\right]_a^b \qquad (8.85)$$

Hence the functions $\phi_n(x)$ and $\phi_m(x)$, which correspond to two different values (λ_n and λ_m) of the constant, will be orthogonal if the right-hand side of equation (8.85) is zero. It is possible for the function on the right-hand side of (8.85) to have the same value at both limits and thus cancel itself out, but it is more usual for the right-hand side to vanish independently at both limits. The term is zero when $p(x) = 0$ at the limit or when the boundary conditions are suitable. When the same boundary conditions apply to every equation of the set, independent of the value of λ, there are three types of boundary condition which are suitable.

(i) $\qquad\qquad$ at $\quad x = a, \quad y = 0$

$$\therefore \quad \phi_n = \phi_m = 0$$

(ii) $\qquad\qquad$ at $\quad x = a, \quad \frac{dy}{dx} = 0$

$$\therefore \quad \frac{d\phi_n}{dx} = \frac{d\phi_m}{dx} = 0$$

(iii) $\qquad\qquad$ at $\quad x = a, \quad \frac{dy}{dx} = \beta y$

$$\therefore \quad \frac{d\phi_n}{dx} = \beta\phi_n, \quad \frac{d\phi_m}{dx} = \beta\phi_m$$

Therefore, if both boundary conditions are of the above three types (with duplication allowed), then the set of functions $\phi_n(x)$ will be orthogonal with respect to the weighting function $r(x)$ which occurs in equation (8.83) in conjunction with λ.

It should be noted that the conditions given above for the solutions to be orthogonal are more restrictive than the three types of boundary conditions listed in Section 8.4. Condition (i) states that the boundary value of the dependent variable must be zero and cannot be a function of another independent variable. This condition can be relaxed a little to allow the dependent variable to have a constant finite value at the boundary by moving the origin so that the new dependent variable will be zero at the boundary. If condition (i) is used at both boundaries, the origin cannot be moved to two different places, but one method of overcoming this difficulty will be described in Example 2 of Section 8.7. Thus most type 1 boundary conditions will give rise to an orthogonal set of solutions to a Sturm–Liouville equation, provided that there is no dependence upon another independent variable.

There is no easy way of extending condition (ii) to include a constant gradient at the boundary, hence only a vanishing derivative of type 2 boundary conditions will give orthogonal solutions. The same remarks apply to condition (iii) as to condition (i). That is, any type 3 boundary condition is suitable provided that only one movement of the origin is required to bring the condition at both boundaries to one of the forms (i), (ii), or (iii). Condition (iii) does include conditions (i) and (ii) as the special cases $\beta = \infty$, and $\beta = 0$ respectively.

Finally, the integro-differential boundary condition given in Section 8.4.4 is not suitable for obtaining orthogonal solutions to a Sturm–Liouville equation.

8.7. METHOD OF SEPARATION OF VARIABLES

When a dependent variable, say T, is related to two independent variables x and t by means of a partial differential equation, the solution will be of the form

$$T = F(x, t) \tag{8.86}$$

The method of separation of variables looks for solutions of the form

$$T = f(x) . g(t) \tag{8.87}$$

where the functional dependence upon x is separated from the dependence upon t. A glance at the particular solutions (8.74) or (8.78) of equation (8.47) is sufficient to show that they cannot be expressed in the form (8.87), and thus any solution which is obtained by the method of separation of variables will be a new one.

The procedure to follow in this method is to substitute equation (8.87) into the differential equation and separate the terms dependent upon x from those dependent upon t by the equality sign. The equation thus takes the form of a function of x only, being equal to a function of t only, which can only be satisfied by both sides of the equation having the same constant value. Thus, two ordinary differential equations, each involving an unknown parameter, are derived from one partial differential equation. The procedure can be better understood in reference to particular examples, and special applications of the time-dependent Laplace equation will be used in the following subsections because it occurs so frequently in chemical engineering. The three most common coordinate systems will be used and the description will be related to simplified forms of the heat conduction equation

$$\alpha \nabla^2 T = \frac{\partial T}{\partial t} \tag{8.88}$$

for convenience. With a suitable interchange of symbols and interpretation of boundary conditions, the solutions could apply equally well to the mass diffusion equation

$$D \nabla^2 c = \frac{\partial c}{\partial t} \tag{8.89}$$

or any other equation of the same form.

8.7.1. *Unsteady-state Linear Heat Conduction*

Equation (8.88) in cartesian coordinates takes the familiar form

$$\alpha \frac{\partial^2 T}{\partial x^2} = \frac{\partial T}{\partial t} \tag{8.47}$$

Assuming that the solution can be expressed by

$$T = f(x) . g(t) \tag{8.87}$$

then

$$\frac{\partial T}{\partial t} = f(x) . g'(t) \tag{8.90}$$

and

$$\frac{\partial^2 T}{\partial x^2} = f''(x) . g(t) \tag{8.91}$$

where dashes denote differentiation with respect to the argument of the function. Since $f(x)$ is a function of x only, the differentiation can only be performed with respect to x; and similarly, $g(t)$ can only be differentiated with respect to t. It is customary to drop the arguments at this stage and refer back to equation (8.87) if necessary. Substituting equations (8.90) and (8.91) into (8.47) gives

$$\alpha f'' g = f g' \tag{8.92}$$

The variable dependence can be separated by dividing by $\alpha f g$, thus

$$\frac{f''}{f} = \frac{g'}{\alpha g} \tag{8.93}$$

The left-hand side of equation (8.93) is independent of t and the ratio f''/f must therefore be constant for all time. Similarly, the ratio $g'/\alpha g$ must be constant at all positions because it is independent of x. Hence each half of equation (8.93) must be equal to the same constant denoted by $-\lambda$, thus

$$f'' + \lambda f = 0 \tag{8.94}$$

and

$$g' + \lambda \alpha g = 0 \tag{8.95}$$

Equations (8.94) and (8.95) are both linear ordinary differential equations with constant coefficients and the solutions are

$$f = A \sin \sqrt{(\lambda)}\,x + B \cos \sqrt{(\lambda)}\,x \tag{8.96}$$

and
$$g = C e^{-\lambda \alpha l} \tag{8.97}$$

if λ is not zero; or

$$f = A_0 x + B_0 \tag{8.98}$$

and
$$g = C_0 \tag{8.99}$$

if λ is zero. Putting equations (8.96) to (8.99) into (8.87) gives the particular solutions

$$T = A_0 x + B_0 \tag{8.100}$$

and
$$T = [A_\lambda \sin \sqrt{(\lambda)}\,x + B_\lambda \cos \sqrt{(\lambda)}\,x]\,e^{-\lambda \alpha l} \tag{8.101}$$

where the arbitrary constant C_0 has been "absorbed" into A_0 and B_0, and the subscripts λ denote that the arbitrary constants in equation (8.101) will depend upon the choice of λ.

A_λ, B_λ, and λ must be chosen to satisfy the boundary conditions and the simplest assumption will be made that

$$\text{at} \quad x = 0, \quad T = 0 \tag{8.102}$$

and
$$\text{at} \quad x = L, \quad T = 0 \tag{8.103}$$

To satisfy condition (8.102), B_λ must be zero, and condition (8.103) then gives $A_0 = 0$ and

$$0 = A_\lambda [\sin \sqrt{(\lambda)}\,L]\,e^{-\lambda \alpha l}$$

which means that $A_\lambda = 0$ unless

$$\sqrt{(\lambda)}\,L = n\pi \tag{8.104}$$

Hence, the only solutions of equation (8.47) having the form (8.101) which also satisfy the boundary conditions (8.102) and (8.103) are given by

$$T = A_n \sin \frac{n\pi x}{L} e^{-n^2 \pi^2 \alpha l / L^2} \tag{8.105}$$

where n is an integer and the subscript has been altered for convenience. Thus the most general solution of this type can be found by superposing the particular solutions (8.105).

$$T = \sum_{n=1}^{\infty} A_n \sin \frac{n\pi x}{L} e^{-n^2 \pi^2 \alpha l / L^2} \tag{8.106}$$

The final boundary condition which is usually imposed on solutions of equation (8.47) is one specifying the initial temperature distribution when $t = 0$. Thus, let

$$T = f_0(x) \quad \text{when} \quad t = 0 \tag{8.107}$$

and substitute into equation (8.106). Therefore

$$f_0(x) = \sum_{n=1}^{\infty} A_n \sin \frac{n\pi x}{L} \qquad (8.108)$$

Choosing the constants A_n to satisfy equation (8.108) is accomplished by using the properties of orthogonal functions. Comparing equation (8.94) with the general Sturm–Liouville equation (8.83) shows that if $p(x) = r(x) = 1$, $q(x) = 0$, then equation (8.94) is of Sturm–Liouville type. The boundary conditions (8.102) and (8.103) are of the suitable form (i) and thus the functions $\sin(n\pi x/L)$ are orthogonal with respect to unity in the range 0 to L. This property is utilized by multiplying both sides of equation (8.108) by any other function from the set $\sin(m\pi x/L)$ and integrating the result from 0 to L. Thus

$$\int_0^L f_0(x) \sin \frac{m\pi x}{L} dx = \sum_{n=1}^{\infty} \int_0^L A_n \sin \frac{n\pi x}{L} \sin \frac{m\pi x}{L} dx$$

$$= \int_0^L A_m \left(\sin \frac{m\pi x}{L} \right)^2 dx \qquad (8.109)$$

by using the orthogonality property (8.82). Evaluating the final integral in equation (8.109) gives

$$A_m = \frac{2}{L} \int_0^L f_0(x) \sin \frac{m\pi x}{L} dx \qquad (8.110)$$

which can be evaluated from the known initial distribution $f_0(x)$. Substitution of the value given by equation (8.110) into equation (8.106) gives the final solution

$$T = \sum_{n=1}^{\infty} \frac{2}{L} \sin \frac{n\pi x}{L} \int_0^L f_0(x) \sin \frac{n\pi x}{L} dx \, e^{-n^2\pi^2\alpha t/L^2} \qquad (8.111)$$

It is convenient to introduce the nomenclature of the method in relationship to the above calculation. The method relies upon the feasibility of separating the variable dependence as in equation (8.93) so that a "separation constant" $(-\lambda)$ can be introduced. This constant can be zero, positive, negative or even complex, and the choice is usually made on a physical basis. The λ used in the above calculation has to be real and positive to make the solution of equation (8.95) physically reasonable. Thus equation (8.97) gives the time dependence of the particular solution as a decaying exponential for positive values of λ. If λ were complex, its real part would have to be positive to control the temperature at large times, and its imaginary part would introduce hyperbolic functions into equation (8.101). The hyperbolic cosine term would be removed by condition (8.102) and the hyperbolic sine term by condition (8.103). Hence only real positive values of λ are permitted by the boundary conditions. It is not always possible to substitute the right form for the constant initially, but the right form is usually apparent at a later stage. In most of the illustrations of the method in this book, the most convenient form will be chosen at the start for ease of presentation, although in an actual problem this form would not be known.

Any other constant (such as α in the above problem) is normally transferred to the simpler side of the separated equation (8.93) for algebraic convenience. When the equivalent of equation (8.104) is reached, the separation constant is usually restricted to an infinite number of discrete values which can be labelled by an integer suffix. Thus

$$\lambda_n = (n\pi/L)^2 \tag{8.112}$$

and these are called the "eigen-values" of the problem because they are the only permitted values of the separation constant. The functions arising from the eigen-values are called the "eigen-functions".

By comparing the ordinary differential equation and its boundary conditions which determine the eigen-values with the conditions given in Section 8.6.1, the orthogonality of the eigen-functions can be proved without the necessity of performing the integrations. Hence, the last set of arbitrary constants is determined by the orthogonality property.

8.7.2. *Steady-state Heat Transfer with Axial Symmetry*

Laplace's equation in spherical polar coordinates with axial symmetry was derived in Section 8.3.2 for the analogous process of mass transfer in the more general case in which fluid velocity was allowed. Putting $u = v = 0$, removing the time derivative from equation (8.52), and replacing c by T gives

$$\frac{\partial}{\partial r}\left(r^2 \frac{\partial T}{\partial r}\right) + \frac{1}{\sin\theta}\frac{\partial}{\partial\theta}\left(\sin\theta\frac{\partial T}{\partial\theta}\right) = 0$$

or
$$r^2\frac{\partial^2 T}{\partial r^2} + 2r\frac{\partial T}{\partial r} + \frac{\partial^2 T}{\partial\theta^2} + \cot\theta\frac{\partial T}{\partial\theta} = 0 \tag{8.113}$$

Looking for solutions of the form

$$T = f(r)g(\theta) \tag{8.114}$$

by substituting into equation (8.113) gives

$$r^2 f''g + 2rf'g + fg'' + \cot\theta\, fg' = 0 \tag{8.115}$$

Dividing by fg and separating the variables gives

$$\frac{r^2 f'' + 2rf'}{f} = -\frac{g'' + \cot\theta\, g'}{g} = l(l+1) \tag{8.116}$$

The conventional choice of separation constant in this case is $l(l+1)$ for convenience of presentation. λ can be chosen, but the algebra becomes formidable and the above form of separation constant is then derived in the argument leading to equation (8.132). The first part of equation (8.116) is

$$r^2\frac{d^2 f}{dr^2}+2r\frac{df}{dr}-l(l+1)f = 0 \qquad (8.117)$$

which is linear and homogeneous with a solution of the form

$$f = Ar^n \qquad (8.118)$$

$$\therefore \quad n(n-1)Ar^n+2nAr^n-l(l+1)Ar^n = 0$$

$$\therefore \quad n^2+n-l^2-l = 0$$

$$\therefore \quad (n-l)(n+l+1) = 0$$

$$\therefore \quad f = Ar^l+Br^{-l-1} \qquad (8.119)$$

which partially explains the choice of separation constant.

The second part of equation (8.116) can be written

$$\frac{d^2 g}{d\theta^2}+\cot\theta\frac{dg}{d\theta}+l(l+1)g = 0 \qquad (8.120)$$

Because of the product $\sin\theta\,d\theta$ in the denominator of the second term, it is convenient to change the dependent variable by putting

$$m = \cos\theta \quad \text{and} \quad dm = -\sin\theta\,d\theta \qquad (8.121)$$

Hence,

$$(1-m^2)\frac{d^2 g}{dm^2}-2m\frac{dg}{dm}+l(l+1)g = 0 \qquad (8.122)$$

which is known as "Legendre's equation of order l". This has to be solved by the method of Frobenius given in Chapter 3. Thus let

$$g = \sum_0^\infty a_n m^{n+c}$$

$$\therefore \quad (1-m^2)\sum_0^\infty (n+c)(n+c-1)a_n m^{n+c-2}$$

$$-2m\sum_0^\infty (n+c)a_n m^{n+c-1}+l(l+1)\sum_0^\infty a_n m^{n+c} = 0 \qquad (8.123)$$

The indicial equation is found by equating coefficients of m^{c-2}, hence

$$c(c-1)a_0 = 0$$

$$\therefore \quad c = 0 \quad \text{or} \quad 1 \qquad (8.124)$$

Equating coefficients of m^{c-1} gives

$$(c+1)ca_1 = 0$$

$$\therefore \quad c = 0 \quad \text{or} \quad a_1 = 0 \qquad (8.125)$$

The recurrence relation obtained by equating coefficients of m^{c+s} is

$$(s+c+2)(s+c+1)a_{s+2} = (s+c-l)(s+c+l+1)a_s \qquad (8.126)$$

The roots of the indicial equation differ by an integer, but since a_1 is indeterminate when $c = 0$, the solution falls into Case IIIb and the method yields two normal ascending power series solutions. As stated in Section 3.4.5, the complete solution can be obtained in this case by putting $c = 0$ and leaving a_0 and a_1 as the arbitrary constants. Thus, the recurrence relation (8.126) becomes

$$(s+2)(s+1)a_{s+2} = (s-l)(s+l+1)a_s \qquad (8.127)$$

Putting $s = 2p$ will give the recurrence relation governing the series multiplied by a_0, and putting $s = 2p+1$ will give the other series multiplied by a_1. Thus

$$(2p+2)(2p+1)a_{2p+2} = (2p-l)(2p+l+1)a_{2p} \qquad (8.128)$$

and

$$(2p+3)(2p+2)a_{2p+3} = (2p-l+1)(2p+l+2)a_{2p+1} \qquad (8.129)$$

If the solutions are to be useful, they must be convergent. Comparing equation (8.122) with the standard form (3.20) shows that equations (3.21, 3.22) become

$$F(m) = \frac{-2m^2}{1-m^2} \qquad (8.130)$$

$$G(m) = \frac{l(l+1)m^2}{1-m^2} \qquad (8.131)$$

the series expansions of which have a radius of convergence $R = 1$. Clearly, when $m = \pm 1$, the expansions of (8.130, 8.131) are not convergent so the series solutions will not be convergent at $m = \pm 1$, i.e. along the axis $\theta = 0$. However, there are special values of l which are exceptions to the above analysis, and these provide the solutions.

Suppose $l = 1$. Then equation (8.129) with $p = 0$ gives

$$(3)(2)a_3 = (0)(3)a_1$$

that is, $$a_3 = 0$$

$$\therefore \quad a_5 = a_7 = a_9 = \ldots = 0$$

Hence, when $l = 1$, $g = a_1 m$ is a solution of equation (8.122). This can easily be verified by substitution, and the solution is finite when $m = 1$. The solution is thus physically reasonable.

It can readily be seen that if l is an odd integer, there will be a value of p making the right-hand side of equation (8.129) zero and thus giving a polynomial of finite degree. Similarly, if l is any even integer, there will be a value of p making the right-hand side of equation (8.128) zero with similar effect. These solutions are called the "Legendre polynomials" and are given the symbol $P_l(m)$. The solution of equation (8.122) is thus

$$g = AP_l(m) + BQ_l(m) \qquad (8.132)$$

where $Q_l(m)$ denotes the infinite series. It is as well to remember that this solution exists mathematically, but it is invariably unreasonable on physical grounds and it will be rejected at this stage.

Combining equations (8.119) and (8.132) according to equation (8.114) gives the particular solution

$$T = (A_l r^l + B_l r^{-l-1}) P_l(\cos \theta) \qquad (8.133)$$

Again, any number of these solutions can be superposed for integer values of l giving

$$T = \sum_{l=0}^{\infty} (A_l r^l + B_l r^{-l-1}) P_l(\cos \theta) \qquad (8.134)$$

Equation (8.122) can be written in the form

$$\frac{d}{dm} \left[(1 - m^2) \frac{dg}{dm} \right] + l(l+1) g = 0 \qquad (8.135)$$

which shows that it is of Sturm–Liouville type with unit weighting function. The function $p(m) = 1 - m^2$ is zero at both limits of integration (-1 and $+1$) and thus the solutions in this range will be orthogonal for any boundary conditions as shown by equation (8.85). Hence the Legendre polynomials are orthogonal with respect to unity in the range from -1 to $+1$.

A convenient way of deriving the polynomials is given by Rodrigue's formula

$$P_n(x) = \frac{1}{2^n n!} \frac{d^n}{dx^n} (x^2 - 1)^n \qquad (8.136)$$

which gives

$$\left. \begin{array}{ll} P_0(x) = 1, & P_2(x) = \frac{1}{2}(3x^2 - 1) \\ P_1(x) = x, & P_3(x) = \frac{1}{2}(5x^3 - 3x) \\ & \text{etc.} \end{array} \right\} \qquad (8.137)$$

for the first few polynomials.

Example 1. Two concentric spherical metallic shells of radii a and b (where $b > a$), are separated by a solid thermal insulator. The inner shell is kept at a constant uniform temperature T_1 by a steady distributed source of heat. The temperature on the outer shell is found experimentally to vary with angular distribution from a fixed point on the surface. The distribution can be specified as

$$T = T_2 + T_3 \cos \theta + T_4 \cos^2 \theta$$

How should T_1, T_2, T_3, and T_4 be plotted against heat input so that the thermal conductivity of the insulator can be found from a straight-line plot of results for different rates of heat input?

The relevant differential equation is Laplace's equation with axial symmetry in spherical polar coordinates (8.113). This equation has just been solved in terms of Legendre polynomials to give the solution

$$T = \sum_{l=0}^{\infty} (A_l r^l + B_l r^{-l-1}) P_l(\cos \theta) \tag{8.134}$$

The boundary conditions given in the problem are:

at $r = a$, $T = T_1 P_0(\cos \theta)$ I

at $r = b$, $T = T_2 + T_3 \cos \theta + T_4 \cos^2 \theta$

$\qquad = \tfrac{2}{3} T_4 P_2(\cos \theta) + T_3 P_1(\cos \theta) + (T_2 + \tfrac{1}{3} T_4) P_0(\cos \theta)$ II

by using equations (8.137). Putting these boundary conditions into equation (8.134) gives

$$\sum_{l=0}^{\infty} (A_l a^l + B_l a^{-l-1}) P_l(\cos \theta) = T_1 P_0(\cos \theta) \tag{III}$$

$$\sum_{l=0}^{\infty} (A_l b^l + B_l b^{-l-1}) P_l(\cos \theta)$$
$$\qquad = (T_2 + \tfrac{1}{3} T_4) P_0(\cos \theta) + T_3 P_1(\cos \theta) + \tfrac{2}{3} T_4 P_2(\cos \theta) \tag{IV}$$

Multiplying both sides of equations III and IV by $P_n(\cos \theta)\,d(\cos \theta)$, integrating over the range $0 \leqslant \theta \leqslant \pi$, and using the orthogonality property of the Legendre polynomials amounts to the same thing as equating coefficients of the same polynomials in equations III and IV. For example, multiplying by $P_1(\cos \theta)\,d(\cos \theta)$ and integrating would remove all terms from both sides except those which already contained $P_1(\cos \theta)$. Hence,

$$A_0 + B_0 a^{-1} = T_1 \qquad\qquad\qquad\qquad \text{V}$$
$$A_l a^l + B_l a^{-l-1} = 0 \quad (\text{for } l > 0) \qquad\qquad \text{VI}$$
$$A_0 + B_0 b^{-1} = T_2 + \tfrac{1}{3} T_4 \qquad\qquad\qquad \text{VII}$$
$$A_1 b + B_1 b^{-2} = T_3 \qquad\qquad\qquad\qquad \text{VIII}$$
$$A_2 b^2 + B_2 b^{-3} = \tfrac{2}{3} T_4 \qquad\qquad\qquad\qquad \text{IX}$$
$$A_l b^l + B_l b^{-l-1} = 0 \quad (\text{for } l > 2) \qquad\qquad \text{X}$$

Although there are an infinite number of equations and unknowns in the above set, they can be solved in pairs. Thus equations V and VII can be solved for A_0 and B_0, and equation VI with $l = 1$ and equation VIII can be solved for A_1 and B_1. Rather than complete the algebra at this stage, it is more efficient to find the required answer in terms of A_l and B_l, then solve sufficient of equations V to X for the constants required.

The rate at which the solid conducts heat away from the sphere of radius a is given by

$$Q = -k \int_0^\pi 2\pi a^2 \left(\frac{\partial T}{\partial r}\right)_{r=a} \sin \theta \, d\theta \qquad \text{XI}$$

Differentiating equation (8.134) with respect to r and substituting into equation XI gives

$$Q = -2\pi ka^2 \int_0^\pi \sum_{l=0}^\infty [lA_l a^{l-1} - (l+1) B_l a^{-l-2}] P_l(\cos \theta) \sin \theta \, d\theta$$

$$= -2\pi ka^2 \sum_{l=0}^\infty [lA_l a^{l-1} - (l+1) B_l a^{-l-2}] \int_{-1}^{+1} P_l(m) \, dm \qquad \text{XII}$$

But the integral occurring in equation XII is the orthogonality integral if $P_0(m) = 1$ is inserted into it. Therefore, all terms after the first are zero in the series in equation XII.

$$\therefore \quad Q = 4\pi kB_0 \qquad \text{XIII}$$

and hence only B_0 need be determined from the set of equations V to X. Thus

$$A_0 + B_0 a^{-1} = T_1$$

$$A_0 + B_0 b^{-1} = T_2 + \tfrac{1}{3} T_4$$

$$\therefore \quad B_0(a^{-1} - b^{-1}) = T_1 - T_2 - \tfrac{1}{3} T_4$$

$$\therefore \quad B_0 = \frac{ab}{b-a}(T_1 - T_2 - \tfrac{1}{3} T_4) \qquad \text{XIV}$$

Substituting into equation XIII gives

$$Q = \frac{4\pi kab}{b-a}(T_1 - T_2 - \tfrac{1}{3} T_4) \qquad \text{XV}$$

Therefore, if Q is plotted as ordinate and $(T_1 - T_2 - \tfrac{1}{3} T_4)$ as abscissa, for a variety of heat supply rates, the gradient will be $4\pi kab/(b-a)$ from which the thermal conductivity (k) can be evaluated.

8.7.3. *Equations Involving Three Independent Variables*

The steady-state flow of heat in a cylinder is governed by Laplace's equation in cylindrical polar coordinates (7.125)

$$\frac{\partial^2 T}{\partial \omega^2} + \frac{1}{\omega} \frac{\partial T}{\partial \omega} + \frac{1}{\omega^2} \frac{\partial^2 T}{\partial \theta^2} + \frac{\partial^2 T}{\partial z^2} = 0 \qquad (8.138)$$

where there are three independent variables ω, θ, and z. This equation has to be solved in stages and solutions are first sought in the form

$$T = f(\omega, \theta) \cdot g(z) \qquad (8.139)$$

Substituting this form into equation (8.138) gives

$$\frac{\partial^2 f}{\partial \omega^2} g + \frac{1}{\omega} \frac{\partial f}{\partial \omega} g + \frac{1}{\omega^2} \frac{\partial^2 f}{\partial \theta^2} g + f g'' = 0$$

Dividing by fg and separating the terms dependent upon z,

$$\frac{1}{f} \left[\frac{\partial^2 f}{\partial \omega^2} + \frac{1}{\omega} \frac{\partial f}{\partial \omega} + \frac{1}{\omega^2} \frac{\partial^2 f}{\partial \theta^2} \right] = \frac{-g''}{g} = -\nu^2 \qquad (8.140)$$

where $-\nu^2$ is the chosen separation constant. Each side of equation (8.140) must still be constant because a function of ω and θ is equated to a function of z only.

The second half of equation (8.140) gives

$$\frac{d^2 g}{dz^2} - \nu^2 g = 0 \qquad (8.141)$$

which has the solution

$$g = A_\nu e^{\nu z} + B_\nu e^{-\nu z} \qquad (8.142)$$

The other half of equation (8.140) gives the partial differential equation

$$\omega^2 \frac{\partial^2 f}{\partial \omega^2} + \omega \frac{\partial f}{\partial \omega} + \frac{\partial^2 f}{\partial \theta^2} + \omega^2 \nu^2 f = 0 \qquad (8.143)$$

which has only two independent variables. Applying the method of separation of variables again, by looking for solutions of the type

$$f(\omega, \theta) = F(\omega) . G(\theta) \qquad (8.144)$$

$$\therefore \quad \omega^2 F'' G + \omega F' G + F G'' + \omega^2 \nu^2 F G = 0 \qquad (8.145)$$

Dividing by FG and separating the variables gives

$$\frac{\omega^2 F'' + \omega F'}{F} + \omega^2 \nu^2 = \frac{-G''}{G} = k^2 \qquad (8.146)$$

Solving the second part as before gives

$$G = A_k \cos k\theta + B_k \sin k\theta \qquad (8.147)$$

The first part of equation (8.146) gives

$$\omega^2 F'' + \omega F' + (\omega^2 \nu^2 - k^2) F = 0$$

Changing the variable by substituting

$$\omega \nu = x \qquad (8.148)$$

$$x^2 \frac{d^2 F}{dx^2} + x \frac{dF}{dx} + (x^2 - k^2) F = 0 \qquad (8.149)$$

which is the standard form of Bessel's equation (3.35).

The most convenient form has been chosen for both separation constants, and k must be an integer for the trigonometrical functions in equation (8.147) to be cyclic with period 2π. The boundary conditions usually make ν real as chosen. Equation (8.149) has the solution

$$F = AJ_k(x) + BY_k(x) \tag{8.150}$$

but at the axis of the cylinder, $Y_k(\nu w)$ is infinite and thus $B = 0$ on physical grounds. Combining equations (8.142, 8.147, 8.150) gives the particular solution

$$T = (A_\nu e^{\nu z} + B_\nu e^{-\nu z}) \cos(k\theta + \varepsilon) J_k(\nu w) \tag{8.151}$$

Solutions of this type can be superposed for integer values of k, and values of ν permitted by the boundary conditions, which involves a double summation.

Bessel's equation (8.149) can be rewritten as

$$\frac{d}{dx}\left(x\frac{dF}{dx}\right) + (x - k^2 x^{-1})F = 0$$

which is of Sturm–Liouville type with a weighting function $(1/x)$. The range of integration is dependent upon ν because of equation (8.148) and it is not useful to proceed further with the general case. However, for suitable boundary conditions, the eigen-functions will be orthogonal because equation (8.149) is of Sturm–Liouville type.

8.7.4. *General Use of the Orthogonality Property*

If the separated partial differential equation is of Sturm–Liouville type (8.83) and the boundary conditions are suitable, the solution of the complete partial differential equation takes the form

$$T = a_0 f_0(t)\phi_0(x) + a_1 f_1(t)\phi_1(x) + \ldots + a_n f_n(t)\phi_n(x) + \ldots \tag{8.152}$$

where t is the other independent variable and $\phi_n(x)$ are orthogonal. The other boundary condition will be given at a fixed value of t [say $t = 0$, $T = f(x)$]. Therefore

$$f(x) = A_0\phi_0(x) + A_1\phi_1(x) + \ldots + A_n\phi_n(x) + \ldots \tag{8.153}$$

Multiplying equation (8.153) by $r(x)\phi_m(x)$ and integrating to generate the orthogonality integrals gives

$$\int_a^b r(x) f(x) \phi_m(x)\,dx = \int_a^b \sum_{n=0}^\infty A_n \phi_n(x) r(x) \phi_m(x)\,dx \tag{8.154}$$

But all terms in the infinite sum are zero except the term when $n = m$. Hence

$$\int_a^b r(x) f(x) \phi_m(x)\,dx = \int_a^b A_m r(x) [\phi_m(x)]^2\,dx$$

or rearranging,

$$A_m = \int_a^b r(x) f(x) \phi_m(x)\,dx \bigg/ \int_a^b r(x) [\phi_m(x)]^2\,dx \tag{8.155}$$

8.7.5. *Eigen-functions not Orthogonal*

Sometimes the variables will separate in the partial differential equation to give an equation of Sturm–Liouville type, but the boundary conditions are such that the resulting eigen-functions are not orthogonal. If an attempt is made to use the orthogonality relationships, the general equation (8.154) is obtained, but the terms in the infinite sum are not zero. Instead of each equation (8.154) being solvable for A_m, each equation contains all of the A_ms and thus an infinite set of equations, each containing an infinite number of unknowns is obtained. Such a system of equations is normally insoluble and an approximate solution has to be found by assuming that $A_7, A_8, ...,$ etc. are all zero and solving the first six equations for the first six coefficients. The accuracy can be improved by taking an increasing number of equations and coefficients, but the effort is rarely worth while and six equations is the normal practical limit. There are, however, still two ways of finding an exact solution and these are illustrated below.

Example 2. A solid rectangular slab at a uniform temperature T_0 has its four edges thermally insulated. The temperature of one exposed face is raised to and maintained at T_1, whilst the temperature of the other exposed face is held constant at T_0. Find how the temperature distribution varies with time.

Because the edges are insulated, heat will only flow in one direction and equation (8.47) will describe the process. The solution follows the method given in Section 8.7.1 identically until the particular solutions (8.100) and (8.101) arise. The boundary conditions can be written:

$$\text{at} \quad x = 0, \quad T = T_0 \qquad\qquad \text{I}$$

and
$$\text{at} \quad x = L, \quad T = T_1 \qquad\qquad \text{II}$$

These do not satisfy the conditions of orthogonality given in Section 8.6.

Substituting equations I and II into the particular solution (8.100) gives

$$T_0 = B_0 \qquad\qquad \text{III}$$

$$T_1 = A_0 L + B_0 \qquad\qquad \text{IV}$$

Hence, the particular solution is

$$T = T_0 + (T_1 - T_0) x/L \qquad\qquad \text{V}$$

and this should be recognized as the steady-state solution.

Equation V is now used to define a new variable Z thus,

$$Z = T - T_0 - (T_1 - T_0) x/L \qquad\qquad \text{VI}$$

Substituting into the differential equation (8.47) gives the new equation

$$\alpha \frac{\partial^2 Z}{\partial x^2} = \frac{\partial Z}{\partial t} \qquad\qquad \text{VII}$$

which is the same as equation (8.47) with T replaced by Z. The boundary conditions I and II become:

$$\text{at} \quad x = 0, \quad Z = 0 \qquad\qquad \text{VIII}$$

$$\text{and at} \quad x = L, \quad Z = 0 \qquad\qquad \text{IX}$$

The equation and boundary conditions in terms of Z are now completely identical with the corresponding equations in Section 8.7.1. Hence, the solution is given by equation (8.111) with T replaced by Z. In the present problem, the final boundary condition that the temperature is T_0 at all points initially becomes

$$Z = -(T_1 - T_0)\, x/L \quad \text{when} \quad t = 0 \qquad\qquad \text{X}$$

Comparing equation X with equation (8.107) shows that

$$f_0(x) = -(T_1 - T_0)\, x/L \qquad\qquad \text{XI}$$

$$\therefore \quad \int_0^L f_0(x) \sin\frac{n\pi x}{L}\, dx = \frac{T_1 - T_0}{n\pi}\left[x\cos\frac{n\pi x}{L} - \int \cos\frac{n\pi x}{L}\, dx \right]_0^L$$

$$= \frac{T_1 - T_0}{n\pi}\left[(-1)^n L \right] \qquad\qquad \text{XII}$$

by integrating by parts and simplifying. Substituting equation XII into the solution (8.111) gives

$$Z = \sum_{n=0}^{\infty} \frac{2}{n\pi}(-1)^n(T_1 - T_0)\sin\frac{n\pi x}{L} e^{-n^2\pi^2\alpha t/L^2} \qquad\qquad \text{XIII}$$

Returning to the original variables by using equation VI,

$$\frac{T - T_0}{T_1 - T_0} = \frac{x}{L} + \sum_{n=0}^{\infty} \frac{2}{n\pi}(-1)^n \sin\frac{n\pi x}{L} e^{-n^2\pi^2\alpha t/L^2} \qquad\qquad \text{XIV}$$

The above method, which should be adopted when the boundary conditions I and II will not give an orthogonal set of eigen-functions, involves finding any particular solution (such as V) which satisfies the conditions; and defining a new variable (Z) so that the new solution consists of an orthogonal set of eigen-functions. In this example, the first eigen-function is given by equation V and provided it is removed from the others an orthogonal set remains and the method can be pursued as if it were normal.

Example 3. In the study of flow distribution in a column packed with woven-wire gauze,† it has been observed that the liquid tends to aggregate at the walls. If the column is a cylinder of radius b m and the feed to the column is distributed within a central core of radius a m with velocity U_0 m/s, determine the fractional amount of liquid on the walls as a function of distance from the inlet in terms of the parameters of the system.

† Porter, K. E. and Jones, M. C. *Trans. Instn. Chem. Engrs.*, **41**, 240 (1963).

The problem is illustrated in Fig. 8.7 where the vertical velocity U m/s is considered to be a function of the radial coordinate r and the vertical coordinate z measured from the inlet. It is assumed that a normal diffusion equation

$$V = -D\frac{\partial U}{\partial r} \qquad\qquad\qquad \text{I}$$

applies to the velocity distribution. V m/s is the horizontal component of fluid velocity and D m is the coefficient governing the equalization of velocities. D will be a property of the packing and of the particular fluid used.

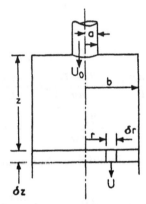

FIG. 8.7. Flow in a packed column

Taking a material balance over the element shown in Fig. 8.7 gives

$$\text{Input} = 2\pi r U\,\delta r - D\frac{\partial U}{\partial r}2\pi r\,\delta z$$

$$\text{Output} = 2\pi r U\,\delta r + \frac{\partial}{\partial z}(2\pi r U\,\delta r)\delta z - D\frac{\partial U}{\partial r}2\pi r\,\delta z$$

$$-\frac{\partial}{\partial r}\left(D\frac{\partial U}{\partial r}2\pi r\,\delta z\right)\delta r$$

Accumulation $= 0$

$$\therefore \quad \frac{\partial}{\partial z}(2\pi r U\,\delta r)\delta z - \frac{\partial}{\partial r}\left(D\frac{\partial U}{\partial r}2\pi r\,\delta z\right)\delta r = 0 \qquad\qquad \text{II}$$

$$\therefore \quad r\frac{\partial U}{\partial z} = D\frac{\partial}{\partial r}\left(r\frac{\partial U}{\partial r}\right)$$

$$\therefore \quad \frac{\partial U}{\partial z} = D\left(\frac{\partial^2 U}{\partial r^2} + \frac{1}{r}\frac{\partial U}{\partial r}\right) \qquad\qquad \text{III}$$

The boundary conditions are as follows:

$$\text{at} \quad z = 0, \quad \text{if } r < a, \quad U = U_0 \quad\Big\}$$
$$\text{if } r > a, \quad U = 0 \quad\Big\} \qquad \text{IV}$$

$$\text{and at} \quad r = 0, \quad U \text{ must be finite} \qquad \text{V}$$

The final boundary condition at $r = b$ must allow fluid to accumulate on the walls and a reasonable assumption is that there is a capacity $2\pi bk$ m²/unit length at $r = b$. It is further assumed that fluid which has diffused to the wall is equally distributed in this capacity, so that the same flow rate exists in this fictitious layer on the wall as in the packing in contact with the wall. This condition can be stated symbolically thus,

$$\text{at} \quad r = b, \quad -D \frac{\partial U}{\partial r} 2\pi b = 2\pi bk \frac{\partial U}{\partial z}$$

that is,
$$-D \frac{\partial U}{\partial r} = k \frac{\partial U}{\partial z} \qquad \text{VI}$$

The problem is thus to solve equation III, satisfy the initial condition IV, and satisfy the boundary conditions V and VI. Trying the method of separation of variables by putting

$$U = f(r)g(z) \qquad \text{VII}$$
$$\therefore \quad fg' = D(f''g + f'g/r)$$

Dividing by Dfg gives

$$\frac{g'}{Dg} = \frac{f'' + f'/r}{f} = -\alpha^2 \qquad \text{VIII}$$

where $-\alpha^2$ is the chosen separation constant. The first part of the equation gives

$$\frac{dg}{dz} + \alpha^2 Dg = 0$$

which has the solution

$$g = A' e^{-\alpha^2 Dz} \qquad \text{IX}$$

indicating that α is real and has been chosen correctly.

The second part of equation VIII gives

$$r^2 \frac{d^2 f}{dr^2} + r \frac{df}{dr} + \alpha^2 r^2 f = 0 \qquad \text{X}$$

Equation X can be transformed to the standard form of Bessel's equation (3.35) by putting $x = \alpha r$. The solution is thus

$$f = AJ_0(\alpha r) + BY_0(\alpha r) \qquad \text{XI}$$

if $\alpha \neq 0$, or

$$f = A_0 + B_0 \ln r \qquad \text{XII}$$

if $\alpha = 0$.

Since $Y_0(0) = -\infty$ and $\ln 0 = -\infty$, boundary condition V asserts that $B_0 = B = 0$. The solutions of equation III are thus

$$\text{if} \quad \alpha = 0, \quad U = A_0 \qquad \text{XIII}$$

$$\text{if} \quad \alpha \neq 0, \quad U = A_\alpha J_0(\alpha r) e^{-\alpha^2 Dz} \qquad \text{XIV}$$

Boundary condition VI is satisfied by solution XIII, and putting equation XIV into equation VI gives

$$DA_\alpha \alpha J_1(\alpha b) e^{-\alpha^2 Dz} = -kA_\alpha \alpha^2 D J_0(\alpha b) e^{-\alpha^2 Dz} \qquad \text{XV}$$

by using equations (3.75) and (3.84) to differentiate the Bessel function. Simplifying equation XV gives

$$J_1(\alpha_n b) + k\alpha_n J_0(\alpha_n b) = 0 \qquad \text{XVI}$$

which determines the eigen-values α_n. There are an infinite number of solutions of equation XVI, the first few of which have been tabulated.† The most general solution of equation III is therefore obtained by adding all permissible particular solutions together. Thus

$$U = A_0 + \sum_{n=1}^{\infty} A_n J_0(\alpha_n r) e^{-\alpha_n^2 Dz} \qquad \text{XVII}$$

where the eigen-functions are zero-order Bessel functions of the first kind.

It can be shown by rearranging equation X that it is of Sturm–Liouville type with a weighting function r. Boundary condition VI is not one of the types listed in Section 8.6 and hence the eigen-functions will not be orthogonal. Nevertheless, the usual method can be followed by putting $z = 0$ and multiplying equation XVII by $rJ_0(\alpha_m r)$ and integrating. This gives

$$\int_0^b U|_{z=0} rJ_0(\alpha_m r)\,dr = \int_0^b A_0 rJ_0(\alpha_m r)\,dr + \sum_{n=1}^{\infty} A_n \int_0^b J_0(\alpha_n r) J_0(\alpha_m r) r\,dr$$

$$\text{XVIII}$$

It is simpler to evaluate the terms of equation XVIII separately. Thus the first term becomes

$$\int_0^b U|_{z=0} rJ_0(\alpha_m r)\,dr = \int_0^a U_0 rJ_0(\alpha_m r)\,dr$$

$$= U_0 aJ_1(\alpha_m a)/\alpha_m \qquad \text{XIX}$$

† Crank, J. "The Mathematics of Diffusion." Oxford University Press (1956). Carslaw, H. S. and Jaeger, J. C. "Conduction of Heat in Solids." Oxford University Press (1974).

by using equation (3.75) with $k = 1$. Similarly, the next term gives

$$\int_0^b A_0 r J_0(\alpha_m r)\, dr = A_0 b J_1(\alpha_m b)/\alpha_m$$

$$= - A_0 k b J_0(\alpha_m b) \qquad\qquad \text{XX}$$

Taking a general term with $n \neq m$ from the infinite sum and using equation (3.79) gives

$$A_n \int_0^b J_0(\alpha_n r) J_0(\alpha_m r) r\, dr = \frac{A_n b}{\alpha_n{}^2 - \alpha_m{}^2} [\alpha_n J_0(\alpha_m b) J_1(\alpha_n b)$$

$$- \alpha_m J_0(\alpha_n b) J_1(\alpha_m b)]$$

$$= \frac{A_n b}{\alpha_n{}^2 - \alpha_m{}^2} [-k\alpha_n{}^2 J_0(\alpha_n b) J_0(\alpha_m b)$$

$$+ k\alpha_m{}^2 J_0(\alpha_n b) J_0(\alpha_m b)]$$

because α_n must satisfy equation XVI. Therefore

$$A_n \int_0^b J_0(\alpha_n r) J_0(\alpha_m r) r\, dr = -k A_n b J_0(\alpha_n b) J_0(\alpha_m b) \qquad \text{XXI}$$

For the special term in the series when $n = m$, equation (3.80) must be used. Hence

$$A_m \int_0^b [J_0(\alpha_m r)]^2 r\, dr = \tfrac{1}{2} b^2 A_m [J_0{}^2(\alpha_m b) + J_1{}^2(\alpha_m b)]$$

$$= \tfrac{1}{2} b^2 A_m (1 + k^2 \alpha_m{}^2) J_0{}^2(\alpha_m b) \qquad \text{XXII}$$

by using equations (3.84) and XVI.

Combining equations XIX, XX, XXI, and XXII into equation XVIII,

$$\frac{U_0 a}{\alpha_m} J_1(\alpha_m a) = - A_0 k b J_0(\alpha_m b) + \tfrac{1}{2} b^2 A_m (1 + k^2 \alpha_m{}^2) J_0{}^2(\alpha_m b)$$

$$- \sum_{n \neq m} k A_n b J_0(\alpha_n b) J_0(\alpha_m b) \qquad \text{XXIII}$$

Completing the infinite series by inserting the terms $k A_m b J_0{}^2(\alpha_m b)$ twice with opposite signs, therefore

$$\frac{U_0 a}{\alpha_m} J_1(\alpha_m a) = - A_0 b k J_0(\alpha_m b) + \tfrac{1}{2} A_m (b^2 + 2kb + b^2 k^2 \alpha_m{}^2) J_0{}^2(\alpha_m b)$$

$$- k b J_0(\alpha_m b) \sum_{n=1}^{\infty} A_n J_0(\alpha_n b) \qquad \text{XXIV}$$

But when $z = 0$ and $r = b$, equations IV and XVII give

$$0 = A_0 + \sum_{n=1}^{\infty} A_n J_0(\alpha_n b) \qquad\qquad \text{XXV}$$

Because all of the terms in equation XXV occur in equation XXIV with a factor $-kbJ_0(\alpha_m b)$, then

$$\frac{U_0 a}{\alpha_m} J_1(\alpha_m a) = \tfrac{1}{2} A_m (b^2 + 2bk + b^2 k^2 \alpha_m{}^2) J_0{}^2(\alpha_m b) \qquad \text{XXVI}$$

which determines A_m. It is both fortunate and unusual that equation XXV eliminates all coefficients except A_m from equation XXIV. The solution by putting A_m from equation XXVI into equation XVII is thus

$$U = A_0 + \sum_{n=1}^{\infty} \frac{2U_0 a}{\alpha_n} \frac{J_1(\alpha_n a) J_0(\alpha_n r)}{(b^2 + 2bk + b^2 k^2 \alpha_n{}^2) J_0{}^2(\alpha_n b)} e^{-\alpha_n{}^2 Dz} \qquad \text{XXVII}$$

Taking an overall material balance at large values of z,

$$\pi a^2 U_0 = (\pi b^2 + 2\pi bk) A_0$$

$$\therefore \quad a^2 U_0 = b(b + 2k) A_0 \qquad \text{XXVIII}$$

which determines A_0, the equilibrium velocity in the packing.
The fraction of total fluid in the wall layer is given by

$$F = \frac{2\pi bk U}{\pi a^2 U_0}$$

$$= \frac{2k}{b + 2k} + \sum_{n=1}^{\infty} \frac{4bk}{a\alpha_n} \frac{J_1(\alpha_n a) e^{-\alpha_n{}^2 Dz}}{(b^2 + 2bk + b^2 k^2 \alpha_n{}^2) J_0(\alpha_n b)} \qquad \text{XXIX}$$

8.8. THE LAPLACE TRANSFORM METHOD

It was shown in Chapter 6 that a Laplace transformation would remove the derivatives from an ordinary differential equation. It is thus fairly obvious that the same technique can be used to remove all derivatives with respect to one independent variable from a partial differential equation. However, a particular independent variable can only be removed if it has an open range, because the Laplace transformation involves integrating with respect to the variable from zero to infinity. Since the boundary conditions can only be used at the extremities of the range of integration, they must all be imposed at zero. Hence the variable to be transformed must have an open range as described in Section 8.4.5.

If the partial differential equation has two independent variables the removal of one of them yields an ordinary differential equation which can be solved by the methods of Chapter 2. An added complication in the study of partial differential equations is the occurrence of arbitrary constants in the solution of the transformed equation. These constants must be evaluated before the solution is inverted because they are normally functions of the transform parameter as will be seen in the examples. Hence the boundary conditions which are not used to transform the equation must themselves be transformed.

These ideas are better illustrated by solving specific equations, and in the first three examples below, the one-dimensional heat conduction equation will be solved again for the same selection of boundary conditions as used in the previous sections of this chapter. This duplicate derivation emphasizes the versatility of the Laplace transformation by determining all of the solutions by one technique instead of using a different method for each specific example. The later examples illustrate the solutions of different partial differential equations.

Example 1. Solutions of the Linear Heat Conduction Equation in a Semi-infinite Medium. Taking the now familiar form of the conduction equation (8.47),

$$\alpha \frac{\partial^2 T}{\partial x^2} = \frac{\partial T}{\partial t} \qquad\qquad \text{I}$$

and using the initial condition that

$$\text{at} \quad t = 0, \quad T = T_0 \qquad\qquad \text{II}$$

the Laplace transformation can be used to remove the variable t as follows.
Equation (6.11) from Chapter 6 gives

$$\mathscr{L}\left[\frac{\partial T}{\partial t}\right] = s\bar{T} - T_0$$

Because x and t are independent variables, then

$$\mathscr{L}\left[\frac{\partial^2 T}{\partial x^2}\right] = \frac{\mathrm{d}^2 \bar{T}}{\mathrm{d}x^2}$$

Hence the transform of equation I is

$$\alpha \frac{\mathrm{d}^2 \bar{T}}{\mathrm{d}x^2} = s\bar{T} - T_0 \qquad\qquad \text{III}$$

Equation III is a second order linear ordinary differential equation having the solution

$$\bar{T} = A\,\mathrm{e}^{-qx} + B\,\mathrm{e}^{qx} + T_0/s \qquad\qquad \text{IV}$$

where $\qquad\qquad q^2 = s/\alpha \qquad\qquad$ V

has been introduced for simplicity of presentation.

(a) The first problem, solved in Section 8.5.1, involved maintaining the exposed face of the slab at a fixed temperature, i.e.

$$\text{at} \quad x = 0, \quad T = T_1 \qquad\qquad \text{VI}$$

Transforming equation VI gives

$$\text{at} \quad x = 0, \quad \bar{T} = T_1/s \qquad\qquad \text{VII}$$

As x approaches infinity, T must remain finite, and hence $\bar{T}$ must remain finite. Therefore $B = 0$ and condition VII substituted into equation IV gives

$$\frac{T_1}{s} = A + \frac{T_0}{s}$$

Eliminating A from equation IV gives

$$\bar{T} = (T_1 - T_0)(e^{-\sqrt{(s/\alpha)}x}/s) + T_0/s \qquad \text{VIII}$$

Inverting by using transforms 83 and 1 in the table gives

$$T = (T_1 - T_0)\,\text{erfc}\,[x/2\sqrt{(\alpha t)}] + T_0 \qquad \text{IX}$$

which can be compared with equation (8.74).

(b) In Section 8.5.2 it was shown that equation (8.78) was the solution expressing the movement of a fixed quantity of heat within a solid body when the heat is initially concentrated at the surface. This solution will now be derived from equation IV.

If all of the heat is at the surface initially, then $T_0 = 0$; and since $B = 0$ to satisfy conditions at infinity, equation IV becomes

$$\bar{T} = A\,e^{-qx} \qquad \text{X}$$

If the total heat content (H) of the body is constant,

$$H = \int_0^\infty \rho C_p T\,dx \qquad (8.79)$$

Transforming equation (8.79) gives

$$\frac{H}{\rho C_p s} = \int_0^\infty \bar{T}\,dx \qquad \text{XI}$$

Using equation X for $\bar{T}$ and integrating, therefore

$$\frac{H}{\rho C_p s} = \frac{A}{q} = A\sqrt{\frac{\alpha}{s}} \qquad \text{XII}$$

Eliminating A from equation X gives

$$\bar{T} = \frac{H}{\rho C_p \sqrt{\alpha}}\frac{e^{-\sqrt{(s/\alpha)}x}}{\sqrt{s}} \qquad \text{XIII}$$

This can be inverted by using transform 84 to give

$$T = \frac{H}{\rho C_p \sqrt{(\alpha\pi)}}t^{-\frac{1}{2}}e^{-x^2/4\alpha t} \qquad \text{XIV}$$

which is identical with equation (8.78) with H given by equation (8.81).

(c) In Example 1 from Section 8.5.2, heat was suppled at a fixed rate to the exposed face and this boundary condition can be expressed by the equation:

$$\text{at} \quad x = 0, \quad Q = -k\frac{\partial T}{\partial x} \qquad\qquad \text{XV}$$

Taking the uniform initial temperature as the datum and applying the physical limitation on the temperature at infinite depth, equation IV simplifies to

$$\bar{T} = A\,e^{-qx} \qquad\qquad \text{XVI}$$

Transforming the boundary condition XV,

$$\text{at} \quad x = 0, \quad \frac{-Q}{ks} = \frac{\partial \bar{T}}{\partial x} \qquad\qquad \text{XVII}$$

Substituting equation XVI into equation XVII gives

$$\frac{-Q}{ks} = -Aq\,e^{-qx} = -Aq$$

$$\therefore \quad A = Q/kqs \qquad\qquad \text{XVIII}$$

$$\therefore \quad \bar{T} = \frac{Q\sqrt{\alpha}}{k}s^{-\frac{3}{2}}e^{-\sqrt{(s/\alpha)}x} \qquad\qquad \text{XIX}$$

Because integrating a function with respect to t is equivalent to dividing its transform by s, as shown by equation (6.21), equation XIX can be inverted by integrating the function given in transform 84 in the tables. Thus

$$T = \frac{Q\sqrt{\alpha}}{k}\int_0^t \frac{1}{\sqrt{(\pi\tau)}}e^{-x^2/4\alpha\tau}\,d\tau \qquad\qquad \text{XX}$$

which is the same result as that given by equation III in the original example. At the surface ($x = 0$), equation XX gives

$$T_s = \frac{2Q}{k}\sqrt{\left(\frac{\alpha t}{\pi}\right)} \qquad\qquad \text{XXI}$$

as in Section 8.5.2, Example 1, equation IV.

Example 2. Heat Conduction between Parallel Planes. In Section 8.7.5, the second example considered the flow of heat between parallel planes maintained at different temperatures. The boundary conditions were

$$\text{at} \quad t = 0, \quad T = T_0 \qquad\qquad \text{I}$$

$$\text{at} \quad x = 0, \quad T = T_0 \qquad\qquad \text{II}$$

$$\text{at} \quad x = L, \quad T = T_1 \qquad\qquad \text{III}$$

Because condition I is the same, the Laplace transformation of equation (8.47) is as above, and the general solution is

$$\bar{T} = A\,e^{-qx} + B\,e^{qx} + T_0/s \qquad\qquad \text{IV}$$

where

$$q^2 = s/\alpha \qquad\qquad \text{V}$$

Transforming the boundary conditions II and III gives

$$\text{at}\quad x = 0,\quad \bar{T} = T_0/s \qquad\qquad \text{VI}$$

$$\text{at}\quad x = L,\quad \bar{T} = T_1/s \qquad\qquad \text{VII}$$

Using these conditions, solving for A and B, and substituting into equation IV gives

$$\bar{T} = \frac{T_1 - T_0}{s}\,\frac{e^{qx} - e^{-qx}}{e^{qL} - e^{-qL}} + \frac{T_0}{s} \qquad\qquad \text{VIII}$$

The first term on the right-hand side of equation VIII must be inverted by the method of residues (see Section 6.8) as follows. The denominator is zero when either

$$s = 0 \qquad\qquad \text{IX}$$

or

$$q_n L = in\pi \quad (\text{i.e. } s_n = -\alpha n^2 \pi^2/L^2) \qquad\qquad \text{X}$$

and all of the above poles are simple poles. The residue at $s = 0$ is

$$(T_1 - T_0)\lim_{q \to 0}\left[\frac{e^{qx} - e^{-qx}}{e^{qL} - e^{-qL}}\right]$$

which can be evaluated by L'Hôpital's rule to give:

$$\text{residue at}\quad s = 0\quad \text{is}\quad (T_1 - T_0)\,x/L \qquad\qquad \text{XI}$$

The simplest method of evaluating the other residues is to express equation VIII in terms of hyperbolic functions and use equation (4.90) from Section 4.14.1. Thus

$$\bar{T} = \frac{T_1 - T_0}{s}\,\frac{\sinh[\sqrt{(s/\alpha)}\,x]}{\sinh[\sqrt{(s/\alpha)}\,L]} + \frac{T_0}{s} \qquad\qquad \text{XII}$$

The residue at s_n is

$$\frac{T_1 - T_0}{s_n}\,\frac{\sinh[\sqrt{(s_n/\alpha)}\,x]}{\cosh[\sqrt{(s_n/\alpha)}\,L]}\,\frac{2\sqrt{(s_n\,\alpha)}}{L}\,\exp(s_n t)$$

or using equation X:

$$\frac{2(T_1 - T_0)}{in\pi}\,\frac{\sinh(in\pi x/L)}{\cosh(in\pi)}\,e^{-\alpha n^2\pi^2 t/L^2} = \frac{2}{n\pi}(T_1 - T_0)(-1)^n \sin\left(\frac{n\pi x}{L}\right)e^{-\alpha n^2\pi^2 t/L^2}$$

Adding all of the residues together, and inverting the last term by transform 1 in the tables, gives

$$\frac{T-T_0}{T_1-T_0} = \frac{x}{L} + \sum_{n=0}^{\infty} \frac{2}{n\pi}(-1)^n \sin\frac{n\pi x}{L} e^{-n^2\pi^2\alpha t/L^2} \qquad \text{XIII}$$

as in Example 2, Section 8.7.5, equation XIV.

Example 3. *Symmetrical Heat Conduction between Parallel Planes.* The example in Section 8.5.3 again involved the linear heat conduction equation with the boundary conditions:

$$\text{at} \quad t = 0, \quad T = 0 \qquad \text{I}$$

$$\text{at} \quad x = 0, \quad T = T_0 \qquad \text{II}$$

$$\text{at} \quad x = L, \quad \frac{\partial T}{\partial x} = 0 \qquad \text{III}$$

Solving as in Example 1 above gives

$$\bar{T} = A e^{-qx} + B e^{qx} \qquad \text{IV}$$

where

$$q^2 = s/\alpha \qquad \text{V}$$

The transformed boundary conditions are

$$\text{at} \quad x = 0, \quad \bar{T} = \frac{T_0}{s} \qquad \text{VI}$$

$$\text{at} \quad x = L, \quad \frac{\partial \bar{T}}{\partial x} = 0 \qquad \text{VII}$$

Using conditions VI and VII to determine A and B in equation IV gives the transformed solution

$$\bar{T} = \frac{T_0}{s} \frac{(e^{2qL}-1)e^{-qx} + (1-e^{-2qL})e^{qx}}{e^{2qL}-e^{-2qL}} \qquad \text{VIII}$$

This could be inverted by the method of residues as in the last example, but to obtain the error function form given in Section 8.5.3, it is necessary to expand equation VIII in terms of exponential functions.

$$\frac{(e^{2qL}-1)e^{-qx}}{e^{2qL}-e^{-2qL}} = e^{-qx}(1-e^{-2qL})(1-e^{-4qL})^{-1}$$

$$= e^{-qx}(1-e^{-2qL})(1+e^{-4qL}+e^{-8qL}+...)$$

$$= e^{-qx}(1-e^{-2qL}+e^{-4qL}-e^{-6qL}+e^{-8qL}-...)$$

$$= \sum_{n=0}^{\infty}(-1)^n e^{-(2nL+x)q} \qquad \text{IX}$$

Similarly, it can be shown that

$$\frac{(1-e^{-2qL})e^{qx}}{e^{2qL}-e^{-2qL}} = \sum_{n=0}^{\infty}(-1)^n e^{-(2nL+2L-x)q} \qquad \text{X}$$

Substituting equations IX and X into VIII, using V and inverting each term by using transform 83 in the tables, gives

$$T = T_0 \sum_{n=0}^{\infty}(-1)^n\{\operatorname{erfc}[(x+2nL)/2\sqrt{(\alpha t)}] + \operatorname{erfc}[(2nL+2L-x)/2\sqrt{(\alpha t)}]\} \qquad \text{XI}$$

which agrees with its counterpart in Example 2, Section 8.5.3.

The above solutions of the linear heat conduction equation and the analogous linear diffusion equation are all contained in Crank or Carslaw and Jaeger,† together with many other solutions for different combinations of the four types of boundary condition in the three common coordinate systems. As pointed out by those authors, the error function type of solution arising from the Laplace transform is useful for determining the behaviour of the system in the early stages, but the summation of eigen-functions is more useful for determining the approach to equilibrium in the later stages of the process.

Example 4. *Exploitation of an oilfield.*   An extensive shallow oilfield is to be exploited by removing product at a constant rate from one well. How will the pressure distribution in the formation vary with time?

Taking a radial coordinate r measured from the base of the well system it is known that the pressure (p) obeys the normal diffusion equation

$$\frac{\partial^2 p}{\partial r^2} + \frac{1}{r}\frac{\partial p}{\partial r} = \frac{1}{\eta}\frac{\partial p}{\partial t} \qquad \text{I}$$

where η is the hydraulic diffusivity.

The initial pressure in the formation is assumed to be uniform at p_0. If oil is removed at a constant rate q, then

$$q = \lim_{r \to 0}\frac{2\pi rkh}{\mu}\frac{\partial p}{\partial r} \qquad \text{II}$$

where k is the permeability, h is the thickness of the formation, and μ is the coefficient of viscosity.

Transforming equation I by the same procedure as given in Example 1

$$\frac{d^2\bar{p}}{dr^2} + \frac{1}{r}\frac{d\bar{p}}{dr} - \frac{s}{\eta}\bar{p} = -\frac{p_0}{\eta} \qquad \text{III}$$

Putting
$$x = r\sqrt{(s/\eta)}. \qquad \text{IV}$$

$$x^2\frac{d^2\bar{p}}{dx^2} + x\frac{d\bar{p}}{dx} - x^2\bar{p} = -\frac{x^2 p_0}{s} \qquad \text{V}$$

† Crank, J. "The Mathematics of Diffusion." Oxford University Press (1956). Carslaw, H. S. and Jaeger, J. C. "Conduction of Heat in Solids." Oxford University Press (1974).

which is the modified form of Bessel's equation (3.51) having the solution

$$\bar{p} = AI_0(x) + BK_0(x) + p_0/s \qquad\qquad \text{VI}$$

Equation (3.69) shows that as $x \to \infty$, $I_0(x) \to \infty$ and hence $A = 0$ and

$$\bar{p} = BK_0(x) + p_0/s \qquad\qquad \text{VII}$$

Transforming boundary condition II gives

$$\frac{q}{s} = \lim_{r \to 0} \frac{2\pi rkh}{\mu} \frac{d\bar{p}}{dr} \qquad\qquad \text{VIII}$$

Using equation (3.58) that $I_0(0) = 1$, and taking the limit of equation (3.55) as $x \to 0$ gives

$$K_0(x) \to -\ln x \qquad\qquad \text{IX}$$

Therefore, near the base of the well, equation VII becomes

$$\bar{p} \doteqdot -B\ln x = -B\ln [r\sqrt{(s/\eta)}] \qquad\qquad \text{X}$$

Substituting into equation VIII, therefore

$$\frac{q}{s} = \lim_{r \to 0} -\frac{2\pi rkh}{\mu} \frac{B}{r} = -\frac{2\pi khB}{\mu} \qquad\qquad \text{XI}$$

$$\therefore \quad B = \frac{-\mu q}{2\pi khs} = -\frac{\alpha}{s} \qquad\qquad \text{XII}$$

Hence the transformed solution VII becomes

$$\bar{p} = -\alpha \frac{K_0[r\sqrt{(s/\eta)}]}{s} + \frac{p_0}{s} \qquad\qquad \text{XIII}$$

Since integration of a function with respect to t is equivalent to dividing its transform by s, as shown by equation (6.21), transform number 117 in the tables enables equation XIII to be inverted, thus

$$p = -\int_0^t \frac{\alpha}{2t} e^{-r^2/4\eta t}\, dt + p_0 \qquad\qquad \text{XIV}$$

Putting

$$t = r^2/4\eta z \qquad\qquad \text{XV}$$

and hence

$$dt = -\frac{r^2\, dz}{4\eta z^2}$$

$$\therefore \quad p = \int_\infty^{z_0} \frac{\alpha}{2z} e^{-z}\, dz + p_0$$

$$= -\tfrac{1}{2}\alpha \int_{z_0}^\infty \frac{e^{-z}}{z}\, dz + p_0 \qquad\qquad \text{XVI}$$

where $z_0 = r^2/4\eta t$. Comparing the integral in equation XVI with equation (5.35), the solution can be written in terms of a tabulated function. Thus

$$p = \frac{\mu q}{4\pi k h} Ei(-r^2/4\eta t) + p_0 \qquad\qquad \text{XVII}$$

Example 5. Gas absorption accompanied by a first order reaction in a falling film. A wetted wall column is to be used for the absorption of a gas (A). The gas is consumed in the liquid phase by a pseudo first order reaction in terms of the gas concentration. Develop expressions giving the point absorption rate and effective penetration depth as a function of distance from the liquid inlet.

Figure 8.8 shows a section through the liquid film where z is the distance from the liquid inlet and x is a coordinate measured inwards from the surface of the film which is assumed to be of uniform thickness L. Although it is a poor assumption which depends upon inlet conditions, the parabolic velocity distribution $v(x)$ will be assumed to exist right at the inlet and remain unchanged throughout the length of the column.

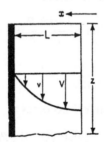

FIG. 8.8. Velocity distribution in a falling film

The gas is assumed pure so that there will be no gas phase resistance and the surface concentration of A in the liquid will be constant at c_0. For any element of the liquid ($\delta x\,\delta z$) to be in equilibrium, the amount of A carried in the z direction by bulk flow, plus the amount diffusing in the x direction, must be sufficient to supply the chemical reaction proceeding within the element. Expressing this fact symbolically,

Input rate $\qquad = vc\,\delta x - D\dfrac{\partial c}{\partial x}\delta z$

Output rate $\qquad = vc\,\delta x + \dfrac{\partial}{\partial z}(vc\,\delta x)\,\delta z - D\dfrac{\partial c}{\partial x}\delta z + \dfrac{\partial}{\partial x}\left(-D\dfrac{\partial c}{\partial x}\delta z\right)\delta x$

Consumption rate $= kc\,\delta x\,\delta z$

where c is the concentration of A in the liquid, D is its diffusivity, and k is the specific reaction rate constant. Completing the balance and cancelling appropriate terms gives

$$-v\frac{\partial c}{\partial z}+D\frac{\partial^2 c}{\partial x^2}=kc \qquad\qquad \text{I}$$

Transforming equation I by the method previously explained gives

$$-sv\bar{c}+D\frac{\mathrm{d}^2\bar{c}}{\mathrm{d}x^2}=k\bar{c}$$

$$\therefore \quad D\frac{\mathrm{d}^2\bar{c}}{\mathrm{d}x^2}-(k+sv)\,\bar{c}=0 \qquad\qquad \text{II}$$

But $v=V(1-x^2/L^2)$ where V is the interfacial velocity, and equation II would have to be integrated using the method of Frobenius to obtain an accurate solution. However, the penetration of A into the liquid will be small for small values of z and as a first approximation it will be assumed that $x\ll L$ so that v can be replaced by V in equation II. Thus

$$D\frac{\mathrm{d}^2\bar{c}}{\mathrm{d}x^2}-(k+sV)\,\bar{c}=0 \qquad\qquad \text{III}$$

which is a linear second order differential equation with constant coefficients having the solution

$$\bar{c}=A'\exp\left[x\sqrt{\left(\frac{k+sV}{D}\right)}\right]+B\exp\left[-x\sqrt{\left(\frac{k+sV}{D}\right)}\right] \qquad \text{IV}$$

As x increases, the concentration of A must decrease to zero, and hence $A'=0$. At $x=0$, the boundary condition gives $c=c_0$

$$\therefore \quad \bar{c}=c_0/s$$

$$\therefore \quad \bar{c}=\frac{c_0}{s}\exp\left\{-x\sqrt{\left[\frac{V}{D}\left(s+\frac{k}{V}\right)\right]}\right\} \qquad \text{V}$$

Using transform 82 and the shifting theorem (Section 6.2.2),

$$\mathscr{L}\left[\frac{x}{2}\sqrt{\left(\frac{V}{D\pi z^3}\right)}\exp\left(\frac{-x^2 V}{4Dz}\right)\exp\left(\frac{-kz}{V}\right)\right]=\exp\left\{-x\sqrt{\left[\frac{V}{D}\left(s+\frac{k}{V}\right)\right]}\right\} \qquad \text{VI}$$

Applying the property given in equation (6.21) to equation VI and inverting equation V gives

$$c=c_0\int_0^z\frac{x}{2}\sqrt{\left(\frac{V}{D\pi z^3}\right)}\exp\left(-\frac{x^2 V}{4Dz}-\frac{kz}{V}\right)\mathrm{d}z \qquad \text{VII}$$

Putting $\qquad\qquad\qquad\qquad \alpha^2=x^2 V/4D$

and $\qquad\qquad\qquad\qquad\qquad \beta^2=k/V \qquad\qquad\qquad\qquad\qquad\qquad \text{VIII}$

to simplify the algebra gives

$$c = c_0 \int_0^z \frac{\alpha}{\sqrt{(\pi z^3)}} \exp\left(-\frac{\alpha^2}{z} - \beta^2 z\right) dz \qquad\qquad \text{IX}$$

It is difficult to evaluate this integral, but its form can be changed by putting

$$r = \frac{\alpha}{\sqrt{z}} - \beta\sqrt{z} \qquad\qquad \text{X}$$

and hence

$$r^2 = \frac{\alpha^2}{z} + \beta^2 z - 2\alpha\beta$$

and

$$dr = \left(-\frac{\alpha}{2z^{\frac{3}{2}}} - \frac{\beta}{2z^{\frac{1}{2}}}\right) dz$$

into equation IX to give

$$c = c_0 \int_0^z \frac{2}{\sqrt{\pi}} \left(\frac{\alpha}{2z^{\frac{3}{2}}} + \frac{\beta}{2z^{\frac{1}{2}}} - \frac{\beta}{2z^{\frac{1}{2}}}\right) \exp\left(-r^2 - 2\alpha\beta\right) dz$$

$$= \frac{-2c_0}{\sqrt{\pi}} \int_\infty^z e^{-2\alpha\beta} e^{-r^2} dr - c_0\beta \int_0^z \frac{1}{\sqrt{(\pi z)}} \exp\left(-\frac{\alpha^2}{z} - \beta^2 z\right) dz$$

$$\therefore \quad c = c_0 e^{-2\alpha\beta} \operatorname{erfc}\left(\frac{\alpha}{\sqrt{z}} - \beta\sqrt{z}\right) - c_0\beta \int_0^z \frac{1}{\sqrt{(\pi z)}} \exp\left(-\frac{\alpha^2}{z} - \beta^2 z\right) dz \qquad \text{XI}$$

If the sign of β were reversed in equation IX, the integral would not be altered in any way. Hence the sign of β may be reversed in equation XI without altering the result. Thus

$$c = c_0 e^{2\alpha\beta} \operatorname{erfc}\left(\frac{\alpha}{\sqrt{z}} + \beta\sqrt{z}\right) + c_0\beta \int_0^z \frac{1}{\sqrt{(\pi z)}} \exp\left(-\frac{\alpha^2}{z} - \beta^2 z\right) dz \qquad \text{XII}$$

Adding equations XI and XII together, dividing the result by 2 and returning to the original symbols through equation VIII yields the final result

$$\frac{c}{c_0} = \tfrac{1}{2}\exp\left(-x\sqrt{\frac{k}{D}}\right) \operatorname{erfc}\left(\frac{x}{2}\sqrt{\frac{V}{Dz}} - \sqrt{\frac{kz}{V}}\right)$$

$$+ \tfrac{1}{2}\exp\left(x\sqrt{\frac{k}{D}}\right) \operatorname{erfc}\left(\frac{x}{2}\sqrt{\frac{V}{Dz}} + \sqrt{\frac{kz}{V}}\right) \qquad \text{XIII}$$

The point absorption rate per unit area is given by

$$R = -D\frac{\partial c}{\partial x}\bigg|_{x=0} \qquad\qquad \text{XIV}$$

Transforming equation XIV gives

$$\bar{R} = -D\frac{\partial \bar{c}}{\partial x}\bigg|_{x=0} \qquad\qquad \text{XV}$$

and substituting for $\bar{c}$ from equation V gives

$$\bar{R} = \frac{Dc_0}{s}\sqrt{\left[\frac{V}{D}\left(s+\frac{k}{V}\right)\right]}$$

$$\therefore \quad \bar{R} = c_0\sqrt{(DV)}\sqrt{\frac{s+k/V}{s}} \qquad\qquad \text{XVI}$$

Inverting equation XVI by using transform 38 and the shifting theorem (Section 6.2.2), therefore

$$R = c_0\sqrt{(DV)}\exp(-kz/V)\left[\frac{1}{\sqrt{(\pi z)}}+\sqrt{\frac{k}{V}}\exp(kz/V)\,\text{erf}\sqrt{(kz/V)}\right]$$

$$\therefore \quad R = c_0\sqrt{(kD)}\left[\text{erf}\sqrt{(kz/V)}+\sqrt{(V/\pi kz)}\exp(-kz/V)\right] \qquad \text{XVII}$$

In order to determine the depth of penetration of the solute, some arbitrary limit must be set for its concentration to be negligible. This limit is chosen as $c/c_0 = 0.002$, and equation XIII gives an implicit expression for the penetration depth x as a function of film length z. The relationship is simple enough to solve by trial and error except for small values of z but an approximate solution can be found for this region as follows.

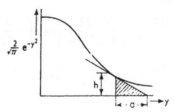

FIG. 8.9. Approximation for erfc y

Figure 8.9 shows $(2/\sqrt{\pi})e^{-y^2}$ plotted against y so that erfc y is the shaded area. If y is large it is reasonable to suppose that erfc y is some constant multiple of the area of the triangle formed as shown by the tangent to the curve. Putting

$$h = \frac{2}{\sqrt{\pi}}e^{-y^2} \qquad\qquad \text{XVIII}$$

$$\therefore \quad \frac{dh}{dy} = \frac{-4y}{\sqrt{\pi}}e^{-y^2} = -2yh$$

but from the triangle,

$$\frac{dh}{dy} = \frac{-h}{a}$$

$$\therefore \quad a = 1/2y \qquad\qquad \text{XIX}$$

The approximation can be written

$$\text{erfc}\, y = \tfrac{1}{2}Kha \qquad\qquad \text{XX}$$

where K is some proportionality constant greater than unity. Substituting for h and a from equations XVIII and XIX gives

$$\text{erfc}\, y = \frac{K}{2y\sqrt{\pi}}\, e^{-y^2} \qquad\qquad \text{XXI}$$

Evaluating K by putting $y = 1.5$ on both sides of equation XXI gives $K = 1.71$. Hence

$$\text{erfc}\, y \simeq (0.484/y)\, e^{-y^2} \qquad\qquad \text{XXII}$$

If $y = 2.5$, the error in equation XXII is less than $\tfrac{1}{2}\%$, and if $y = 1$, $\text{erfc}\, y = 0.178$ whereas $(0.484/y)\, e^{-y^2} = 0.157$. Hence equation XXII should only be used for $y > 1$.

Introducing dimensionless coordinates into equation XIII by putting

$$X = x\sqrt{(k/D)} \quad \text{and} \quad Z^2 = kz/V \qquad\qquad \text{XXIII}$$

$$\frac{c}{c_0} = \tfrac{1}{2}e^{-X}\text{erfc}\left(\frac{X}{2Z} - Z\right) + \tfrac{1}{2}e^{X}\,\text{erfc}\left(\frac{X}{2Z} + Z\right)$$

Using equation XXII gives

$$\frac{c}{c_0} \simeq \left[\frac{0.242}{(X/2Z) - Z} + \frac{0.242}{(X/2Z) + Z}\right]\exp\left(-\frac{X^2}{4Z^2} - Z^2\right)$$

$$\simeq \frac{0.242X}{Z[(X^2/4Z^2) - Z^2]}\exp\left(-\frac{X^2}{4Z^2} - Z^2\right)$$

which is valid for

$$\frac{X}{2Z} > 1 + Z \qquad\qquad \text{XXIV}$$

The penetration depth is thus given by

$$0.002Z\left(\frac{X^2}{4Z^2} - Z^2\right) = 0.242X\exp\left(-\frac{X^2}{4Z^2} - Z^2\right) \qquad\qquad \text{XXV}$$

Solving by trial and error using equation XXV for $Z < 1$ and equation XIII for $Z \geqslant 1$ gives:

Z	0·1	0·5	1·0	1·5	2·0	3·0
X	0·435	2·130	4·00	5·33	6·00	6·215

which is plotted in Fig. 8.10.

8.9. Other Transforms

In the last section, the Laplace transform was shown to be quite versatile, and although the transform involved introducing an exponential function, this did not influence the type of solution obtained. The chief restriction on its use is that the problem must be of initial value type, but provided the dependent variable and its derivative remain finite† as the transformed variable tends to infinity, no analytical difficulties arise. The inversion process is often tedious if the method of residues has to be adopted, but in principle the inversion process can always be completed.

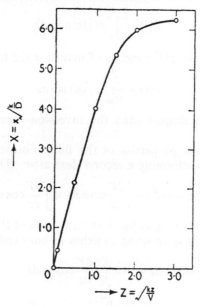

FIG. 8.10. Relation between penetration of solute and length of wetted wall

No reason has been given for using the function e^{-st} as the "kernel" of the Laplace transform and attempts have been made to use other functions as kernels thus giving rise to other transforms. The purpose of the present section is to give the names and formulae for these other transforms and indicate their range of applicability.

8.9.1. Finite Fourier Transforms

The finite Fourier sine transform can be derived from the general formula developed in Section 8.7.1 for linear heat conduction. Replacing the boundary condition (8.107) into the general solution (8.111) and dropping the suffix from $f_0(x)$ gives the identity

$$f(x) = \sum_{n=1}^{\infty} \frac{2}{L} \sin \frac{n\pi x}{L} \int_0^L f(x) \sin \frac{n\pi x}{L} dx \qquad (8.156)$$

† The method is still applicable when the dependent variable or its derivative become infinite to a finite exponential order, but physical variables never actually take infinite values so there is no restriction.

which is universally true and no longer related to the specific problem of Section 8.7.1. To obtain the conventional notation, put $z = \pi x/L$.

$$\therefore \quad f(z) = \frac{2}{\pi} \sum_{n=1}^{\infty} \sin nz \int_0^{\pi} f(z) \sin nz \, dz \tag{8.157}$$

The finite Fourier sine transform is defined as

$$\bar{f}(n) = \int_0^{\pi} f(z) \sin nz \, dz \tag{8.158}$$

Equation (8.157) provides the means of inverting the transform. Thus

$$f(z) = \frac{2}{\pi} \sum_{n=1}^{\infty} \bar{f}(n) \sin nz \tag{8.159}$$

which is very much simpler than the inversion formula for the Laplace transform (6.31).

The most important properties of the finite Fourier sine transform can be determined by transforming a second derivative. Thus,

$$\int_0^{\pi} \frac{\partial^2 V}{\partial z^2} \sin nz \, dz = \left[\frac{\partial V}{\partial z} \sin nz - n \int \frac{\partial V}{\partial z} \cos nz \, dz \right]_0^{\pi} \tag{8.160}$$

Provided the first derivative is finite at both ends of the range (as it will be in practical cases) and since $\sin nz$ vanishes at both ends of the range, then

$$\int_0^{\pi} \frac{\partial^2 V}{\partial z^2} \sin nz \, dz = -n \int_0^{\pi} \frac{\partial V}{\partial z} \cos nz \, dz$$

$$= -n \left[V \cos nz + n \int V \sin nz \, dz \right]_0^{\pi}$$

$$= -n^2 \bar{V} + n[V_0 - (-1)^n V_\pi] \tag{8.161}$$

Thus the value of the dependent variable must be known at both ends of the range. Little further investigation is needed to show the following severe restrictions on the use of the finite Fourier sine transform.

S(a) The variable to be transformed must have a closed range.
S(b) Type 1 boundary conditions must be given at both ends of the range.
S(c) All derivatives to be transformed must be of even order.

Similarly, a finite Fourier cosine transform can be defined by

$$\bar{f}(n) = \int_0^{\pi} f(z) \cos nz \, dz \tag{8.162}$$

and inverted by

$$f(z) = \frac{1}{\pi} \bar{f}(0) + \frac{2}{\pi} \sum_{n=1}^{\infty} \bar{f}(n) \cos nz \tag{8.163}$$

but again the following restrictions apply.

C(a)   The variable to be transformed must have a closed range.
C(b)   Type 2 boundary conditions must be given at both ends of the range.
C(c)   All derivatives to be transformed must be of even order.

8.9.2. *Fourier Transforms*

Equation (8.156) can be written

$$f(x) = \sum_{n=1}^{\infty} 2 \sin \frac{n\pi x}{L} \int_0^L f(x) \sin \frac{n\pi x}{L} dx \, \delta\left(\frac{n}{L}\right) \qquad (8.164)$$

where $\delta n = 1$. Putting $p = n\pi/L$ and taking the limit as $L \to \infty$, then $\delta(n\pi/L) \to dp$ and the sum becomes an integral in terms of the "continuous" variable p. Thus

$$f(x) = \frac{2}{\pi} \int_0^\infty \sin px \left(\int_0^\infty f(x) \sin px \, dx \right) dp \qquad (8.165)$$

The Fourier sine transform is defined by

$$\bar{f}(p) = \int_0^\infty f(x) \sin px \, dx \qquad (8.166)$$

and equation (8.165) gives the inversion formula

$$f(x) = \frac{2}{\pi} \int_0^\infty \bar{f}(p) \sin px \, dp \qquad (8.167)$$

This transform is mathematically elegant because the same formula (except for a constant $2/\pi$) is used both for forming the transform and for inverting it. Although the Fourier sine transform is less restricted than the finite transform, it does not overcome all of the difficulties. If an attempt is made to remove a second derivative as in equation (8.160) with the upper limit changed, the term $(\partial V/\partial z) \sin pz$ oscillates boundedly as z increases indefinitely unless $(\partial V/\partial z) \to 0$ as $z \to \infty$. Hence the Fourier sine transform has the following limitations.

(i) The variable to be transformed must have an open range.
(ii) A type 1 boundary condition must be given at the lower limit.
(iii) The dependent variable and its derivative must vanish as $z \to \infty$.
(iv) All derivatives to be transformed must be of even order.

Similarly, a Fourier cosine transform can be defined by

$$\bar{f}(p) = \int_0^\infty f(x) \cos px \, dx \qquad (8.168)$$

and inverted by

$$f(x) = \frac{2}{\pi} \int_0^\infty \bar{f}(p) \cos px \, dp \qquad (8.169)$$

but the same restrictions as given above for the Fourier sine transform apply, except that the boundary conditions at the lower limit must be type 2.

The more general complex Fourier transform is defined by

$$\bar{f}(p) = \int_{-\infty}^{\infty} f(x) e^{ipx} \, dx \tag{8.170}$$

and inverted by

$$f(x) = \frac{1}{2\pi} \int_{-\infty}^{\infty} \bar{f}(p) e^{ipx} \, dp \tag{8.171}$$

This transform is not so severely restricted in application because derivatives of any order can be removed by its use, but it does not allow boundary conditions to be used at all! It is thus useful for general problems, but for specific problems it is virtually useless.

8.9.3. *Hankel and Mellin Transforms*

When the next two transforms are used, it is usually to remove the radial coordinate from a system of cylindrical polar coordinates, and because the integrals range from zero to infinity, they can only be applied to axisymmetrical disturbances in infinite media.

The Hankel transform is defined by

$$\bar{f}(p) = \int_{0}^{\infty} f(x) \, x J_n(px) \, dx \tag{8.172}$$

and inverted by the symmetrical formula

$$f(x) = \int_{0}^{\infty} \bar{f}(p) \, p J_n(px) \, dp \tag{8.173}$$

where $J_n(px)$ is the Bessel function of the first kind of order n.

The Mellin transform is defined by

$$\bar{f}(p) = \int_{0}^{\infty} f(x) \, x^{p-1} \, dx \tag{8.174}$$

and inverted by

$$f(x) = \frac{1}{2\pi i} \int_{\gamma-i\infty}^{\gamma+i\infty} \bar{f}(p) \, x^{-p} \, dp \tag{8.175}$$

which has certain similarities with the inversion formula for the Laplace transform.

8.10. CONCLUSIONS

The methods given in Section 8.5 are useful for finding isolated particular solutions of partial differential equations, but these solutions are unlikely to satisfy the desired boundary conditions. If the boundary conditions are of type 1, or vanishing derivative type 2, then the method of images can be very useful for generating a more general solution. The most powerful methods for finding solutions are undoubtedly the separation of variables (Section 8.7) and the Laplace transform (Section 8.8). Due to its versatility, the Laplace transform should be tried whenever a variable has an open range, and the method of separation of variables should be used in all other cases, including those in which the Laplace transform has been tried and found to be inconvenient. There are many partial differential equations which cannot be solved by either method, and the numerical methods described in Chapter 11 are recommended for these.

The other transform methods of Section 8.9 are specialized in that the four real Fourier transforms are only capable of removing derivatives of even order, and cannot solve any problem in which there is either a mixture of type 1 and type 2 boundary conditions or a type 3 condition. The complex Fourier transform can be taken without reference to boundary conditions and is thus only suitable for problems in which a variable can alter throughout its entire range from $-\infty$ to $+\infty$ without crossing a boundary. The use of other transforms such as the Hankel or Mellin transform is only advantageous when cylindrical polar coordinates would normally be used for stating the problem. It is not efficient for an engineer to try to learn the characteristics of all of these transforms when most problems requiring their use can be solved conveniently by using the Laplace transform. The reader is therefore recommended to familiarize himself with the Laplace transform, and turn to the other transforms later if he so desires.

Chapter 9

FINITE DIFFERENCES

9.1. INTRODUCTION

CHEMICAL processes involving separation, purification and chemical reactions are frequently carried out in a series of stages, and these processes are often analysed by stage-to-stage calculations or in the simpler examples by graphical methods. Thus distillation, gas absorption, and extraction operations performed in plate columns involve stepwise changes in the concentration of each phase as it passes from plate to plate. In particular, in a distillation column, the vapours leaving any plate differ in composition from the vapours entering the same plate by a finite amount, thus an abrupt change in the vapour composition occurs on the plate. Therefore, in contrast with a packed distillation column where the vapour composition changes continuously throughout the packed height and where the height of the packing is estimated from the solution of a differential equation, a plate column should be analysed in such a way that these abrupt finite changes enter into the analysis. This is possible by finite difference calculus and in particular by the solution of finite difference equations; as will be shown in this chapter.

9.2. THE DIFFERENCE OPERATOR, Δ

Considering an independent variable x which can take a number of discrete values x_0; $(x_0 + h)$; $(x_0 + 2h)$; etc., then any function $y = f(x)$ will take a corresponding set of values

$$y_0; y_1; y_2; \ldots; y_n$$

From this series of values of the dependent variable y, a series of differences can be constructed as follows.

$$y_1 - y_0; y_2 - y_1; y_3 - y_2; \ldots; y_n - y_{n-1}$$

This series can be written in the form

$$\Delta y_0; \Delta y_1; \Delta y_2; \ldots \Delta y_{n-1}$$

where
$$\Delta y_0 = y_1 - y_0 = f(x_0 + h) - f(x_0) = \Delta f(x_0) \tag{9.1}$$

Δy is called the "first difference" of y with respect to h. The symbol "Δ" is called the difference operator, and has meaning only when applied to a function of a given independent variable with reference to a specific increment h. In fact writing Δ before a variable or function takes a lot on trust. A more precise representation would be $\Delta_{x,h}^1$ implying that the differencing operation has been applied once to a function of an independent variable x through an increment h. However, it is seldom necessary to make these additions to Δ as both x and h would be clearly understood in a particular problem. Since it is possible to repeat the differencing operation, the number just above and on the right of Δ should be inserted for all differences greater than the first. Thus Δ^2 would be the second difference of a variable or function, and Δ^n would be the nth difference. These will be discussed more fully below.

9.2.1. *Properties of the Difference Operator* Δ

Let there be another function of the independent variable x that has discrete values at x_0; (x_0+h); etc., and let this function be expressed as

$$z = \phi(x) \tag{9.2}$$

Then the series of first differences of z can be written

$$z_1-z_0;\ z_2-z_1;\ z_3-z_2;\ \ldots;\ z_n-z_{n-1}$$

or in difference form

$$\Delta z_0;\ \Delta z_1;\ \Delta z_2;\ \Delta z_3\ \ldots;\ \Delta z_{n-1}$$

where

$$\Delta z_0 = z_1-z_0 = \phi(x_0+h)-\phi(x_0) \tag{9.3}$$

Adding equations (9.1) and (9.3) gives

$$\Delta y_0+\Delta z_0 = (y_1-y_0)+(z_1-z_0) = (y_1+z_1)-(y_0+z_0) = \Delta(y_0+z_0) \tag{9.4}$$

or more generally

$$\Delta(y_n+z_n) = \Delta y_n+\Delta z_n \tag{9.5}$$

That is, Δ obeys the "distributive law of algebra".

If k is a finite constant

$$\Delta(ky_n) = ky_{n+1}-ky_n = k\Delta y_n \tag{9.6}$$

so that k and Δ can be interchanged without altering the result. That is, Δ obeys the "commutative law of algebra" with regard to constants.

Finally, Δ obeys the "index law of algebra". That is

$$\Delta^{p+q} = \Delta^p \Delta^q = \Delta^q \Delta^p \tag{9.7}$$

This will be further explained below.

9.2.2. *Differences of Second and Higher Orders*

In Section 9.2 a sequence of first differences of the function $y = f(x)$ was constructed thus

$$(y_1-y_0);\ (y_2-y_1);\ \ldots;\ (y_n-y_{n-1}) = \Delta y_0;\ \Delta y_1;\ \ldots;\ \Delta y_{n-1} \tag{9.8}$$

The differencing process described by equation (9.8) can be repeated to give

$$\Delta y_1-\Delta y_0 = \Delta^2 y_0 \tag{9.9}$$

But from the relationship of Δy to y it can be seen that

$$\Delta^2 y_0 = \Delta y_1 - \Delta y_0 = (y_2 - y_1) - (y_1 - y_0) = y_2 - 2y_1 + y_0 \qquad (9.10)$$

Second difference relationships of the type given in equation (9.10) are employed to replace second derivatives in differential equations by finite differences prior to using a numerical method of solution. This will be illustrated later in Section 11.4.1 *et seq.*

The procedure developed in the above paragraph to obtain an expression for the second difference of a variable y can be extended to differences of higher order. For example, the third difference of y_0 is

$$\Delta^3 y_0 = y_3 - 3y_2 + 3y_1 - y_0 \qquad (9.11)$$

and the fourth difference is

$$\Delta^4 y_0 = y_4 - 4y_3 + 6y_2 - 4y_1 + y_0 \qquad (9.12)$$

In general terms for any variable y_n in a sequence, the mth difference is

$$\Delta^m y_n = y_{m+n} - \binom{m}{1} y_{m+n-1} + \binom{m}{2} y_{m+n-2} - \ldots (-1)^m y_n \qquad (9.13)$$

In equation (9.13) the coefficients of the expansion on the right of the equation are the same as those obtainable from a binomial expansion of $(1-x)^m$.

9.2.3. *Difference Tables*

The most convenient way of expressing the first, second and higher order differences of a function is by means of difference tables. Such a table is illustrated below for a function $y = f(x)$.

TABLE 9.1. The Difference Table

Variable	Function	Δ	Δ^2	Δ^3	Δ^4
x_0	y_0				
		Δy_0			
x_1	y_1		$\Delta^2 y_0$		
		Δy_1		$\Delta^3 y_0$	
x_2	y_2		$\Delta^2 y_1$		$\Delta^4 y_0$
		Δy_2		$\Delta^3 y_1$	
x_3	y_3		$\Delta^2 y_2$		
		Δy_3			
x_4	y_4				

In the above table, the second column gives the value of the function at specified values of the variable x. The third column having the heading Δ is constructed from the values of y in the second column by subtracting y_0 from y_1, y_1 from y_2, and so on to give the column of first differences. The fourth column of second differences is obtained from the first differences in the third column by subtracting the entry above from the one below. The difference table is completed by repeating the procedure outlined above. This will be illustrated by the following example.

Example. Construct a difference table of $y = f(x)$ from the following data and state what conclusions can be drawn from the result.

x	1·0	1·1	1·2	1·3	1·4	1·5	1·6
y	7·567	12·159	17·399	23·339	30·031	37·527	45·879

The difference table of these data is

x	y	Δ^1	Δ^2	Δ^3	Δ^4
1·0	7·567				
		4·592			
1·1	12·159		0·648		
		5·240		0·052	
1·2	17·399		0·700		0
		5·940		0·052	
1·3	23·339		0·752		0
		6·692		0·052	
1·4	30·031		0·804		0
		7·496		0·052	
1·5	37·527		0·856		
		8·352			
1·6	45·879				

Inspection of this difference table shows that all the third differences have the same value which from the elementary properties of a polynomial suggests that the data represent a function of the third degree.

Difference tables are very useful for assessing scientific data whether it be interpolation, extrapolation or fitting an equation to a curve of experimental data. These applications will be illustrated in Section 9.4 after more of the background has been established. However, certain properties of difference tables will be stated at this point. These are:

(i) If the value of a function is known at equally spaced values of the independent variable of the function, or if the difference between successive values of the function is known, all the columns of higher differences can be constructed.

(ii) If one complete column of differences of any order, and one item of each column on the left of the complete column is known, all the columns of the lower differences and the values of the function itself can be determined within the range of the independent variable.

(iii) If the members of the column of nth differences have the same constant value, the data from which the difference table was constructed are derived from a polynomial of degree n.

These rules can be tested by considering the example given above. Thus $\Delta^3 y = 0\cdot052$ and the last member of the Δ^2 column is $0\cdot856$. Therefore the last but one member will be $(0\cdot856 - 0\cdot052) = 0\cdot804$. In a similar manner each item in the Δ^2 column can be inserted and thereafter the Δ^1 column can be completed if one of its members is given.

9.3. Other Difference Operators

The differences presented in the above sections are, to be precise, called "forward differences" and are the ones most generally used for the solution of engineering problems. Consequently they are simply referred to as the first, second or nth differences and the prefix "forward" is omitted when it is clearly understood which type of difference is meant. However, in certain interpolation formulae the terms "backward difference" and "central difference" are employed and therefore their meanings must be understood. In actual fact there is no basic difference between a forward difference and a backward difference since both express the difference between the higher and next lower value of the function, and the significance between these terms is best understood by reference to Table 9.1. In that table it will be seen that the first entry in each column refers to y_0 and the various differences are obtained from successive members of the series starting from y_0. These are the forward differences and they are shown by the solid arrow. On the other hand, the last entry in each column is the first, second and third forward difference of each successively lower member of the series as shown by the dotted arrow. These entries in the table can be relabelled as backward differences by using the following definition.

$$\nabla y_n = y_n - y_{n-1} = \Delta y_{n-1} \tag{9.14}$$

The second backward difference is written similarly thus

$$\nabla^2 y_n = y_n - 2y_{n-1} + y_{n-2} = \Delta^2 y_{n-2} \tag{9.15}$$

It is now clear that all of the backward differences shown by the dotted arrow in Table 9.1 refer to y_4.

A third type of difference known as the "central difference" is symbolized by prefixing δ to the function. Thus

$$\delta y_n = y_{n+\frac{1}{2}} - y_{n-\frac{1}{2}} \tag{9.16}$$

Central differences would be represented by a horizontal line in Table 9.1.

To sum up, forward differences should be used when data from the initial parts of a table are to be assessed and backward differences used when data from the end of a table are to be considered. This is particularly true for interpolation and extrapolation purposes. Finally, for interpolation near the centre of a collection of data central differences should be employed.

There is another important difference operator which will be described in the following sub-section.

9.3.1. *The operator E*

The operation of changing the value of a function to correspond with a change of one increment in the independent variable is denoted by E. That is,

$$Ey_0 = y_1 \tag{9.17}$$

In a similar manner

$$y_n = Ey_{n-1} = E(Ey_{n-2}) = E^2 y_{n-2} = \ldots = E^n y_0 \tag{9.18}$$

Equation (9.18) shows that E^n operates on y_0 n times to increase the value of the function corresponding to an increase of n increments in the independent variable.

The exponent n in E^n can take any positive or negative integer value. Thus

$$E^{-1}f(x_0) = f(x_0 - h) \tag{9.19}$$

where the exponent (-1) signifies that the function is reduced to the next lower value of the independent variable.

The operator E has significance only when it is placed before a function or variable, and in this respect it is similar to other operators. Also, E can be treated like an algebraic symbol in that it obeys the distributive, commutative, and index laws of algebra. The reader is asked to verify these for himself.

Since E and Δ are both finite difference operators that obey the same laws of algebra, it would be expected that they must be inter-related. The relationship between them can be shown as follows.

$$y_{n+1} = Ey_n \tag{9.20}$$

$$y_{n+1} - y_n = \Delta y_n \tag{9.21}$$

Substituting equation (9.20) into (9.21) gives

$$(E-1)y_n = \Delta y_n \tag{9.22}$$

which can be written as an identity between operators thus,

$$E = 1 + \Delta \tag{9.23}$$

This simple relationship is most important because it often allows algebraic expressions in terms of the one operator to be simplified in terms of the other by using ordinary algebraic manipulations. Thus, if x is the independent variable that is capable of taking the values x_0; $(x_0 + h)$; $(x_0 + 2h)$; etc., then

$$f(x_0 + h) = Ef(x_0) \tag{9.24}$$

By Taylor's theorem (see Section 3.3.6)

$$Ef(x_0) = f(x_0 + h) = f(x_0) + h\,Df(x_0) + \frac{h^2}{2!}D^2 f(x_0) + \frac{h^3}{3!}D^3 f(x_0) + \ldots \tag{9.25}$$

where D is the differential operator d/dx. From equation (9.25) it can be seen that:

$$\left[1+h\,\mathrm{D}+\frac{(h\,\mathrm{D})^2}{2!}+\frac{(h\,\mathrm{D})^3}{3!}+\ldots\right]f(x_0) = Ef(x_0) = (1+\Delta)f(x_0) \quad (9.26)$$

The terms in the square bracket form the expansion of the exponential function $e^{h\mathrm{D}}$, and therefore equation (9.26) provides a relationship between the differential operator and the finite difference operators. That is

$$\mathrm{E} = 1+\Delta = e^{h\mathrm{D}} \quad (9.27)$$

The three symbols in equation (9.27) are all operators and they can be treated like algebraic quantities. In the mathematical relations developed to show the relationship between them, the analysis has been confined to the first order of the operator. This was done for convenience only, and for any value of m

$$\mathrm{E}^m = (1+\Delta)^m = e^{mh\mathrm{D}} \quad (9.28)$$

In addition, any polynomial of these symbols having constant coefficients represents an operation on a function. Thus

$$\mathrm{E}^m f(x_0) = (1+\Delta)^m f(x_0)$$

$$= f(x_0)+\binom{m}{1}\Delta f(x_0)+\binom{m}{2}\Delta^2 f(x_0)+\binom{m}{3}\Delta^3 f(x_0)+\ldots$$

$$= \left[1+\binom{m}{1}\Delta+\binom{m}{2}\Delta^2+\binom{m}{3}\Delta^3+\ldots\right]f(x_0) \quad (9.29)$$

Equation (9.29) is known as "Newton's forward difference formula". The polynomial within the square bracket is a polynomial operator and m is any number, integer or otherwise. This polynomial operator only has significance when it operates on a function such as $f(x)$ in equation (9.29). This result finds extensive use in interpolation (Section 9.4).

In a similar manner it is possible to produce an expansion of E^m in terms of the operator ∇. Thus by equation (9.23)

$$\mathrm{E}^m = \left(\frac{\mathrm{E}}{\mathrm{E}-\Delta}\right)^m = (1-\Delta\,\mathrm{E}^{-1})^{-m} \quad (9.30)$$

and if the last expression in equation (9.30) is expanded and made to operate on $f(x_0)$ the result is

$$\mathrm{E}^m f(x_0) = (1-\Delta\mathrm{E}^{-1})^{-m}f(x_0)$$

$$= f(x_0)+\binom{m}{1}\Delta f(x_0-h)+\binom{m+1}{2}\Delta^2 f(x_0-2h)+\ldots\binom{m+r-1}{r}$$

$$\times \Delta^r f(x_0-rh)+\ldots \quad (9.31)$$

Using equations (9.14, 9.15) to introduce backward differences,

$$E^m f(x_0) = f(x_0) + \binom{m}{1} \nabla f(x_0) + \binom{m+1}{2} \nabla^2 f(x_0) + \ldots \binom{m+r-1}{r} \nabla^r f(x_0) + \ldots$$

$$= \left[1 + \binom{m}{1} \nabla + \binom{m+1}{2} \nabla^2 + \binom{m+2}{3} \nabla^3 + \ldots \right] f(x_0) \qquad (9.32)$$

Equation (9.32) is known as "Newton's backward difference formula" and like equation (9.29) is used for interpolation. Again m can be any number, integer or otherwise.

9.4. INTERPOLATION AND EXTRAPOLATION

It frequently happens that the engineer is confronted with a collection of data relating two variables, such as the pressure inside a vessel corresponding to the temperature of its contents. The pressure is thus some unknown function of the temperature, and he is required to predict the pressure at some temperature in between or outside the limits of the data he possesses. If the pressure is required within the limits of the data, the mathematical procedure is called "interpolation"; whereas if the point lies outside the range of the data, the mathematical process is called "extrapolation".

Interpolation or extrapolation depends on the assumption that the functional relationship between the two variables is continuous over the range of the independent variable being considered. Therefore in order to be able to interpolate or extrapolate, the engineer must obtain this functional relationship or some other functional relationship that approximates to it. For this purpose the trends in the data must be considered from the point of view of the experimental results and the mathematical analysis. For example, the chemical engineer would be familiar with the general shape of a vapour pressure–temperature curve and its position on a graph from the experimental data. The mathematician is able to assign an approximate equation to this curve and thereafter interpolation is straightforward.

Usually the experimental data are approximated by a polynomial and the degree of the polynomial can often be estimated by preparing a difference table from the collected data. The difference column that gives an approximately constant value gives the degree of the polynomial which can be fitted to the data by selection of the constants.

9.4.1. *Newton's Formulae*

When the data points are available at equal intervals of the independent variable, Newton's difference formulae can be employed and the procedure is best illustrated by the following example.

Example 1. Using the data given in the table below estimate the vapour pressure and density of ammonia vapour at 75 °C. The latent heat of ammonia is 1265 kJ/kg.

Temperature °C	20	25	30	35	40	45	50	55	60
Pressure (kN/m²)	805	985	1170	1365	1570	1790	2030	2300	2610

The following difference table is prepared from the above data

t	p	Δp	$\Delta^2 p$	$\Delta^3 p$	$\Delta^4 p$
20	805				
		180			
25	985		5		
		185		5	
30	1170		10		-5
		195		0	
35	1365		10		$+5$
		205		5	
40	1570		15		0
		220		5	
45	1790		20		$+5$
		240		10	
50	2030		30		0
		270		10	
55	2300		40		
		310			
60	2610				

The fourth differences oscillate between ± 5, therefore the calculations will be restricted to the third differences. Inspection of the table shows that the last entry in the $\Delta^3 p$ column (i.e. 10) corresponds to $p = 1790$ kN/m². Therefore 45 °C will be taken as the base temperature. Then

$$p = f(t) = f(t_0 + mh)$$

where $\qquad t = 75\,°C, \quad t_0 = 45\,°C, \quad \text{and} \quad h = 5\,°C$

$$\therefore \quad 75 = 45 + 5m \quad \text{or} \quad m = 6 \cdot 0$$

Newton's forward difference formula (9.29) gives

$$p = p_0 + \binom{m}{1}\Delta p_0 + \binom{m}{2}\Delta^2 p_0 + \binom{m}{3}\Delta^3 p_0 \qquad\qquad \text{I}$$

$$= 1790 + \frac{6 \cdot 0}{1} \times 240 + \left(\frac{6 \cdot 0 \times 5 \cdot 0}{1 \times 2}\right)30 + \left(\frac{6 \cdot 0 \times 5 \cdot 0 \times 4 \cdot 0}{1 \times 2 \times 3}\right)10$$

$$= 1790 + 1440 + 450 + 200$$

$$= 3880 \text{ kN/m}^2.$$

The value extrapolated from Perry[†] is 3860 kN/m².

† Perry, J. H. (Ed.) "Chemical Engineers Handbook," 5rd edition. McGraw-Hill (1976).

In order to find the density of the ammonia vapour it is convenient to use the Clausius Clapeyron equation

$$\frac{dp}{dT} = \frac{\delta H}{T(V_v - V_L)} \qquad\qquad \text{II}$$

where T is the temperature in degrees Kelvin. Assuming that the volume of the liquid is negligible compared with the volume of the vapour gives

$$\rho = \frac{1}{V_v} = \frac{T}{\delta H}\frac{dp}{dT} \qquad\qquad \text{III}$$

dp/dT can be evaluated from the difference table by taking the logarithm of equation (9.27). Thus

$$h\mathrm{D} = \ln(1+\Delta) = \Delta - \tfrac{1}{2}\Delta^2 + \tfrac{1}{3}\Delta^3 - \ldots \qquad\qquad \text{IV}$$

but Δp, $\Delta^2 p$ and $\Delta^3 p$ are required at 75 °C. This can be done by applying Newton's interpolation formula to the differences as above for the pressure. Thus with

$$m = 6{\cdot}0, \quad \Delta p_0 = 240, \quad \Delta^2 p_0 = 30 \quad \text{and} \quad \Delta^3 p_0 = 10,$$

$$\Delta p = 240 + \left(\frac{6{\cdot}0}{1}\right)30 + \left(\frac{6{\cdot}0 \times 5{\cdot}0}{1 \times 2}\right)10 = 570$$

$$\Delta^2 p = 30 + \left(\frac{6{\cdot}0}{1}\right)10 = 90$$

and $\qquad\qquad \Delta^3 p = 10.$

Therefore $\qquad\qquad \rho = \dfrac{348}{5 \times 1265}\left(570 - \dfrac{90}{2} + \dfrac{10}{3}\right)$

$$= 29{\cdot}0 \text{ kg/m}^3$$

This value is more reliable than that obtained from the ideal gas laws.

9.4.2. Bessel's Interpolation Formula

This formula is for use with central differences along a horizontal line in the difference table. It can be seen in Table 9.1 that values are only available in alternate columns. The gaps are filled by taking an arithmetical average of the two values adjacent to the one required in the same column and denoting the value by $\mu\delta$, $\mu\delta^2$, ..., etc., depending upon the order of the difference. It is customary to choose a fractional value of m which is less than unity and use the formula

$$E^m f(x_0) = f(x_0) + m\delta f(x_0 + \tfrac{1}{2}h) + \frac{m(m-1)}{2!}\left[\mu\delta^2 f(x_0 + \tfrac{1}{2}h)\right.$$

$$\left. + \frac{m - \tfrac{1}{2}}{3}\delta^3 f(x_0 + \tfrac{1}{2}h)\right] + \frac{(m+1)m(m-1)(m-2)}{4!}$$

$$\times \left[\mu\delta^4 f(x_0 + \tfrac{1}{2}h) + \frac{m - \tfrac{1}{2}}{5}\delta^5 f(x_0 + \tfrac{1}{2}h)\right] + \ldots \quad \text{etc.} \quad (9.33)$$

9.4.3. *Lagrange's Interpolation Formula*

Quite often, the data are available at unequal intervals of the independent variable and this precludes the use of a normal difference table. Lagrange's method fits a polynomial to the data but otherwise has no connection with finite difference calculus; it is included in this chapter merely for completeness within this section concerning interpolation.

The fourth degree polynomial which passes through the four points $(x_0, y_0), (x_1, y_1), (x_2, y_2), (x_3, y_3)$ can be written

$$y = y_0 \frac{(x-x_1)(x-x_2)(x-x_3)}{(x_0-x_1)(x_0-x_2)(x_0-x_3)} + y_1 \frac{(x-x_0)(x-x_2)(x-x_3)}{(x_1-x_0)(x_1-x_2)(x_1-x_3)}$$

$$+ y_2 \frac{(x-x_0)(x-x_1)(x-x_3)}{(x_2-x_0)(x_2-x_1)(x_2-x_3)} + y_3 \frac{(x-x_0)(x-x_1)(x-x_2)}{(x_3-x_0)(x_3-x_1)(x_3-x_2)} \quad (9.34)$$

and this form is known as "Lagrange's interpolation formula". The extension to more than four points is fairly obvious; the numerator and denominator of each fraction are extended to include the further values of x_n and the complete equation is extended by adding further terms to include all values of y_n. It is to be noted that the term corresponding to the appropriate y_n is omitted from each numerator and its counterpart from the denominator. The use of many more than four points is not recommended due to the computational effort.

Example 2. Find the thermal conductivity of propane at $1 \cdot 013 \times 10^4$ kN/m² and 99 °C from the data given by Leng and Comings.†

T (°C)	P (kN/m²)	k (W/m °K)
68	$9 \cdot 7981 \times 10^3$	$0 \cdot 0848$
	$13 \cdot 324 \times 10^3$	$0 \cdot 0897$
87	$9 \cdot 0078 \times 10^3$	$0 \cdot 0762$
	$13 \cdot 355 \times 10^3$	$0 \cdot 0807$
106	$9 \cdot 7981 \times 10^3$	$0 \cdot 0696$
	$14 \cdot 277 \times 10^3$	$0 \cdot 0753$
140	$9 \cdot 6563 \times 10^3$	$0 \cdot 0611$
	$12 \cdot 463 \times 10^3$	$0 \cdot 0651$

From the graph given in the paper, other results show that in this region the conductivity varies approximately linearly with pressure so that a linear interpolation is adequate in this direction.

A set of values for conductivity and temperature at a pressure of $1 \cdot 013 \times 10^4$ kN/m² can be found from the above values as follows. At 68 °C and

† Leng, D. E. and Comings, E. W. *I.E.C.* **49** (ii), 2042 (1957).

$1 \cdot 013 \times 10^4$ kN/m²

$$k = 0 \cdot 0848 + \frac{(10 \cdot 13 - 9 \cdot 7981)(0 \cdot 0897 - 0 \cdot 0848) \times 10^3}{(13 \cdot 324 - 9 \cdot 7981) \times 10^3}$$

$$= 0 \cdot 0853 \text{ W/m }^{\circ}\text{K}$$

Hence the following table can be compiled

T	68	87	106	140
k	0·0853	0·0774	0·0699	0·0618

Substituting these values into equation (9.34) gives the thermal conductivity of propane at $1 \cdot 013 \times 10^4$ kN/m² and 99 °C. Thus

$$k = 0 \cdot 0853 \frac{(99-87)(99-106)(99-140)}{(68-87)(68-106)(68-140)} + 0 \cdot 0774 \frac{(99-68)(99-106)(99-140)}{(87-68)(87-106)(87-140)}$$

$$+ 0 \cdot 0699 \frac{(99-68)(99-87)(99-140)}{(106-68)(106-87)(106-140)}$$

$$+ 0 \cdot 0618 \frac{(99-68)(99-87)(99-106)}{(140-68)(140-87)(140-106)}$$

$$\therefore \quad k = 0 \cdot 0725 \text{ W/m }^{\circ}\text{K}$$

9.5. FINITE DIFFERENCE EQUATIONS

A finite difference equation is a relationship between an independent variable that can take discrete values only, a dependent variable and successive differences of the dependent variable. Thus an equation of the type

$$a_0 y_0 + a_1 y_1 + a_2 y_2 + \dots + a_n y_n = \phi(x) \tag{9.35}$$

is a finite difference equation if $y = f(x)$ and x can take the values $x_0, x_0 + h,$ $\dots, x_0 + nh$ only. In equation (9.35), $a_0, a_1, a_2, \dots, a_n$ can be constants or functions of the independent variable, $\phi(x)$ could be a constant or a function of x. Equation (9.35) can therefore be written

$$a_0 f(x_0) + a_1 f(x_0 + h) + \dots + a_n f(x_0 + nh) = \phi(x)$$

or

$$(a_n E^n + a_{n-1} E^{n-1} + \dots + a_0) f(x_0) = \phi(x) \tag{9.36}$$

because

$$f(x_0 + rh) = E^r f(x_0).$$

But

$$E = 1 + \Delta,$$

$$\therefore \quad (p_n \Delta^n + p_{n-1} \Delta^{n-1} + \dots + p_0) f(x_0) = \phi(x) \tag{9.37}$$

where the coefficients p_r are obtained from a_r. Equations (9.35) to (9.37) are different forms of the same finite difference equation.

Finite difference equations are similar to differential equations in that they can be classified by "order" and "degree". Thus the order of a finite difference equation is determined by the order of the highest difference. For example, the equation

$$y_{n+1} - ay_n = 0 \qquad (9.38)$$

is of order 1 because it involves only the first difference, i.e.

$$(E - a)y_n = [\Delta - (a - 1)]y_n = 0 \qquad (9.39)$$

On the other hand, the equation

$$y_n - ay_{n-1} + by_{n-2} = 0 \qquad (9.40)$$

is of order 2 because the highest difference is of order 2, thus

$$[\Delta^2 + (2 - a)\Delta + (1 - a + b)]y_{n-2} = 0 \qquad (9.41)$$

It should be noted from equations (9.39) and (9.41) that it is the difference between the highest and lowest members of the dependent variable that determines the order of the equation, and not the highest member as indicated by the suffix to the dependent variable.

The degree of a finite difference equation is the highest degree of the dependent variable, or any of its differences in the equation. Equations containing y_n, Δy_n, and $\Delta^n y_n$ are of the first degree and are said to be linear, whereas equations containing products of the type $y_n y_{n-1}$ or their differences are of degree 2. Linear difference equations are by far the most common and they will be treated initially.

9.6. Linear Finite Difference Equations

Equation (9.35) is a linear finite difference equation of order n and the coefficients $a_0, a_1, \ldots, a_n$ may be functions of the independent variable or constants.

The solution of this type of equation is similar to a linear differential equation in that the complete solution is made up of a complementary solution and a particular solution. Proof of this can be obtained in a manner very similar to that shown in Section 2.5.1 for differential equations, and therefore will not be repeated. However, each part of the complete solution will be discussed.

9.6.1. *The Complementary Solution*

For ease of manipulation, consider the second order linear finite difference equation with constant coefficients.

$$y_{n+2} - Ay_{n+1} + By_n = \phi(n) \qquad (9.42)$$

In the same way as for differential equations, the complementary solution is the solution of equation (9.42) when $\phi(n)$ is zero, i.e.

$$y_{n+2} - Ay_{n+1} + By_n = 0 \qquad (9.43)$$

But equation (9.43) can be written in the form

$$(E^2 - AE + B)y_n = 0 \qquad (9.44)$$

Since the polynomial within the brackets can be treated as an algebraic quantity, it can be factorized.

$$\therefore \quad (E - \rho_1)(E - \rho_2)y_n = 0 \qquad (9.45)$$

The two factors could be in either order since the operator E will commute with constants such as ρ_1 and ρ_2; hence there are two independent solutions of equation (9.43) satisfying

(i) $$\qquad\qquad (E - \rho_1)y_n = 0$$

or (ii) $$\qquad\qquad (E - \rho_2)y_n = 0$$

It can be seen by substitution that the solution of (i) is

$$y_n = A'\rho_1{}^n$$

and the solution of (ii) is

$$y_n = B'\rho_2{}^n$$

Hence the general solution of equation (9.43) is given by

$$y_n = A'\rho_1{}^n + B'\rho_2{}^n \qquad (9.46)$$

where A' and B' are two arbitrary constants that would be evaluated by the boundary conditions of the problem.

It will be noticed that equation (9.46) is the complementary solution of the original second order equation and that it contains two arbitrary constants. In a similar manner it can be shown that an nth order equation will provide a complementary solution containing n arbitrary constants; which again shows the similarity to the complementary function of a linear differential equation.

Equation (9.44) is known as the "characteristic equation" of the finite difference equation (9.42). Its roots may be different or equal, real, or complex. The case of different roots is given above, and if ρ_1 and ρ_2 are real, equation (9.46) is the final complementary solution. If the roots are complex, they will occur in conjugate pairs, thus

$$\rho_1 = \alpha + i\beta = r\,e^{i\phi}; \quad \rho_2 = \alpha - i\beta = r\,e^{-i\phi} \qquad (9.47)$$

In this case, the solution (9.46) can be written in the form

$$y_n = r^n(A'\cos n\phi + B'\sin n\phi) \qquad (9.48)$$

This is the complementary solution when the roots of the characteristic equation are complex and is similar to the solution of the corresponding differential equation.

Finally, when the roots of the characteristic equation are equal, the complementary solution takes the form

$$y_n = (A' + B'n)\rho^n \tag{9.49}$$

again like the comparable form of the differential equation (Section 2.5.2b).

9.6.2. *The Particular Solution*

This can invariably be found either by the method of undetermined coefficients or the method of inverse operators; both of which are very similar to the corresponding method for linear differential equations. Since these methods were described in Section 2.5, they will only be repeated briefly here with reference to the second order finite difference equation

$$(E^2 - AE + B)y_n = \phi(n) \tag{9.42}$$

A particular solution of this equation is

$$y_n = \left(\frac{1}{E^2 - AE + B}\right)\phi(n) \tag{9.50}$$

by the method of inverse operators. The operator can be factorized as in the above sub-section, and separated into partial fractions. Thus

$$\frac{1}{E^2 - AE + B} = \frac{1}{(E - \rho_1)(E - \rho_2)} = \frac{\alpha}{(E - \rho_1)} - \frac{\alpha}{(E - \rho_2)} \tag{9.51}$$

where $\alpha = 1/(\rho_1 - \rho_2)$. Each partial fraction on the right-hand side of equation (9.51) can be expanded as follows

$$\frac{\alpha}{E - \rho_1} = \frac{\alpha}{(1 - \rho_1)}\left[1 + \frac{\Delta}{1 - \rho_1}\right]^{-1}$$

$$= \frac{\alpha}{(1 - \rho_1)}\left[1 - \frac{\Delta}{1 - \rho_1} + \frac{\Delta^2}{(1 - \rho_1)^2} - \cdots\right] \tag{9.52}$$

and

$$\frac{\alpha}{E - \rho_2} = \frac{\alpha}{(1 - \rho_2)}\left[1 - \frac{\Delta}{1 - \rho_2} + \frac{\Delta^2}{(1 - \rho_2)^2} - \cdots\right] \tag{9.53}$$

The forms (9.52, 9.53) of the operators are most convenient for finding the particular solution if $\phi(n)$ is a polynomial in n, because only a finite number of terms will be needed in the expansions.

When $\phi(n) = ka^n$, an alternative procedure is advantageous. Thus

$$Ea^n = a^{n+1} = a.a^n$$

and

$$E^m a^n = a^{m+n} = a^m.a^n$$

In general,

$$f(E)a^n = f(a)a^n \tag{9.54}$$

provided $f(E)$ can be expanded as a polynomial in E. Equation (9.54) is the key to a particular solution when $\phi(n) = ka^n$. The usual difficulties can arise when $f(a)$ is infinite, but these can be overcome by techniques similar to those given in Section 2.5.5. It should be noted, however, that the equation analogous to equation (2.79) takes the form

$$f(E)(a^n y_n) = a^n f(aE) y_n \qquad (9.55)$$

where a and E appear as a product instead of a sum on the right-hand side. Equation (9.55) is of little practical value since it does not lead to any simplification in the form of the operator.

Hence, rewriting equation (9.42) with $\phi(n) = ka^n$.

$$(E^2 - AE + B)y_n = ka^n \qquad (9.56)$$

and using equation (9.54), the particular solution is

$$y_n = \frac{ka^n}{a^2 - Aa + B} \qquad (9.57)$$

provided $a^2 - Aa + B \neq 0$. This exceptional case can be treated in a similar manner to the corresponding differential equation by using equation (9.55) and introducing the difference operator Δ using equation (9.23).

In concluding the sub-section on the determination of particular solutions of finite difference equations, it should be pointed out that a particular solution can be obtained in most cases when $\phi(n)$ takes the form of a polynomial or power by the method of undetermined coefficients. Here there is no difference in technique between finite difference equations and differential equations; the trial solutions take exactly the same form in the corresponding cases.

Finally, the complete solution is the sum of the complementary solution and a particular solution in a manner similar to differential equations. If possible the operational method should be used to evaluate both parts of the solution, but on occasion the method of undetermined coefficients may be more suitable.

Example 1. G_{N+1} kg moles/h of a wet gas containing Y_{N+1} kg moles/mole of solute are fed into the base of a plate absorption column where the solute is to be stripped from the gas by absorption in L_0 kg moles/h of lean oil which is fed into the top of the column. If the solute in the entering oil is X_0 kg moles/mole of lean oil and the solute in the exit gas is Y_1 kg moles of solute/mole of wet gas, show that the performance of the absorber can be expressed in terms of the absorption factor $A = (L_0/KG_{N+1})$ and the number of ideal stages by the Kremser–Brown equation

$$\frac{Y_{N+1} - Y_1}{Y_{N+1} - Y_0} = \frac{A^{N+1} - A}{A^{N+1} - 1}$$

where K_m is the equilibrium constant. That is

$$y_m = K_m x_m$$

where y_m and x_m are the mole fractions of the solute in the gas and liquid phases respectively.

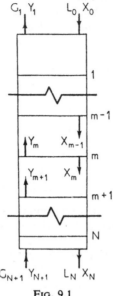

FIG. 9.1

Considering Fig. 9.1 and taking a material balance over plate m gives

$$L_0(X_m - X_{m-1}) = G_{N+1}(Y_{m+1} - Y_m) \qquad \text{I}$$

But

$$y_m G_m = Y_m G_{N+1} \quad \text{and} \quad x_m L_m = X_m L_0 \qquad \text{II}$$

$$\therefore \quad Y_m \frac{G_{N+1}}{G_m} = K_m \frac{X_m L_0}{L_m}$$

or

$$X_m = \left(\frac{G_{N+1}}{L_0}\right)\left(\frac{L_m}{K_m G_m}\right) Y_m \qquad \text{III}$$

Substituting equation III into equation I to eliminate X_m and X_{m-1} gives

$$Y_m = \frac{Y_{m+1} + (L_{m-1}/K_{m-1} G_{m-1}) Y_{m-1}}{1 + (L_m/K_m G_m)} \qquad \text{IV}$$

But $(L_{m-1}/K_{m-1} G_{m-1}) = A_{m-1}$, the absorption factor for plate $m-1$; and $(L_m/K_m G_m) = A_m$, the absorption factor for plate m. Therefore equation IV can be written as

$$Y_m = \frac{Y_{m+1} + A_{m-1} Y_{m-1}}{1 + A_m} \qquad \text{V}$$

Taking a mean value of the absorption factor as suggested by Edmister[†] or Horton and Franklin[‡] equation V becomes

$$Y_{m+1} - (1 + A) Y_m + A Y_{m-1} = 0 \qquad \text{VI}$$

where A is the mean value of the absorption factor between the bottom and top of the column.

Putting equation VI into operational form and factorizing gives

$$[E^2 - (1 + A) E + A] Y_m = (E - 1)(E - A) Y_m = 0 \qquad \text{VII}$$

which has the solution

$$Y_m = C_1 + C_2 A^m \qquad \text{VIII}$$

where C_1 and C_2 are arbitrary constants.

Now
$$Y_0 = K X_0 \qquad \text{IX}$$

since A is assumed constant; and equation VIII gives

$$Y_0 = C_1 + C_2 \qquad \text{X}$$

Similarly,
$$Y_{N+1} = C_1 + C_2 A^{N+1} \qquad \text{XI}$$

$$\therefore \quad Y_{N+1} - Y_0 = C_2(A^{N+1} - 1) \qquad \text{XII}$$

and also
$$Y_{N+1} - Y_1 = C_2(A^{N+1} - A) \qquad \text{XIII}$$

Dividing equation XIII by equation XII gives

$$\frac{Y_{N+1} - Y_1}{Y_{N+1} - Y_0} = \frac{A^{N+1} - A}{A^{N+1} - 1} \qquad \text{XIV}$$

which is the well-known Kremser equation.

9.6.3. *Simultaneous Linear Difference Equations*

The analysis presented in the sections above has been confined to one dependent variable. However, when two or more dependent variables are present in a system of difference equations it is usually possible to eliminate all of these variables except one by the normal rules of algebra. Thus consider the following examples.

Example 2. In a proposed chemical process 5000 kg/h of a pure liquid A is to be fed continuously to the first of a battery of two equal-sized stirred tank reactors operating in series. If both vessels are to be maintained at the same constant temperature so that the chemical reaction

$$A \xrightarrow{\ k_1\ } B \xrightarrow{\ k_2\ } C \qquad \text{I}$$

---

† Edmister, W. C. *I.E.C.* **35**, 837 (1943).
‡ Horton, G. and Franklin, W. B. *I.E.C.* **32**, 1384 (1940).

will take place, estimate the size of the vessels that will give the maximum yield of the product B. The specific reaction rate constant $k_1 = 6{\cdot}0$ h^{-1} and $k_2 = 3{\cdot}0$ h^{-1} at the temperature of the reactors, and the fluid density is constant at 960 kg/m^3.

Let the concentration of A leaving any vessel n be $C_{A,n}$, the concentration of B be $C_{B,n}$ and of C be $C_{C,n}$. Then a material balance over stage n gives for:

component A,
$$C_{A,n-1} - C_{A,n} = k_1 C_{A,n}\, \theta \qquad\qquad\qquad \text{II}$$

where θ h is the nominal holding time in the reactor;

component B,
$$C_{B,n-1} - C_{B,n} = k_2 C_{B,n}\, \theta - k_1 C_{A,n}\, \theta \qquad\qquad \text{III}$$

Solution of equation II is obtained in the usual way. Thus let $k_1 \theta = \alpha$ and $k_2 \theta = \beta$. Then

$$(1+\alpha) C_{A,n} - C_{A,n-1} = 0 \qquad\qquad\qquad \text{IV}$$

$$[E(1+\alpha) - 1] C_{A,n} = 0 \qquad\qquad\qquad \text{V}$$

or
$$C_{A,n} = K_1 \rho_1{}^n \qquad\qquad\qquad \text{VI}$$

where K_1 is an arbitrary constant and $\rho_1 = 1/(1+\alpha)$. Substituting for $C_{A,n}$ from equation VI into equation III and rearranging gives

$$[(1+\beta) E - 1] C_{B,n-1} = \alpha K_1 \rho_1{}^n \qquad\qquad \text{VII}$$

The complementary solution of equation VII is

$$C_{B,n} = K_2 \rho_2{}^n \qquad\qquad\qquad \text{VIII}$$

where K_2 is the arbitrary constant and $\rho_2 = 1/(1+\beta)$.

The particular solution of equation VII can be written

$$C_{B,n-1} = \left[\frac{1}{(1+\beta) E - 1}\right] \alpha K_1 \rho_1{}^n$$

$$= \frac{\alpha K_1 \rho_1{}^n}{(1+\beta) \rho_1 - 1}$$

$$= \frac{\alpha K_1 \rho_1{}^{n-1}}{\beta - \alpha} \qquad\qquad\qquad \text{IX}$$

by using equation (9.54). The complete solution of equation VII is thus

$$C_{B,n} = K_2 \rho_2{}^n + \frac{\alpha K_1}{\beta - \alpha} \rho_1{}^n \qquad\qquad \text{X}$$

Since the initial feed to the reactors is pure A, at $n = 0$, $C_A = C_{A,0}$ and $C_B = 0$.

$$\therefore \quad K_1 = C_{A,0} \qquad\qquad\qquad \text{XI}$$

and
$$K_2 = \frac{\alpha}{\alpha - \beta} C_{A,0} \qquad\qquad\qquad \text{XII}$$

Substituting these values into equation X gives

$$C_{B,n} = \frac{\alpha C_{A,0}}{\beta - \alpha}(\rho_1{}^n - \rho_2{}^n) \qquad \text{XIII}$$

The condition required is that $C_{B,n}$ must be a maximum for $n = 2$ and some value of θ. Hence, by putting $n = 2$,

$$\rho_1 = \frac{1}{1 + \alpha} = \frac{1}{1 + 6\theta} \qquad \text{and} \qquad \rho_2 = \frac{1}{1 + \beta} = \frac{1}{1 + 3\theta},$$

$$C_{B,2} = \frac{6C_{A,0}}{(3 - 6)}\left\{\frac{1}{(1 + 6\theta)^2} - \frac{1}{(1 + 3\theta)^2}\right\}$$

$$= 2C_{A,0}\left\{\frac{1}{(1 + 3\theta)^2} - \frac{1}{(1 + 6\theta)^2}\right\} \qquad \text{XIV}$$

Differentiating equation XIV and equating to zero,

$$\frac{dc_{B,2}}{d\theta} = 2C_{A,0}\left\{\frac{-6}{(1 + 3\theta)^3} - \frac{-12}{(1 + 6\theta)^3}\right\} = 0 \qquad \text{XV}$$

$$\therefore \quad 6(1 + 6\theta)^3 = 12(1 + 3\theta)^3$$

$$\therefore \quad \frac{1 + 6\theta}{1 + 3\theta} = 2^{1/3}$$

and hence

$$\theta = \frac{2^{1/3} - 1}{3(2 - 2^{1/3})} = \frac{0 \cdot 2599}{2 \cdot 2203} = 0.117 \qquad \text{XVI}$$

$$\frac{\text{Reactor volume}}{\text{Volumetric feed rate}} = \frac{960\,v}{5000} = 0 \cdot 117$$

$\therefore$ Optimum volume of each reactor is $0 \cdot 61$ m^3.

Example 3. 3875 kg/h of an animal fat are to be hydrolysed and extracted in a spray column using 1705 kg/h of water.† If the column is to operate under counter-current flow conditions, the percentage of hydrolysable glycerine in the fat is 8·53% by weight, and the glycerine in fatty acid leaving the tower is 0·24% by weight, calculate the number of theoretical stages in the column.

A mass balance will give a glycerine concentration in the sweetwater of 18·8%, and the total weight of fat phase held up in the column is 5535 kg. The distribution ratio of glycerine between water and fat is 10·32, and the reaction rate constant is 10·2 h^{-1}.

† Jeffreys, G. V., Jenson, V. G. and Miles, F. R. *Trans. Instn. Chem. Eng.*, 39, 389 (1961).

Solution

In order to determine N, the number of theoretical stages, the plate column illustrated in Fig. 9.2 is considered and the following symbols are used:

L kg/h fat phase rising through the column.

G kg/h water phase descending through the column.

H kg fat phase held up per stage.

x weight fraction of glycerine in raffinate.

y weight fraction of glycerine in extract.

z weight fraction of unreacted fat in raffinate.

w kg fat required to produce 1·0 kg of glycerine.

k pseudo first order reaction rate constant expressed in terms of fat concentration (h^{-1}).

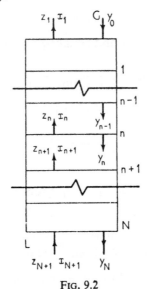

FIG. 9.2

A glycerine balance over stage n gives

$$Gy_{n-1}+Lx_{n+1}+\frac{kH}{w}z_n = Gy_n+Lx_n \qquad\qquad \text{I}$$

and an equivalent glycerine balance between plate n and the base of the column gives

$$\frac{L}{w}z_{N+1}+Gy_{n-1} = Gy_N+L\left(x_n+\frac{z_n}{w}\right) \qquad\qquad \text{II}$$

The equilibrium relation is

$$y_n = mx_n \qquad\qquad \text{III}$$

where m is the equilibrium ratio.

Substitution of (z_n/w) from equation I into equation II, and elimination of x by the use of equation III gives

$$\frac{mG}{L}(y_{n-1}-y_n)+y_{n+1}-y_n+\frac{kH}{L}\left(\frac{m}{w}z_{N+1}+\frac{mG}{L}y_{n-1}-\frac{mG}{L}y_N-y_n\right)=0 \quad \text{IV}$$

Letting $mG/L = \alpha$, $kH/L = \beta$ and rearranging IV gives

$$y_{n+1}-(\alpha+\beta+1)y_n+(\alpha+\alpha\beta)y_{n-1} = \beta\left(\alpha y_N-\frac{m}{w}z_{N+1}\right) \quad \text{V}$$

Equation V is a linear finite difference equation. The subsidiary equation is

$$[E^2-(\alpha+\beta+1)E+\alpha(1+\beta)]y_n = 0 \quad \text{VI}$$

and the complementary solution is

$$y_n = A\alpha^n + B(1+\beta)^n \quad \text{VII}$$

The particular solution is

$$y_n = \frac{1}{(E-\alpha)(E-1-\beta)}C \quad \text{VIII}$$

where

$$C = \beta\left(\alpha y_N-\frac{m}{w}z_{N+1}\right)$$

Because C is independent of n and $E = 1+\Delta$, E can be replaced by unity. Hence equation VIII becomes

$$y_n = \frac{-C}{\beta(1-\alpha)}$$

$$= \frac{\alpha y_N - mz_{N+1}/w}{\alpha-1} \quad \text{IX}$$

The complete solution of equation V is then

$$y_n = A\alpha^n + B(1+\beta)^n + \frac{\alpha y_N - mz_{N+1}/w}{\alpha-1} \quad \text{X}$$

with the boundary conditions that $n = 0$, $y = 0$; and at $n = N+1$, $x = 0$. Substituting these boundary conditions into equation X gives

$$A = \left(\frac{\alpha y_N - mz_{N+1}/w}{\alpha-1}\right)\left[\frac{(1+\beta)^{N+1}-1}{\alpha^{N+1}-(1+\beta)^{N+1}}\right]$$

and

$$B = -\left(\frac{\alpha y_N - mz_{N+1}/w}{\alpha-1}\right)\left[\frac{\alpha^{N+1}-1}{\alpha^{N+1}-(1+\beta)^{N+1}}\right]$$

The final solution after simplifying is

$$y_N = \frac{mz_{N+1}}{w(\alpha-1)}\left[\alpha^N+\frac{(\alpha-1)\alpha^N-\beta(1+\beta)^N}{(1-\alpha+\beta)(1+\beta)^N}\right] \quad \text{XI}$$

Substituting the data supplied into equation XI gives

$$\alpha = \frac{mG}{L} = \frac{10\cdot32 \times 1705}{3875} = 4\cdot54$$

$$\beta = \frac{kH}{L} = \frac{10\cdot2 \times 5535}{3875 \times N} = \frac{14\cdot6}{N}$$

where $5535/N = H$, the hold-up per stage. Then from equation XI

$$0\cdot188 = \left(\frac{10\cdot32 \times 0\cdot0853}{4\cdot54^{N+1} - 1}\right)\left[4\cdot54^N + \frac{3\cdot54 \times 4\cdot54^N - (14\cdot6/N)(1 + 14\cdot6/N)^N}{[(14\cdot6/N) - 3\cdot54][1 + (14\cdot6/N)]^N}\right]$$

gives $N = 2\cdot8$ (by trial and error)

9.7. NON-LINEAR FINITE DIFFERENCE EQUATIONS

These types of equation occasionally arise in the solution of engineering problems and they are difficult and sometimes impossible to solve. However, first order non-linear equations can be solved graphically, and some second order equations can be solved by special substitutions. These methods will be discussed separately below.

9.7.1. *Graphical Solution*

Consider the first order non-linear difference equation

$$y_{n+1} - y_n^2 + Ay_n + B = 0 \tag{9.58}$$

where A and B are constants, and rearrange it to separate the terms according to their subscripts. Thus

$$y_{n+1} = y_n^2 - Ay_n - B \tag{9.59}$$

By selecting arbitrarily a suitable set of values for y_n, a corresponding set of values of y_{n+1} can be calculated from equation (9.59). Hence a table of values of y_n and y_{n+1} can be compiled and the results plotted on rectangular coordinates in the form y_n vs y_{n+1}. The curve AB in Fig. 9.3 illustrates equation (9.59). In order to find the solution, the diagonal $y_n = y_{n+1}$ shown by the line PQ is constructed. Starting at the boundary condition y_0 at A (of coordinates y_0, y_1) and drawing the ordinate, the point C is located on the diagonal where $y_0 = y_1$. From C, a horizontal line is drawn back to meet the curve AB at the point D. The coordinates of D are (y_1, y_2) by equation (9.59) and hence y_2 is evaluated. By continuing this stepwise procedure to N steps, the value of y_{N+1} can be obtained. Thus the value of y corresponding to any value of n can be obtained.

This method will work for all first order equations provided that they can be separated into a simple form such as equation (9.59). It is usually easier to choose values for the variable on the more complicated side of the equation and solve for the other variable. Thus it is simpler to choose a set of values for y_n and solve equation (9.59) for y_{n+1} than assume values for y_{n+1} and solve the quadratic for y_n.

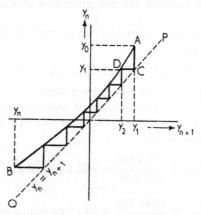

FIG. 9.3. Graphical solution of a finite difference equation

Example 1. 500 kg/h of ethyl alcohol is to be esterified by reacting with 425 kg/h of acetic acid in a battery of continuously stirred tank reactors, each of 0·935 m³ capacity and maintained at 100 °C. If the equilibrium relation is such that 75·2% of the acid will be esterified, estimate the number of reactors required for 60% conversion.

At 100 °C the specific reaction rate for esterification is 0·0285 m³/kg mole h and that for the hydrolysis of the ester is 0·0098 m³/kg mole h. The density of the reaction mixture can be assumed constant at 865 kg/m³.

In a battery of N stirred tank reactors, consider a material balance on reactor m. That is,

$$qC_{A,m-1} - qC_{A,m} = rV \qquad\qquad \text{I}$$

where $C_{A\,m}$ is the concentration of reactant A leaving vessel m, r is the rate of reaction, q is the volumetric flow rate and V is the volume of one reactor.

For a second order reaction

$$A + B \;\rightleftharpoons\; C + D \qquad\qquad \text{II}$$

the rate of chemical reaction is

$$r = k_1 C_A C_B - k_2 C_C C_D \qquad\qquad \text{III}$$

and if the agitation is assumed to be perfect, the concentration of the components in the effluent from the vessel will be the same as the contents of the vessel. Therefore equation I becomes

$$C_{A,m-1} - C_{A,m} = (k_1 C_A C_B - k_2 C_C C_D)_m \, \theta \qquad\qquad \text{IV}$$

where θ is the nominal holding time in the vessel.

If the concentration of reactant B exceeds that of reactant A by an amount c initially, this difference will be maintained throughout the system. That is, in any reactor m the concentration of reactant B is $(C_{A,m}+c)$. By the stoichiometry of the reaction the concentration of each product is $(C_{A,0}-C_{A,m})$.

Substituting these concentrations of components B, C, and D into equation IV and rearranging gives

$$C_{A,m-1} = C_{A,m} + [k_1 C_{A,m}(C_{A,m}+c) - k_2(C_{A,0}-C_{A,m})^2]\,\theta \qquad \text{V}$$

Equation V is a first order non-linear finite difference equation which will be solved graphically for the total number of reactors required to convert 60% of the acetic acid. Thus assuming the feed is uniformly mixed, the initial concentration of acetic acid is

$$C_{A,0} = \frac{425 \times 865}{925 \times 60} = 6.62 \text{ kg moles/m}^3$$

Similarly,
$$C_{B,0} = \frac{500 \times 865}{925 \times 46} = 10.17 \text{ kg moles/m}^3$$

$$\therefore \quad c = 10.17 - 6.62 = 3.55 \text{ kg moles/m}^3$$

$$\theta = \frac{0.935 \times 865}{925} = 0.875 \text{ h,}$$

Substituting these values into equation V gives

$$C_{A,m-1} = C_{A,m} + [0.0285\,C_{A,m}(C_{A,m}+3.55) - 0.0098(6.62 - C_{A,m})^2]\,(0.875)$$

$$= 0.0164\,C_{A,m}^2 + 1.202C_{A,m} - 0.376 \qquad \text{VI}$$

Arbitrary values of $C_{A,m}$ given in the table were selected and substituted into equation VI to give the corresponding values of $C_{A,m-1}$.

$C_{A,\,m}$	$C_{A,\,m-1}$
6·6	8·27
6·0	7·43
5·0	6·05
4·0	4·70
3·0	3·38
2·0	2·09

$C_{A,m-1}$ is plotted against $C_{A,m}$ in Fig. 9.4; and starting with the feed composition of 6.62 kg moles/m³, the number of reactors is "stepped off" as shown.

The required conversion is 60%,

$\therefore$ acid in final effluent $= 0.40 \times 6.62 = 2.65 \text{ kg moles/m}^3$;

$\therefore$ number of reactors to produce this effluent is 7.

9.7.2. *Analytical Solution*

At the present time only a limited number of non-linear finite difference equations can be solved analytically and these are solvable because they can be transformed into linear equations. Thus consider the following example.

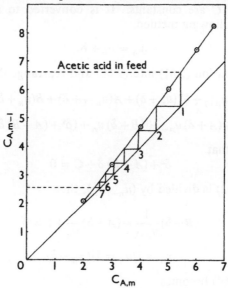

FIG. 9.4. Graphical reactor analysis

Example 2. Solve the equation

$$y_{n+2}y_n = y_{n+1}^2 \qquad\qquad\qquad\text{I}$$

Converting equation I to logarithmic form,

$$\log y_{n+2} + \log y_n = 2\log y_{n+1} \qquad\qquad\qquad\text{II}$$

Letting
$$u_n = \log y_n$$

equation II becomes the second order linear equation

$$u_{n+2} - 2u_{n+1} + u_n = 0 \qquad\qquad\qquad\text{III}$$

Solution of equation III by the methods already explained is

$$u_n = C_1 + C_2 n \qquad\qquad\qquad\text{IV}$$

or
$$\log y_n = C_1 + C_2 n \qquad\qquad\qquad\text{V}$$

Now let
$$C_1 = \log A \quad \text{and} \quad C_2 = \log B$$

then
$$y_n = AB^n \qquad\qquad\qquad\text{VI}$$

The solution of the above equation depended on the linearization of equation I and unless such a transformation can be found it is not possible to solve the equation.

A non-linear finite difference equation of second order that frequently appears in engineering problems is the Riccati finite difference equation.

$$y_{n+1}y_n + Ay_{n+1} + By_n + C = 0 \tag{9.60}$$

where A, B, and C are constants. It is converted to a linear difference equation by the following method.

Let
$$y_n = u_n + \delta \tag{9.61}$$

and substitute equation (9.61) into (9.60) and rearrange thus

$$(u_{n+1} + \delta)(u_n + \delta) + A(u_{n+1} + \delta) + B(u_n + \delta) + C = 0$$

$$u_{n+1}u_n + (A + \delta)u_{n+1} + (B + \delta)u_n + (\delta^2 + (A + B)\delta + C] = 0 \tag{9.62}$$

If δ is chosen so that
$$\delta^2 + (A + B)\delta + C = 0 \tag{9.63}$$

and equation (9.62) is divided by $(u_{n+1}u_n)$, then

$$(B + \delta)\frac{1}{u_{n+1}} + (A + \delta)\frac{1}{u_n} + 1 = 0 \tag{9.64}$$

Putting
$$x_n = 1/u_n = 1/(y_n - \delta) \tag{9.65}$$

then equation (9.64) becomes

$$x_{n+1} + Px_n + Q = 0 \tag{9.66}$$

which is a first order linear finite difference equation with

$$P = \frac{A + \delta}{B + \delta} \quad \text{and} \quad Q = \frac{1}{B + \delta}$$

Equation (9.66) can be solved by the methods already given, to yield the final solution

$$\frac{1}{y_n - \delta} = K\left[-\frac{A + \delta}{B + \delta}\right]^n - \frac{1}{A + B + 2\delta} \tag{9.67}$$

where the constant K is to be evaluated from the boundary conditions of the problem. This is illustrated in the example below.

Example 3. A benzene–toluene feed containing 60 mole per cent of benzene is fed continuously to a distillation column. If there are nine plates between the reboiler and the feed plate, and the top product contains 98 mole per cent benzene whilst the liquid leaving the base of the column contains 2 mole per cent benzene, estimate the overall plate efficiency of the column. The feed enters the column at its boiling point and the relative volatility of benzene to toluene can be considered to be constant at 2·3. The reflux ratio is 3.0.

Consider 100 kg moles feed and let D = moles distillate; W = moles residue.

$$\therefore \quad 100 = D + W \qquad\qquad\qquad\qquad\qquad \text{I}$$

and

$$60 = 0.98D + 0.02(100 - D) \qquad\qquad\qquad \text{II}$$

from which $D = 60.4$ kg moles and $W = 39.6$ kg moles.

A mass balance between the bottom of the column and some plate n in the stripping section is

$$Lx_{n+1} = Gy_n + Wx_w \qquad\qquad\qquad\qquad \text{III}$$

where L is the molar reflux down the column and G is the molar vapour rate up the column. Since the relative volatility is constant, the equilibrium relationship between x and y on any plate is

$$y_n = \frac{\alpha x_n}{1 + (\alpha - 1)x_n} \qquad\qquad\qquad\qquad \text{IV}$$

Substituting equation IV into equation III gives

$$Lx_{n+1} = \frac{\alpha Gx_n}{1 + (\alpha - 1)x_n} + Wx_w \qquad\qquad\qquad \text{V}$$

or

$$x_{n+1}x_n + \frac{x_{n+1}}{\alpha - 1} - \left[\frac{\alpha G + (\alpha - 1)Wx_w}{L(\alpha - 1)}\right]x_n - \frac{Wx_w}{L(\alpha - 1)} = 0 \qquad \text{VI}$$

Now $L = F + RD = (100 + 3 \times 60.4) = 281.2$ kg moles/h and letting

$$A = \frac{1}{\alpha - 1} = \frac{1}{1.3} = 0.769$$

$$B = \frac{\alpha G + (\alpha - 1)Wx_w}{L(\alpha - 1)} = \frac{(2.3 \times 4 \times 60.4) + (1.3 \times 39.6 \times 0.02)}{281.2 \times 1.3}$$

$$= 1.523$$

$$C = \frac{Wx_w}{L(\alpha - 1)} = \frac{39.6 \times 0.02}{281.2 \times 1.3} = 0.0022$$

equation VI becomes the Riccati equation

$$x_{n+1}x_n + Ax_{n+1} - Bx_n - C = 0 \qquad\qquad\qquad \text{VII}$$

The solution of equation VII is obtained by following the procedure set out in the above text. Thus let

$$z_n = \frac{1}{x_n - \delta} \qquad\qquad\qquad\qquad\qquad \text{VIII}$$

where δ is obtained from the equation

$$\delta^2 + (A - B)\delta - C = 0$$

or

$$\delta^2 - 0.754\delta = 0.0022 \qquad\qquad\qquad\qquad \text{IX}$$

The roots of this equation are

$$\delta = 0.757 \quad \text{or} \quad -0.003 \qquad \qquad \text{X}$$

The complete solution of equation VII is

$$z_n = \frac{1}{x_n - \delta} = K\left[\frac{A+\delta}{B-\delta}\right]^n - \frac{1}{A-B+2\delta} \qquad \text{XI}$$

Selecting the root $\delta = -0.003$ and inserting the boundary condition that when $n = 0$, $x_n = 0.02$ gives

$$K = 42.2$$

The value of n corresponding to $x_n = 0.60$ is given by

$$1.66 = 42.2\left[\frac{0.766}{1.526}\right]^n + 1.32 \qquad \text{XII}$$

as $n = 7.0$ ideal stages. Subtracting the reboiler from this number gives the overall plate efficiency as $6.0/9 = 67\%$

9.8. Differential-difference Equations

It has been shown in the above sections that chemical engineering equipment consisting of a number of stages can be analysed by means of finite difference equations when the operating conditions are steady. However, when this type of equipment is subjected to a step change in operating conditions, or is started up, or is being shut down, the compositions of the streams passing through these stages change with time. This results in the presence of differential terms in addition to the finite difference terms in what is known as a "differential-difference" equation. This type of equation will be treated in this section, but it should be pointed out that on many occasions the final equation describing the process is too complex for analytical solution, and it is necessary to make a "stage-to-stage" analysis with the aid of a computer. The solution of such problems will be treated in Chapter 11.

The analytical solution of differential–difference equations is accomplished by converting this type of equation into a finite difference form by means of the Laplace transformation. The transformed equation is solved and inverted to show how the conditions of the process streams passing through the equipment vary with time during the transient period. The technique to apply, where possible, is illustrated by the following examples.

Example 1. A system consists of N stirred tanks each of volume vm^3 arranged in cascade. If each tank initially contains pure water and a salt stream of concentration x_0 kg/m³ is fed to the first tank at a rate Rm^3/h, calculate what the output concentration from the last tank should be as a function of time if the stirring is 100% efficient.

Use the result to compare the transient behaviour of all systems of total volume $Nv = V$ (constant).

Solution

With the normal convention regarding subscripts, a salt balance over the nth stage gives

$$Rx_{n-1} - Rx_n = v\frac{dx_n}{dt} \qquad \text{I}$$

Taking the Laplace transformation of equation I by using equation (6.11) to remove the derivative, and remembering that the system initially contains no salt,

$$\therefore \quad R\bar{x}_{n-1} - R\bar{x}_n = vs\bar{x}_n \qquad \text{II}$$

Solving this linear finite difference equation by the method of Section 9.6 gives

$$\bar{x}_n = A\left(\frac{R}{R+vs}\right)^n \qquad \text{III}$$

Because the feed composition is constant,

$$\bar{x}_0 = x_0/s \qquad \text{IV}$$

$$\therefore \quad \bar{x}_n = \frac{x_0}{s}\left(\frac{R}{R+vs}\right)^n \qquad \text{V}$$

$$\therefore \quad \bar{x}_N = \frac{x_0}{s}\left(\frac{R}{R+vs}\right)^N \qquad \text{VI}$$

From transform number 10 in the table at the end of the book,

$$\mathscr{L}\left[\frac{t^{N-1}e^{-at}}{(N-1)!}\right] = \frac{1}{(s+a)^N}$$

Hence, using equation (6.21),

$$\mathscr{L}^{-1}\left[\frac{1}{s(s+a)^N}\right] = \int_0^t \frac{t^{N-1}e^{-at}}{(N-1)!}\,dt \qquad \text{VII}$$

Integrating the right-hand side of equation VII by parts $(N-1)$ times gives

$$\mathscr{L}^{-1}\left[\frac{1}{s(s+a)^N}\right] = \frac{1}{(N-1)!}\left[\frac{-t^{N-1}}{a}e^{-at} + \frac{N-1}{a}\int t^{N-2}e^{-at}\,dt\right]_0^t$$

$$= \frac{-t^{N-1}e^{-at}}{a(N-1)!} + \frac{1}{a(N-2)!}\int_0^t t^{N-2}e^{-at}\,dt$$

$$= \cdots$$

$$= \frac{-t^{N-1}e^{-at}}{a(N-1)!} - \frac{t^{N-2}e^{-at}}{a^2(N-2)!} - \cdots - \frac{e^{-at}}{a^N} + \frac{1}{a^N} \qquad \text{VIII}$$

Inverting equation VI by using equation VIII,

$$x_N = x_0\left[1-e^{-Rt/v}-\frac{Rt}{v}e^{-Rt/v}-\dots-\frac{(Rt/v)^{N-1}}{(N-1)!}e^{-Rt/v}\right]$$

$$\therefore \quad x_N = x_0-x_0 e^{-Rt/v}\left[1+\frac{Rt}{v}+\frac{(Rt/v)^2}{2!}+\dots+\frac{(Rt/v)^{N-1}}{(N-1)!}\right] \qquad \text{IX}$$

which is the solution to the first part of the problem.

For the second part, if $V = Nv$, equation IX becomes

$$x_N = x_0-x_0 e^{-NRt/V}\left[1+\frac{NRt}{V}+\frac{(NRt/V)^2}{2!}+\dots+\frac{(NRt/V)^{N-1}}{(N-1)!}\right] \qquad \text{X}$$

In particular, if $N = 1$,

$$x_N = x_0-x_0 e^{-Rt/V} \qquad \text{XI}$$

and if $N = 4$

$$x_N = x_0-x_0 e^{-4Rt/V}\left[1+\frac{4Rt}{V}+8\left(\frac{Rt}{V}\right)^2+\frac{32}{3}\left(\frac{Rt}{V}\right)^3\right] \qquad \text{XII}$$

It is difficult to evaluate equation X in the limit as $N\to\infty$, but putting $V = Nv$ into equation VI gives

$$\bar{x}_N = \frac{x_0}{s}\left(1+\frac{Vs}{RN}\right)^{-N} \qquad \text{XIII}$$

which can be expanded by the binomial theorem thus,

$$\bar{x}_N = \frac{x_0}{s}\left[1-\frac{Vs}{R}+\frac{N+1}{2N}\left(\frac{Vs}{R}\right)^2-\frac{(N+1)(N+2)}{3!N^2}\left(\frac{Vs}{R}\right)^3+\dots\right]$$

If N is large, the expression in the square brackets can be approximated to give

$$\bar{x}_N = \frac{x_0}{s}\left[1-\frac{Vs}{R}+\frac{1}{2}\left(\frac{Vs}{R}\right)^2-\frac{1}{3!}\left(\frac{Vs}{R}\right)^3+\dots\right]$$

$$= \frac{x_0}{s}e^{-Vs/R} \qquad \text{XIV}$$

The inversion of equation XIV is given by number 61 in the table at the end of the book. Hence as $N\to\infty$

$$x_N\to x_0 S_k(t) \qquad \text{XV}$$

where
$$k = V/R$$

Equations X, XI, XII, and XIV have been plotted in Fig. 9.5 to show how the system responds for different values of N. The important result emerges that a large number of small stirred tanks in series has approximately the same response to a step change as a single tube of the same volume with plug flow.

The converse is also true, that flow in a tube with a small amount of back-mixing can be represented by a chain of stirred tanks of the same total volume.

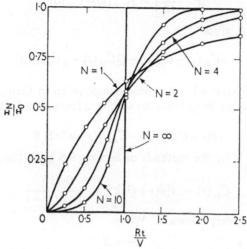

FIG. 9.5. Output from an N tank system of total volume V

Example 2. Acetic anhydride is to be hydrolysed in a laboratory stirred tank reactor battery consisting of three vessels of equal size. 1800 cm³ of anhydride solution of concentration 0·21 kg moles/m³ is charged into each vessel at 40 °C, and 600 cm³/min of a solution containing 0·137 kg moles/m³ is continuously fed to the first reactor. Estimate the time required for the reactor system to settle down to steady-state operation.

The specific reaction rate constant for the hydrolysis at 40 °C is 0·38 min⁻¹.

Initially let there be a battery of n reactors in series and consider a material balance over reactor m. Let q be the volumetric flow rate; V be the volume of each reaction vessel; C be the concentration of anhydride. If the density of the liquid and the temperature of each reaction vessel are considered to be constant, then

$$qC_{m-1} - qC_m = r_m V + V\frac{dC_m}{dt} \qquad\qquad \text{I}$$

Now $V/q = \theta$, the holding time, and $r_m = kC_m$.

$$\therefore \quad \frac{dC_m}{dt} = \frac{C_{m-1}}{\theta} - C_m\left(\frac{1+k\theta}{\theta}\right) \qquad\qquad \text{II}$$

Since all the vessels are to be of the same size and the battery is to be operated isothermally,

$$\frac{1+k\theta}{\theta} = Q \quad \text{(constant)} \qquad\qquad \text{III}$$

Thus carrying out the Laplace transformation on equation II, by making use of

$$\mathscr{L}\left[\frac{dC_m}{dt}\right] = s\bar{C}_m(s) - C_m(0)$$

from Section 6.2.1, gives

$$s\bar{C}_m(s) - C_m(0) + Q\bar{C}_m(s) - \frac{1}{\theta}\bar{C}_{m-1}(s) = 0 \qquad \text{IV}$$

But the vessels all start with the same composition in them, hence $C_m(0) = 0\cdot21$. Equation IV can be put into the standard form for a first order difference equation, thus

$$\theta(s+Q)\,\bar{C}_m(s) - \bar{C}_{m-1}(s) = 0\cdot21\,\theta \qquad \text{V}$$

Solving equation V by the methods of Section 9.6 gives the general solution

$$\bar{C}_m(s) = A[\theta(s+Q)]^{-m} + \frac{0\cdot21\,\theta}{\theta(s+Q)-1} \qquad \text{VI}$$

where A is an arbitrary constant. When

$$m = 0$$

$$C_0 = 0\cdot137$$

$$\therefore \quad \bar{C}_0(s) = \frac{0\cdot137}{s}$$

$$\therefore \quad \frac{0\cdot137}{s} = A + \frac{0\cdot21\,\theta}{\theta(s+Q)-1} \qquad \text{VII}$$

which determines A. Substituting into equation VI and rearranging gives

$$\bar{C}_m(s) = \frac{0\cdot137}{s\theta^m(s+Q)^m} + \frac{0\cdot21\,\theta}{\theta(s+Q)-1}\left[1 - \frac{1}{\theta^m(s+Q)^m}\right]$$

$$= \frac{0\cdot137}{s\theta^m(s+Q)^m} + 0\cdot21\,\theta\left[\frac{1}{\theta(s+Q)} + \frac{1}{\theta^2(s+Q)^2} + \dots + \frac{1}{\theta^m(s+Q)^m}\right]$$

$$\text{VIII}$$

by algebraic division.

Each term in the square brackets can be inverted directly by using the tables at the end of the book (transform number 10), and the first term can be inverted by using the property of the transform given by equation (6.21). Hence the solution is

$$C_m = \frac{0\cdot137}{(\theta Q)^m} - \frac{0\cdot137}{(\theta Q)^m}\left[1 + Qt + \frac{(Qt)^2}{2!} + \dots + \frac{(Qt)^{n-1}}{(n-1)!}\right]e^{-Qt}$$

$$+ 0\cdot21\left[1 + \frac{t}{\theta} + \frac{1}{2!}\left(\frac{t}{\theta}\right)^2 + \dots + \frac{(t/\theta)^{n-1}}{(n-1)!}\right]e^{-Qt} \qquad \text{IX}$$

Equation IX is a general expression giving the concentration of the effluent from a stirred tank reactor battery containing n vessels of equal size, each starting from the same initial concentration. Expressions of this type have been derived for different kinds of reaction by Mason and Piret[†] and the reader is advised to consult these authors for a more extensive analysis.

In the special example considered here, the last vessel to reach steady state will be the third, and since

$$\theta = \frac{V}{q} = \frac{1800}{600} = 3{\cdot}0$$

$$Q = \frac{1+k\theta}{\theta} = \tfrac{1}{3}+0{\cdot}38 = 0{\cdot}713$$

$$\theta Q = 2{\cdot}14$$

$$\therefore \quad C_3 = 0{\cdot}014 - 0{\cdot}014(1+0{\cdot}713t+0{\cdot}254t^2)\,e^{-0{\cdot}713t}$$

$$+0{\cdot}21(1+0{\cdot}333t+0{\cdot}056t^2)\,e^{-0{\cdot}713t}$$

$$\therefore \quad C_3 = 0{\cdot}014 + (0{\cdot}196+0{\cdot}060t+0{\cdot}0082t^2)\,e^{-0{\cdot}713t} \qquad \text{X}$$

Equation X which gives C_3, the composition of the effluent from the final reactor as a function of time, has been plotted in Fig. 9.6. This shows that the laboratory reactor battery has reached steady-state conditions within the limits of experimental detection in about 15 min.

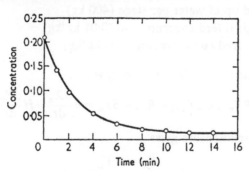

FIG. 9.6. Time to reach steady state

Example 3. 1000 kg/h of a solution containing 0·01 kg nicotine/kg of water is being extracted by 1200 kg/h of kerosine in a counter-current extractor containing eight stages. If the concentration of the feed liquor is suddenly changed to 0·02 kg nicotine/kg of water, how long will it take for the extractor to settle down to steady-state operation under the new conditions?

The equilibrium relation is

$$Y = 0{\cdot}86X$$

† Mason, D. R. and Piret, E. L. *I.E.C.* **42**, 817 (1950).

where Y is kg nicotine/kg of kerosine, and X is kg nicotine/kg of water. The hold-up per stage can be taken to be constant at 400 kg water and 200 kg kerosine.

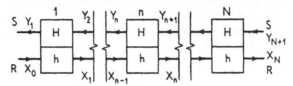

FIG. 9.7. Stagewise counterflow system

Solution

The system is illustrated in Fig. 9.7 and the symbols have the following meanings.

$\qquad R$ flow rate of water phase (1000 kg/h)
$\qquad S$ flow rate of kerosine phase (1200 kg/h)
$\qquad N$ total number of stages (8)
$\qquad X_n$ nicotine concentration in water phase in nth stage
$\qquad Y_n$ nicotine concentration in kerosine phase in nth stage
$\qquad m$ distribution ratio (0·86)
$\qquad h$ hold-up of kerosine per stage (200 kg)
$\qquad H$ hold-up of water per stage (400 kg)
$\qquad X_0'$ original feed concentration (0·01 kg/kg)
$\qquad X_F$ new feed concentration (0·02 kg/kg)

A nicotine balance over the nth stage gives

$$RX_{n-1} + SY_{n+1} - RY_n - SY_n = h\frac{dX_n}{dt} + H\frac{dY_n}{dt} \qquad\qquad \text{I}$$

The equilibrium relationship gives

$$Y_n = mX_n \qquad\qquad \text{II}$$

Eliminating Y_n, Y_{n+1} between equations I and II and rearranging,

$$\therefore \quad \alpha X_{n+1} - (\alpha + 1)X_n + X_{n-1} = \beta\frac{dX_n}{dt} \qquad\qquad \text{III}$$

where $\qquad\qquad \alpha = mS/R, \quad \beta = (h + mH)/R \qquad\qquad \text{IV}$

The plant is initially operating at steady state, and the concentration in each stage must be determined for these conditions first. Thus, putting $dX_n/dt = 0$, equation III becomes

$$\alpha X_{n+1} - (\alpha + 1)X_n + X_{n-1} = 0 \qquad\qquad \text{V}$$

The method of Section 9.6 gives the general solution

$$X_n = A + B\alpha^{-n} \qquad \text{VI}$$

The boundary conditions for this part of the problem are:-

$$\left.\begin{array}{ll} \text{at} \quad n = 0, & X_0 = X_0' \\ \text{at} \quad n = N+1, & Y_{N+1} = mX_{N+1} = 0 \end{array}\right\} \qquad \text{VII}$$

Hence, $\qquad A = \dfrac{-X_0'}{\alpha^{N+1}-1} \quad \text{and} \quad B = \dfrac{\alpha^{N+1} X_0'}{\alpha^{N+1}-1} \qquad$ VIII

Therefore, equation VI gives the initial state of the system where A and B are determined by equation VIII.

For the unsteady-state part of the problem, it is necessary to take the Laplace transformation of equation III. Thus

$$\alpha \bar{X}_{n+1} - (\alpha+1)\,\bar{X}_n + \bar{X}_{n-1} = \beta(s\bar{X}_n - A - B\alpha^{-n}) \qquad \text{IX}$$

where equation VI has been used for the initial conditions. Rearranging,

$$\alpha \bar{X}_{n+1} - (\alpha+1+s\beta)\,\bar{X}_n + \bar{X}_{n-1} = -A\beta - B\beta\alpha^{-n}$$

or, in operator form,

$$[\alpha E^2 - (\alpha+1+s\beta)\,E + 1]\,\bar{X}_{n-1} = -A\beta - B\beta\alpha^{-n} \qquad \text{X}$$

Equation X is linear so that its solution can be expressed as the sum of a complementary solution and a particular solution. The complementary solution is given by

$$\bar{X}_{n-1} = C\rho_1^{n-1} + D\rho_2^{n-1} \qquad \text{XI}$$

where $\qquad\qquad \rho_1 + \rho_2 = (\alpha+1+s\beta)/\alpha \qquad\qquad$ XII

and $\qquad\qquad\qquad \rho_1\rho_2 = 1/\alpha \qquad\qquad\qquad$ XIII

The particular solutions for the separate terms on the right-hand side of equation X can be found by the inverse operator method described in Section 9.6.2. Thus for the first term

$$\bar{X}_{n-1} = \frac{-A\beta}{\alpha - \alpha - 1 - s\beta + 1} = \frac{A}{s} \qquad \text{XIV}$$

and for the second term

$$\bar{X}_{n-1} = \frac{-B\beta\alpha^{-n}}{\alpha^{-1} - 1 - \alpha^{-1} - s\beta\alpha^{-1} + 1} = \frac{B}{s}\alpha^{-n+1} \qquad \text{XV}$$

Combining equations XI, XIV, and XV gives the complete solution of equation X, which is

$$\bar{X}_{n-1} = C\rho_1^{n-1} + D\rho_2^{n-1} + \frac{A + B\alpha^{-n+1}}{s}$$

or
$$\bar{X}_n = C\rho_1{}^n + D\rho_2{}^n + \frac{A + B\alpha^{-n}}{s} \qquad\qquad \text{XVI}$$

The boundary conditions can also be transformed thus,

$$\left.\begin{array}{lll} \text{at} \quad n = 0, & X_0 = X_F, \quad \therefore & \bar{X}_0 = X_F/s \\[2mm] \text{at} \quad n = N+1, & X_{N+1} = 0, \quad \therefore & \bar{X}_{N+1} = 0 \end{array}\right\} \qquad \text{XVII}$$

Substituting these boundary values into equation XVI and using equations VIII to eliminate A and B, gives

$$\frac{X_F}{s} = C + D + \frac{A+B}{s} = C + D + \frac{X_0{}'}{s} \qquad\qquad \text{XVIII}$$

and
$$0 = C\rho_1{}^{N+1} + D\rho_2{}^{N+1} + \frac{A + B\alpha^{-N-1}}{s}$$

$$= C\rho_1{}^{N+1} + D\rho_2{}^{N+1} \qquad\qquad \text{XIX}$$

Solving equations XVIII and XIX for C and D, and substituting into equation XVI gives the solution

$$\bar{X}_n = \frac{A + B\alpha^{-n}}{s} + \frac{X_F - X_0{}'}{s}\left[\frac{\rho_2{}^{N+1}\rho_1{}^n - \rho_1{}^{N+1}\rho_2{}^n}{\rho_2{}^{N+1} - \rho_1{}^{N+1}}\right] \qquad \text{XX}$$

In order to invert equation XX it must be remembered that ρ_1 and ρ_2 are determined by equations XII and XIII as functions of s. Therefore, the method of residues must be used to invert the last term in equation XX.

To simplify the nomenclature, put

$$f(s) = \frac{\rho_2{}^{N+1}\rho_1{}^n - \rho_1{}^{N+1}\rho_2{}^n}{s(\rho_2{}^{N+1} - \rho_1{}^{N+1})} \qquad\qquad \text{XXI}$$

The numerator on the right-hand side of equation XXI could become infinite if $|\rho_2|$ became infinite. But if $|\rho_2|$ is large, equation XIII ensures that $|\rho_1|$ is small. Hence it can be shown that as $|\rho_2| \to \infty$, $f(s) \to 0$; and the only singularities of $f(s)$ arise from the zeros of the denominator. These zeros occur at $s = 0$ and when $\rho_1{}^{N+1} = \rho_2{}^{N+1}$. The second solution shows that $|\rho_1| = |\rho_2|$, and since ρ_1 and ρ_2 are the roots of a quadratic equation with real coefficients, they must be equal or mutual complex conjugates. Therefore, putting

$$\rho_1 = r\,\mathrm{e}^{-i\theta}, \quad \rho_2 = r\,\mathrm{e}^{i\theta} \qquad\qquad \text{XXII}$$

$$\therefore \quad \rho_2{}^{N+1} - \rho_1{}^{N+1} = r^{N+1}[\mathrm{e}^{(N+1)i\theta} - \mathrm{e}^{-(N+1)i\theta}]$$

$$= 2ir^{N+1}\sin(N+1)\theta \qquad\qquad \text{XXIII}$$

Hence $\bar{f}(s)$ has singularities when $s = 0$, and when

$$(N+1)\theta = k\pi \qquad\qquad \text{XXIV}$$

where k is an integer.

Putting equations XXII into equations XIII and XII gives

$$r^2 = 1/\alpha \qquad\qquad \text{XXV}$$

$$2\alpha r \cos\theta = \alpha + 1 + s\beta \qquad\qquad \text{XXVI}$$

$$\therefore \quad s = -\frac{\alpha + 1 - 2\sqrt{\alpha}\cos[k\pi/(N+1)]}{\beta} \qquad\qquad \text{XXVII}$$

Equation XXVII gives a distinct set of values of s for $k = 0, 1, ..., (N+1)$, but repeated values for integer values of k exceeding $(N+1)$. When $k = 0$, or $(N+1)$, equations XXII show that $\rho_1 = \rho_2$, and an application of L'Hôpital's rule to equation XXI shows that $\bar{f}(s)$ remains finite as $\rho_2 \to \rho_1$. Therefore $\bar{f}(s)$ has $(N+1)$ simple poles:

one at $s = 0$,

and N given by equation XXVII with $k = 1, 2, ..., N$.

Using equation (4.90) to find the residue of $\bar{f}(s)$ at $s = 0$ gives at

$$s = 0, \quad \rho_1 = 1, \quad \rho_2 = \alpha^{-1}$$

$$B_0 = \frac{\alpha^{-N-1} - \alpha^{-n}}{\alpha^{-N-1} - 1} = \frac{\alpha^{N+1-n} - 1}{\alpha^{N+1} - 1} \qquad\qquad \text{XXVIII}$$

Substituting equations XXII into equation XXI gives

$$\bar{f}(s) = \frac{r^n}{s} \frac{\sin(N+1-n)\theta}{\sin(N+1)\theta} \qquad\qquad \text{XXIX}$$

$$\therefore \quad e^{st}\bar{f}(s) = \frac{r^n}{s} \frac{\sin(N+1-n)\theta}{\sin(N+1)\theta} e^{st} \qquad\qquad \text{XXX}$$

Remembering that θ is a function of s given by equation XXVI, and using equation (4.90) to find the residue of $e^{st}\bar{f}(s)$ gives

$$B_k = \frac{r^n}{s} \frac{\sin(N+1-n)\theta}{(N+1)\cos(N+1)\theta} \frac{e^{st}}{(d\theta/ds)}$$

$$= \frac{-r^n}{s(N+1)} \frac{\sin(N+1-n)\theta}{\cos(N+1)\theta} \frac{2\alpha r \sin\theta \, e^{st}}{\beta} \qquad\qquad \text{XXXI}$$

Using equations XXIV, XXV, and XXVII to simplify XXXI,

$$B_k = \frac{2(-1)^k \alpha^{-(n-1)/2} \sin\theta \sin(N+1-n)\theta}{(N+1)(\alpha+1-2\sqrt{\alpha}\cos\theta)} \exp\left[\frac{-(\alpha+1-2\sqrt{\alpha}\cos\theta)t}{\beta}\right]$$

Combining the residues as given by equation (6.43) and inverting the remainder of equation XX,

$$X_n = X_0'\left[\frac{\alpha^{N+1-n}-1}{\alpha^{N+1}-1}\right] + (X_F - X_0')\left[\frac{\alpha^{N+1-n}-1}{\alpha^{N+1}-1}\right] + (X_F - X_0')$$

$$\times \sum_{k=1}^{N} \frac{2(-1)^k \alpha^{-(n-1)/2}\sin\theta\sin(N+1-n)\theta}{(N+1)(\alpha+1-2\sqrt{\alpha}\cos\theta)}e^{st}$$

$$\therefore\quad X_n = X_F\left[\frac{\alpha^{N+1-n}-1}{\alpha^{N+1}-1}\right] + (X_F - X_0')$$

$$\times \sum_{k=1}^{N} \frac{2(-1)^k \alpha^{-(n-1)/2}\sin\theta\sin(N+1-n)\theta}{(N+1)(\alpha+1-2\sqrt{\alpha}\cos\theta)}e^{st}\qquad \text{XXXII}$$

Equation XXXII is the complete solution to the problem where θ is given by equation XXIV and s by equation XXVII.

Since the final stage will be the last to reach the new steady-state conditions, a graph should be plotted showing X_8 as a function of time. From equation XXXII,

$$X_8 = X_F\left[\frac{\alpha-1}{\alpha^9-1}\right] + (X_F - X_0')\sum_{k=1}^{8} \frac{2(-1)^k \alpha^{-7/2}\sin^2\theta}{9(\alpha+1-2\sqrt{\alpha}\cos\theta)}e^{st}\qquad \text{XXXIII}$$

Using the numerical values given in the problem,

$$\theta = k\pi/9 = 20k°$$

$$\alpha = mS/R = 1\cdot032$$

$$\beta = \frac{h+mH}{R} = 0\cdot544$$

$$\therefore\quad \alpha^9 = 1\cdot3278,\quad \alpha^{-7/2} = 0\cdot8956$$

k	θ	$2\sqrt{\alpha}\cos\theta$	$-s$	$\sin^2\theta$	$\dfrac{2\alpha^{-7/2}\sin^2\theta}{9(\alpha+1-2\sqrt{\alpha}\cos\theta)}$
1	20	1·9092	0·2257	0·1170	0·1896
2	40	1·5564	0·8743	0·4132	0·1729
3	60	1·0159	1·868	0·7500	0·1469
4	80	0·3528	3·087	0·9699	0·1149
5	100	−0·3528	4·384	0·9699	0·0809
6	120	−1·0159	5·603	0·7500	0·0489
7	140	−1·5564	6·596	0·4132	0·0230
8	160	−1·9092	7·245	0·1170	0·0059

$$\therefore\quad 100X_8 = 0\cdot195 - 0\cdot1896\,e^{s_1t} + 0\cdot1729\,e^{s_2t} - 0\cdot1469\,e^{s_3t}$$

$$+0\cdot1149\,e^{s_4t} - 0\cdot0809\,e^{s_5t} + 0\cdot0489\,e^{s_6t} - 0\cdot0230\,e^{s_7t} + 0\cdot0059\,e^{s_8t}$$

$$\text{XXXIV}$$

Equation XXXIV has been plotted in Fig. 9.8 from the following calculated values.

t	X_8	t	X_8
0	0·0972	10	0·1752
1	0·0975	12	0·1824
2	0·1012	14	0·1870
3	0·1107	16	0·1899
4	0·1232	18	0·1917
6	0·1470	20	0·1929
8	0·1640	24	0·1942

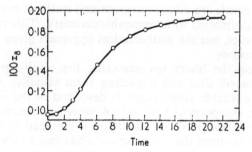

FIG. 9.8. Nicotine concentration in extract

It can be seen that it takes about 24 h for the system to settle down to the new conditions.

Chapter 10

TREATMENT OF EXPERIMENTAL RESULTS

10.1 INTRODUCTION

THE analysis of many chemical engineering problems is concerned with using experimental results to verify theories or develop empirical formulae. Some of the data points are often inaccurate and methods must be found for eliciting reliable information with reasonable certainty. This subject falls into the realm of statistics, but the mathematical approach given in this chapter yields valid conclusions.

A great deal can be learnt by presenting the results graphically with suitably chosen coordinates and inspecting them visually; and the use of various kinds of available graph paper is described in the first part of the chapter. With the exception of Section 10.4, where a contour representation of three variables is described, all discussion is confined to systems of two variables; it being assumed that all other variables can be kept constant and treated as parameters.

Experimental results can also be used to determine an average value of a variable, or a set of data can be integrated to find some total value. For example, the determination of an average velocity across the cross-section of a pipe or piece of equipment involves averaging data points; and the prediction of an accumulated composition from a set of values of point composition necessitates numerical or graphical integration. Methods for performing these calculations are given in the final part of this chapter.

10.2. GRAPH PAPER

Results can be presented in the form of a table or as a graph. Although the tabular form is capable of showing greater accuracy, the graphical presentation is more readily assimilated, because it gives a clearer picture of the behaviour of the variables. The graphical method is usually sufficiently accurate for the presentation of experimental points, but for theoretical points calculated from a numerical method (see Chapter 11) a tabular form is preferred.

For a particular piece of equipment where the number of adjustable parameters is small, a direct plot of the variables is the most useful, but if comparison with other work on similar equipment is envisaged, the appropriate dimensionless groups should be used along the axes. If the dimensionless groups are chosen correctly, each group should only contain one principal variable and the rest of its constituent parts should be parameters. No purpose would be served by plotting a dimensionless group against a dimensional variable and this form of presentation should be avoided. Any mathematical equation must be dimensionally consistent and hence if one term is dimensionless, so must all other terms be dimensionless. Disregard of this important fact can lead to apparent anomalies, particularly when a logarithmic plot is used.

There are many types of graph paper which are described below and the correct type should be chosen for each problem. "The most desirable shape for a curve is a straight line", and the scales should be chosen in such a way that the major part of the curve is inclined at 45° to each axis. This ensures equal accuracy in all regions of the graph both in plotting the curve and reading results from it.

10.2.1. *Linear Graph Paper*

The most common and generally used graph paper has its axes divided into equal intervals so that any range of numerical values of a variable can be accommodated by choosing a suitable scale. Most normal types of graph paper are subdivided into ten divisions, and it is customary to count in a decimal system; hence factors of three should be avoided when choosing the scales, so that each subdivision will represent an easily written number.

The equation of a straight line is usually written

$$y = mx + c \tag{10.1}$$

where m is the gradient and c (the intercept) is the value of y corresponding to $x = 0$. Because c is defined in this manner, the origin should not be suppressed.

For a general curve on linear scales, the equation

$$y = f(x) \tag{10.2}$$

is satisfied. By differentiating equation (10.2) the gradient of the tangent at a point can be found. Conversely, if a tangent is drawn to the curve then a numerical value can be found for the derivative at that point. The most accurate way to differentiate a curve graphically is to choose a gradient and construct a set of parallel lines with that gradient which intersect the curve as shown in Fig. 10.1. If each chord so constructed is bisected and the midpoints are joined with a smooth curve, this curve must intersect the original curve orthogonally at the point where the tangent has the chosen slope.

Integration of equation (10.2) with respect to x between the fixed limits $x = a$ and $x = b$ is represented graphically by the area bounded by the curve $y = f(x)$, the ordinates $x = a$ and $x = b$, and the x axis. This can readily be visualized by dividing the area into vertical strips of equal width and considering the integral as the limit of the sum of the areas of these strips as their width approaches zero. This area can be evaluated by counting the squares on the graph paper which form the desired area or by using a numerical method from Section 10.7.

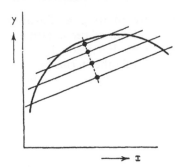

FIG. 10.1. Graphical differentiation

10.2.2. *Semi-logarithmic Graph Paper*

In many practical cases, particularly those involving the approach to steady-state conditions of a process, the dependent variable is a decaying exponential function of the independent variable. For example, in Section 1.6, equation (1.29) is of the form

$$y = A\,e^{-bx} \tag{10.3}$$

after a change of origin. If an experiment is designed to determine the coefficients A and b from a series of measurements of x and y, it is more convenient to use the logarithmic form of equation (10.3) which is

$$\ln y = \ln A - bx \tag{10.4}$$

Defining a new dependent variable by

$$Y = \ln y \tag{10.5}$$

equation (10.4) becomes

$$Y = -bx + \ln A \tag{10.6}$$

By comparison of equations (10.1) and (10.6) it can be seen that $-b$ is the gradient of the curve of Y against x, and $\ln A$ is the intercept. Because such plots are frequently needed, semi-logarithmic graph paper is available which has one scale divided logarithmically, as on a slide rule. Hence, the values of y can be used directly instead of consulting log tables to evaluate Y, the logarithms being intrinsically taken on the graph paper itself. It must be remembered, however, that the gradient $(-b)$ must be determined from the linear measurement of Y and not from reading the scales which give values of y. The intercept can be measured as a linear value of $\ln A$ when $x = 0$, or read directly from the scale of y as A, which is simpler and more usual.

It should be noted that the logarithms are all taken to the base e and the gradient must be corrected for the scales of the axes. To do this, the measured gradient must be multiplied by the length of the unit on the linear scale and divided by the length of the unit on the logarithmic scale. The logarithmic unit is the distance between 1 and 2·718 on the scale of y. Graph paper is usually printed with these two units approximately equal, but a 5% discrepancy is quite common and should be corrected.

10.2.3. *Logarithmic Graph Paper*

Dimensional analysis frequently indicates an empirical equation of the form

$$y = Cx^n \tag{10.7}$$

where x and y are dimensionless groups and C and n are constants to be determined experimentally. Because the groups are usually made dimensionless by using the physical properties of the materials used, the values of x and y are likely to vary over very wide ranges. To combine many results into a single correlation applicable to such a wide range, a logarithmic scale is superior to a linear scale. Also, the logarithmic form of equation (10.7) is

$$Y = nX + c \tag{10.8}$$

where $Y = \ln y$, $X = \ln x$, and $c = \ln C$; and this is the equation of a straight line. Therefore, for plotting such equations as (10.8), graph paper with both axes subdivided logarithmically is available. It is normal for the logarithmic unit to be the same along each axis so that the gradient n can be found by direct linear measurement. The intercept c is the value of Y when $X = 0$. This can be found more easily as the value $y = C$ when $x = 1$.

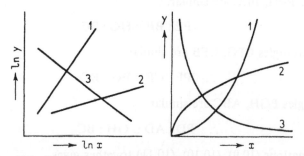

FIG. 10.2. Equivalent curves on log–log and linear graph paper

The origin of linear scales can never be plotted on logarithmic graph paper. If any point is referred to as the origin, the point $x = 1$, $y = 1$, corresponding to $X = 0$, $Y = 0$, is usually meant. The intercept as described above is determined relative to axes through this origin.

Inspection of equation (10.7) shows that if n is positive, then the curve will pass through the origin of linear scales. If n is negative, the curve will be asymptotic to both axes. It is therefore useless trying to fit an equation of the form (10.7) to any set of points which obviously intersect either axis, unless the point of intersection can be transferred to the origin. The three types of curve which give rise to straight-line plots on log–log paper are illustrated in Fig. 10.2.

10.2.4. *Triangular Graph Paper*

In the study of liquid–liquid extraction systems where three components are present in two phases, a convenient graphical representation of the composition of a phase is needed. The data can be presented on a triangular diagram by using the geometrical property of a triangle which is proved below.

Figure 10.3 shows any triangle ABC and any interior point P. The lines AP, BP, CP are produced to meet BC, AC, AB at D, E, F respectively. PG is drawn parallel to AB, and PH is drawn parallel to AC so that G and H lie on BC.

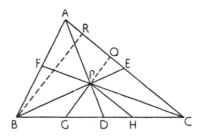

FIG. 10.3. The triangular diagram

Triangles BPH, BEC are similar.

$$\therefore \quad PE : BE :: HC : BC \qquad (10.9)$$

Similarly, triangles CPG, CFB are similar.

$$\therefore \quad PF : CF :: BG : BC \qquad (10.10)$$

Also, triangles PGH, ABC are similar

$$\therefore \quad PD : AD :: GH : BC \qquad (10.11)$$

Adding equations (10.9), (10.10), (10.11) together gives

$$\frac{PE}{BE} + \frac{PF}{CF} + \frac{PD}{AD} = 1 \qquad (10.12)$$

Dropping perpendiculars from P and B on to AC at Q and R shows that

$$\frac{PE}{BE} = \frac{PQ}{BR} \qquad (10.13)$$

Hence, if lines are constructed parallel to AC and subdividing BR into equal intervals, the concentration of one component of the ternary mixture can be read against the scale as the ratio PQ/BR which is the same as PE/BE. Similarly, the other two components can be represented by the ratios PD/AD and PF/CF. Equation (10.12) then shows that the three separate concentrations add up to the whole mixture, for any point P inside any triangle ABC.

Although the above proof is valid for any triangle, it is normal practice to use an equilateral triangle for a symmetrical representation of the three components. If printed triangular paper is not available, however, it is more convenient to use rectangular graph paper and construct diagonals so that an isoceles right-angle triangle can be used.

10.2.5. *Special Graph Pagers*

There are many other types of graph paper available for special applications. For example, the axes can be subdivided to a quadratic scale or a square root scale for orifice calibrations. Alternatively, a reciprocal scale can be used to automatically change a flow rate into a residence time. Other special graph papers are provided for thermometer calibrations, and then the borderline between special graph papers and charts for chart recorders is rather ill-defined. Any of the above graph papers can aid the understanding of a phenomenon by providing a pictorial representation, and anything which does this can be loosely described as a graph.

10.3. THEORETICAL PROPERTIES

Whenever a set of experimental data is plotted against linᵣar scales, a decision has to be taken regarding the type of curve which best summarizes the results. Experimental data are always subject to error and tɦe assessment of these errors will be discussed in Section 10.5. Hence any fundamental properties which the curve must possess on theoretical grounds are invaluable aids to the correct choice of representative curve. There are three points which are always worth considering and these will be discussed separately.

10.3.1. *The Origin*

It is often possible to state with certainty the value of the dependent variable when the independent variable is zero. For example, if the pressure drop across a length of pipe is being determined for various flow rates, it is certainly true that the pressure drop is zero for a zero flow rate. Although all other points on the graph are subject to experimental error and the curve need only pass close to them, the origin is one point which the curve must pass through. Hence the origin is more important than any single experimental point.

When time is the independent variable, the intercept at zero time is frequently set as a boundary condition instead of being a measured value, and this point is thus more accurately determined than any other. If a chemical reaction is being studied by plotting the concentration of one of the products as a function of time, the origin must be a point on the curve.

There are many other examples, and a few moments' thought should always indicate whether or not a curve should pass through the origin.

10.3.2. *Starting Gradient*

In most problems involving axial symmetry, the gradient of the curve tends to zero as the axis is approached. Examples of this are the velocity profiles for flow inside a cylindrical pipe, temperature profiles within a cylindrical conductor, or the variation of local mass transfer coefficient near the front pole of a dissolving sphere. Not all dependent variables have a vanishing derivative on the axis of a symmetrical system as may be seen from equation XV, Example 1, Section 7.10.1, which gives the vorticity distribution for slow flow past a sphere. When $\theta = 0$, $\zeta = 0$, but $\partial \zeta / \partial \theta \neq 0$.

If the concentration of the final product of two or more consecutive chemical reactions is being plotted against time from mixing the reactants, the curve must pass through the origin and have a vanishing derivative when it does so. This is because the rate of formation of the product is proportional to the concentration of the intermediate product which is initially zero.

10.3.3. *Efficiencies*

A lot depends upon the context, but most efficiency curves either approach a limit asymptotically or pass through a maximum value before decaying to zero. One thing is certain, however; an efficiency curve cannot be a straight line over an appreciable range. Any straight line, except a horizontal one, must exceed 100% and fall below zero for some values of the independent variable, and both types of behaviour are inconsistent with a properly defined efficiency.

10.4. Contour Plots

It is often difficult to isolate the independent variables in a system so that one variable can be altered through a range of values whilst all other independent variables are held constant. When only two independent variables are involved, it is possible to establish a family of curves by presenting the results in a manner analogous to a contour map which shows height as a function of two independent coordinates. This method has been used to illustrate the dependence of start-up time on boil-up rate and feed composition for a distillation column.† The results are shown in Fig. 10.4 in the form of start-up time varying with feed composition, using the boil-up rate as a parameter. It will be seen that very few experiments were performed at the exact boil-up rate, the actual rates being indicated adjacent to the points. However, the three curves shown are almost as accurately determined as they would have been from the same number of experiments performed at the three exact boil-up rates. There is no reason why the positions of the three variables should not be interchanged, but it is customary to plot one of the independent variables as abscissa.

10.5. Propagation of Errors

Any experimentally determined quantity is subject to error, and hence any calculated result which is based on experimental evidence is also limited in

† Barker, P. E., Jenson, V. G. and Rustin, A. *Inst. Pet. J.* **49**, 316 (1963).

accuracy. The determination of the derived error from the observed error is a calculation of some importance since it indicates the position where experimental technique should be improved. The way in which errors accumulate is governed by different rules depending upon the type of calculation performed (i.e. addition, subtraction, multiplication, etc.), and these will be considered separately.

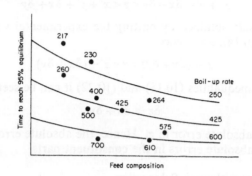

FIG. 10.4. Contour plot of three variables

10.5.1. *Propagation through Addition*

If a quantity z is calculated from

$$z = x + y \qquad (10.14)$$

where x and y are two measured variables, then any error in determining x or y will result in an error in z. Equation (10.14) is assumed to be exactly true and satisfied identically by the exact values of x, y, and z. But any experimental reading (x') for the variable (x) will only approximate to the true value and this is usually stated in the form

$$x - \delta x < x' < x + \delta x \qquad (10.15)$$

i.e. the error in the determination of x is given by

$$|x' - x| < \delta x \qquad (10.16)$$

δx is called the "absolute error" in x, has the same dimensions as x, and is usually assigned by the observer with due regard to the manner in which the measurement was made.

A similar argument leads to the definition of δy, the absolute error in y.

Thus
$$y - \delta y < y' < y + \delta y \qquad (10.17)$$

The inequalities (10.15) and (10.17) can be written in the alternative forms

$$x' - \delta x < x < x' + \delta x \qquad (10.18)$$

$$y' - \delta y < y < y' + \delta y \qquad (10.19)$$

Adding y to each term in inequality (10.18) and using (10.19) gives

$$x'+y'-\delta x-\delta y<x'-\delta x+y<x+y<x'+\delta x+y<x'+y'+\delta x+\delta y$$
$$(10.20)$$

Remembering that x and y cannot be determined except by means of x', y', δx, and δy, inequality (10.20) can be combined with equation (10.14) to give

$$x'+y'-\delta x-\delta y<z<x'+y'+\delta x+\delta y \qquad (10.21)$$

But z' is the result obtained by putting the experimental values (x' and y') into equation (10.14), and hence

$$z'-(\delta x+\delta y)<z<z'+(\delta x+\delta y) \qquad (10.22)$$

Comparing the inequalities (10.18) and (10.22) it can be seen that

$$\delta z = \delta x+\delta y \qquad (10.23)$$

where δz is the absolute error in z. Hence, the absolute error in the result is the sum of the absolute errors in the constituent parts.

10.5.2. *Propagation through Subtraction*

Assuming that
$$z = x-y \qquad (10.24)$$

and using the same terminology as above, the following set of inequalities can be obtained from (10.18) and (10.19).

$$x'-\delta x-(y'+\delta y)<x'-\delta x-y<x-y<x'+\delta x-y<x'+\delta x-(y'-\delta y)$$
$$(10.25)$$

But
$$z' = x'-y'$$

$$\therefore \quad z'-(\delta x+\delta y)<z<z'+(\delta x+\delta y) \qquad (10.26)$$

$$\therefore \quad \delta z = \delta x+\delta y \qquad (10.27)$$

Hence, the absolute error in the difference of two quantities is the sum of the absolute errors in those quantities.

The above two results can be generalized to any number of operations involving addition and subtraction only. Thus, the absolute error in the result of any calculation involving only addition and subtraction is the sum of the absolute errors in the constituent parts.

It should be noted that each operation increases the absolute error, but any subtraction reduces the result. Hence the "relative error", which is the ratio of the absolute error to the result, will always increase during subtraction. The relative error is usually expressed as a percentage and is dimensionless.

10.5.3. *Propagation through Multiplication and Division*

Using the same terminology as before but letting z be determined from

$$z = xy \qquad (10.28)$$

the inequality

$$(x' - \delta x)(y' - \delta y) < z < (x' + \delta x)(y' + \delta y) \qquad (10.29)$$

is valid provided that x' and y' are both positive.

Defining the relative errors by

$$x_r = \delta x / x' \quad \text{and} \quad y_r = \delta y / y'$$

inequality (10.29) can be rewritten

$$x'y'(1 - x_r)(1 - y_r) < z < x'y'(1 + x_r)(1 + y_r) \qquad (10.30)$$

Provided the errors are small, products of two errors can be neglected by comparison with the error itself. Since z' is the value calculated by substituting x' and y' into equation (10.28), (10.30) can be written in the form

$$z'(1 - x_r - y_r) < z < z'(1 + x_r + y_r)$$

which shows that

$$z_r = x_r + y_r \qquad (10.31)$$

Thus, the relative error in a product is the sum of the relative errors in the constituent parts.

Starting with the inequality

$$\frac{x' - \delta x}{y' + \delta y} < \frac{x}{y} < \frac{x' + \delta x}{y' - \delta y} \qquad (10.32)$$

it can be shown that the relative error in a quotient is also the sum of the relative errors in the constituent parts.

10.5.4. *Propagation through a General Functional Relationship*

It can be seen from the above calculations that the error in the calculated result is greatest when the individual readings are at the ends of the error ranges. The terminology can be simplified by writing

$$x' = x + \delta x \qquad (10.33)$$

with the interpretation that x' is the poorest acceptable reading obtained for the unknown true value x; and δx can be either positive or negative. If z is to be determined from the general expression

$$z = f(x, y) \qquad (10.34)$$

by using the values x' and y' for x and y, then the value obtained (z') will be in error as follows. Applying equation (8.9) to equation (10.34) gives

$$dz = \frac{\partial f}{\partial x} dx + \frac{\partial f}{\partial y} dy \qquad (10.35)$$

Provided the errors δx and δy are small, dx and dy can be replaced by them in equation (10.35). Thus

$$\delta z = \frac{\partial f}{\partial x} \delta x + \frac{\partial f}{\partial y} \delta y \qquad (10.36)$$

Equation (10.36) can also be obtained from the generalized form of Taylor's theorem by neglecting products of errors in comparison with the errors themselves. The equation can be applied to functions of any number of variables and it is the general law for propagation of errors.

Example 1. If $z = xy^n$ with n known and x and y determined experimentally, determine the relative error in z in terms of the relative errors in x and y.

$$f(x, y) = xy^n \qquad\qquad\qquad \text{I}$$

$$\therefore \quad \frac{\partial f}{\partial x} = y^n, \quad \frac{\partial f}{\partial y} = nxy^{n-1} \qquad\qquad \text{II}$$

Substituting into equation (10.36),

$$\delta z = y^n\, \delta x + nxy^{n-1}\, \delta y$$

$$= z\left(\frac{\delta x}{x} + \frac{n\delta y}{y}\right)$$

$$\therefore \quad z_r = x_r + ny_r \qquad\qquad\qquad \text{III}$$

The special case $n = 1$ reduces to equation (10.31) for the product. Because δx and δy can have either sign, if n is negative equation III should be written

$$z_r = x_r - ny_r \qquad\qquad\qquad \text{IV}$$

Hence, if $n = -1$ the relative error in a quotient is the sum of the relative errors in the constituent parts.

Example 2. It was shown in the first problem in Section 2.5.6 that if a chemical reaction

$$A \rightarrow B$$

has a first order reaction rate constant $k\ s^{-1}$, the concentration of A leaving a tubular reactor of length L m with velocity u m/s is given by

$$c = c_0 e^{-kL/u}$$

where c_0 is the initial concentration of A, diffusion has been neglected and plug flow has been assumed. How accurately must k be known and the flow rate be steady for it to be possible to design a reactor to give $94\frac{1}{2}$–$95\frac{1}{2}\%$ completion?

When the reactor is designed, L will be known very accurately; and (c/c_0) can be considered as a single variable (f) since the limit is on completion.

$$f = e^{-kL/u} \qquad\qquad\qquad \text{I}$$

Applying equation (10.36),

$$\therefore \quad \delta f = -\frac{L}{u} f\, \delta k + \frac{kL}{u^2} f\, \delta u$$

which can be rearranged using equation I to give

$$\frac{-\delta f}{f \ln f} = \frac{\delta u}{u} - \frac{\delta k}{k} \qquad \text{II}$$

Since $f < 1$ the left-hand side will be positive. The errors can be of either sign and the worst case must be considered, therefore

$$\frac{-\delta f}{f \ln f} = \frac{\delta u}{u} + \frac{\delta k}{k} = u_r + k_r \qquad \text{III}$$

Using the numerical values given,

$$u_r + k_r = \frac{0 \cdot 005}{0 \cdot 95 \times 0 \cdot 0513} = 0 \cdot 10 \qquad \text{IV}$$

Equation IV shows that if the relative error in k exceeds 10% it will be impossible to design the reactor to contain the completion within the given limits. If the flow rate is likely to surge by 5%, then k must be known to an accuracy of 5% otherwise the design is impossible.

10.5.5 Sources of Error

There are many sources of error which should be assessed before quoting the accuracy of any determination. They can be classified as follows.

    (i) Errors of measurement.
    (ii) Precision errors.
    (iii) Errors of method.

The first type of error is due to physical limitations of reading a scale. Without a vernier attachment a length of 2 cm could easily be in error by $\frac{1}{2}$ mm which is $2\frac{1}{2}$%. Precision errors are the "built-in" errors of the apparatus. For example, the scales on a mercury-in-glass thermometer assume that the bore is uniform, so that a reading of 28 °C could easily be 1 °C in error due to an uncalibrated scale. Errors of method include such faults as neglecting heat losses, assuming constant molal overflow, or neglecting back-mixing in a tubular reactor.

All of the above errors can only be estimated and rarely measured, consequently an error is usually given to only one or two significant figures. A "significant figure" is any digit, except zero when it is positioning the decimal point. It is usual to imply the accuracy of a reading by the number of significant figures quoted. That is, if a reading is known to be in error by about 2%, three significant figures are barely justified; whilst an error of 10% will only justify two significant figures. To quote too many significant figures is misleading since it implies a false sense of accuracy; whilst quoting too few squanders the accuracy which has been dearly bought. One extra significant figure should always be carried through a calculation to smother the calculation errors due to rounding-off.

10.6. Curve Fitting

If a set of experimental data is to be represented by an empirical equation, the equation must possess the correct theoretical properties as discussed in Section 10.3. Assuming that an equation of the form

$$y = a + bx^n \tag{10.37}$$

is known to fit a set of results for x and y, values are required for a, b, and n. The best way of finding them graphically is to estimate a value for a and plot $(y-a)$ against x on log–log paper. If the value of a is correct, the result will be a straight line of slope n and intercept $\ln b$ as described in Section 10.2.3, but this is unlikely. Too large a value for a will give a curve on log–log paper which is convex upwards, and too small a value for a gives a curve which is concave upwards as illustrated in Fig. 10.5.

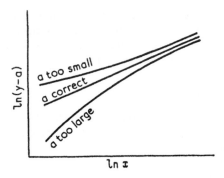

Fig. 10.5. Log–log plot of $y = a + bx^n$

The approach to curve fitting in the earlier sections of this chapter has been of a visual or graphical nature, but there are two analytical methods available. However, it is still necessary to choose the appropriate type of curve, but instead of determining the parameters graphically, they can be found analytically by the method of averages or the method of least squares.

10.6.1. *Method of Averages*

Having chosen the type of curve which it is desired to fit to the experimental data, the problem becomes one of determining the values of certain parameters in the equation so that the "best fit" is obtained. In graphical methods, the best fit is a matter of opinion, but in the present method a set of average points is defined and the best curve is the one passing through the average points.

Suppose that eight experimental values of a variable y are available at eight different known values of x, and the best curve of the type

$$y = A + B \sin x + C \cos x \tag{10.38}$$

is to be chosen to represent them. Because the values of y are experimental, it is unlikely that any pair of values (x_n, y_n) will satisfy the best fit curve (10.38) exactly. Therefore, the experimental results are substituted into equation (10.38) and an extra term is included for the unknown error. Thus

$$\left.\begin{array}{c} y_1 - A - B \sin x_1 - C \cos x_1 = R_1 \\ y_2 - A - B \sin x_2 - C \cos x_2 = R_2 \\ y_3 - A - B \sin x_3 - C \cos x_3 = R_3 \\ \vdots \qquad\qquad \vdots \\ y_8 - A - B \sin x_8 - C \cos x_8 = R_8 \end{array}\right\} \qquad (10.39)$$

where R_n are the error terms.

In order to determine A, B, and C analytically, three equations are required, and these can be obtained from the set of eight equations (10.39) by assuming relationships between the error terms. Assuming that

$$R_1 + R_2 + R_3 = 0 \qquad (10.40)$$

and adding together the first three of equations (10.39) gives

$$\sum_{n=1}^{3} y_n - 3A - B \sum_{n=1}^{3} \sin x_n - C \sum_{n=1}^{3} \cos x_n = 0 \qquad (10.41)$$

Dividing by the number of points used (3) gives

$$\frac{y_1 + y_2 + y_3}{3} - A - B \frac{\sin x_1 + \sin x_2 + \sin x_3}{3} - C \frac{\cos x_1 + \cos x_2 + \cos x_3}{3} = 0 \qquad (10.42)$$

Equation (10.42) can be obtained directly from equation (10.38) by using an "average point" to represent the three actual points. It should be noted that the average point is determined in a special manner which depends upon the functions arising in the type of equation (10.38) chosen.

Two further equations, similar to (10.42), can be obtained by arranging the remaining points in two groups. Thus

$$\frac{y_4 + y_5}{2} - A - B \frac{\sin x_4 + \sin x_5}{2} - C \frac{\cos x_4 + \cos x_5}{2} = 0 \qquad (10.43)$$

$$\frac{y_6 + y_7 + y_8}{3} - A - B \frac{\sin x_6 + \sin x_7 + \sin x_8}{3} - C \frac{\cos x_6 + \cos x_7 + \cos x_8}{3} = 0 \qquad (10.44)$$

Because every symbol with a suffix is a known experimental result, only A, B, and C are unknown in the three equations (10.42), (10.43), (10.44) and hence they can be determined.

In applications of the method of averages, the following points must be observed.

   (i) Points must be arranged in ascending values of x.
  (ii) The number of groups must equal the number of unknown parameters.
 (iii) Groups should contain approximately equal numbers of points.
 (iv) Each experimental point should be used once only.
  (v) The appropriate average must be taken.

Example 1. The thermal conductivity of graphite varies with temperature according to the equation

$$k = k_0 - \alpha T \qquad\qquad \text{I}$$

Experimentally, it is only possible to obtain a mean conductivity over a temperature range. It is required to find the point conductivity from the mean conductivity given below.†

$T(^\circ C)$	390	500	1000	1500
k_m	141	138	119	115

k_m is determined between T and $25\,^\circ C$ in W/m $^\circ C$.

Solution

   It is first necessary to determine the relationship between k_m and the point values of k. Considering unit cross-sectional area of a conductor of length L and temperatures T_1 and T_2, an elementary heat balance yields the equation

$$-k\frac{\mathrm{d}T}{\mathrm{d}x} = k_m \frac{T_1 - T_2}{L} = K \qquad\qquad \text{II}$$

which defines the constant K. Substituting equation I into equation II gives

$$-(k_0 - \alpha T)\,\mathrm{d}T = K\,\mathrm{d}x$$

$$\therefore \quad -k_0 T + \tfrac{1}{2}\alpha T^2 = Kx + \beta \qquad\qquad \text{III}$$

Using the boundary conditions that

$$\text{at} \quad x = 0, \quad T = T_1 \qquad\qquad \text{IV}$$

$$\text{at} \quad x = L, \quad T = T_2 \qquad\qquad \text{V}$$

$$\therefore \quad -k_0 T_1 + \tfrac{1}{2}\alpha T_1^2 = \beta \qquad\qquad \text{VI}$$

and $\qquad\qquad -k_0 T_2 + \tfrac{1}{2}\alpha T_2^2 = \beta + KL \qquad\qquad \text{VII}$

Subtracting equation VII from VI gives

$$k_0(T_2 - T_1) - \tfrac{1}{2}\alpha(T_2 - T_1)(T_2 + T_1) = -KL = k_m(T_2 - T_1) \qquad \text{VIII}$$

$$\therefore \quad k_m = k_0 - \alpha T_m \qquad\qquad \text{IX}$$

† *I.C.T.* **5**, 86. McGraw-Hill, New York (1937).

where T_m is the arithmetic mean temperature between the end points.

There are four experimental points and two parameters k_0 and α. Therefore, the points must be split into two groups of two, thus

$$k_0 - 207 \cdot 5\alpha - 141 = R_1$$
$$k_0 - 262 \cdot 5\alpha - 138 = R_2$$

$$2k_0 - 470\alpha \quad -279 = 0 \qquad \qquad \text{X}$$

and
$$k_0 - 512 \cdot 5\alpha - 119 = R_3$$
$$k_0 - 762 \cdot 5\alpha - 115 = R_4$$

$$2k_0 - 1275\alpha \quad -234 = 0 \qquad \qquad \text{XI}$$

Solving equations X and XI for k_0 and α gives

$$\alpha = 0 \cdot 056$$

$$k_0 = 152 \cdot 6$$

hence,
$$k = 152 \cdot 6 - 0 \cdot 056T \qquad \qquad \text{XII}$$

Substituting the experimental temperatures into equation IX gives the following comparison.

$T(^{\circ}\text{C})$	390	500	1000	1500
k_m (experimental)	141	138	119	115
k_m (equation IX)	141	138	124	110

This result was used in Example 4, Section 2.4.3.

10.6.2. *Method of Least Squares*

More often than not, this method is used to fit the best straight line to a set of data. Hence the formulae will be derived determining the values of m and c which give the least mean squares fit of the equation

$$y = mx + c \qquad \qquad (10.1)$$

to N pairs of results for x and y.

Following the method of the previous section, each pair of results is substituted into equation (10.1) and an extra term is introduced to allow for the unknown error.

$$y_n - mx_n - c = R_n \qquad \qquad (10.45)$$

R_n may be either positive or negative; therefore, to obtain a positive representation for the errors, equation (10.45) is squared thus,

$$y_n^2 + m^2 x_n^2 + c^2 - 2mx_n y_n - 2cy_n + 2cmx_n = R_n^2 \qquad \qquad (10.46)$$

Equations (10.46) for the individual experimental points are now added together to give

$$\sum_{n=1}^{N} y_n^2 + m^2 \sum_{n=1}^{N} x_n^2 + Nc^2 - 2m \sum_{n=1}^{N} x_n y_n - 2c \sum_{n=1}^{N} y_n + 2cm \sum_{n=1}^{N} x_n = \sum_{n=1}^{N} R_n^2$$

(10.47)

Equation (10.47) gives an expression for the sum of the squares of the error terms. The "method of least squares" defines the best straight line as the one for which the sum of the squares of the error terms is a minimum.

It will be shown in Section 13.3 that the left-hand side of equation (10.47) will have a turning value when its partial derivatives with respect to both m and c are zero. Thus

$$2m \sum_{n=1}^{N} x_n^2 - 2 \sum_{n=1}^{N} x_n y_n + 2c \sum_{n=1}^{N} x_n = 0$$

(10.48)

$$2cN - 2 \sum_{n=1}^{N} y_n + 2m \sum_{n=1}^{N} x_n = 0$$

(10.49)

Solving equations (10.48) and (10.49) gives the result

$$\left. \begin{aligned} m &= \frac{N \sum x_n y_n - \sum x_n \sum y_n}{N \sum x_n^2 - (\sum x_n)^2} \\[2mm] c &= \frac{\sum x_n^2 \sum y_n - \sum x_n y_n \sum x_n}{N \sum x_n^2 - (\sum x_n)^2} \end{aligned} \right\}$$

(10.50)

where all sums must be taken over all experimental points.

The best way to use the above formula is to construct a table with four columns headed $x_n, y_n, x_n^2, x_n y_n$ and complete the table for the experimental results. The sums of the columns will then give all of the values required by equations (10.50). If a desk calculating machine is available, the four sums can be obtained from two cumulative calculations without constructing the table.

The following example shows how the method can be extended to a third variable by fitting the best plane to the three-dimensional plot.

Example 2. It has been proposed† that the second order chemical reaction

$$CO + Cl_2 \rightarrow COCl_2$$

proceeds on the surface of an activated carbon catalyst after adsorption of the two reactants. Each of the three substances is adsorbed to a different extent, but the number of sites occupied by carbon monoxide is small compared with sites otherwise occupied. Assuming that the process is

† Potter, C. and Barron, S. *Chem. Eng. Prog.* **47**, 473 (1951).

controlled by the surface reaction, which is irreversible, find the best values of the adsorption coefficients from the following experimental results.

Partial Pressures			Reaction Rate g mols/h gm catalyst
CO	Cl$_2$	COCl$_2$	
0·406	0·352	0·226	0·00414
0·396	0·363	0·231	0·00440
0·310	0·320	0·356	0·00241
0·287	0·333	0·367	0·00245
0·253	0·218	0·522	0·00157
0·610	0·118	0·231	0·00390
0·179	0·608	0·206	0·00200

Solution

The following nomenclature is used:

C number of sites in the state specified
K adsorption coefficient
k specific reaction rate constant
p partial pressure
r rate of reaction

Subscripts denote the following:

a   carbon monoxide
b   chlorine
c   phosgene
v   vacant site
t   total number of sites

The equilibrium of the three components between the catalyst and the vapour phase can be expressed thus,

$$C_a = K_a p_a C_v \qquad\qquad \text{I}$$

$$C_b = K_b p_b C_v \qquad\qquad \text{II}$$

$$C_c = K_c p_c C_v \qquad\qquad \text{III}$$

The reaction rate is given by

$$r = k C_a C_b \qquad\qquad \text{IV}$$

The total number of active sites must be equal to the sum of the number of sites in each state. Therefore,

$$C_t = C_v + C_a + C_b + C_c \qquad\qquad \text{V}$$

But C_a is known to be small compared with C_b and C_c, hence it can be ignored in equation V. Using equations II and III to eliminate C_b and C_c from equation V,

$$C_t = C_v(1 + K_b p_b + K_c p_c) \qquad\qquad \text{VI}$$

Eliminating C_a and C_b from equation IV by using equations I and II,

$$r = kK_a K_b C_v^2 p_a p_b \qquad \text{VII}$$

Eliminating C_v between equations VI and VII and rearranging gives

$$1 + K_b p_b + K_c p_c = C_t \sqrt{(kK_a K_b)} \sqrt{(p_a p_b/r)} \qquad \text{VIII}$$

Equation VIII is of the form

$$1 + \alpha x_n + \beta y_n - \gamma z_n = R_n \qquad \text{IX}$$

where x_n, y_n, and z_n are the experimental quantities p_b, p_c, and $\sqrt{(p_a p_b/r)}$, α, β, and γ are the constants to be determined, and R_n is the experimental error.

Squaring equation IX and summing over the N experimental points gives

$$N + \alpha^2 \sum x_n^2 + \beta^2 \sum y_n^2 + \gamma^2 \sum z_n^2 + 2\alpha \sum x_n + 2\beta \sum y_n - 2\gamma \sum z_n + 2\alpha\beta \sum x_n y_n$$
$$- 2\alpha\gamma \sum x_n z_n - 2\beta\gamma \sum y_n z_n = \sum R_n^2 \qquad \text{X}$$

Differentiating equation X partially with respect to α, β, and γ in turn and equating to zero gives the three equations:

$$\alpha \sum x_n^2 + \sum x_n + \beta \sum x_n y_n - \gamma \sum x_n z_n = 0 \qquad \text{XI}$$

$$\beta \sum y_n^2 + \sum y_n + \alpha \sum x_n y_n - \gamma \sum y_n z_n = 0 \qquad \text{XII}$$

$$-\gamma \sum z_n^2 + \sum z_n + \alpha \sum x_n z_n + \beta \sum y_n z_n = 0 \qquad \text{XIII}$$

In order to solve equations XI, XII, and XIII for α, β, and γ it is necessary to evaluate the nine sums over the given experimental results. Inspection of the given data indicates that there is a much higher level of inert gas in the sixth result. This is considered unlikely and the result should be rejected. However, the data are sparse and it has been assumed that an error has been made which did not affect the proportions of the components present. This result has therefore been scaled to the same order of inerts as the other readings.

The preliminary calculations are shown in Table 10.1. Equations XI, XII, and XIII now become

$$0 \cdot 9008\alpha + 0 \cdot 6673\beta - 14 \cdot 595\gamma + 2 \cdot 315 = 0$$

$$0 \cdot 6673\alpha + 0 \cdot 7370\beta - 12 \cdot 890\gamma + 2 \cdot 145 = 0$$

$$14 \cdot 595\alpha + 12 \cdot 890\beta - 256 \cdot 3\gamma + 41 \cdot 99 = 0$$

Solving for α, β, and γ by successive elimination or by matrix inversion (see Section 12.10) gives

$$\alpha = 2 \cdot 61, \quad \beta = 1 \cdot 60, \quad \gamma = 0 \cdot 393$$

Returning to the original symbols, this means that

$$K_b = 2 \cdot 61, \quad K_c = 1 \cdot 60, \quad C_t \sqrt{(kK_a)} = 0 \cdot 243$$

TABLE 10.1. Least Squares Fit of Reaction Data

$r(\times 10^3)$	p_a	$x = p_b$	$y = p_c$	$z = \sqrt{\dfrac{p_a p_b}{r}}$	x^2	y^2	z^2	xy	yz	zx	z (calculated)
4·14	0.406	0.352	0.226	5·88	0·1239	0·0511	34·52	0·0796	1·329	2·070	5·80
4·40	0.396	0.363	0.231	5·72	0·1318	0·0534	32·67	0·0838	1·321	2·076	5·89
2·41	0.310	0.320	0.356	6·42	0·1024	0·1267	41·16	0·1139	2·285	2·054	6·12
2·45	0.287	0.333	0.367	6·25	0·1109	0·1347	39·01	0·1222	2·294	2·081	6·25
1·57	0.253	0.218	0.522	5·93	0·0475	0·2725	35·13	0·1138	3·096	1·293	6·11
3·90	0.626	0.121	0.237	4·41	0·0146	0·0562	19·42	0·0287	1·045	0·534	4·31
2·0	0.179	0.608	0.206	7·38	0·3697	0·0424	54·42	0·1253	1·520	4·487	7·42
—	—	2.315	2.145	41·99	0·9008	0·7370	256·3	0·6673	12·890	14·595	—

Putting these values into equation VIII, the final column in Table 10.1 can be calculated to show the agreement between the experimental results and the empirical equation.

It can be seen that the greatest error in z is 5% which corresponds to a 10% error in predicting the rate of reaction.

The methods of averages and least squares are based on different definitions of the best curve which fits the data. If a desk calculating machine is available, the method of least squares should always be used since it is more soundly based. Without the aid of a machine, however, the calculation is rather tedious and liable to arithmetical error and the method of averages, which is shorter, should be used to give an adequate result.

10.7. NUMERICAL INTEGRATION

It is sometimes necessary to perform a calculation which involves integration. For example, the volumetric flow rate of a gas through a duct can be determined from the linear velocity distribution by evaluating a suitable integral. If an average value is required for a particular quantity, it is frequently obtained by integration. For instance, an average surface temperature can be determined by integrating the measured temperature distribution over the surface and dividing by the area. Also, if a representative sample of a process stream is required during a steady-state investigation, it is much more reliable to use a composite sample than a single sample. The intervals at which the parts of the composite sample are taken, and the proportions in which they should be mixed, will be determined theoretically in the following work. Occasionally, a difficult integral arises in a theoretical prediction, and this has to be evaluated numerically for each individual case.

Many of the problems mentioned above are treated graphically, by plotting the variables against one another and evaluating the area enclosed by two ordinates and the parts of the abscissa and curve which join the ordinates. In this section, numerical methods for calculating this area will be presented as an alternative to the graphical method.

One way of integrating a set of data is by fitting an empirical equation to the points, using one of the methods from the previous section and then integrating the equation analytically. Unless the empirical equation is needed for some other purpose, this procedure is not usually worth the effort. All of the methods to be described involve fitting a polynomial to a set of points so that the polynomial passes through the points. Since the points may be subject to experimental error, the curve obtained in this way will not be a good representation of the data and would certainly give false results for a derivative, but for purposes of integration the method is quite satisfactory. The above statement is illustrated in Fig. 10.6 where a cubic has been fitted to four points. For general purposes, the straight line is as good a fit to the four points as is justified by their accuracy, and if the slope of the curve is required, the cubic fit would give misleading results. It is quite clear, however, that the area beneath both curves is approximately the same. Although the cubic is the more complex curve, it is the easier to fit because it passes through the points. Thus polynomials are used to fit data for integration purposes.

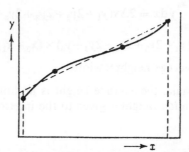

FIG. 10.6. Polynomial fit for integration

10.7.1. *The Trapezium Rule*

The simplest polynomial, consisting of two terms, is the linear equation

$$y = a_0 + a_1 x \tag{10.51}$$

Since it only contains two parameters (a_0 and a_1) it gives the equation of the straight line joining two points. Taking two points (x_1, y_1) and (x_2, y_2), a_0 and a_1 can be evaluated as

$$a_0 = \frac{x_2 y_1 - x_1 y_2}{x_2 - x_1}, \quad a_1 = \frac{y_2 - y_1}{x_2 - x_1} \tag{10.52}$$

Integrating equation (10.51) with respect to x between $x = x_1$ and $x = x_2$ gives

$$I = \int_{x_1}^{x_2} y \, dx = [a_0 x + \tfrac{1}{2} a_1 x^2]_{x_1}^{x_2}$$

$$= a_0(x_2 - x_1) + \tfrac{1}{2} a_1 (x_2 - x_1)(x_2 + x_1) \tag{10.53}$$

Using the values for a_0 and a_1 given in equations (10.52) gives

$$I = x_2 y_1 - x_1 y_2 + \tfrac{1}{2}(x_2 y_2 - x_2 y_1 + x_1 y_2 - x_1 y_1)$$

$$= \tfrac{1}{2}(x_2 y_2 + x_2 y_1 - x_1 y_2 - x_1 y_1)$$

$$= \tfrac{1}{2}(y_2 + y_1) \times (x_2 - x_1)$$

$$= \text{average height} \times \text{width} \tag{10.54}$$

The value of the integral is thus proved to be the arithmetic average of the two ordinates multiplied by the distance between them. Figure 10.7 illustrates a repeated use of equation (10.54). It can be seen that if the curve is convex upwards, the trapezium rule under-estimates the area by neglecting the segments shown. Conversely, if the curve is convex downwards, the trapezium rule over-estimates the area. The error can be reduced by subdividing the intervals still further.

If the intervals Δx are all equal, and there are four of them, the integral can be written

$$I = \int_{x_1}^{x_5} y\, dx = 2\Delta x(y_1 + 2y_2 + 2y_3 + 2y_4 + y_5)$$

$$= \tfrac{1}{8}(y_1 + 2y_2 + 2y_3 + 2y_4 + y_5) \times (x_5 - x_1) \qquad (10.55)$$

$$= \text{average height} \times \text{width}$$

In this combined formula, the average height is no longer a plain arithmetic average, but twice as much weight is given to the interior points as to the end points.

10.7.2. Simpson's Rule

The next polynomial which contains three terms is the quadratic

$$y = a_0 + a_1 x + a_2 x^2 \qquad (10.56)$$

and this can be fitted to three points. However, it will now be shown that all cubic equations of the type

$$y = a_0 + a_1 x + a_2 x^2 + a_3 x^3 \qquad (10.57)$$

which pass through any three chosen points at equally spaced values of x, all have the same area beneath them between the end ordinates. The three ordinates and the area required are illustrated in Fig. 10.8, where h is the increment of x.

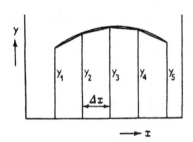

FIG. 10.7. The trapezium rule   FIG. 10.8. Simpson's rule

The algebra in the following derivation is simpler if the variable x is changed to a new variable z according to the formula

$$z = (x - x_2)/h \qquad (10.58)$$

which transforms the required integral as follows

$$I = \int_{x_1}^{x_3} y\, dx = h \int_{-1}^{1} y\, dz \qquad (10.59)$$

and the three ordinates y_1, y_2, and y_3 are now at $z = -1$, 0, and 1.

The polynomial (10.57) can be transformed using (10.58) to a polynomial of the same order in z, thus

$$y = b_0 + b_1 z + b_2 z^2 + b_3 z^3 \tag{10.60}$$

where the b_ns are functions of the a_ns. Following the same method as for the trapezium rule, the three values of z are substituted into equation (10.60) with the corresponding ordinates. Thus

$$y_1 = b_0 - b_1 + b_2 - b_3 \tag{10.61}$$

$$y_2 = b_0 \tag{10.62}$$

$$y_3 = b_0 + b_1 + b_2 + b_3 \tag{10.63}$$

Substituting equation (10.60) into equation (10.59) and evaluating the integral gives

$$I = h \int_{-1}^{1} (b_0 + b_1 z + b_2 z^2 + b_3 z^3) \, dz$$

$$= h[b_0 z + \tfrac{1}{2} b_1 z^2 + \tfrac{1}{3} b_2 z^3 + \tfrac{1}{4} b_3 z^4]_{-1}^{1}$$

$$= h(2b_0 + \tfrac{2}{3} b_2) \tag{10.64}$$

Adding equations (10.61) and (10.63) together gives

$$y_1 + y_3 = 2b_0 + 2b_2 \tag{10.65}$$

and using equation (10.62) shows that

$$y_1 - 2y_2 + y_3 = 2b_2 \tag{10.66}$$

Substituting equations (10.62) and (10.66) into (10.64) gives

$$I = h[2y_2 + \tfrac{1}{3}(y_1 - 2y_2 + y_3)]$$

$$= \tfrac{1}{6}(y_1 + 4y_2 + y_3) \times 2h \tag{10.67}$$

$$= \text{average height} \times \text{width}$$

The average height in this case is found by adding the end values to four times the central value and dividing the result by six.

The same result is obtained for all cubics through the three points because the step from equations (10.61) and (10.63) to equation (10.65) eliminates two parameters (b_1 and b_3) instead of one, which is normally the case.

If the range of integration is subdivided into equal intervals by using any odd number of ordinates (say 7), Simpson's rule can be applied to each group of three points and the result of all integrations added together. Thus

$$I = \int_{x_1}^{x_7} y \, dx = \tfrac{1}{18}(y_1 + 4y_2 + 2y_3 + 4y_4 + 2y_5 + 4y_6 + y_7) \times (x_7 - x_1)$$

$$= \text{average height} \times \text{width} \tag{10.68}$$

The average height in equation (10.68) is found by adding four times the even ordinates, and twice the interior odd ordinates, to the end ordinates; the sum being divided by the total number of times an ordinate has been used.

Formula (10.68) gives twice as much weight to some of the interior points than to others, and is not really satisfactory for this reason. It implies that some of the interior points should be measured more accurately than others. Nevertheless, the results obtained from the use of the formula are quite accurate if sufficient sub-intervals are used.

The methods used above can be extended by fitting polynomials of higher degree to a greater number of points, but this is inconvenient. The increased complexity arising from the greater number of points is not justified by a significant improvement in accuracy. A slightly different approach, due to Gauss, enables a polynomial of degree $(2n-1)$ to be fitted to n points, resulting in a much more efficient process.

10.7.3. *Gauss' Method*

The derivation of the formula for fitting a quintic equation to three points will be described below in order to illustrate the method.

It is always possible to transform an integral over a finite range to an integral between the limits -1 and 1, as shown in the derivation of Simpson's rule. In this derivation, it will be assumed that this change of variable has been made by considering the integral

$$I = \int_{-1}^{1} y \, dx = 2y_m \tag{10.69}$$

which defines y_m, the average height.

Assuming that the quintic

$$y = a_0 + a_1 x + a_2 x^2 + a_3 x^3 + a_4 x^4 + a_5 x^5 \tag{10.70}$$

adequately represents the shape of the curve y vs x between $x = -1$ and $x = 1$, an expression for y_m can be obtained by substituting equation (10.70) into equation (10.69) and integrating. Thus

$$2y_m = [a_0 x + \tfrac{1}{2}a_1 x^2 + \tfrac{1}{3}a_2 x^3 + \tfrac{1}{4}a_3 x^4 + \tfrac{1}{5}a_4 x^5 + \tfrac{1}{6}a_5 x^6]_{-1}^{1}$$

$$= 2a_0 + \tfrac{2}{3}a_2 + \tfrac{2}{5}a_4$$

or $\qquad y_m = a_0 + \tfrac{1}{3}a_2 + \tfrac{1}{5}a_4 \tag{10.71}$

Referring back to the derivation of Simpson's rule, equations (10.61), (10.62), (10.63) contained four unknowns and could not be solved completely. However, the coefficients b_0 and b_2 could be evaluated due to b_1 and b_3 being eliminated together. It will now be shown that in Gauss' method, the three coefficients a_0, a_2, a_4 occurring in equation (10.71) can be evaluated from only three ordinates, if three special values of the abscissa are chosen. Denoting these three values by x_1, x_2, x_3, and the corresponding ordinates by y_1, y_2, y_3, each pair of values must satisfy equation (10.70). Hence

$$y_1 = a_0 + a_1 x_1 + a_2 x_1{}^2 + a_3 x_1{}^3 + a_4 x_1{}^4 + a_5 x_1{}^5 \tag{10.72}$$

$$y_2 = a_0 + a_1 x_2 + a_2 x_2{}^2 + a_3 x_2{}^3 + a_4 x_2{}^4 + a_5 x_2{}^5 \tag{10.73}$$

$$y_3 = a_0 + a_1 x_3 + a_2 x_3{}^2 + a_3 x_3{}^3 + a_4 x_3{}^4 + a_5 x_3{}^5 \tag{10.74}$$

It is now assumed that the above three equations can be added together in the proportions $K_1 : K_2 : K_3$ so that a_1, a_3, a_5 are simultaneously eliminated, and the coefficients of a_0, a_2, a_4 correspond with equation (10.71). Thus

$$
\begin{aligned}
K_1 y_1 + K_2 y_2 + K_3 y_3 = {}&(K_1 + K_2 + K_3)\,a_0 + (K_1 x_1 + K_2 x_2 + K_3 x_3)\,a_1 \\
&+ (K_1 x_1{}^2 + K_2 x_2{}^2 + K_3 x_3{}^2)\,a_2 \\
&+ (K_1 x_1{}^3 + K_2 x_2{}^3 + K_3 x_3{}^3)\,a_3 \\
&+ (K_1 x_1{}^4 + K_2 x_2{}^4 + K_3 x_3{}^4)\,a_4 \\
&+ (K_1 x_1{}^5 + K_2 x_2{}^5 + K_3 x_3{}^5)\,a_5
\end{aligned}
\tag{10.75}
$$

and hence

$$
\left.
\begin{aligned}
K_1 + K_2 + K_3 &= 1 \\
K_1 x_1 + K_2 x_2 + K_3 x_3 &= 0 \\
K_1 x_1{}^2 + K_2 x_2{}^2 + K_3 x_3{}^2 &= \tfrac{1}{3} \\
K_1 x_1{}^3 + K_2 x_2{}^3 + K_3 x_3{}^3 &= 0 \\
K_1 x_1{}^4 + K_2 x_2{}^4 + K_3 x_3{}^4 &= \tfrac{1}{5} \\
K_1 x_1{}^5 + K_2 x_2{}^5 + K_3 x_3{}^5 &= 0
\end{aligned}
\right\}
\tag{10.76}
$$

(10.76) is a set of six equations in the six unknowns $x_1, x_2, x_3, K_1, K_2, K_3$ and can thus be solved. The solution is tedious and proceeds according to the following pattern.

(i) Eliminate K_1 by multiplying each equation by x_1 and subtracting it from the following equation.

(ii) Eliminate K_2 from the new set of five equations by multiplying each by x_2 and subtracting from the following equation.

(iii) Eliminate K_3 in a similar manner.

(iv) The three remaining equations show that x_1, x_2, x_3 can be interpreted as the roots of the cubic equation

$$5x^3 - 3x = 0 \tag{10.77}$$

(v) Therefore $x_3 = -x_1 = \sqrt{\tfrac{3}{5}}$, and $x_2 = 0$.

(vi) Substituting these values back into various parts of the calculation it can be shown that $K_2 = \tfrac{4}{9}$, and $K_1 = K_3 = \tfrac{5}{18}$.

Equations (10.71) and (10.75) are identical, and hence

$$\tfrac{1}{2}\int_{-1}^{1} y\,dx = \frac{5y_1 + 8y_2 + 5y_3}{18} \tag{10.78}$$

where y_1, y_2, y_3 are evaluated at $x = -0.7746$, 0, and 0.7746 respectively.

Gauss' two-point method is seldom used since it shows little advantage over Simpson's rule, but Gauss' four-point method should be used for improved accuracy. The formulae are:

$$\tfrac{1}{2} \int_{-1}^{1} y \, dx = \tfrac{1}{2}(y_1 + y_2) \tag{10.79}$$

where y_1 and y_2 are evaluated at $x = \pm 0.5773$; and

$$\tfrac{1}{2} \int_{-1}^{1} y \, dx = 0.1739(y_1 + y_4) + 0.3261(y_2 + y_3) \tag{10.80}$$

where y_1, y_2, y_3, y_4 are evaluated at $x = -0.8611$, -0.3400, 0.3400, and 0.8611 respectively,

Example. Evaluate $\int_0^4 (1 + x^2)^{-\frac{1}{2}} \, dx$ by using the following methods:
(a) Analytical.
(b) Trapezium rule (three points).
(c) Trapezium rule (nine points).
(d) Simpson's rule (three points).
(e) Simpson's rule (nine points).
(f) Gauss three-point.
(g) Gauss four-point.
Compare all other results with the analytical solution to determine their accuracy in representing the above function.

Solution

(a) Put $x = \sinh z$, and hence $dx = \cosh z \, dz$.

$$\therefore \quad I = \int_0^4 (1 + x^2)^{-\frac{1}{2}} \, dx = \int_0^\alpha \frac{\cosh z \, dz}{\sqrt{(1 + \sinh^2 z)}}$$

$$= [z]_0^\alpha$$

$$= \sinh^{-1} 4$$

$$= 2.0947$$

x	$1 + x^2$	$y = (1 + x^2)^{-\frac{1}{2}}$
0.0	1.00	1.00000
0.5	1.25	0.89445
1.0	2.00	0.70711
1.5	3.25	0.55475
2.0	5.00	0.44722
2.5	7.25	0.37138
3.0	10.00	0.31623
3.5	13.25	0.27473
4.0	17.00	0.24254

(b) From the trapezium rule for three points

$$I = y_1 + 2y_2 + y_3$$
$$= 1 \cdot 00000 + 2(0 \cdot 44722) + 0 \cdot 24254$$
$$= 2 \cdot 1369$$

(c) From the trapezium rule for nine points,

$$I = \tfrac{1}{4}(y_1 + 2y_2 + 2y_3 + 2y_4 + 2y_5 + 2y_6 + 2y_7 + 2y_8 + y_9)$$
$$= \tfrac{1}{4}(1 \cdot 00000 + 0 \cdot 24254) + \tfrac{1}{2}(0 \cdot 89445 + 0 \cdot 70711 + 0 \cdot 55475 + 0 \cdot 44722$$
$$+ 0 \cdot 37138 + 0 \cdot 31623 + 0 \cdot 27473)$$
$$= 2 \cdot 0936$$

(d) Simpson's rule for three points (10.67) gives

$$I = \tfrac{1}{6}(y_1 + 4y_2 + y_3) \times 4$$
$$= \tfrac{2}{3}(1 \cdot 00000 + 1 \cdot 78888 + 0 \cdot 24254)$$
$$= 2 \cdot 0209$$

(e) Simpson's rule for nine points gives

$$I = \tfrac{1}{24}(y_1 + 4y_2 + 2y_3 + 4y_4 + 2y_5 + 4y_6 + 2y_7 + 4y_8 + y_9) \times 4$$
$$= \tfrac{1}{6}(1 \cdot 00000 + 0 \cdot 24254) + \tfrac{1}{3}(0 \cdot 70711 + 0 \cdot 44722 + 0 \cdot 31623)$$
$$+ \tfrac{2}{3}(0 \cdot 89445 + 0 \cdot 55475 + 0 \cdot 37138 + 0 \cdot 27473)$$
$$= 2 \cdot 0941$$

(f) In Gauss' three-point method (10.78), if the range of integration is -1 to 1, then the ordinates are required at 0 and $\pm 0 \cdot 7746$. Therefore, if the range of integration is 0–4, then ordinates are required at $x = 2$ and $x = 2 \pm 2(0 \cdot 7746) = 3 \cdot 5492$ and $0 \cdot 4508$.

$$I = \tfrac{4}{18}[5(0 \cdot 91165 + 0 \cdot 27118) + 8(0 \cdot 44722)]$$
$$= 2 \cdot 1093$$

(g) In Gauss' four-point method, the ordinates will be required at $x = 0 \cdot 2778$, $1 \cdot 3200$, $2 \cdot 6800$, and $3 \cdot 7222$. Equation (10.80) then gives the average height within the range of integration.

$$I = 4[0 \cdot 1739(0 \cdot 96351 + 0 \cdot 25945) + 0 \cdot 3261(0 \cdot 60386 + 0 \cdot 34959)]$$
$$= 2 \cdot 0944$$

The above results can be compared in Table 10.2.

It can be seen that Gauss' method is the superior one for three points, and that Gauss' four-point method is better than either of the other nine-point methods. The reason why Simpson's rule is so poor for three points is

TABLE 10.2. *Comparison of Integration Formulae*

Method	Result	Absolute error	Relative error (%)
Trapezium rule (3 points)	2·1369	0·0422	2·1
Trapezium rule (9 points)	2·0936	−0·0011	0·05
Simpson's rule (3 points)	2·0209	0·0738	3·5
Simpson's rule (9 points)	2·0941	−0·0006	0·03
Gauss 3-point	2·1093	0·0146	0·7
Gauss 4-point	2·0944	−0·0003	0·01

illustrated in Fig. 10.9, where the true curve and the parabolic fit can be compared. The cause of the bad fit is the point of inflection at $x = 0·7071$ which cannot be represented by a quadratic equation.

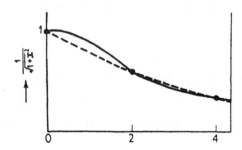

FIG. 10.9. Parabolic approximation for Simpson's rule

10.7.4. *Comparison of Methods*

In most cases, the four-point method of Gauss gives an accuracy equivalent to the use of Simpson's rule with nine points equally spaced across the same range. Gauss' method is thus more economical in the use of data, but the calculations are a little more complicated because of the more difficult weighting factors.

Gauss' method should have many applications to the investigation of the behaviour of full-scale production equipment, where it is necessary to compress all readings into as short a time as possible due to plant fluctuations. A reduction in the number of readings required without loss of accuracy is an obvious advantage. Another example is obtaining a representative sample of a steady-state process stream. It is much more efficient to take a set of samples at the Gauss intervals, mix them in the Gauss proportions, and analyse the composite sample than to take twice as many samples at equal intervals and analyse each one separately or mix them all together in equal proportions.

An interesting point has been noted by McDermott† concerning the integration of experimental results. Figure 10.10 illustrates the true relationship between two variables x and y as the straight line AB. The two groups of three equally spaced experimental points (a), (b) have been chosen so that the total error in each group is zero, and these are the only two possible groups which satisfy this condition (except for their mirror images in the line AB). It is obvious from the group (a) that the parabolic fit seriously underestimates the area, whilst the two straight lines joining the points give a closer approximation. For the group (b), the difference between the straightline area and the parabolic area is small, but does show a marginal advantage in favour of the trapezium rule. It is thus apparent that the trapezium rule is more satisfactory than Simpson's rule. This conclusion is in direct contrast to the result demonstrated in Table 10.2, where the higher polynomials gave the better results.

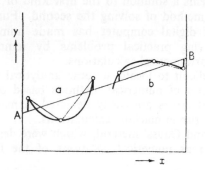

FIG. 10.10. Comparison of the trapezium rule and Simpson's rule

These two conflicting conclusions can be reconciled as follows. For evaluating analytical integrals or integrating accurate experimental data, the methods of Gauss and Simpson's rule should be used; but if the data are inherently inaccurate, the trapezium rule should be used. The question of how accurate the data must be to justify the use of Gauss' method or Simpson's rule is a very difficult one, and the answer must depend upon the type of curve being fitted. For linear relationships the trapezium rule appears to be the best for all experimental data, and the simplicity of the method is a further recommendation. However, this statement does not detract from the value of Gauss' method or Simpson's rule for integrating accurate experimental or theoretical data; nor does it invalidate a planned experiment to sample a process at the Gauss intervals.

† McDermott, C. M.Sc. Thesis, Chemical Engineering Department, University of Birmingham (1962).

Chapter 11

NUMERICAL METHODS

11.1. INTRODUCTION

THERE are many problems in mathematics for which no analytical solution is known. There are also others, for which the analytical solution is tedious and the answer may be in the form of an infinite series that can only be interpreted after much computational effort. A numerical method is the only one which yields a solution to the first kind of problem, and may be the most efficient method of solving the second. Furthermore, the advent of the high-speed digital computer has made numerical methods more attractive for solving practical problems by removing the tedium of repetitive manual arithmetical calculations.

Mistakes are difficult to locate in some analytical solutions whereas the self-checking features of numerical methods based on successive approximation ensure that errors are easily detected in manual calculations and errors should not arise in machine calculations.

Simpson's rule and Gauss' method, which were described in Chapter 10, can be interpreted as numerical solutions of the first order differential equation

$$\frac{dy}{dx} = f(x) \tag{11.1}$$

These solutions are valuable if $f(x)$ is not a simple function. The purpose of this present chapter is to present numerical solutions for more complicated types of ordinary differential equations, to locate roots of algebraic and transcendental equations, and to solve partial differential equations numerically.

11.2. FIRST ORDER ORDINARY DIFFERENTIAL EQUATIONS

In Section 2.3 many first order differential equations were solved, but if the equation is not linear and the variables will not separate, none of the methods given can be applied. The problem is thus to solve

$$\frac{dy}{dx} = f(x, y) \tag{11.2}$$

Two kinds of method are commonly used, one due to Runge and Kutta based on satisfying as many terms as possible in the Taylor series expansion of the function by an explicit method, and the other kind, called predictor-corrector methods, which are implicit.

378

11.2.1. *The Runge–Kutta Method*

The method to be described is a composite one, using the principles discovered separately by Runge and by Kutta. The analytical details of the modern method differ from the original development and the algebra is rather tedious; however, the general principles can be outlined by reference to Fig. 11.1. Assuming that equation (11.2) is satisfied, and an initial condition ($y = y_0$ at $x = x_0$) is given, it is desired to find the value of y when $x = x_0 + h$ where h is some given constant.

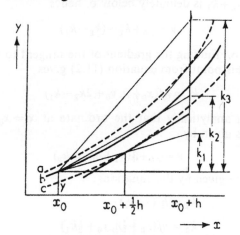

FIG. 11.1. Third order Runge–Kutta process

The general solution of equation (11.2) will consist of a family of curves, each curve having a particular value of the constant of integration. Three specially chosen curves of this family (a, b, c) are illustrated in Fig. 11.1; the curve b passing through the point (x_0, y_0). The gradient of the tangent to the curve at (x_0, y_0) can be calculated from the equation (11.2) thus determining the length k_1 by

$$k_1 = hf(x_0, y_0) \qquad (11.3)$$

$y_0 + k_1$ is obviously a poor approximation to the desired answer unless the curve happens to be a straight line, and a better approximation can be obtained from the following reasoning. The gradient of the chord to the curve between x_0 and $x_0 + h$ should be approximately equal to the gradient of the tangent to the curve at $x_0 + \frac{1}{2}h$. Unfortunately, at this stage of the calculation, the position of the curve at $x_0 + \frac{1}{2}h$ is not known; however, an approximate value for y at $x = x_0 + \frac{1}{2}h$ is given by $y = y_0 + \frac{1}{2}k_1$. This point $(x_0 + \frac{1}{2}h, y_0 + \frac{1}{2}k_1)$ lies on a different curve of the family as illustrated by c in Fig. 11.1. The gradient of the tangent to curve c at $(x_0 + \frac{1}{2}h, y_0 + \frac{1}{2}k_1)$ can be calculated from equation (11.2) and the length k_2 constructed by drawing a line parallel to this tangent through (x_0, y_0). Thus

$$k_2 = hf(x_0 + \frac{1}{2}h, y_0 + \frac{1}{2}k_1) \qquad (11.4)$$

For the type of curve illustrated, k_2 is also an under-estimate. The method of obtaining an even closer estimate for the end ordinate is to find some average gradient of three tangents, one at the start of the interval, one in the centre, and one at the end. All three tangents should be taken to the correct curve b but this is not possible. The tangents so far, are to the correct curve b, and to the curve c which lies below b. To compensate for the effect of the latter tangent, the final tangent must be taken to a curve lying above b. On the ordinate at $x = x_0 + h$, $y = y_0 + k_2$ is approximately on curve b, $y = y_0 + k_1$ is definitely below b, hence

$$y = y_0 + k_2 + (k_2 - k_1)$$

should be above b. Finding the gradient of the tangent to the curve a which passes through this point from equation (11.2) gives

$$k_3 = hf(x_0 + h, y_0 + 2k_2 - k_1) \tag{11.5}$$

It can be shown analytically that the ordinate at $x = x_0 + h$ to the curve through (x_0, y_0) is given by

$$y = y_0 + \tfrac{1}{6}(k_1 + 4k_2 + k_3) \tag{11.6}$$

where k_1, k_2, k_3 are given by the equations:

$$k_1 = hf(x_0, y_0) \tag{11.3}$$

$$k_2 = hf(x_0 + \tfrac{1}{2}h, y_0 + \tfrac{1}{2}k_1) \tag{11.4}$$

$$k_3 = hf(x_0 + h, y_0 + 2k_2 - k_1) \tag{11.5}$$

This formula (11.6) is known as the "third order Runge–Kutta formula" because the term corresponding to the third derivative term in the Taylor series for y expanded about (x_0, y_0) is correct.

To illustrate the use of this method, it will be applied to the following example.

Example 1. If $y = 1$ when $x = 1$, and the equation

$$\frac{dy}{dx} = \frac{x^2 - y}{x} \qquad\qquad \text{I}$$

is satisfied, find the value of y when $x = 2$.

Using the above symbols, $h = 1$; hence equation (11.3) gives

$$k_1 = f(1, 1) = 0 \qquad\qquad \text{II}$$

Equation (11.4) now gives

$$k_2 = f(1\tfrac{1}{2}, 1) = \frac{(\tfrac{9}{4}) - 1}{\tfrac{3}{2}}$$

$$= \tfrac{5}{6} \qquad\qquad \text{III}$$

Putting equations II and III into (11.5) gives

$$k_3 = f(2, 2\tfrac{2}{3}) = \frac{4 - 2\tfrac{2}{3}}{2}$$

$$= \tfrac{2}{3} \qquad\qquad \text{IV}$$

Substituting equations II, III, and IV into (11.6) gives the answer

$$y = 1 + \tfrac{1}{6}(0 + 3\tfrac{1}{3} + \tfrac{2}{3})$$

$$= 1\tfrac{2}{3} \qquad\qquad \text{V}$$

An analytical solution is available for this problem because equation I is linear. Solving by the method of Section 2.3.4 gives

$$y = \tfrac{1}{3}x^2 + \tfrac{2}{3}x^{-1} \qquad\qquad \text{VI}$$

so that at $x = 2$,

$$y = 1\tfrac{2}{3} \qquad\qquad \text{VII}$$

The exact value obtained in this example is fortuitous but the method is usually quite reliable. If the range of integration is large, it can be sub-divided just as for Simpson's rule, with consequent increase in accuracy. This will be illustrated by an example in the next sub-section 11.3.

Another way of increasing accuracy is by averaging the gradients of four or more tangents instead of three, but this complication is usually only adopted for digital computer solutions. In the program library of most computer centres there is an integrating routine which implements the Runge–Kutta method. A popular version is the more accurate fourth order method which consists of evaluating the following gradients in turn.

$$k_1 = hf(x_0, y_0) \qquad\qquad (11.3)$$

$$k_2 = hf(x_0 + \tfrac{1}{2}h, y_0 + \tfrac{1}{2}k_1) \qquad\qquad (11.4)$$

$$k_3 = hf(x_0 + \tfrac{1}{2}h, y_0 + \tfrac{1}{2}k_2) \qquad\qquad (11.7)$$

$$k_4 = hf(x_0 + h, y_0 + k_3) \qquad\qquad (11.8)$$

The value of y at $x_0 + h$ is then given by

$$y = y_0 + \tfrac{1}{6}(k_1 + 2k_2 + 2k_3 + k_4) \qquad\qquad (11.9)$$

Further modifications, possibly involving the evaluation of a fifth gradient, to estimate the residual error in equation (11.9) are often provided auto-matically by the library program. These details are usually specific to the computer program and will not be discussed here.

11.2.2 Predictor-Corrector Methods

As their name implies these methods use one formula to predict an approximate value for y at the end of the interval of integration, followed by a more accurate formula which improves or corrects this value by successive approximation. The simplest and obvious method of this type is to predict the terminal value by using the initial gradient, thus

$$y_p = y_0 + hf(x_0, y_0) \tag{11.10}$$

and correct by taking the average of the gradients at each end of the interval, viz.

$$y_o^{(n)} = y_0 + \tfrac{1}{2}h[f(x_0, y_0) + f(x_0 + h, y_o^{(n-1)})] \tag{11.11}$$

where $y_o^{(n)}$ is the nth approximation to the solution, y_p having been used for the first. The previous example will be repeated to illustrate the method.

Example 2. If $y = 1$ when $x = 1$, and the equation

$$\frac{dy}{dx} = \frac{x^2 - y}{x} \qquad\qquad \text{I}$$

is satisfied, find the value of y when $x = 2$.

$$\text{At} \quad x = 1, \quad y = 1 \quad \text{and} \quad \frac{dy}{dx} = f(1, 1) = 0 \qquad\qquad \text{II}$$

Equation (11.10) thus predicts

$$y^{(1)} = 1 \qquad\qquad \text{III}$$

The gradient at the end of the interval where $x = 2$ is

$$f(2, 1) = \frac{4 - 1}{2} = 1 \cdot 5 \qquad\qquad \text{IV}$$

Substitution into equation (11.11) gives

$$y^{(2)} = 1 + \tfrac{1}{2}(0 + 1 \cdot 5) = 1 \cdot 75 \qquad\qquad \text{V}$$

Repeated substitution gives

$$f(2, 1 \cdot 75) = \frac{4 - 1 \cdot 75}{2} = 1 \cdot 125$$

$$y^{(3)} = 1 + \tfrac{1}{2}(0 + 1 \cdot 125) = 1 \cdot 562$$

$$y^{(4)} = 1 + \frac{1}{2}\left(0 + \frac{4 - 1 \cdot 562}{2}\right) = 1 \cdot 6095$$

$$y^{(5)} = 1 + \frac{1}{2}\left(0 + \frac{4 - 1 \cdot 6095}{2}\right) = 1 \cdot 5976$$

$$y^{(6)} = 1 + \frac{1}{2}\left(0 + \frac{4 - 1 \cdot 5976}{2}\right) = 1 \cdot 6006$$

$$y^{(7)} = 1 + \frac{1}{2}\left(0 + \frac{4 - 1 \cdot 6006}{2}\right) = 1 \cdot 600 \qquad\qquad \text{VI}$$

The solution obtained is not very accurate but the gradient has changed rather rapidly from zero to $1 \cdot 200$ across the interval. A more accurate solution can be obtained by covering the interval in two steps instead of one by using $h = 0 \cdot 5$. The intermediate and final answers are:

$$\text{at } x = 1 \cdot 5, \quad y = 1 \cdot 1786, \quad f(1 \cdot 5, 1 \cdot 1786) = 0 \cdot 7143$$

and
$$\text{at } x = 2, \quad y = 1 \cdot 6508, \quad f(2, 1 \cdot 6508) = 1 \cdot 1745$$

These answers are more acceptable but a further subdivision of the interval is really needed.

At this point it is worth noting that a simpler method with a small step length gives similar accuracy to a more complex method with a larger step length but the total amount of work involved is normally not significantly changed. However, the more sophisticated methods are generally more reliable.

Equation (11.11) is the second order corrector in the more general series of correctors used in the Adams–Bashforth method. The general formula can be derived as follows from the backward finite difference methods described in Chapter 9.

Operating on equation (9.14) with E gives

$$E \nabla y_n = \Delta y_n \tag{11.12}$$

and using equation (9.27),

$$E \nabla = \Delta = E - 1$$

$$\therefore \quad E = (1 - \nabla)^{-1} = e^{hD} \tag{11.13}$$

Taking logarithms

$$hD = -\ln(1 - \nabla) \tag{11.14}$$

Applying these operators to y_{n+1} gives

$$h D y_{n+1} = -\ln(1 - \nabla) y_{n+1}$$

which can be rearranged to

$$y_{n+1} = \frac{-1}{\ln(1 - \nabla)} h D y_{n+1} \tag{11.15}$$

Taking the backward difference of both sides and expanding the logarithm in series,

$$\nabla y_{n+1} = \frac{\nabla}{\nabla + \frac{1}{2}\nabla^2 + \frac{1}{3}\nabla^3 \ldots} h D y_{n+1} \tag{11.16}$$

$$= \frac{1}{1 + (\frac{1}{2}\nabla + \frac{1}{3}\nabla^2 + \ldots)} h D y_{n+1}$$

Expanding the operator on the right-hand side by the binomial theorem gives

$$\nabla y_{n+1} = (1 - \tfrac{1}{2}\nabla - \tfrac{1}{12}\nabla^2 - \tfrac{1}{24}\nabla^3 - \tfrac{21}{720}\nabla^4 - \ldots) h D y_{n+1} \tag{11.17}$$

Writing the original differential equation (11.2) in the form

$$Dy_n = f_n(x_n, y_n) = f_n \qquad (11.18)$$

where

$$x_n = x_0 + nh \qquad (11.19)$$

allows equation (11.17) to be written in the form

$$y_{n+1} = y_n + h(1 - \tfrac{1}{2}\nabla - \tfrac{1}{12}\nabla^2 - \tfrac{1}{24}\nabla^3 - \tfrac{21}{720}\nabla^4 - ...)f_{n+1} \qquad (11.20)$$

Taking just the first two terms in the operator gives equation (11.11). Equation (11.20) is the Adams–Bashforth corrector. The corresponding predictor can be obtained by replacing $n+1$ with n in equation (11.17) then operating on both sides with E or its equivalent $(1 - \nabla)^{-1}$. The result is

$$y_{n+1} = y_n + h(1 + \tfrac{1}{2}\nabla + \tfrac{5}{12}\nabla^2 + \tfrac{3}{8}\nabla^3 + \tfrac{249}{720}\nabla^4 + ...)f_n \qquad (11.21)$$

Although the Adams–Bashforth method gives good results it suffers from many disadvantages. The coefficients of the higher order-differences do not decrease very rapidly so that errors due to truncation of the series are serious unless a small interval (h) is used, the method is implicit involving iterative substitutions until agreement is obtained for y_{n+1}, and many previous values of the derivatives must be available for evaluating the differences thus presenting problems on the first step of the integration. The method is therefore most useful when it can be truncated at about the second difference and used with a small step length. In most applications, if a computer is available, the Runge–Kutta subroutine should be used.

11.3. HIGHER ORDER DIFFERENTIAL EQUATIONS (INITIAL VALUE TYPE)

It is a simple matter to convert an nth order differential equation to n simultaneous first order equations. It is only necessary to define $(n-1)$ new variables by

$$\left. \begin{aligned} w_1 &= \frac{dy}{dx} \\[2mm] w_2 &= \frac{d^2 y}{dx^2} = \frac{dw_1}{dx} \\ &\;\;\vdots \\ w_{n-1} &= \frac{d^{n-1} y}{dx^{n-1}} = \frac{dw_{n-2}}{dx} \end{aligned} \right\} \qquad (11.22)$$

and remove all derivatives from the original differential equation except $(d^n y/dx^n)$ which is replaced by (dw_{n-1}/dx).

The solution of the set of equations (11.22) with the original differential equation is now very similar to the solution of a single first order equation, provided that the boundary conditions are of initial value type (Section 8.4.5). In this case, at some value $x = x_0$, values for $y, w_1, w_2 \ldots, w_{n-1}$, will be available and an extension of the Runge–Kutta process can be used. The third order formulae for the second order differential equation

$$\frac{d^2 y}{dx^2} = F\left(x, y, \frac{dy}{dx}\right) \qquad (11.23)$$

are as follows. Putting

$$\frac{dy}{dx} = w \qquad (11.24)$$

equation (11.23) becomes

$$\frac{dw}{dx} = F(x, y, w) \qquad (11.25)$$

From the initial values (x_0, y_0, w_0) and the interval (h) in x, k_1 and K_1 can be defined as before. Thus

$$k_1 = hw_0 \qquad (11.26)$$
$$K_1 = hF(x_0, y_0, w_0) \qquad (11.27)$$

where k_1 is the first approximation to the change in y and K_1 is the corresponding change in w. Proceeding in a similar manner to the first order case,

$$k_2 = h(w_0 + \tfrac{1}{2}K_1) \qquad (11.28)$$
$$K_2 = hF(x_0 + \tfrac{1}{2}h, y_0 + \tfrac{1}{2}k_1, w_0 + \tfrac{1}{2}K_1) \qquad (11.29)$$
$$k_3 = h(w_0 + 2K_2 - K_1) \qquad (11.30)$$
$$K_3 = hF(x_0 + h, y_0 + 2k_2 - k_1, w_0 + 2K_2 - K_1) \qquad (11.31)$$

The values of y and w at the end of the interval are given by

$$y = y_0 + \tfrac{1}{6}(k_1 + 4k_2 + k_3) \qquad (11.32)$$
$$w = w_0 + \tfrac{1}{6}(K_1 + 4K_2 + K_3) \qquad (11.33)$$

Example 1. Given the differential equation

$$8\frac{d^2 y}{dx^2} - x^2\frac{dy}{dx} + 2y^2 = 0$$

and the initial conditions that at $x = 0$, $y = 1$ and $dy/dx = 0$, find the value of y when $x = 1$.

Solution

Putting

$$\frac{dy}{dx} = w \qquad \qquad \text{I}$$

$$\therefore \quad \frac{dw}{dx} = \frac{x^2 w - 2y^2}{8} \qquad \qquad \text{II}$$

Since $h = 1$, $x_0 = 0$, $y_0 = 1$, $w_0 = 0$, equations (11.26) to (11.31) can be used in succession. Thus

$$k_1 = 0 \qquad \text{III}$$

$$K_1 = 1\left(\frac{0-2}{8}\right) = -0.2500 \qquad \text{IV}$$

$$k_2 = 1(0 - 0.1250) = -0.1250 \qquad \text{V}$$

$$K_2 = 1\left[\frac{(0.5)^2(-0.125) - 2(1)^2}{8}\right]$$

$$= -0.25391 \qquad \text{VI}$$

$$2k_2 - k_1 = -0.2500 \qquad \text{VII}$$

$$2K_2 - K_1 = -0.25782 \qquad \text{VIII}$$

$$k_3 = -0.25782 \qquad \text{IX}$$

$$K_3 = 1\left[\frac{(1)^2(-0.25782) - 2(0.7500)^2}{8}\right]$$

$$= -0.17285 \qquad \text{X}$$

Equation (11.32) now gives the solution

$$y = 1 - \tfrac{1}{8}(0 + 0.500000 + 0.25782)$$

$$\therefore \quad y = 0.8737 \qquad \text{XI}$$

There is a large relative variation between k_1, k_2, and k_3 in this example and the above solution may not be very accurate. To check this, the integration can be performed in two stages by choosing $h = \tfrac{1}{2}$ and applying the formulae (11.26) to (11.33) twice. The intermediate and final results are

$$\text{at} \quad x = 0.5, \quad y = 0.9687, \quad w = -0.1230$$

and $\qquad\qquad$ at $\quad x = 1.0, \quad y = 0.8779 \qquad \text{XII}$

The answers given by equations XI and XII agree to better than 1%, and the solution can be written

$$y = 0.878$$

with reasonable certainty that the third significant figure is within two units of the correct solution.

11.4. HIGHER ORDER DIFFERENTIAL EQUATIONS
(BOUNDARY VALUE TYPE)

With second and higher order differential equations, the boundary conditions may be specified at two different values of the independent variable. In such cases, there is insufficient data available at $x = x_0$ for the above method to be used directly; however, a trial and error method is practicable for second order equations. Thus, consider the equation

$$\frac{d^2 y}{dx^2} = F\left(x, y, \frac{dy}{dx}\right) \tag{11.23}$$

again, with the boundary conditions:

$$\text{at} \quad x = x_0, \quad y = y_0 \tag{11.34}$$

and
$$\text{at} \quad x = x_N, \quad y = y_N \tag{11.35}$$

where it has been assumed that N integration steps are necessary to cover the distance between the boundaries. If a value is assumed for $y'(x_0)$, equation (11.23) can be solved as an initial value problem, thus generating a value for y at $x = x_N$. This value is unlikely to equal y_N, therefore a second choice is made for $y'(x_0)$, resulting in a second value for y at $x = x_N$. The value of y at $x = x_N$ can thus be evaluated as a function of $y'(x_0)$, and the correct value of $y'(x_0)$ can be found by using an interpolation formula from Chapter 10. The value of $y'(x_0)$ obtained in this way can then be checked by a final solution of an initial value problem.

If this method is adopted, the early trials are usually made with a small value of N and widely spaced values of $y'(x_0)$. As the solution is refined, the value of N is increased and the assumed values of $y'(x_0)$ are contained within a narrower band. The calculated internal point values then give the shape of the solution curve (y vs x).

Example 1. In Example 4, Section 2.4.3, the problem of cooling a graphite electrode was considered, and the problem was stated in mathematical terms as follows. Find the rate of flow of heat into the water cooler at $x = 30$ cm when the temperature distribution satisfies

$$(k_0 - \alpha T)\frac{d^2 T}{dx^2} - \alpha\left(\frac{dT}{dx}\right)^2 - \beta(T - T_0) = 0 \qquad \text{I}$$

where $T_0 = 20$, $k_0 = 152 \cdot 6$, $\alpha = 0 \cdot 056$, and $\beta = 45 \cdot 3$.
The boundary conditions are

$$\text{at} \quad x = 0, \quad T = 1500 \qquad \text{II}$$

$$\text{at} \quad x = 0 \cdot 3, \quad T = 150 \qquad \text{III}$$

and the rate of flow of heat into the cooler is given by

$$H = -2 \cdot 55 \frac{dT}{dx}\bigg|_{x=0 \cdot 3} \qquad \text{IV}$$

Solution

Since dT/dx is required at $x = 0.3$, it is convenient to change the independent variable by putting

$$z = 0.3 - x \qquad\qquad \text{V}$$

$$\therefore \quad (k_0 - \alpha T)\frac{d^2 T}{dz^2} - \alpha\left(\frac{dT}{dz}\right)^2 - \beta(T - T_0) = 0 \qquad\qquad \text{VI}$$

and the boundary conditions become,

$$\text{at} \quad z = 0, \quad T = 150 \qquad\qquad \text{VII}$$

and

$$\text{at} \quad z = 0.3, \quad T = 1500 \qquad\qquad \text{VIII}$$

From equations VII and VIII the average temperature gradient is 4500, but the gradient at $z = 0$ should be smaller than this because less heat reaches the water cooler than leaves the furnace, and the conductivity is higher at the lower temperature. Therefore, as a first approximation, put

$$T' = 3500 \quad \text{at} \quad z = 0 \qquad\qquad \text{IX}$$

Putting

$$\frac{dT}{dz} = w \qquad\qquad \text{X}$$

equation VI becomes

$$\frac{dw}{dz} = \frac{\alpha w^2 + \beta(T - T_0)}{k_0 - \alpha T} \qquad\qquad \text{XI}$$

Taking one interval and using the method of Section 11.3, the following values can be calculated:

$$k_1 = 1050 \qquad K_1 = 1439$$
$$k_2 = 1266 \qquad K_2 = 2682$$
$$2k_2 - k_1 = 1482 \quad 2K_2 - K_1 = 3925$$
$$k_3 = 2227 \qquad K_3 = 15\,492$$

$$\therefore \quad T = 150 + \tfrac{1}{6}(1050 + 5064 + 2227) = 1540 \qquad\qquad \text{XII}$$

$$\therefore \quad w = 3500 + \tfrac{1}{6}(1439 + 10\,728 + 15\,492) = 8110 \qquad\qquad \text{XIII}$$

The answer given by equation XIII is suspect because K_3 is so different from K_1 and K_2. It appears that $T'(0)$ has been chosen 3% too large by comparing equations VIII and XII. Therefore $T'(0)$ is reduced to 3400 for the second trial, and the interval is halved to reduce the discrepancy between K_3 and K_2.

The calculation for the first interval gives

$$k_1 = 510 \qquad K_1 = 679.5$$
$$k_2 = 561 \qquad K_2 = 924.6$$
$$k_3 = 685.5 \qquad K_3 = 1642.2$$

$$\therefore \quad T(0.15) = 150 + \tfrac{1}{6}(510 + 2244 + 685) = 723 \qquad\qquad \text{XIV}$$

$$\therefore \quad w(0.15) = 3400 + \tfrac{1}{6}(679.5 + 3698.4 + 1642.2) = 4403 \qquad\qquad \text{XV}$$

For the second interval,

$$k_1 = \quad 660 \cdot 5 \quad K_1 = 1495$$

$$k_2 = \quad 772 \cdot 6 \quad K_2 = 2455$$

$$k_3 = 1172 \cdot 7 \quad K_3 = 8374$$

$$\therefore \quad T(0 \cdot 3) = 723 + \tfrac{1}{6}(660 \cdot 5 + 3090 \cdot 4 + 1172 \cdot 7) = 1544 \qquad \text{XVI}$$

$$\therefore \quad w(0 \cdot 3) = 4403 + \tfrac{1}{6}(1495 + 9820 + 8374) = 7684 \qquad \text{XVII}$$

Equation XVI shows that the temperature drop across the electrode is now 1394 °K compared with the true value 1350 °K. Hence the assumed value $T'(0) = 3400$ is probably still $3\tfrac{1}{4}\%$ too high. Therefore, putting $dT/dx = -3290$ at $x = 0 \cdot 3$ into equation IV gives the result

$$H = 8390 \text{ W} \qquad \text{XVIII}$$

This answer is considered to be sufficiently accurate for the purpose of the problem. It could be improved further by taking more intervals and choosing $T'(0)$ more accurately, but this is not justified by the accuracy of the original data.

11.4.1. *Finite Difference Methods*

An alternative method of solving higher order ordinary differential equations of boundary value type is to replace the differential equation with an equivalent finite difference equation. This technique is most useful for second order equations, or higher order equations replaced by simultaneous second order equations.

There are many bases for the method, but the derivation from Taylor's theorem is probably the easiest to follow. The closed range of the independent variable $a \leqslant x \leqslant b$ is divided into equal intervals and the points are labelled x_n, where $x_0 = a$ and $x_N = b$. The dependent variable (y) can be expressed in the neighbourhood of any point (x_n) in terms of x and its derivatives at x_n by means of Taylor's theorem (Section 3.3.6), and if h denotes the increment of x between neighbouring points,

$$y_{n+1} = f(x_{n+1}) = f(x_n) + hf'(x_n) + \tfrac{1}{2}h^2 f''(x_n) + \dots \qquad (11.36)$$

$$y_n = f(x_n) \qquad (11.37)$$

$$y_{n-1} = f(x_{n-1}) = f(x_n) - hf'(x_n) + \tfrac{1}{2}h^2 f''(x_n) - \dots \qquad (11.38)$$

If h is chosen sufficiently small, terms in h^3 and higher degree can be neglected in equations (11.36) and (11.38). These three equations relate neighbouring point values to the derivatives at the central point (x_n); and they can be solved to determine the derivatives in terms of the point values. Thus

$$f(x_n) = y_n \tag{11.39}$$

$$f'(x_n) = \frac{y_{n+1} - y_{n-1}}{2h} \tag{11.40}$$

$$f''(x_n) = \frac{y_{n+1} - 2y_n + y_{n-1}}{h^2} = \frac{\Delta^2 y_{n-1}}{h^2} \tag{11.41}$$

These equations allow any second order differential equation to be replaced by an equivalent second order difference equation.

The calculation details and treatment of boundary conditions are more easily explained with reference to a particular example.

Example 2. The cylindrical combustion chamber of a rocket motor has coolant ducts of lenticular cross-section at frequent evenly spaced intervals within the cylindrical wall. The dimensions of a typical duct are illustrated in Fig. 11.2. Radiant heat is received uniformly by the internal surface of

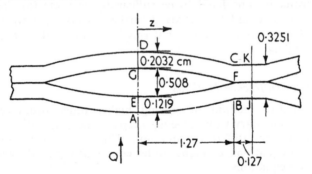

FIG. 11.2. Rocket motor cooling duct

the combustion chamber at a rate $Q = 57$ cal/s cm² and this heat is conducted into the liquid coolant at a bulk temperature $T_B = 130\,°C$ according to a variable heat transfer coefficient h cal/s cm² °C. h can be determined from the Dittus–Boelter equation expressed for gases as $h = \alpha u^{0.8}$, where u is the local centre line velocity and $\alpha = 0.00626$ for the coolant used. Spikins[†] has determined the velocity distribution as

$$u = U_0 \left[1 - \left(\frac{z}{1.27} \right)^3 \right]$$

where z is the coordinate measured as shown in Fig. 11.2, and U_0, the axial velocity, has a value 100 cm/s. If the metal conductivity $k = 0.107$ cal/ s cm² °C cm⁻¹, find the temperature distribution within the chamber wall.

† Spikins, D. J., Ph.D. Thesis, Chemical Engineering Department, University of Birmingham, 1958.

Solution

The shape of the duct is rather complicated, so it was idealized as shown in Fig. 11.3. There are lines of symmetry at AE, DG, and JK, and a typical section has been straightened out by distorting the junction CKJBF. In the idealized situation, radiant heat enters through AB and BC which is chosen to equal BJ, negligible heat passes through the external surface CD, and the coolant flows over EF and F′G. All areas and heat paths are correct in this model, except for the curved path from BF to CF which is only approximately correct. Neglecting the temperature drop through the metal wall with respect to the temperature drop within the coolant film reduces the problem to the determination of the metal temperature (T) as a function of a single coordinate (x).

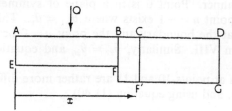

FIG. 11.3. Idealized metal section

Taking a heat balance over an element of metal in the section ABFE gives

$$\frac{d^2 T}{dx^2} - \frac{h}{ks}(T - T_B) + \frac{Q}{ks} = 0 \qquad\qquad \text{I}$$

or inserting the data

$$\frac{d^2 T}{dx^2} - \frac{0.00626 \times 100^{0.8}}{0.107 \times 0.1219}\left[1 - \left(\frac{x}{1.27}\right)^3\right]^{0.8}(T - T_B) + \frac{57}{0.107 \times 0.1219} = 0 \qquad \text{II}$$

To simplify the arithmetic, it is convenient to put

$$x = 1.27X \qquad\qquad \text{III}$$

and

$$\theta = T - T_B \qquad\qquad \text{IV}$$

$$\therefore \quad \frac{d^2 \theta}{dX^2} - 30.8(1 - X^3)^{0.8}\,\theta + 7048 = 0 \qquad\qquad \text{V}$$

Similarly, in the section BCF′F, where no coolant flows but heat is still absorbed in a thicker section,

$$\frac{d^2 \theta}{dX^2} + 4228 = 0 \qquad\qquad \text{VI}$$

Finally, in the section $1.1 \leqslant X \leqslant 2.1$,

$$\frac{d^2 \theta}{dX^2} - 18.5[1 - (2.1 - X)^3]^{0.8}\,\theta = 0 \qquad\qquad \text{VII}$$

In the finite difference representation, AB and CD are each divided into ten equal intervals, BC is one interval, and the point temperatures are labelled θ_n where $n = 10X$. The finite difference form of equation V can be found from equations (11.39), (11.40), (11.41) as

$$\theta_{n+1} - [2 + 0 \cdot 308(1 - X_n{}^3)^{0 \cdot 8}]\,\theta_n + \theta_{n-1} + 70 \cdot 5 = 0 \qquad \text{VIII}$$

which is valid for $1 \leqslant n \leqslant 9$. The finite difference form of equation VII is

$$\theta_{n+1} - \{2 + 0 \cdot 185[1 - (2 \cdot 1 - X_n)^3]^{0 \cdot 8}\}\,\theta_n + \theta_{n-1} = 0 \qquad \text{IX}$$

which is valid for $12 \leqslant n \leqslant 20$.

The four points 0, 10, 11, 21 are influenced by the boundary conditions in the following manner. Point 0 is in a plane of symmetry, and it can be assumed that a point $n = -1$ exists where $\theta_{-1} = \theta_1$. This ensures that the gradient is zero at the boundary and the point $n = 0$ becomes an ordinary point of equation VIII. Similarly, $\theta_{22} = \theta_{20}$ and equation IX is valid for $n = 21$.

The conditions at points 10 and 11 are rather more difficult but returning to first principles, and using equation (11.40),

$$\frac{\theta_{10} - \theta_9}{1 \cdot 27 \delta X} \quad \text{is the temperature gradient at } n = 9\tfrac{1}{2}$$

$$\frac{\theta_{11} - \theta_{10}}{1 \cdot 27 \delta X} \quad \text{is the temperature gradient at } n = 10\tfrac{1}{2}$$

Taking a heat balance over an element of width $\delta X = 0 \cdot 1$ astride point 10 as illustrated in Fig. 11.4 gives

$$\frac{-0 \cdot 1219k}{1 \cdot 27 \delta X}(\theta_{10} - \theta_9) + 1 \cdot 27 Q \delta X = \frac{-0 \cdot 2032k}{1 \cdot 27 \delta X}(\theta_{11} - \theta_{10})$$

$$+ \tfrac{1}{4}(\theta_{10} + \theta_9)\,1 \cdot 27 \delta X h|_{X=0 \cdot 95} \qquad \text{X}$$

which simplifies to

$$0 \cdot 3666\theta_9 - \theta_{10} + 0 \cdot 6213\theta_{11} + 26 \cdot 3 = 0 \qquad \text{XI}$$

A similar balance at point 11 gives

$$0 \cdot 4976\theta_{10} - \theta_{11} + 0 \cdot 4927\theta_{12} + 10 \cdot 5 = 0 \qquad \text{XII}$$

There are thus ten equations of type VIII, ten of type IX, plus equations XI and XII which are to be solved for θ_n where $0 \leqslant n \leqslant 21$. The simplest method of solution is to estimate values of θ_n for odd values of n and solve the equations directly for θ_n at even values of n. From the calculated even values, improved odd values can be calculated. By repetition, the solution can be obtained as accurately as required.

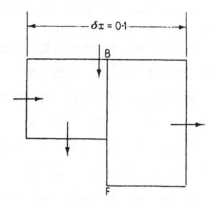

FIG. 11.4. Heat balance at BF

Equations VIII and IX are first solved for θ_n and the coefficients of θ_{n+1}, θ_{n-1} evaluated at each point as shown in Table 11.1.

TABLE 11.1. *Coefficients in the Difference Equations*

n	Coefficient of		plus	n	Coefficient of		plus
	θ_{n-1}	θ_{n+1}			θ_{n-1}	θ_{n+1}	
0	—	0·8665	30·5	1	0·4333	0·4333	30·5
2	0·4336	0·4336	30·6	3	0·4345	0·4345	30·6
4	0·4363	0·4363	30·8	5	0·4392	0·4392	31·0
6	0·4438	0·4438	31·3	7	0·4504	0·4504	31·8
8	0·4601	0·4601	32·4	9	0·4743	0·4743	33·4
10	0·3666	0·6213	26·3	11	0·4976	0·4927	10·5
12	0·4842	0·4842	—	13	0·4752	0·4752	—
14	0·4690	0·4690	—	15	0·4646	0·4646	—
16	0·4616	0·4616	—	17	0·4597	0·4597	—
18	0·4585	0·4585	—	19	0·4579	0·4579	—
20	0·4577	0·4577	—	21	0·9154	—	—

To start the calculation shown in Table 11.2, values must be estimated for θ_n with n odd. Neglecting the change of heat transfer coefficient with velocity, and assuming that θ_n is constant for $n \leqslant 10$ gives $\theta_n = 227$ by equation VIII; hence $\theta_n = 250$ will be a fair approximation for odd values of n between 0 and 10. An approximate solution of equation VII by the methods of Chapter 2 after neglecting the term $(2 \cdot 1 - X)^3$ gives

$$\theta = 250 \exp\left[-4 \cdot 25(X - 1 \cdot 1)\right] \qquad \text{XIII}$$

Evaluating equation XIII at the chosen points completes the estimate of the temperature distribution.

TABLE 11.2. *Temperature Distribution Around a Cooling Duct*

0	1	2	3	4	5	6	7	8	9	10	11	12	13	14	15	16	17	18	19	20	21
	250		250		250		250		250		250		107		46		20		8		4
247	245	247	246	249	251	253	264	262	287	273	232	173	116	72	47	30	20	13	8	5	5
243	241	243	244	248	254	260	278	286	300	276	231	169	116	76	50	31	20	13	9	6	5
239	238	241	243	248	257	267	286	298	308	280	233	168	117	78	51	32	21	13	9	6	5
237	237	239	243	249	260	272	292	306	313	284	235	169	118	79	52	33	22	14	9	6	5
236	236	239	243	250	262	276	296	311	317	287	238	171	119	80	53	34	22	14	9	6	5
235	235	238	243	251	264	279	299	314	320	290	240	173	121	81	54	35	23	14	9	6	5
	233		244		268		305		326		244		125		56		24		9		5
232	234	237	244	254	268	286	306	323	327	297	246	179	125	85	57	37	24	15	10	6	5
233	235	238	244	254	268	286	307	324	329	299	248	180	126	85	57	37	24	16	11	7	6
	236		245		269		309		332		251		127		57		24		11		6
235	235	239	245	255	269	288	309	327	333	304	252	183	128	86	57	37	24	16	11	8	7
234		239	245	255	269	288	310	330	337	309	256	187	131	89	59	39	26	17	11	8	7

The equations given in the left-hand half of Table 11.1 can be used to calculate the values in the second row of Table 11.2. The calculation is thus started and alternate use of the left- and right-hand halves of Table 11.1 gives the solution. A quicker way to the solution is by making a fresh estimate at any stage, and this has been done on two rows of Table 11.2. The solution given in the last row was obtained from eight further rows of calculation.

This solution will be fairly accurate in the metal immediately surrounding the duct, but not very good in the weld section CKJBF. Taking a heat balance over an element within this section gives

$$\frac{d^2 T}{dz^2} + \frac{Q}{ks} = 0 \qquad\qquad \text{XIV}$$

or

$$\frac{d^2 T}{dz^2} + 1640 = 0 \qquad\qquad \text{XV}$$

Solving equation XV and remembering that $z = 1\cdot397$ is a plane of symmetry gives

$$T = A - 820(1\cdot397 - z)^2 \qquad\qquad \text{XVI}$$

But the temperature at point F $(z = 1\cdot270)$ is

$$T = T_B + \theta = 130 + \tfrac{1}{2}(309 + 256) = 412\tfrac{1}{2}$$

$$\therefore \quad A = 426 \qquad\qquad \text{XVII}$$

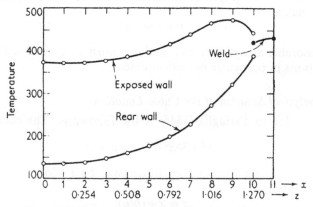

FIG. 11.5. Temperature distribution in cooling duct

These results have been plotted in Fig. 11.5, which shows that the hottest part of the duct is about 1·5 mm from the corner of the coolant channel along the heated wall, and not at the centre of the uncooled weld. Because the temperature drop across the metal wall has been neglected, the results presented should approximate to the average temperature within the metal at the point in question, but the exposed surface temperature could be about 35 °C higher than the figure given.

It is possible to modify the calculation procedure with advantage in more complicated problems, by working in terms of temperature changes instead of temperatures. To do this, the first row of estimated temperatures is subtracted from the first row of calculated temperatures at the same points. All subsequent calculations are then performed to give correction terms, and the final temperature at each point becomes the sum of the values in that column. The result obtained in this way should then be checked by one cycle of the basic calculation in terms of the temperatures.

11.5. ALGEBRAIC EQUATIONS

In many problems it is necessary to solve a quadratic equation, and this is readily accomplished by means of the formula:

Roots of
$$ax^2 + bx + c = 0 \tag{11.42}$$

are
$$x = \frac{-b \pm \sqrt{(b^2 - 4ac)}}{2a} \tag{11.43}$$

Less frequently, a cubic equation needs to be solved, and this can also be done analytically without too much difficulty. An analytical solution of a quartic equation is also feasible, but not very profitable in most practical cases, but polynomial equations of higher degree are insoluble analytically, hence numerical or graphical methods must be used. Transcendental equations such as

$$\tan x = kx \tag{11.44}$$

are also insoluble analytically and methods must be established for solving such equations in particular numerical cases.

11.5.1. *Analytical Solution of the Cubic Equation*

A method due to Tartaglia (1545) is the following. The cubic equation

$$ax^3 + 3bx^2 + 3cx + d = 0 \tag{11.45}$$

will have three roots, the sum of which will be $(-3b/a)$. Changing the variable by putting
$$z = x + (b/a) \tag{11.46}$$
gives the equation
$$z^3 + 3Hz + G = 0 \tag{11.47}$$

which has zero for the sum of its roots, where

$$a^2 H = ac - b^2$$

and
$$a^3 G = a^2 d - 3abc + 2b^3$$

To solve equation (11.47), a further substitution is made. That is,

$$z = u+v \qquad (11.48)$$

where v is a function of u which is still to be determined. Hence

$$u^3+v^3+3(uv+H)(u+v)+G = 0 \qquad (11.49)$$

Choosing $v = -H/u$ will simplify equation (11.49) to

$$u^6-H^3+Gu^3 = 0 \qquad (11.50)$$

Equation (11.50) is a quadratic in u^3 which has the solution

$$u^3 = -\tfrac{1}{2}G \pm \tfrac{1}{2}\sqrt{(G^2+4H^3)} \qquad (11.51)$$

Equation (11.51) determines six values for u and since $uv = -H$, six values of v are determined. Hence equation (11.48) appears to give six roots to the cubic (11.47). However, there are only three distinct values for z because each pair of values for u and v occurs twice, once as $u+v$ and again as $v+u$. Equation (11.46) transforms the values of z into the corresponding values of x which satisfy equation (11.45).

Example 1. Solve the equation

$$x^3-4x^2+7x-5 = 0 \qquad \text{I}$$

Removing the second term by putting

$$z = x-1\tfrac{1}{3} \qquad \text{II}$$

gives

$$27z^3+45z-11 = 0 \qquad \text{III}$$

Putting

$$z = u+v \qquad \text{IV}$$

$$\therefore \quad 27u^3+27v^3+(81uv+45)(u+v)-11 = 0$$

Assuming that

$$81uv+45 = 0$$

$$\therefore \quad v = \frac{-5}{9u} \qquad \text{V}$$

$$\therefore \quad 729u^6-297u^3-125 = 0 \qquad \text{VI}$$

The solution of equation VI is obtained from (11.43) as

$$u^3 = \frac{297 \pm \sqrt{(88\ 209+364\ 500)}}{1458}$$

$$= 0{\cdot}6652$$

$$\therefore \quad u = 0{\cdot}8729 \qquad \text{VII}$$

From equation V

$$v = -0{\cdot}6364 \qquad \text{VIII}$$

Hence, from equations IV and II,

$$x = 1\cdot5698 \qquad\qquad \text{IX}$$

is one root of the cubic equation I.

The simplest way of finding the other roots is to divide equation I by $(x - 1\cdot5698)$ algebraically to obtain

$$x^2 - 2\cdot4302x + 3\cdot1851 = 0 \qquad\qquad \text{X}$$

which has the roots

$$x = 1\cdot2151 \pm 2\cdot6143i \qquad\qquad \text{XI}$$

In general, if the cubic equation (11.45) has three real roots, $G^2 + 4H^3 < 0$ and equation (11.50) has complex roots. The cube root of equation (11.51) is best evaluated in the Argand diagram in this case. When the cubic has only one real root, as in Example 1, the quadratic (11.50) will always have real roots. Hence the method is most useful for finding a single real root of a cubic equation.

11.5.2. *Graphical Location of Roots*

In all numerical methods it is useful to know an approximate answer to the problem; this can then be successively improved. This was the basic technique in Example 2 of Section 11.4.1, where a very approximate temperature distribution was assumed. It can be seen that a very good first approximation will yield an accurate solution quickly, but even a poor first approximation will not extend the subsequent calculation unduly. Graphical methods are very useful for determining a first approximation. If the graph is drawn accurately, the graphical solution may be acceptable as the final solution; but even a rough sketch will give the approximate location of the root of an equation. Thus consider the following example.

Example 2. Find the approximate location of the roots of the equation

$$\tan x = kx \qquad\qquad \text{I}$$

where k is a constant.

Equation I can be resolved into two equations

$$y = \tan x \qquad\qquad \text{II}$$

and

$$y = kx \qquad\qquad \text{III}$$

which must be satisfied simultaneously. Taking a set of values for x, y can be evaluated from each of equations II and III, and both equations can be plotted graphically as shown in Fig. 11.6.

The principal branch of $y = \tan x$ has asymptotes at $x = \pm\frac{1}{2}\pi$ and passes through the origin. This shape is then repeated indefinitely at intervals of π along the x axis. Equation III is just the equation of a straight line and the solutions of equation I are the points of intersection of the two curves. The gradient of equation II at the origin is unity and therefore two possibilities arise. If $k = k_1 < 1$, there is only one root on the principal branch, but if $k = k_2 > 1$, there are three roots on the principal branch. Altogether there are an infinite number of roots which occur approximately at odd multiples of $\frac{1}{2}\pi$. As already mentioned, the first few roots near the origin are a fair distance from the asymptotes and there may be one or three roots near the origin, depending upon the value of k.

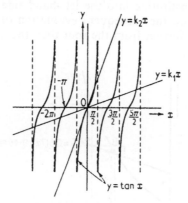

FIG. 11.6. Graphical solution of $\tan x = kx$

11.5.3. *Improvement of Roots by Successive Approximation*

The equation to be solved can always be rearranged, as in the above example, to give a simple function on at least one side of the equation. After a graphical investigation to find an approximate root, the value obtained can be substituted into one side of the equation, and a second value obtained from the other side of the equation. This process of substituting into one side and estimating a new value from the other can be repeated until two consecutive results agree. Only one side of the equation can be used for the evaluation step because the one side will give a convergent solution whereas the other side gives a divergent solution. The method is illustrated by reference to the problem arising in Section 6.9.3, equation VII.

Example 3. Find the root of equation

$$15\alpha^3 - 68\alpha^2 + 95\alpha - 47 = 0 \qquad\qquad \text{I}$$

which is close to $\alpha = 2.5$.

It is always useful when solving cubic equations to eliminate the second term by using the substitution (11.46). Thus putting

$$z = \alpha - \tfrac{68}{45} = \alpha - 1.5111 \qquad\qquad \text{II}$$

into equation I and rearranging,

$$z^3 = 0.5170z + 0.4641 \qquad\qquad \text{III}$$

Each side of equation III has been plotted in Fig. 11.7 to show the location of the root. The mechanism of the numerical method is also illustrated by the stepped path ABCD. Starting from the approximate value $z = z_1$, the right-hand side of equation III is evaluated to give the ordinate of A. The point where the curve $y = z^3$ has the same ordinate value at B determines a new approximation for the root ($z = z_2$). Repeating the process gives the points C and D and a closer approximation to the root. It is obvious from the diagram that substituting into the left-hand side of equation III would give the point A', and the subsequent evaluation of z from the right-hand side of III would be further from the root than the original estimate.

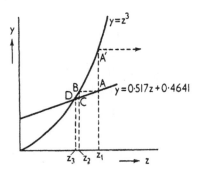

FIG. 11.7. Solution of a cubic equation

The approximate root $\alpha = 2 \cdot 5$ corresponds to $z = 1 \cdot 0$ from equation II. Hence the calculation proceeds,

$$0 \cdot 5170 z_1 + 0 \cdot 4641 = 0 \cdot 9811 = z_2{}^3$$

$$\therefore \quad z_2 = 0 \cdot 9937$$

$$0 \cdot 5170 z_2 + 0 \cdot 4641 = 0 \cdot 9778 = z_3{}^3$$

$$\therefore \quad z_3 = 0 \cdot 9926$$

$$0 \cdot 5170 z_3 + 0 \cdot 4641 = 0 \cdot 9773 = z_4{}^3$$

$$\therefore \quad z_4 = 0 \cdot 9924$$

$$0 \cdot 5170 z_4 + 0 \cdot 4641 = 0 \cdot 97717 = z_5{}^3$$

$$\therefore \quad z_5 = 0 \cdot 99233$$

$$0 \cdot 5170 z_5 + 0 \cdot 4641 = 0 \cdot 97713 = z_6{}^3$$

$$\therefore \quad z_6 = 0 \cdot 99232$$

Since z_5 and z_6 are approximately equal, z_6 is close enough to the true root. Hence, by equation II,

$$\alpha = 2 \cdot 5034$$

The determination of which side of an equation to substitute into can always be found by rearranging the equation into the standard form

$$x_{n+1} = F(x_n) \tag{11.52}$$

where the known value (x_n) is substituted into the function on the right-hand side to determine the new value (x_{n+1}). If the process is convergent to the root $x = x^*$,

$$x^* = F(x^*) \tag{11.53}$$

Expanding the right-hand side about the point $x = x_n$ by Taylor's theorem, and neglecting second derivative terms and above,

$$x^* = F(x_n) + (x^* - x_n) F'(x_n) \tag{11.54}$$

Using equation (11.52) and rearranging,

$$x^* - x_{n+1} = (x^* - x_n) F'(x_n) \tag{11.55}$$

Hence the new approximation (x_{n+1}) will be closer to the true root (x^*) if

$$|F'(x_n)| < 1 \tag{11.56}$$

Returning to the above example, it was seen that the convergent process was represented by

$$z_{n+1} = (0 \cdot 5170 z_n + 0 \cdot 4641)^{\frac{1}{3}} = F(z_n)$$

$$\therefore \quad F'(z_n) = \tfrac{1}{3} \times 0 \cdot 5170 \times (0 \cdot 5170 z_n + 0 \cdot 4641)^{-\frac{2}{3}}$$

Since $z_n \simeq 1$, $F'(z_n) \simeq 0 \cdot 18$ and the method is proved to be convergent.

However, the reverse substitution process is represented by

$$z_{n+1} = (z_n{}^3 - 0 \cdot 4641)/0 \cdot 5170 = F(z_n)$$

$$\therefore \quad F'(z_n) = 3 z_n{}^2 / 0 \cdot 5170$$

Since $z_n \simeq 1$, $F'(z_n) \simeq 6$ and the method is proved to be divergent.

11.5.4. *Newton's Method*

In this method, Taylor's theorem is used directly to improve an approximate root. If a root of the equation

$$f(x) = 0 \tag{11.57}$$

is required, and $x = x_1$ is an approximate root, the function in equation (11.57) can be expanded in the neighbourhood of x_1, thus

$$f(x_1 + \delta x) = f(x_1) + \delta x f'(x_1) \tag{11.58}$$

where the higher order terms have been neglected. If $x_2 = x_1 + \delta x$ is to be a better approximation to the solution of equation (11.57) then (11.58) gives

$$f(x_1) + (x_2 - x_1) f'(x_1) = 0$$

or

$$x_2 = x_1 - f(x_1)/f'(x_1) \tag{11.59}$$

Further approximations can be found by repeated application of equation (11.59). Thus

$$x_3 = x_2 - f(x_2)/f'(x_2), \quad \text{etc.}$$

This method can be employed to solve non-linear simultaneous equations and the following is an example of such use.

Example 4. In a study of the settling of suspensions of spherical particles, Oliver† has suggested the correlation

$$V_s = V_0(1 - kc)(1 - Kc^{\frac{1}{3}}) \qquad\qquad \text{I}$$

where V_s is the settling velocity of the suspension,
$\quad V_0$ is the Stokes' velocity of a single particle,
$\quad c$ is the volume concentration of the solids,
$\quad k, K$ are empirical constants.
From the following data given in the paper, find the best values for k, K, and V_0 by the method of least squares.

c	V_s (cm/s)	c	V_s (cm/s)
0	0·0943	0·10	0·0492
0·00333	0·0815	0·15	0·0397
0·00666	0·0761	0·20	0·0316
0·010	0·0740	0·25	0·0234
0·020	0·0688	0·30	0·0169
0·050	0·0612	0·35	0·0125

Solution

Rearranging equation I and introducing an error term to accommodate the experimental error,

$$V_s - V_0 + V_0 \alpha c^{\frac{1}{3}} + \beta c - \alpha\beta c^{\frac{4}{3}} = R \qquad\qquad \text{II}$$

where $\qquad\qquad \alpha = K \quad \text{and} \quad \beta = V_0 k \qquad\qquad \text{III}$

Following the method described in Section 10.6.2 by squaring equation II and equating to zero the partial derivatives of R^2 with respect to V_0, α, and β gives

$$V_0(N + \alpha^2 \sum c^{2/3} - 2\alpha \sum c^{1/3})$$
$$= \sum V_s - \alpha \sum V_s c^{1/3} + \beta \sum c - 2\alpha\beta \sum c^{4/3} + \alpha^2 \beta \sum c^{5/3} \qquad \text{IV}$$

$$\alpha(2\beta V_0 \sum c^{5/3} - \beta^2 \sum c^{8/3} - V_0^2 \sum c^{2/3})$$
$$= V_0 \sum V_s c^{1/3} - \beta \sum V_s c^{4/3} - V_0^2 \sum c^{1/3} + 2\beta V_0 \sum c^{4/3} - \beta^2 \sum c^{7/3} \qquad \text{V}$$

$$\beta(\sum c^2 + \alpha^2 \sum c^{8/3} - 2\alpha \sum c^{7/3})$$
$$= \alpha \sum V_s c^{4/3} + V_0 \sum c - 2\alpha V_0 \sum c^{4/3} + \alpha^2 V_0 \sum c^{5/3} - \sum V_s c \qquad \text{VI}$$

† Oliver, D. R. *Chem. Eng. Sci.* **15**, 230, 1961.

where $N = 12$ is the number of experimental points, and each sum is taken over all points. These sums, which are evaluated in Table 11.3, can be substituted into equations IV, V, and VI to give:

$$V_0(12 + 2 \cdot 4949\alpha^2 - 9 \cdot 5544\alpha)$$

$$= 62 \cdot 92 - 18 \cdot 093\alpha + 1 \cdot 440\beta - 1 \cdot 7516\alpha\beta + 0 \cdot 5488\alpha^2\beta \qquad \text{VII}$$

$$\alpha(1 \cdot 0976V_0\beta - 0 \cdot 14854\beta^2 - 2 \cdot 4949V_0^2)$$

$$= 18 \cdot 093V_0 - 2 \cdot 1103\beta - 4 \cdot 7772V_0^2 + 1 \cdot 7516V_0\beta - 0 \cdot 2270\beta^2 \qquad \text{VIII}$$

$$\beta(0 \cdot 35055 + 0 \cdot 14854\alpha^2 - 0 \cdot 4540\alpha)$$

$$= 2 \cdot 1103\alpha - 3 \cdot 8445 + 1 \cdot 440V_0 - 1 \cdot 7516\alpha V_0 + 0 \cdot 5488\alpha^2 V_0 \qquad \text{IX}$$

which are to be solved for α, β, and V_0.

The results given in the paper were obtained graphically, and they are

$$\alpha = 0 \cdot 75, \quad \beta = 18 \cdot 98, \quad V_0 = 9 \cdot 35 \qquad \text{X}$$

Using the method of Section 11.5.3 to improve these values by substituting α and β from equations X into equation VII to evaluate V_0, then using this value of V_0 with β from equation X in equation VIII to evaluate α etc., the following results were obtained.

$$\alpha = 0 \cdot 747, \quad \beta = 18 \cdot 67, \quad V_0 = 9 \cdot 21 \qquad \text{XI}$$

Unfortunately, there is no guarantee that the correct equation is being used for each determination and the results tended to oscillate in successive stages of calculation. Hence the values in equations XI can be checked by Newton's method as follows.

Each value is assumed to be in error by a small amount (x, y, or z). That is,

$$\alpha = 0 \cdot 747 + x, \quad \beta = 18 \cdot 67 + y, \quad V_0 = 9 \cdot 21 + z \qquad \text{XII}$$

The higher order functions can be linearized by neglecting products of the small quantities x, y, and z. Thus

$$\alpha^2 = 0 \cdot 55801 + 1 \cdot 494x$$
$$\beta^2 = 348 \cdot 57 + 37 \cdot 37y$$
$$V_0^2 = 84 \cdot 824 + 18 \cdot 42z$$
$$\alpha\beta = 13 \cdot 946 + 18 \cdot 67x + 0 \cdot 747y$$
$$\alpha V_0 = 6 \cdot 880 + 9 \cdot 21x + 0 \cdot 747z$$
$$\beta V_0 = 171 \cdot 95 + 9 \cdot 21y + 18 \cdot 67z$$
$$\alpha^2\beta = 10 \cdot 418 + 27 \cdot 89x + 0 \cdot 558y$$
$$\alpha^2 V_0 = 5 \cdot 1393 + 13 \cdot 76x + 0 \cdot 558z$$
$$\alpha\beta V_0 = 128 \cdot 45 + 172 \cdot 0x + 6 \cdot 880y + 13 \cdot 95z$$
$$\alpha\beta^2 = 260 \cdot 4 + 348 \cdot 6x + 27 \cdot 92y$$
$$\alpha V_0^2 = 63 \cdot 36 + 84 \cdot 82x + 13 \cdot 76z$$

TABLE 11.3. *Least Mean Squares Fit of Settling Data*

V_s^*	$c^{1/3}$	$c^{2/3}$	c	$c^{4/3}$	$c^{5/3}$	c^2	$c^{7/3}$	$c^{8/3}$	$V_s c^{1/3}$	$V_s c^{4/3}$	$V_s c$
9·43	0	0	0	0	0	0	0	0	0	0	0
8·15	0·1494	0·0223	0·0033	0·0005	0·0001	0·00001	0	0	1·218	0·0041	0·0272
7·61	0·1882	0·0354	0·0067	0·0013	0·0002	0·00004	0	0	1·432	0·0096	0·0507
7·40	0·2154	0·0464	0·0100	0·0022	0·0005	0·00010	0·00001	0·00001	1·594	0·0159	0·0740
6·88	0·2714	0·0737	0·0200	0·0054	0·0015	0·00040	0·00002	0·00003	1·867	0·0373	0·1376
6·12	0·3684	0·1357	0·0500	0·0184	0·0068	0·00250	0·00011	0·00034	2·255	0·1127	0·3060
4·92	0·4642	0·2154	0·1000	0·0464	0·0215	0·01000	0·00092	0·00215	2·284	0·2284	0·4920
3·97	0·5313	0·2823	0·1500	0·0797	0·0423	0·02250	0·00464	0·00635	2·109	0·3164	0·5955
3·16	0·5848	0·3420	0·2000	0·1170	0·0684	0·04000	0·01195	0·01368	1·848	0·3696	0·6320
2·34	0·6300	0·3968	0·2500	0·1575	0·0992	0·06250	0·02339	0·02480	1·474	0·3686	0·5850
1·69	0·6694	0·4482	0·3000	0·2008	0·1345	0·09000	0·03938	0·04034	1·131	0·3394	0·5070
1·25	0·7047	0·4967	0·3500	0·2466	0·1738	0·12250	0·06025	0·06084	0·881	0·3083	0·4375
62·92	4·7772	2·4949	1·4400	0·8758	0·5488	0·35055	0·22700	0·14854	18·09?	2·1103	3·8445

* The values of V_s have been multiplied by 100 to simplify the arithmetic.

Putting the above linearized expressions into equations VII, VIII, and IX and simplifying, therefore

$$18 \cdot 19x + 0 \cdot 438y - 6 \cdot 255z = 0 \cdot 029 \qquad \text{XIII}$$

$$74 \cdot 61x - 65 \cdot 82y - 18 \cdot 19z = 0 \cdot 15 \qquad \text{XIV}$$

$$2 \cdot 138x + 0 \cdot 0944y - 0 \cdot 438z = 0 \cdot 0029 \qquad \text{XV}$$

Solving these equations by successive elimination or by matrices gives the solutions

$$x = 0 \cdot 0014, \quad y = -0 \cdot 0006, \quad z = -0 \cdot 0006 \qquad \text{XVI}$$

Hence, the values of α, β, and V_0 are

$$\alpha = 0 \cdot 748, \quad \beta = 18 \cdot 67, \quad V_0 = 9 \cdot 21 \qquad \text{XVII}$$

from equations XII.

Returning to the original constants by equations III, and remembering the factor (100) introduced in Table 11.3, gives the final result

$$V_0 = 0 \cdot 0921 \text{ cm/s}$$

$$k = 2 \cdot 027$$

$$K = 0 \cdot 748$$

which confirms the graphical values given in the original paper.

11.5.5. Search Methods

In order to use Newton's method to solve

$$f(x) = 0 \qquad (11.57)$$

it was necessary to differentiate the function analytically. With some complicated functions this is not convenient, but it is sometimes possible to introduce an approximate derivative which reduces the rate of convergence, or a numerically determined gradient from two consecutive approximations. However, the evaluation of derivatives can be avoided altogether if two values of x can be found, say x_1 and x_2, such that $f(x_1)$ has the opposite sign to $f(x_2)$, a root of the equation (11.57) must lie between x_1 and x_2 (unless there is an asymptote between x_1 and x_2 of course). A third value (x_3) can then be obtained by linear interpolation between x_1 and x_2 using the equation

$$x_3 = \frac{x_2 f(x_1) - x_1 f(x_2)}{f(x_1) - f(x_2)} \qquad (11.60)$$

Evaluating $f(x_3)$ and selecting whichever of $f(x_1)$ and $f(x_2)$ has opposite sign allows a further interpolation between x_3 and either x_1 or x_2. Although this procedure is attractive at first sight, it is rather slow because if $d^2 f/dx^2$

has the same sign at all values of x between x_1 and x_2 then either x_1 or x_2 will be retained in every step of the interpolation process, even when the root is very close to the other point.

A much simpler, yet quite effective, method, which is ideally suited to computer application, is to evaluate the function at reasonably large equal intervals of the variable and to look for a sign change in the function. The interval in which the sign change occurs is then subdivided into ten equal parts (an arbitrary subdivision which suits the base of normal arithmetic) and the sign change of the function is located again. Thus for an average of five function evaluations the uncertainty in the location of the root can be reduced by a factor of ten. The process can be repeated as many times as required. Although it is desirable to start with a fairly large increment to economize on the time taken to locate the root initially, too large an interval may permit two roots to occur within the same interval giving two sign changes and the roots will be missed.

Example 5. Determine the boiling point, at atmospheric pressure, of a mixture of 50 mole % benzene, 30 mole % toluene, and 20 mole % ethyl benzene and also determine the equilibrium vapour composition. The saturation vapour pressure (p_i^0) of each pure component (i) is related to the absolute temperature (T) by

$$\log_{10}(p_i^0) = a_i - b_i/T$$

where p_i^0 is measured in mm Hg and T in °K.

i	Component	a_i	b_i
1	Benzene	7·84135	1750
2	Toluene	8·08840	1985
3	Ethyl benzene	8·11404	2129

Assuming that the system is ideal and denoting liquid mole fractions by x_i, the partial pressures will be given by

$$p_i = p_i^0 x_i$$

and the boiling point will be determined by

$$p_1 + p_2 + p_3 - P = 0 = f(T)$$

or

$$0 \cdot 5 p_1^0 + 0 \cdot 3 p_2^0 + 0 \cdot 2 p_3^0 - 760 = f(T) = 0$$

The search was started at 350 °K, just below the boiling point of the most volatile component, and the first increment was 10 °K. These calculations are summarized in Table 11.4 and the result is 366·33 °K = 93·33 °C.

TABLE 11.4. *Determination of boiling point*

T	p_1^0	p_2^0	p_3^0	$f(T)$
350	694	261	107	−313
360	955	375	159	−138
370	1293	529	229	+91
361	982	389	165	−119
362	1016	403	171	−97
363	1048	417	177	−75
364	1080	432	184	−54
365	1114	447	191	−31
366	1148	462	198	−8
367	1183	478	206	+16
366·1	1151·4	463·9	198·9	−5·3
366·2	1154·9	465·4	199·7	−3·0
366·3	1158·3	467·0	200·4	−0·7
366·4	1161·8	468·6	201·1	+1·7

The vapour composition is obtained as follows:

Partial pressure of benzene $\quad = 0·5 \times 1159·4 = 579·7$

Partial pressure of toluene $\quad = 0·3 \times 467·5 \ = 140·2$

Partial pressure of ethyl benzene $= 0·2 \times 200·6 \ = \ 40·1$

$\qquad\qquad\qquad\qquad\qquad\qquad\qquad\qquad\qquad 760·0$

Hence the vapour mole fractions are 0·763 benzene, 0·184 toluene, and 0·053 ethyl benzene.

11.6. THE INFLUENCE OF COMPUTERS

It is clear from the above manual calculation that the answer could have been obtained much more quickly by using interpolation techniques coupled with intelligent guesswork, and this draws attention to an important influence of computers on numerical techniques. The above procedure, due to its repetitive nature, can be programmed very easily and the execution by a computer would be very rapid, but any attempt to use interpolation complicates the program considerably and intelligent guesswork cannot be programmed at all.

This distinction cannot be over-emphasized and it is as well to look back at the methods described in this chapter to see what impact the availability of a computer will have. Ordinary differential equations can be solved manually with either a third order Runge–Kutta process or a low order predictor–corrector method with short step length whereas the fourth order Runge–Kutta process existing in the program library is likely to be used on a computer. Boundary value problems solved by finite difference methods

(Section 11.4.1) can be handled equally well manually or by machine without modification. However, the ability to make a fresh estimate of the solution at any stage, as was done twice in Table 11.2, is extraordinarily difficult to write into a program because the machine does not readily recognize patterns unless the type of pattern can be mathematically described and anticipated before the solution is known.

Algebraic equations can also be solved manually or by machine with little or no modification to the method but, as pointed out in the last example, human intelligence can often be used manually to find short cuts during a repetitive calculation.

The remainder of this chapter is devoted to the solution of equations containing more than one independent variable and the methods available can usually be divided into those suitable for machine solution and those suitable for manual solution. Because the volume of calculation required is so much greater than in the problems considered so far, manual methods are used less frequently; therefore methods primarily intended for use with computers will be described in the remainder of this chapter. They can also be used manually of course, but usually the work involved is prohibitive unless frequent short cuts can be taken, and the reader would be well advised to consider alternative ways of tackling the problem. These are either to obtain access to a computer, or to seek a modified numerical technique which is more suitable for manual calculation, or even to simplify the problem by taking a less ambitious model.

To avoid the difficulty of introducing a computer programming language, the methods described below will be illustrated by means of logic flow diagrams from which computer programs can readily be written in any selected language. The numerical details can no longer be illustrated fully but the preparation of the problem, the design of the computer program to suit it, and the final numerical solution obtained will be presented for each example. The logical flow diagram for the last example above is shown in Fig. 11.8 and it should be noted that the diagram is specifically designed to generate the same course of action as that already illustrated. The plan is limited in that the original guess for T should be smaller than the eventual result since the logic only permits T to increase from a value $10°$ below the original guess It is also specific to finding the root of a function $f(T)$ for which $f'(T)$ is positive. A more general program to search for roots in both directions and for all function gradients would be considerably more complicated.

11.7. Difference–Differential Equations

The numerical solution of this type of equation follows directly from the method given in Section 11.3. Considering the general form of equation which arises in a chemical engineering stagewise process,

$$\frac{dy_n}{dt} = f(y_{n-1}, y_n, y_{n+1}) \tag{11.61}$$

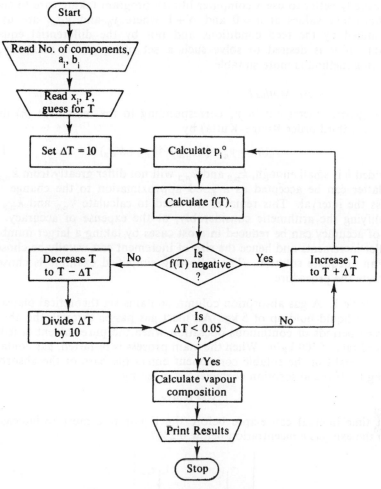

FIG. 11.8. Determination of boiling point

where the values of y_n are given at some value of t (say $t = 0$) and taking an interval (h) in t, then the Runge–Kutta process can be performed on equation (11.61) for each value of n. Thus, evaluating

$$k_{n,1} = hf(y_{n-1}, y_n, y_{n+1}) \tag{11.62}$$

$$k_{n,2} = hf(y_{n-1} + \tfrac{1}{2}k_{n-1,1}, y_n + \tfrac{1}{2}k_{n,1}, y_{n+1} + \tfrac{1}{2}k_{n+1,1}) \tag{11.63}$$

etc., depending on whether the third order, fourth order, or higher is being used, the value of y_n at the end of the interval can be found for all values of n. The whole process is similar to that already described but the calculations are confusing manually and it is all too easy to make mistakes.

It is usually better to use a computer library program taking care to specify the boundary values at $n = 0$ and $N+1$ where y_0 and y_{N+1} are usually determined by the feed conditions and not by the differential equation (11.61). If it is desired to solve such a set of equations manually, the following method is more suitable.

11.7.1. Step-by-step Method

The general increment in y_n corresponding to the increment (h) in t is given (for third order Runge–Kutta) by

$$y_n(h) = y_n + \tfrac{1}{6}(k_{n,1} + 4k_{n,2} + k_{n,3}) \tag{11.64}$$

Provided h is small enough, $k_{n,2}$ and $k_{n,3}$ will not differ greatly from $k_{n,1}$ and the latter can be accepted as a good approximation to the change in y_n across the interval. This removes the need to calculate $k_{n,2}$ and $k_{n,3}$ thus simplifying the arithmetic considerably, at the expense of accuracy. The loss of accuracy can be reduced in most cases by taking a larger number of smaller increments and hence the size of increment can usually be chosen to give an adequate solution. The following simplified example is chosen to illustrate the procedure.

Example 1. A gas absorption column contains six theoretical plates each having a liquid hold-up of 5 kg. An inert gas passes through the absorber between periods of continuous operation when the pure solvent is fed at a reduced rate of 100 kg/h. When the main process is restarted, gas containing 5% by weight of the soluble component enters the base of the absorber at 500 kg/h. If the absorption coefficient is given by

$$y = 0 \cdot 40x \qquad\qquad \text{I}$$

what time interval can elapse before the liquor rate must be increased to keep the exit gas concentration below $0 \cdot 2\%$?

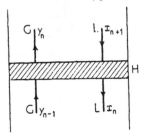

FIG. 11.9. Flow through nth plate of a gas absorber

Solution

Taking a material balance over the nth plate as illustrated in Fig. 11.9,

$$Gy_{n-1} + Lx_{n+1} - Gy_n - Lx_n = H\frac{\mathrm{d}x_n}{\mathrm{d}t} \qquad\qquad \text{II}$$

Because the boundary conditions and concentration limit are expressed in terms of gas composition, it is convenient to use equation I in the form

$$y_n = kx_n \qquad\qquad \text{III}$$

to eliminate x_n and x_{n+1} from equation II. Thus

$$\alpha y_{n-1} + y_{n+1} - \alpha y_n - y_n = \frac{H}{L}\frac{dy_n}{dt} \qquad\qquad \text{IV}$$

where
$$\alpha = kG/L \qquad\qquad \text{V}$$

Taking a time interval Δt and denoting the new gas composition above the nth plate by $y_n{}^*$, the derivative in equation IV can be replaced by a finite difference form. Thus

$$\alpha y_{n-1} + y_{n+1} - (\alpha+1)y_n = \frac{H}{L\Delta t}(y_n{}^* - y_n) \qquad\qquad \text{VI}$$

Solving equation VI for the unknown:

$$y_n{}^* = \frac{L\Delta t\alpha}{H} y_{n-1} + \left[1 - \frac{L\Delta t}{H}(\alpha+1)\right]y_n + \frac{L\Delta t}{H}y_{n+1} \qquad\qquad \text{VII}$$

On physical considerations, the new value $y_n{}^*$ is expected to depend upon the old value y_n on the same plate. Also, a hypothetical increase in the value of y_n at fixed values of y_{n-1} and y_{n+1} should result in an increase in the value of $y_n{}^*$. Hence to keep the problem physically reasonable, Δt must be chosen to satisfy

$$1 - \frac{L\Delta t}{H}(\alpha+1) > 0 \qquad\qquad \text{VIII}$$

or
$$\Delta t < H/L(\alpha+1) \qquad\qquad \text{IX}$$

Inserting the numerical data given in the problem,

$$\alpha = \frac{0.40 \times 500}{100} = 2.0$$

$$\frac{H}{L(\alpha+1)} = \frac{5}{100 \times 3.0} = \frac{1}{60}$$

Hence, Δt is chosen to be $\frac{1}{120}$ to satisfy IX. Equation VII now takes the form

$$y_n{}^* = \tfrac{1}{6}(2y_{n-1} + 3y_n + y_{n+1}) \qquad\qquad \text{X}$$

The boundary conditions are:

$$\text{at} \quad t = 0, \quad y_n = 0 \qquad\qquad \text{XI}$$
$$\text{at} \quad n = 0, \quad y_0 = 0.05 \qquad\qquad \text{XII}$$
$$\text{at} \quad n = 7, \quad y_7 = 0 \qquad\qquad \text{XIII}$$

The third condition results from the pure solvent feed to the top of the absorber (i.e. $x_7 = 0$).

Table 11.5 can now be constructed to receive the results, and the boundary conditions can be inserted in the first row and the end columns (y_0 and y_7). The index (m) has been used to count the time increments. There is a contradiction between conditions XI and XII regarding the composition y_0 at $t = 0$, because of the step change in feed composition. Experience has shown that the best solution usually results from taking a value for y_0 equal to the arithmetic average of the two conflicting requirements.

TABLE 11.5. Gas Compositions in Absorption Column (%)

m	y_0	y_1	y_2	y_3	y_4	y_5	y_6	y_7
0	2·500	0	0	0	0	0	0	0
1	5·000	0·833	0	0	0	0	0	0
2	5·000	2·083	0·278	0	0	0	0	0
3	5·000	2·755	0·833	0·093	0	0	0	0
4	5·000	3·183	1·350	0·324	0·031	0	0	0
5	5·000	3·483	1·790	0·617	0·123	0·010	0	0
6	5·000	3·707	2·159	0·926	0·269	0·046	0·003	0
7	5·000	3·880	2·470	1·227	0·451	0·113	0·017	0
8	5·000	4·018	2·733	1·512	0·653	0·210	0·046	0
9	5·000	4·131	2·958	1·776	0·865	0·330	0·093	0
10	5·000	4·225	3·152	2·018	1·080	0·469	0·156	0
11	5·000	4·304	3·321	2·240	1·291	0·620	0·234	0

The table of results can now be built up, one row at a time, by successive use of equation X. This shows that the exit gas concentration (y_6) exceeds the limit of 0·2% after 11 time increments. Since $\Delta t = \frac{1}{120}$ h $= \frac{1}{2}$ min, the liquor rate must be increased 5 min after the rich gas has entered the absorber.

11.8. PARTIAL DIFFERENTIAL EQUATIONS
(INITIAL VALUE TYPE)

In Sections 11.3 and 11.4 it was shown that some methods of solving ordinary differential equations were unsuitable for boundary value problems and others were unsuitable for initial value problems. It is again necessary to present independent methods of solution for the same classification of problems. Initial value problems were defined in Section 8.4.5 as those problems in which at least one independent variable has an open range; whereas all independent variables in a boundary value problem have closed ranges. Time is the open-range variable of most frequent occurrence in chemical engineering, and coordinates describing a finite space are the most usual closed-range variables, although time is sometimes disguised as the ratio of a coordinate to a velocity as in the falling film of Example 5 in Section 8.8.

The most common partial differential equations in chemical engineering are those which describe heat or mass transfer in stationary or moving fluids in the presence or absence of chemical reactions. All of these situations can be described by second order partial differential equations of one or other of the two types already mentioned. In general, however, second order partial differential equations must be divided into three (rather than two) types. These are often referred to as elliptic, parabolic, or hyperbolic equations. Elliptic equations arise in boundary value problems but either parabolic or hyperbolic equations can arise in initial value problems. Fortunately, the initial value problems in chemical engineering usually involve only first derivatives with respect to time and only one initial value for the dependent variable; these problems are described by parabolic equations. Hyperbolic equations, which require special methods of solution, are rare in chemical engineering and the reader is referred to other texts† for these methods. A common example of a hyperbolic equation is the wave equation

$$c^2 \frac{\partial^2 s}{\partial x^2} = \frac{\partial^2 s}{\partial t^2} \qquad (11.65)$$

where the displacement (s) is a function of the coordinate (x) and time (t). c is the velocity of propagation of the wave. The vibration of a string is described by equation (11.65) and a suitable set of boundary conditions would specify the positions of the two fixed ends of the string and the initial displacement of the string at all points. But a second initial condition is required! This would normally state that the vibration started from rest, viz.

$$\text{at} \quad t = 0, \quad \frac{\partial s}{\partial t} = 0 \qquad (11.66)$$

but a velocity distribution could equally well be prescribed. Thus in solving a hyperbolic equation both the dependent variable and its derivative must be specified at the initial end of the range of the independent variable.

It is worth noting at this point that the subdivision of initial value problems into parabolic or hyperbolic equations was not necessary in Chapter 8 in any of the analytical methods for solving partial differential equations. In particular, the Laplace transform method can be used to solve either parabolic or hyperbolic equations.

11.8.1. *Step-by-step Methods for Initial Value Problems*

The simplest initial value partial differential equation is the time-dependent linear heat conduction equation

$$\frac{\partial T}{\partial t} = \alpha \frac{\partial^2 T}{\partial x^2} \qquad (11.67)$$

† "Modern Computing Methods" (2nd edition). National Physical Laboratory Notes on Applied Science, No. 16, H.M.S.O. London (1961).

A few analytical solutions of equation (11.67) have already been presented in Chapter 8, but if the boundary conditions are functions of time, a numerical solution is often more easily obtained. The basis of the numerical method is to replace the space derivative in equation (11.67) by a finite difference representation, using equation (11.41). Thus

$$\frac{\partial T_m}{\partial t} = \frac{\alpha}{(\Delta x)^2}(T_{m+1} - 2T_m + T_{m-1}) \tag{11.68}$$

Comparison of equations (11.61) and (11.68) shows that the methods suggested for difference-differential equations can now be used to solve equation (11.68). In particular, the step-by-step method can be used by replacing the time derivative by an equivalent finite difference expression. To simplify the symbolism, it is convenient to introduce a second suffix (n) to count the time increments. Hence by using equation (11.40), equation (11.68) becomes

$$\frac{T_{m,n+1} - T_{m,n-1}}{2\Delta t} = \frac{\alpha}{(\Delta x)^2}(T_{m+1,n} - 2T_{m,n} + T_{m-1,n})$$

or $$T_{m,n+1} = T_{m,n-1} + \frac{2\alpha\Delta t}{(\Delta x)^2}(T_{m+1,n} - 2T_{m,n} + T_{m-1,n}) \tag{11.69}$$

By using equations (11.40) and (11.41) to replace the derivatives in equation (11.67), equation (11.69) is accurate to second order differences. Unfortunately, equation (11.69) is unstable, giving a solution showing an unbounded oscillation of temperature as time progresses. The reason for this is that the new point temperature $(T_{m,n+1})$ is calculated from four known temperatures, one of which $(T_{m,n})$ has a negative coefficient. This means that a small positive error in $T_{m,n}$ will lead to a negative error in $T_{m,n+1}$. On the next step in the calculation, this negative error in $T_{m,n+1}$ will augment the positive error in $T_{m,n}$ and generate an increased positive error in $T_{m,n+2}$. Equation (11.69) shows this quite clearly if the subscript n is increased by unity. As a general result, if all known point values are put on to the same side of an equation and any coefficient is negative, then the solution is likely to be unstable.

Since equation (11.69) is unstable, an alternative finite difference expression must be used for the time derivative in equation (11.68). An expression which is only correct to first differences is

$$\frac{\partial T_{m,n}}{\partial t} = \frac{T_{m,n+1} - T_{m,n}}{\Delta t} \tag{11.70}$$

which was effectively used in Section 11.7.1. This leads to the equation

$$T_{m,n+1} = MT_{m+1,n} + (1 - 2M)T_{m,n} + MT_{m-1,n} \tag{11.71}$$

as a finite difference representation of equation (11.67).

$$M = \alpha(\Delta t)/(\Delta x)^2 \tag{11.72}$$

is termed the "modulus" of the equation, and its value determines the "stability" and "convergence" of the method. Formulae do exist to determine which methods are stable and convergent, but the statement above that all coefficients in equation (11.71) should be positive for a stable solution is both simple and effective. Hence

$$M \leqslant \tfrac{1}{2} \qquad (11.73)$$

If the temperature is known at all values of x (or m) for a value of t (or n), then repeated application of equation (11.71) will determine the complete temperature distribution at time $t + \Delta t$ (or $n+1$). Before the method can be started, values must be selected for Δx and Δt such that the condition (11.73) is satisfied. The most convenient procedure is to choose Δx to be a suitable fraction of the total range of x, and then choose M to satisfy (11.73) and give the simplest coefficients in equation (11.71).

The simplest choice of value for M is $\tfrac{1}{2}$ so that equation (11.71) becomes

$$T_{m,n+1} = \tfrac{1}{2}(T_{m+1,n} + T_{m-1,n}) \qquad (11.74)$$

This value of M is on the limit of stability and equation (11.74) cannot be expected to give answers of great accuracy; it does, however, have the merit of simplicity and will give a good approximation to the answer if Δx is chosen sufficiently small. The first known use of equation (11.74) was by Schmidt† who used it as the basis of a graphical construction.

In order to illustrate how the answer emerges from the use of both this method and of the later methods to be developed in this section it is convenient to consider the following simple example.

Example 1. Solve the one-dimensional, unsteady-state heat conduction equation (11.67) subject to the boundary conditions
   (i) at $x = 0$, $T = 100\,^{\circ}\text{C}$ for all t,
   (ii) at $x = 10\,\text{cm}$, $T = 0\,^{\circ}\text{C}$ for all t,
   (iii) at $t = 0$, $T = 0\,^{\circ}\text{C}$ for $0 \leqslant x \leqslant 1$.
The thermal diffusivity $\alpha = 2 \cdot 0\,\text{cm}^2/\text{s}$.

Solution

Following the method described above, the finite difference equation (11.71) is obtained. It is now necessary to choose values for M, Δt, and Δx to satisfy equations (11.72) and (11.73), and the order is important. Δx must be an integral fraction of the total range of the problem and, although it is too large for accurate results, 2 cm is chosen for this illustration. Next, a value of $\tfrac{1}{2}$ is chosen for M so that the simple but stable equation (11.74) can be used. The value of Δt is thus automatically determined by equation (11.72).

$$\Delta t = \frac{M(\Delta x)^2}{\alpha} = \frac{0 \cdot 5 \times 2 \times 2}{2} = 1 \cdot 0\,\text{s} \qquad \text{I}$$

† Schmidt, E. "Foeppls Festschrift", Springer-Verlag OHG, Berlin (1924).

The calculations can now be performed as shown in Table 11.6 where the figures calculated in any column, say $m = 2$, corresponding to $x = 4$ cm give the temperature at that point at equal intervals of time $\Delta t = 1$ s. The boundary conditions determine the temperatures at $n = 0$, $m = 0$, and $m = 5$ but there is a contradiction between conditions (i) and (iii) at $n = 0$ and $m = 0$. It will be seen that this anomaly has been resolved by taking an arithmetic mean of the two alternative values; a device which has been found by experience to be the best way of dealing with a step change occurring at any boundary. An argument supporting this choice is that at $t < 0$, before the problem starts, the temperature must have been uniformly zero otherwise some heat would have penetrated the range $0 \leqslant x \leqslant 10$ at $t = 0$. When $t > 0$, condition (i) states that $T = 100$ at $x = 0$ and if it is assumed that the temperature is raised linearly from 0 to 100 as fast as possible through $t = 0$ as a practical attempt to achieve the step change then the temperature at $x = 0$ would be $50°$ at exactly $t = 0$.

TABLE 11.6. *Temperature distribution (explicit method)*

n \ m	0	1	2	3	4	5
0	50·0	0·0	0·0	0·0	0·0	0·0
1	100·0	25·0	0·0	0·0	0·0	0·0
2	100·0	50·0	12·5	0·0	0·0	0·0
3	100·0	56·2	25·0	6·2	0·0	0·0
4	100·0	62·5	31·2	12·5	3·1	0·0
5	100·0	65·6	37·5	17·2	6·2	0·0
6	100·0	68·8	41·4	21·8	8·6	0·0
7	100·0	70·7	45·3	25·0	10·9	0·0
8	100·0	72·6	47·8	28·1	12·5	0·0

11.8.2. Crank–Nicolson Method

As already mentioned, the finite difference representation given by equation (11.71) is only correct to first order differences in time, and it is clearly desirable to improve the accuracy of the result by devising a formula which is correct to second differences. One such method is best described by reference to Fig. 11.10. The simple step-by-step method described in Section 11.8.1 uses a finite difference equation (11.71) which represents the differential equation (11.67) at point A. The space derivative is adequately represented

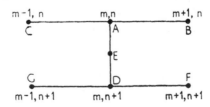

FIG. 11.10. Point values used in iteration method

to second differences at point A but the time derivative is only represented to first differences at point A. In the Crank–Nicolson method, however, the differential equation is expressed correct to second differences in both variables at point E. Considering the time derivative first, equation (11.40) gives

$$\left.\frac{\partial T}{\partial t}\right|_E = \frac{T_{m,n+1} - T_{m,n}}{\Delta t} \tag{11.75}$$

The characteristic assumption of this method is that the second space derivative at E is the average of its values at A and D, viz.

$$\left.\frac{\partial^2 T}{\partial x^2}\right|_E = \frac{1}{2}\left(\left.\frac{\partial^2 T}{\partial x^2}\right|_A + \left.\frac{\partial^2 T}{\partial x^2}\right|_D\right) \tag{11.76}$$

Using equation (11.41) to express the derivatives in finite difference form and substituting equations (11.75) and (11.76) into (11.67) gives

$$\frac{T_{m,n+1} - T_{m,n}}{\Delta t} = \frac{\alpha}{2(\Delta x)^2}(T_{m+1,n} - 2T_{m,n} + T_{m-1,n} + T_{m+1,n+1}$$

$$- 2T_{m,n+1} + T_{m-1,n+1}) \tag{11.77}$$

Introducing

$$M = \alpha\,\Delta t/(\Delta x)^2 \tag{11.72}$$

again and rearranging terms gives

$$(1+M)\,T_{m,n+1} = (1-M)\,T_{m,n} + \tfrac{1}{2}M(T_{m+1,n} + T_{m-1,n})$$

$$+ \tfrac{1}{2}M(T_{m+1,n+1} + T_{m-1,n+1}) \tag{11.78}$$

This is the form in which the equation is used, but the two terms in the last bracket are unknown. The use of equation (11.78) is very similar to the method described in Section 11.5.3 for improving a root of an algebraic equation, except that here, there is a set of simultaneous equations to be solved. Therefore, a first approximation is guessed for $T_{m,n+1}$ for all values of m. The right-hand side of equation (11.78) is next evaluated to provide a better approximation for $T_{m,n+1}$. These new quantities are then used in the right-hand side of equation (11.78) to determine a further approximation and the process is repeated until successive cycles result in negligible changes in the temperature distribution.

The basic mechanism described above is stable for all values of M and is to be recommended for solving initial value problems. There are two further improvements however, both of which lead to a faster solution and a simpler computer program. The first, obvious improvement is to avoid calculating the first three terms on the right-hand side of equation (11.78) more than once at each time step. Thus, by defining a new array $F_{m,n}$ by

$$(1+M)\,F_{m,n} = (1-M)\,T_{m,n} + \tfrac{1}{2}M(T_{m+1,n} + T_{m-1,n}) \tag{11.79}$$

equation (11.78) becomes

$$T_{m,n+1} = F_{m,n} + \frac{M}{2(1+M)}(T_{m+1,n+1} + T_{m-1,n+1}) \tag{11.80}$$

which is the equation to be solved iteratively until the distribution $T_{m,n+1}$ has been determined.

It has been stated above that equation (11.80) is used to obtain the $(r+1)$th approximation from the rth approximation by using the form

$$T_{m,n+1}{}^{r+1} = F_{m,n} + C(T_{m+1,n+1}{}^{r} + T_{m-1,n+1}{}^{r}) \tag{11.81}$$

where $\qquad\qquad C = \tfrac{1}{2}M/(1+M) \tag{11.82}$

The second improvement is to use the latest calculated approximation for all unknown values. For example, when calculating $T_{4,n+1}{}^{r+1}$, the new value $T_{3,n+1}{}^{r+1}$ should be used instead of the old value $T_{3,n+1}{}^{r}$. In general

$$T_{m,n+1}{}^{r+1} = F_{m,n} + C(T_{m+1,n+1}{}^{r} + T_{m-1,n+1}{}^{r+1}) \tag{11.83}$$

which leads to faster convergence on each time step.

Implementation of equations (11.79) and (11.83) in a computer program is much simpler than directly using equation (11.78). The latter requires three one-dimensional arrays; for the known temperatures at the previous time step, the temperatures at the rth iteration, and the new temperatures at the $(r+1)$th iteration, whereas only two arrays are required in the modified method. One array contains the values of $F_{m,n}$ at the previous time step and the other array contains the latest values of $T_{m,n+1}$ which are repeatedly over-written thus ensuring that the latest available values will be used in the right-hand side of equation (11.83). Each temperature must of course be compared with the value it is replacing before over-writing but this is easily accomplished. Once the new temperature distribution has been obtained it can be printed, then used both to calculate the new contents of the array, $F_{m,n+1}$ and left as a first guess for the temperature distribution at the next time step. These ideas are contained in Fig. 11.11 which is a logic flow diagram for solving example 1 again.

A manual calcuiation with $M = \tfrac{2}{3}$, $\Delta t = 1\tfrac{1}{3}$, giving

$$F_{m,n} = 0{\cdot}2(T_{m,n} + T_{m+1,n} + T_{m-1,n})$$

$$T_{m,n+1} = F_{m,n} + 0{\cdot}2(T_{m+1,n+1} + T_{m-1,n+1})$$

is illustrated in Table 11.7. One important point should be noticed on the first row; that is the use of the full step change at $t = 0$, $x = 0$. Referring again to Fig. 11.10, the finite difference equation represents the behaviour of the system around point E which corresponds to the range $0 \leqslant t \leqslant 1\tfrac{1}{3}$. The temperature at $x = 0$ throughout this range is 100 °C, hence it is logical to use this value at $t = 0$ also. Further discussion of this point is given by Pearson† who shows that the proportion of the step change which should

† Pearson, C. E. Mathematical tables and other aids to computation, **19**, 570 (1965).

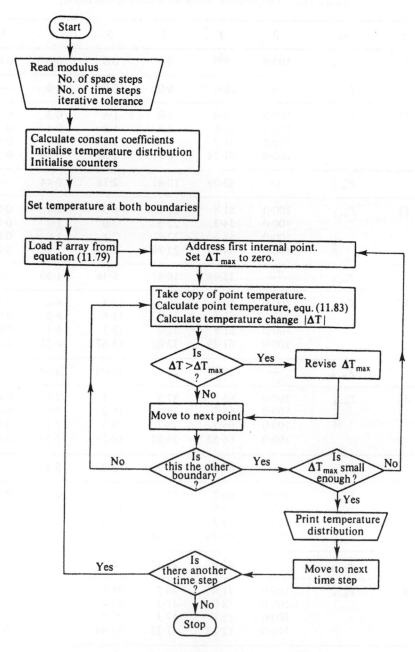

FIG. 11.11. Use of the Crank–Nicolson method

TABLE 11.7. *Temperature distribution* (Crank–Nicolson)

t	m	0	1	2	3	4	5
0	$T_{m,0}$	100·0	0·0	0·0	0·0	0·0	0·0
	$F_{m,0}$	—	20·0	0·0	0·0	0·0	—
$1\frac{1}{3}$	$T_{m,1}$	100·0	40·0	8·0	1·6	0·3	0·0
		100·0	41·6	8·6	1·8	0·4	0·0
		100·0	41·7	8·7	1·8	0·4	0·0
		100·0	41·74	8·71	1·82	0·36	0·0
	$F_{m,1}$	—	30·09	10·45	2·18	0·44	—
$2\frac{2}{3}$	$T_{m,2}$	100·0	51·8	21·2	6·5	1·7	0·0
		100·0	54·3	22·6	7·0	1·8	0·0
		100·0	54·6	22·8	7·1	1·9	0·0
		100·0	54·65	22·80	7·12	1·86	0·0
	$F_{m,2}$	—	35·49	16·91	6·36	1·80	—
4	$T_{m,3}$	100·0	60·0	30·3	12·8	4·4	0·0
		100·0	61·6	31·8	13·6	4·5	0·0
		100·0	61·8	32·0	13·7	4·5	0·0
		100·0	61·89	32·03	13·67	4·53	0·0
	$F_{m,3}$	—	38·78	21·52	10·05	3·64	—
$5\frac{1}{3}$	$T_{m,4}$	100·0	65·2	37·3	18·4	7·3	0·0
		100·0	66·2	38·4	19·2	7·5	0·0
		100·0	66·5	38·7	19·3	7·5	0·0
		100·0	66·52	38·68	19·29	7·50	0·0
	$F_{m,4}$	—	41·04	24·90	13·09	5·36	—
$6\frac{2}{3}$	$T_{m,5}$	100·0	68·8	42·5	23·1	10·0	0·0
		100·0	69·5	43·4	23·8	10·1	0·0
		100·0	69·7	43·6	23·8	10·1	0·0
		100·0	69·76	43·61	23·83	10·13	0·0
	$F_{m,5}$	—	42·67	27·44	15·51	6·79	—
8	$T_{m,6}$	100·0	71·4	46·5	26·8	12·2	0·0
		100·0	72·0	47·2	27·4	12·3	0·0
		100·0	72·1	47·3	27·4	12·3	0·0
		100·0	72·13	47·35	27·44	12·28	0·0
8	Analytical	100·0	72·24	47·47	27·47	12·24	0·0

be imposed is a function of the modulus. For very small values of M the full change should be used, whilst for $M = 1$ the proportion of the step is 83% hence, for $M < 1$ it is sufficiently accurate to use the full step change. The error introduced causes a decaying oscillation so that later parts of the calculation are relatively unaffected. This is demonstrated in the last line of Table 11.7 where the analytical solution

$$T = 100 \cdot 0 - 10 \cdot 0x - \frac{200}{\pi} \sum_{n=1}^{\infty} \frac{1}{n} \sin \frac{n \pi x}{10} \exp \left(\frac{-n^2 \pi^2 \alpha t}{100} \right)$$

has been evaluated at $t = 8$ for comparison.

The backward difference method, which is sometimes used, has no stability or speed advantages over the Crank–Nicolson method and is significantly less accurate. The authors recommend the use of the Crank–Nicolson method with $M < 1$.

11.8.3. *Runge–Kutta Method*

As mentioned in Section 11.8.1, initial value problems can be solved by any of the methods described for difference-differential equations. In particular, the Runge–Kutta method, which is usually supplied as a subroutine in the library of most computers, can be used. Referring to the heat conduction equation (11.67) again, the space derivative is put into finite difference form as before to give

$$\frac{\partial T_m}{\partial t} = \frac{\alpha}{(\Delta x)^2} (T_{m+1} - 2T_m + T_{m-1}) \tag{11.68}$$

Using the same boundary conditions again and five equal increments of $\Delta x = 2$ cm, the set of equations (11.68) consists of four differential equations for T_1, T_2, T_3, and T_4 with T_0 taking the fixed value of 100 and T_5 having the fixed value of 0. It is only necessary to write a very simple computer program to solve the problem in this form and the logic flow diagram is illustrated in Fig. 11.12. The library subroutine always requires the user to write a further subroutine to define the differential equations. In the present example, all of the differential equations are of the form given by equation (11.68) and it is a simple matter to program a loop which increments the value of m from 1 to 4. If Fortran is being used, care must be taken with the first equation to avoid the zero suffix on an array element.

It will be seen that a time increment of 1 s has been chosen; this is the same value as that used for the step-by-step method in Section 11.8.1. The result of running this program is illustrated in the first part of Table 11.8 where it can be seen that the temperature distribution at $t = 8$ s compares very well with the analytical solution given with Crank–Nicolson solution in Table 11.7.

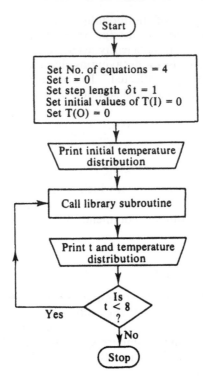

FIG. 11.12. Use of Runge–Kutta library subroutine

TABLE 11.8. *Heat Conduction by Runge-Kutta Method*

t \ x	0	2	4	6	8	10
0	100·00	0	0	0	0	0
1	100·00	32·68	6·77	1·11	0·09	0
2	100·00	47·64	16·74	4·59	0·97	0
3	100·00	56·02	25·14	9·17	2·64	0
4	100·00	61·42	31·76	13·79	4·73	0
5	100·00	65·23	36·97	17·99	6·88	0
6	100·00	68·08	41·13	21·65	8·91	0
7	100·00	70·29	44·50	24·75	10·70	0
8	100·00	72·04	47·23	27·36	12·25	0
0	100·00	0	0	0	0	0
2	100·00	54·17	8·33	10·42	−1·39	0
4	100·00	56·53	39·69	5·87	9·59	0
6	100·00	73·63	32·17	30·60	3·37	0
8	100·00	65·77	57·38	17·21	18·52	0

The second part of Table 11.8 illustrates the result of running the same program with an increased step length of $\delta t = 2$ s. The results are clearly erratic and no reliance can be placed upon them. A further computer run with a reduced step length of $\delta t = 0.5$ s gave results which were barely distinguishable from the first part of Table 11.8. In more detail, the temperatures at $t = 8$ only differed in the fourth decimal place and only one temperature at $t = 1$ differed by slightly more than 0.05 °K.

Thus the step length required for stability in the step-by-step method appears to be suitable for the Runge–Kutta method but attempts to use a larger step length in the Runge–Kutta method result in instability.

11.8.4. Comparison of Methods for Initial Value Problems

The step-by-step method, given in Section 11.8.1, is simple and effective for manual calculations but is not very accurate unless a small step length is used. The Crank–Nicolson method is just about acceptable as a manual method for simple problems and gives much better accuracy and complete stability for the extra effort. If a digital computer is available, it is well worth the effort to use the Crank–Nicolson method for both its stability and efficiency. However, if a computer solution is required with a minimum of programming effort the library subroutine which implements the Runge–Kutta method should be used. This may require a small step length, which will consume computer time, to avoid numerical instability.

11.8.5. Derivative Boundary Conditions

Boundary conditions were classified in Section 8.4. The first type will simply fix the value of the dependent variable at the boundary, and this is the only type considered so far in this section. The second and third types of boundary condition both involve a derivative at the boundary which must be replaced by a finite difference representation. Equation (11.40) is the best one for this purpose, but since the derivative is evaluated at the boundary, point values of the dependent variable are required on both sides of the boundary. This problem has been met already in Example 2, Section 11.4.1, where there were two planes of symmetry at which $dT/dx = 0$. The simple but effective solution was to consider the temperature variation at one extra "fictitious" point beyond the boundary. The temperature at this fictitious point must be the same as the temperature at the corresponding point inside the boundary to maintain the symmetry. The temperature at the point on the boundary can now be calculated from the normal formula since it is an internal point of the extended problem. The same principle can be applied to boundary conditions of the third type by again considering the value of the dependent variable at a fictitious point beyond the boundary. The fictitious point value is calculated from the boundary condition and the value on the boundary is calculated as for an ordinary internal point of the extended problem. The technique is illustrated in the example of a tubular reactor in the next section.

11.9. SIMULTANEOUS PARTIAL DIFFERENTIAL EQUATIONS
(INITIAL VALUE TYPE)

This type of problem is of frequent occurrence in reactor design, and involves the simultaneous solution of two interconnected initial value problems. No new difficulties arise as a result of the inclusion of a second dependent variable, but the calculation is of course about twice as long as a similar calculation with one dependent variable. Although any of the methods presented in Section 11.8 can be tried, the reaction term which links the heat transfer equation and the mass transfer equation is usually non-linear and can be a source of numerical instability. For this reason, the Crank–Nicolson method is preferred and will be applied to the next example.

Example. The dehydrogenation of ethyl benzene has been studied by Wenner and Dybdal† by passing a gaseous mixture of ethyl benzene and steam through a tubular catalytic reactor. They have shown that the rate of reaction can be adequately expressed by

$$r_c = k\left(p_E - \frac{p_S p_H}{K}\right) \text{ kg moles/h kg catalyst} \qquad \text{I}$$

where p_E is the partial pressure of ethyl benzene (bar),
$\quad p_S$ is the partial pressure of styrene (bar),
$\quad p_H$ is the partial pressure of hydrogen (bar).

The reaction rate constant is given by

$$k = 12\,600 \exp\left(-11\,000/T\right) \qquad \text{II}$$

and the equilibrium constant by

$$K = 0.027 \exp\left[0.021(T-773)\right] \qquad \text{III}$$

where T is the temperature in °K.

If the reaction tube has an internal diameter ($2a = 10$ cm) and is to be supplied with 0·069 kg moles/h ethyl benzene and 0·69 kg moles/h steam at 600 °C, the total mass velocity will be 2500 kg/h m². The tube is heated by flue gas which flows counter-current to the reaction mixture at a rate $R = 130$ kg/h and leaves at a temperature of 620 °C. The following data are also supplied.

Bulk density of catalyst (ρ)	= 1440 kg/m³
Operating pressure (P)	= 1·2 bar
Heat of reaction of ethyl benzene (ΔH)	= 140 000 kJ/kg mole
Effective thermal conductivity of bed (k_E)	= 0·45 W/m °K
Ratio of effective diffusivity to linear velocity (D_E/u)	= 0·000427 m
Specific heat of reaction mixture (C_p)	= 2·18 kJ/kg °K
Specific heat of flue gas (C_p')	= 1·0 kJ/kg °K

† Wenner, R. R. and Dybdal, E. C. *Chem. Eng. Prog.* **44**, 275 (1948).

Estimate what length of reactor tube is required to achieve 45% conversion of ethyl benzene.

Solution

In the element of volume shown in Fig. 11.13, ethyl benzene is consumed at a rate

$$2\pi r \, \delta r \, \delta z \, \rho r_c \text{ kg moles/h}$$

and heat is absorbed at a rate

$$2\pi r \, \delta r \, \delta z \, \rho r_c \, \Delta H \text{ kJ/h}$$

Taking a balance of the mass flow rates of ethyl benzene through the element,

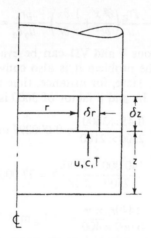

FIG. 11.13. Element of reactor tube

accounting for bulk flow in the axial direction, mass diffusion in the radial direction, and losses by reaction, gives

$$2\pi r \, \delta r \, uc - 2\pi r \, \delta z \frac{D_E}{u} \frac{\partial(uc)}{\partial r} - \left[2\pi r \, \delta r \, uc + \frac{\partial}{\partial z}(2\pi r \, \delta r \, uc) \, \delta z \right]$$

$$+ \left\{ 2\pi r \, \delta z \frac{D_E}{u} \frac{\partial(uc)}{\partial r} + \frac{\partial}{\partial r} \left[2\pi r \, \delta z \frac{D_E}{u} \frac{\partial(uc)}{\partial r} \right] \delta r \right\} = 2\pi r \, \delta r \, \delta z \, \rho r_c$$

where c is the concentration of ethyl benzene (kg moles/m³) and u is the linear gas velocity (m/h). After cancellation of terms, this equation simplifies to

$$\frac{\partial(uc)}{\partial z} - \frac{D_E}{u} \left(\frac{\partial^2(uc)}{\partial r^2} + \frac{1}{r} \frac{\partial(uc)}{\partial r} \right) + \rho r_c = 0 \qquad \text{IV}$$

Similarly, a heat balance for the volume element gives

$$\frac{\partial T}{\partial z} - \frac{k_E}{GC_p}\left(\frac{\partial^2 T}{\partial r^2} + \frac{1}{r}\frac{\partial T}{\partial r}\right) + \frac{\Delta H \rho r_c}{GC_p} = 0 \qquad\qquad \text{V}$$

It is more convenient to work in terms of the fraction of ethyl benzene converted which is denoted by

$$x = \frac{u_0 c_0 - uc}{u_0 c_0} \qquad\qquad \text{VI}$$

where the subscript (0) denotes entry conditions. Differentiating,

$$\therefore \quad d(uc) = -u_0 c_0 \, dx$$

Hence equation IV becomes

$$\frac{\partial x}{\partial z} - \frac{D_E}{u}\left(\frac{\partial^2 x}{\partial r^2} + \frac{1}{r}\frac{\partial x}{\partial r}\right) - \frac{\rho r_c}{u_0 c_0} = 0 \qquad\qquad \text{VII}$$

The coefficients in equations V and VII can be evaluated by using the data given. At this stage in the problem it is also convenient to check that the units used are consistent. Here, for instance, time is measured in hours but the definition of 1 W is 1 J/s and a factor of 3600 is needed. Thus

$$\frac{k_E}{GC_p} = \frac{0 \cdot 45 \times 3600}{2500 \times 2180} = 0 \cdot 000297 \text{ m}$$

$$\frac{\Delta H \rho r_c}{GC_p} = \frac{140\,000 \times 1440 r_c}{2500 \times 2 \cdot 18} = 37\,000 r_c \text{ °K/m}$$

$$\frac{\rho r_c}{u_0 c_0} = \frac{1440 r_c \times \pi}{0 \cdot 069 \times 400} = 164 r_c \text{ m}^{-1}$$

Hence equations V and VII become

$$\frac{\partial T}{\partial z} - 0 \cdot 000297\left(\frac{\partial^2 T}{\partial r^2} + \frac{1}{r}\frac{\partial T}{\partial r}\right) + 37\,000 r_c = 0 \qquad\qquad \text{VIII}$$

$$\frac{\partial x}{\partial z} - 0 \cdot 000427\left(\frac{\partial^2 x}{\partial r^2} + \frac{1}{r}\frac{\partial x}{\partial r}\right) - 164 r_c = 0 \qquad\qquad \text{IX}$$

After the reaction mixture has passed a distance z through the bed, a fraction (x) of the ethyl benzene will have been converted. Thus the reaction mixture will consist of:

10	moles steam
$1-x$	moles ethyl benzene
x	moles styrene
x	moles hydrogen
$\overline{11+x}$	total

Since the total pressure is 1·2 bar, the partial pressures of the reactants are

$$p_E = \frac{1\cdot2(1-x)}{11+x}, \quad p_S = p_H = \frac{1\cdot2x}{11+x} \qquad\qquad \text{X}$$

Substituting equations II and X into equation I gives

$$r_c = 15\ 100 \exp\left(-11\ 000/T\right)\left[\frac{1-x}{11+x} - \frac{1\cdot2x^2}{K(11+x)^2}\right] \qquad \text{XI}$$

where K is given by equation III.

Two of the boundary conditions for this problem are

when $\qquad\qquad\qquad\qquad z = 0, \quad T = 873 \qquad\qquad\qquad\qquad$ XII

and when $\qquad\qquad\qquad r = 0, \quad \dfrac{\partial T}{\partial r} = 0 \qquad\qquad\qquad\qquad$ XIII

The third condition expresses the fact that the flue gas loses heat to the reaction mixture through the wall of the tube. Taking a heat balance over an element of the tube of length δz,

$$\delta z\, RC_p{}'\frac{\partial T}{\partial z} = 2\pi a\, \delta z\, k_E \frac{\partial T}{\partial r}\bigg|_{r=a}$$

or $\qquad\qquad\qquad \dfrac{\partial T}{\partial z} = \dfrac{2\pi a k_E}{RC_p{}'}\dfrac{\partial T}{\partial r}\bigg|_{r=a} \qquad\qquad\qquad$ XIV

Inserting the numerical values, equation XIV becomes

$$\frac{\partial T}{\partial z} = 0\cdot0039\,\frac{\partial T}{\partial r}\bigg|_{r=a} \qquad\qquad \text{XV}$$

The corresponding boundary conditions for the mass transfer equation IX are

when $\qquad\qquad\qquad\qquad z = 0, \quad x = 0 \qquad\qquad\qquad\qquad$ XVI

when $\qquad\qquad\qquad\qquad r = 0, \quad \dfrac{\partial x}{\partial r} = 0 \qquad\qquad\qquad\qquad$ XVII

and when $\qquad\qquad\qquad r = a, \quad \dfrac{\partial x}{\partial r} = 0 \qquad\qquad\qquad\qquad$ XVIII

This completes the mathematical description of the problem and to start the Crank–Nicolson method equations VIII and IX are expressed in finite difference form, correct to second order, at the point E in Fig. 11.10 where $z = n\,\Delta z$ and $r = m\,\Delta r$. Whereas in Section 11.8.2 it was necessary to average two expressions for the second space derivative as in equation (11.76), it is now necessary to take a similar average for all terms except the first in both equation VIII and equation IX. Thus

$$\frac{T_{m,n+1}-T_{m,n}}{\Delta z} - \frac{0\cdot000297}{2}\left[\frac{T_{m+1,n}-2T_{m,n}+T_{m-1,n}}{(\Delta r)^2} + \frac{T_{m+1,n}-T_{m-1,n}}{2m(\Delta r)^2}\right.$$

$$\left. +\frac{T_{m+1,n+1}-2T_{m,n+1}+T_{m-1,n+1}}{(\Delta r)^2} + \frac{T_{m+1,n+1}-T_{m-1,n+1}}{2m(\Delta r)^2}\right]$$

$$+18\ 500[r_c(n+1)+r_c(n)] = 0 \qquad\qquad\text{XIX}$$

and

$$\frac{x_{m,n+1}-x_{m,n}}{\Delta z} - \frac{0\cdot000427}{2}\left[\frac{x_{m+1,n}-2x_{m,n}+x_{m-1,n}}{(\Delta r)^2} + \frac{x_{m+1,n}-x_{m-1,n}}{2m(\Delta r)^2}\right.$$

$$\left. +\frac{x_{m+1,n+1}-2x_{m,n+1}+x_{m-1,n+1}}{(\Delta r)^2} + \frac{x_{m+1,n+1}-x_{m-1,n+1}}{2m(\Delta r)^2}\right]$$

$$-82[r_c(n+1)+r_c(n)] = 0 \qquad\qquad\text{XX}$$

Because the coefficients in the two equations XIX and XX are not equal, a separate modulus needs to be defined for each equation. Thus, define

$$M = 0\cdot000297\Delta z/(\Delta r)^2 \qquad\qquad\text{XXI}$$

$$M' = 0\cdot000427\Delta z/(\Delta r)^2 \qquad\qquad\text{XXII}$$

and rearrange equations XIX and XX to the form

$$(1+M)T_{m,n+1} = \tfrac{1}{2}M\left[\left(1+\frac{1}{2m}\right)(T_{m+1,n}+T_{m+1,n+1})\right.$$

$$\left. +\left(1-\frac{1}{2m}\right)(T_{m-1,n}+T_{m-1,n+1})\right]$$

$$+(1-M)T_{m,n} - 18\ 500[r_c(n+1)+r_c(n)]\Delta z \quad\text{XXIII}$$

$$(1+M')x_{m,n+1} = \tfrac{1}{2}M'\left[\left(1+\frac{1}{2m}\right)(x_{m+1,n}+x_{m+1,n+1})\right.$$

$$\left. +\left(1-\frac{1}{2m}\right)(x_{m-1,n}+x_{m-1,n+1})\right]$$

$$+(1-M')x_{m,n} + 82[r_c(n+1)+r_c(n)]\Delta z \qquad\text{XXIV}$$

Before values are chosen for M and M', it is necessary to inspect equations XXIII and XXIV. Thus it is seen that when $m = 0$, four of the coefficients become infinite, and different equations must be found to represent VIII and IX along the axis of the tube. Boundary conditions XIII and XVII show that an indeterminate fraction arises in each of VIII and IX and L'Hôpital's rule must be used to resolve these terms. Thus

$$\lim_{r\to 0}\frac{\partial T/\partial r}{r} = \frac{\partial^2 T}{\partial r^2} \qquad\qquad\text{XXV}$$

The finite difference equations for use on the axis are thus

$$(1+2M)\,T_{0,n+1} = 2M(T_{1,n}+T_{1,n+1})+(1-2M)\,T_{0,n}$$
$$-18\,500[r_c(n+1)+r_c(n)]\,\Delta z \qquad \text{XXVI}$$

$$(1+2M')\,x_{0,n+1} = 2M'(x_{1,n}+x_{1,n+1})+(1-2M')\,x_{0,n}$$
$$+82[r_c(n+1)+r_c(n)]\,\Delta z \qquad \text{XXVII}$$

where the finite difference representations of boundary conditions XIII and XVII have been used to replace $T_{-1,n}$ with $T_{1,n}$, $x_{-1,n}$ with $x_{1,n}$ and similarly for $T_{-1,n+1}$ and $x_{-1,n+1}$.

Since $M' > M$ it is convenient to choose a suitable small value for M' to ensure that convergence problems will be no more severe on the equation containing M. If the ordinary step-by-step method was being used, the maximum value allowed for M' would be 0·25, so this value will also be chosen here. Therefore, select

$$M' = 0\cdot25, \quad M = 0\cdot174 \qquad \text{XXVIII}$$

Choose five radial increments ($\Delta r = 1$ cm) so that equation XXI or equation XXII gives the consistent value

$$\Delta z = 0\cdot0585 \text{ m} \qquad \text{XXIX}$$

Two new variables, $F_{m,n}$ and $G_{m,n}$, are defined as before to collect together all terms at the time level n. Thus

$$F_{m,n} = 0\cdot0741\left[\left(1+\frac{1}{2m}\right)T_{m+1,n}+\left(1-\frac{1}{2m}\right)T_{m-1,n}\right]$$
$$+0\cdot7036T_{m,n}-922r_c(n) \qquad \text{XXX}$$

$$G_{m,n} = 0\cdot1\left[\left(1+\frac{1}{2m}\right)x_{m+1,n}+\left(1-\frac{1}{2m}\right)x_{m-1,n}\right]$$
$$+0\cdot6x_{m,n}+3\cdot84r_c(n) \qquad \text{XXXI}$$

$$F_{0,n} = 0\cdot258T_{1,n}+0\cdot484T_{0,n}-803r_c(n) \qquad \text{XXXII}$$
$$G_{0,n} = 0\cdot333x_{1,n}+0\cdot334x_{0,n}+3\cdot20r_c(n) \qquad \text{XXXIII}$$

The equations to be solved iteratively (XXIII, XXIV, XXVI, XXVII) thus become

$$T_{m,n+1} = F_{m,n}+0\cdot0741\left[\left(1+\frac{1}{2m}\right)T_{m+1,n+1}+\left(1-\frac{1}{2m}\right)T_{m-1,n+1}\right]$$
$$-922r_c(n+1) \qquad \text{XXXIV}$$

$$x_{m,n+1} = G_{m,n}+0\cdot1\left[\left(1+\frac{1}{2m}\right)x_{m+1,n+1}+\left(1-\frac{1}{2m}\right)x_{m-1,n+1}\right]$$
$$+3\cdot84r_c(n+1) \qquad \text{XXXV}$$

$$T_{0,n+1} = F_{0,n}+0\cdot258T_{1,n+1}-803r_c(n+1) \qquad \text{XXXVI}$$
$$x_{0,n+1} = G_{0,n}+0\cdot333x_{1,n+1}+3\cdot20r_c(n+1) \qquad \text{XXXVII}$$

Boundary condition XVIII takes the form

$$x_{6,n} = x_{4,n} \qquad\qquad \text{XXXVIII}$$

but boundary condition XV presents a new problem. Although the temperature at the wall $(T_{5,n})$ will equal the temperature of the flue gas, the latter is a function of z which is related to the radial temperature gradient in the reactor tube. Introducing a row of "fictitious" points as described in Section 11.8.5 and averaging the finite difference representations of the radial derivatives gives

$$\frac{T_{5,n+1} - T_{5.n}}{\Delta z} = \frac{0\cdot0039}{2}\left[\frac{T_{6,n+1} + T_{6,n} - T_{4,n+1} - T_{4,n}}{2\Delta r}\right] \qquad \text{XXXIX}$$

for the representation of equation XV. Rearranging, and solving for $T_{6.n+1}$ gives

$$T_{6,n+1} = T_{4,n+1} + T_{4,n} - T_{6,n} + \frac{4\Delta r}{0\cdot0039\Delta z}(T_{5,n+1} - T_{5,n}) \qquad \text{XL}$$

Putting $\Delta r = 1$ cm and $\Delta z = 5\cdot85$ cm as before gives the coefficient in front of the last bracket the value 175·3. During the iterative search, the value of $T_{5,n+1}$ will clearly contain an error and this large coefficient (175·3) will magnify the error into $T_{6,n+1}$ via equation XL and the numerical procedure will not converge. The critical coefficient in equation XL should be numerically less than unity for convergence in any other similar problem, but here it is necessary to modify the procedure. The normal intention would be to use equation XXIII with $m = 5$ to calculate $T_{5,n+1}$ from the fictitious value $T_{6,n+1}$ and use equation XL to calculate $T_{6,n+1}$ from $T_{5,n+1}$. It is clear that $T_{6,n+1}$ from equation XL can be substituted algebraically into equation XXIII to avoid this numerical difficulty. Thus

$$(1 + M)\,T_{5,n+1} = \tfrac{1}{2}M\{(1 + \tfrac{1}{10})\,[T_{4,n+1} + T_{4,n} + 175\cdot3(T_{5,n+1} - T_{5,n})]$$
$$+ (1 - \tfrac{1}{10})\,(T_{4,n+1} + T_{4,n})\} + (1 - M)\,T_{5,n}$$
$$- 18\,500[r_o(n+1) + r_o(n)]\,\Delta z \qquad \text{XLI}$$

which can be rearranged to give

$$(95\cdot4M - 1)\,T_{5,n+1} = (97\cdot4M - 1)\,T_{5,n} - M(T_{4,n+1} + T_{4,n})$$
$$+ 18\,500[r_o(n+1) + r_o(n)]\,\Delta z \qquad \text{XLII}$$

Putting in the numerical values gives

$$T_{5,n+1} = 1\cdot0224T_{5,n} - 0\cdot0112(T_{4,n} + T_{4,n+1}) + 69\cdot4[r_o(n+1) + r_o(n)]$$

Redefining the variable $F_{5,n}$ by

$$F_{5,n} = 1\cdot0224T_{5,n} - 0\cdot0112T_{4,n} + 69\cdot4r_o(n) \qquad \text{XLIII}$$

finally gives for the iterative equation

$$T_{5,n+1} = F_{5,n} - 0\cdot0112T_{4,n+1} + 69\cdot4r_o(n+1) \qquad \text{XLIV}$$

The initial conditions given by equations XII and XVI take the form

$$T_{5,0} = 893 \qquad \text{XLV}$$

$$T_{m,0} = 873 \quad (0 \leqslant m \leqslant 4) \qquad \text{XLVI}$$

$$x_{m,0} = 0 \quad (0 \leqslant m \leqslant 6) \qquad \text{XLVII}$$

This completes the conversion of the problem to numerical form and the logic flow diagram on which a computer program can be based is illustrated in Fig. 11.14. It should be appreciated that a fair amount of arithmetic has been done during the above presentation and that a computer program using these numerical values is specific to this one problem. To obtain a more useful, general program the reader should repeat the above development from equation VII, retaining the algebraic coefficients and writing the computer program to evaluate all coefficients from the experimental data. If Δr and Δz are always chosen to make $M' = 0.25$, whatever the numerical value in equation XXII, the only numerical changes will be confined to the coefficients of all reaction terms and changes in the temperature equations due to M differing from 0.174.

The required solution is the length of reactor needed for 45% conversion. The average conversion at any length z is given by

$$\bar{x} = \int_0^a rx \, dr \Big/ \int_0^a r \, dr \qquad \text{XLVIII}$$

Because five increments were chosen, Simpson's rule (Section 10.7.2) cannot be used easily for the integration. The trapezium rule (Section 10.7.1), however, is applicable and gives

$$\bar{x} = \frac{2}{a^2} \frac{a^2}{50} (2x_{1,n} + 4x_{2,n} + 6x_{3,n} + 8x_{4,n} + 5x_{5,n})$$

$$\therefore \quad \bar{x} = \tfrac{1}{25}(2x_{1,n} + 4x_{2,n} + 6x_{3,n} + 8x_{4,n} + 5x_{5,n}) \qquad \text{XLIX}$$

The calculation should thus proceed by taking as many steps of length Δz as are necessary to make $\bar{x}$ as calculated from equation XLIX greater than 0.45.

The solution is presented in Table 11.9. Interpolation between the last two rows gives the required reactor length

$$L = 17.3\Delta z = 1.012 \text{ m} \qquad \text{L}$$

11.10. Partial Differential Equations (Boundary Value Type)

In the previous section it was clear that any of the methods presented in Section 11.3 for initial value ordinary differential equations could be successfully used to solve an initial value partial differential equation. Unfortunately, the methods used for ordinary differential equations of

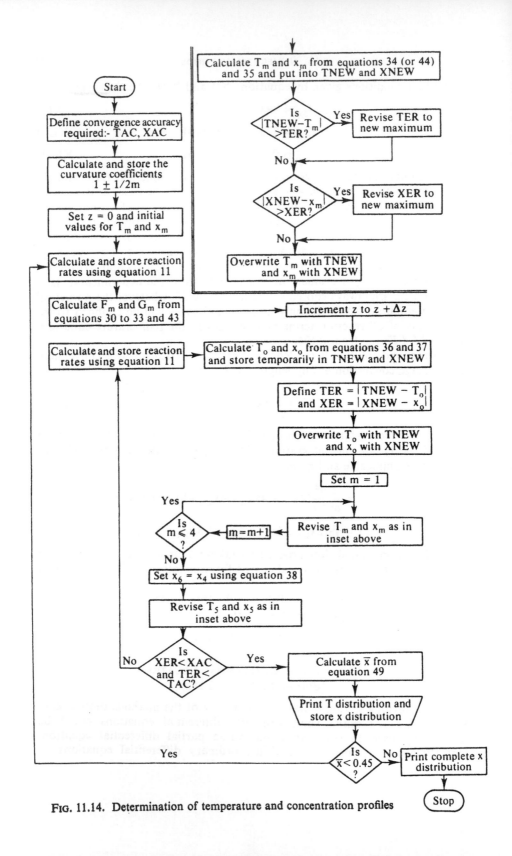

FIG. 11.14. Determination of temperature and concentration profiles

TABLE 11.9. *Iterative Solution for Tubular Reactor*

		Temperatures at $m =$					
n	z	0	1	2	3	4	5
1	0·0585	863·8	863·8	863·8	864·1	867·8	894·3
2	0·1170	856·0	856·1	856·2	857·3	864·3	895·8
3	0·1755	849·4	849·4	849·8	851·9	861·9	897·3
4	0·2340	843·6	843·7	844·5	847·7	860·3	898·8
5	0·2925	838·6	838·8	840·0	844·5	859·4	900·4
6	0·3510	834·2	834·6	836·2	842·0	859·0	902·0
7	0·4095	830·4	830·9	833·1	840·1	859·1	903·7
8	0·4680	827·1	827·7	830·6	838·8	859·4	905·3
9	0·5265	824·2	825·1	828·5	837·9	860·0	907·0
10	0·5850	821·8	822·8	826·9	837·4	860·8	908·7
11	0·6435	819·7	821·0	825·7	837·2	861·7	910·4
12	0·7020	818·1	819·6	824·9	837·3	862·8	912·0
13	0·7605	816·8	818·4	824·3	837·6	864·1	913·7
14	0·8190	815·8	817·6	824·1	838·2	865·4	915·4
15	0·8775	815·1	817·1	824·1	838·9	866·9	917·0
16	0·9360	814·7	816·9	824·3	839·8	868·4	918·7
17	0·9945	814·5	816·9	824·8	840·9	869·9	920·3
18	1·0530	814·6	817·1	825·4	842·1	871·5	921·9

		Conversion at $m =$					
n	$\bar{x}$	0	1	2	3	4	5
1	0·0445	0·0408	0·0408	0·0409	0·0412	0·0433	0·0548
2	0·0840	0·0753	0·0753	0·0756	0·0771	0·0839	0·1026
3	0·1197	0·1050	0·1051	0·1060	0·1096	0·1218	0·1454
4	0·1525	0·1311	0·1315	0·1333	0·1398	0·1570	0·1842
5	0·1829	0·1545	0·1552	0·1585	0·1681	0·1899	0·2198
6	0·2113	0·1758	0·1771	0·1820	0·1949	0·2208	0·2527
7	0·2379	0·1957	0·1975	0·2042	0·2203	0·2498	0·2834
8	0·2632	0·2143	0·2167	0·2254	0·2445	0·2772	0·3120
9	0·2871	0·2321	0·2352	0·2457	0·2677	0·3030	0·3388
10	0·3099	0·2492	0·2529	0·2653	0·2899	0·3276	0·3641
11	0·3316	0·2658	0·2701	0·2842	0·3112	0·3509	0·3880
12	0·3525	0·2820	0·2869	0·3026	0·3316	0·3731	0·4106
13	0·3724	0·2978	0·3033	0·3204	0·3514	0·3943	0·4320
14	0·3916	0·3134	0·3193	0·3378	0·3704	0·4145	0·4524
15	0·4101	0·3286	0·3350	0·3547	0·3887	0·4339	0·4719
16	0·4278	0·3436	0·3504	0·3711	0·4064	0·4525	0·4904
17	0·4450	0·3584	0·3655	0·3872	0·4236	0·4703	0·5082
18	0·4615	0·3729	0·3803	0·4028	0·4402	0·4875	0·5252

boundary value type in Section 11.4 are not all suitable for partial differential equations. In particular, any attempt to use an initial value method (such as Runge–Kutta) with estimates for the unknown initial conditions requires too many values to be estimated and revised. However, the methods presented in Section 11.4.1 can be readily extended.

11.10.1. *Liebmann's Method for Boundary Value Problems*

In this method, equations (11.40) and (11.41) are used to replace all derivatives in the partial differential equation with finite differences. Considering the two-dimensional version of Laplace's equation in cartesian coordinates

$$\frac{\partial^2 T}{\partial x^2} + \frac{\partial^2 T}{\partial y^2} = 0 \qquad (11.84)$$

and using equation (11.41) to replace both derivatives gives

$$\frac{T_{m+1,n} - 2T_{m,n} + T_{m-1,n}}{(\Delta x)^2} + \frac{T_{m,n+1} - 2T_{m,n} + T_{m,n-1}}{(\Delta y)^2} = 0 \qquad (11.85)$$

where $x = m\,\Delta x$ and $y = n\,\Delta y$. Rearrangement of equation (11.85) gives

$$2(1 + k^2)\,T_{m,n} = T_{m+1,n} + T_{m-1,n} + k^2(T_{m,n+1} + T_{m,n-1}) \qquad (11.86)$$

where $$k = \Delta x / \Delta y \qquad (11.87)$$

In Section 11.4.1, Table 11.2 showed that values for the dependent variable at each point were calculated on alternate cycles of the calculation and the same feature is present when the method is applied to partial differential equations. In the earlier example, temperatures were estimated at points where the single suffix (n) was odd and the calculation alternated between the points with even suffix and those with odd suffix. Here, temperature must be estimated at points where the sum of the suffices ($m+n$) is odd and the calculation using equation (11.86) will alternate between the points where the sum of the suffices is even and those where the sum is odd. The main difficulty with this simple method is that convergence is very slow for a manual calculation, and programming for the alternating cycle of calculations is unnecessarily complicated for machine calculation. It is much simpler to write a computer program which will apply equation (11.86) to every point on every cycle of calculation. A further modification is analogous to that developed in the Crank–Nicolson method (Section 11.8.2) where it was found convenient to use the latest values available on the current cycle rather than the values retained from the last cycle. Specifically, equation (11.86) is used in the form

$$2(1 + k^2)\,T_{m,n}^{r+1} = T_{m+1,n}^{r} + T_{m-1,n}^{r+1} + k^2(T_{m,n+1}^{r} + T_{m,n-1}^{r+1})$$

where the superscript r refers to values from the previous cycle and $r+1$ refers to new values available on the current cycle. With this technique it is only necessary to keep one value for the dependent variable at each point in the x–y plane, thus reducing the computer storage requirement.

Example. Find the velocity distribution for steady incompressible laminar flow in a rectangular duct of aspect ratio 2 : 1. Also determine the average velocity in such a duct as a proportion of the average velocity of the same fluid flowing through a cylindrical pipe of the same cross-sectional area with the same pressure gradient.

Solution

The Navier–Stokes equation (7.126) governs the fluid velocity, and taking cartesian coordinates x and y measured along the edges of the rectangular cross section, and z along the axis of the duct, the fluid velocity will only have a z component denoted by u. The velocity will be a function of x and y only, hence all terms arising from $\mathbf{u}.\nabla\mathbf{u}$ will be zero. Since there are no external forces acting, and the fluid flow is steady, equation (7.126) becomes

$$-\frac{1}{\rho}\nabla p + \nu\nabla^2\mathbf{u} = 0 \qquad\qquad \text{I}$$

Because $\mathbf{u}$ only has a z component, the x and y components of equation I become

$$\frac{\partial p}{\partial x} = 0, \quad \frac{\partial p}{\partial y} = 0 \qquad\qquad \text{II}$$

whereas the z component of equation I gives

$$\frac{\partial^2 u}{\partial x^2} + \frac{\partial^2 u}{\partial y^2} + \alpha = 0 \qquad\qquad \text{III}$$

where α is a constant given by

$$\alpha = -\frac{1}{\mu}\frac{\partial p}{\partial z} \qquad\qquad \text{IV}$$

Equation III is "Poisson's equation". Using equation (11.41) to replace the derivatives in equation III gives

$$\frac{u_{m+1,n} - 2u_{m,n} + u_{m-1,n}}{(\Delta x)^2} + \frac{u_{m,n+1} - 2u_{m,n} + u_{m,n-1}}{(\Delta y)^2} + \alpha = 0 \qquad\qquad \text{V}$$

Because the ratio of the lengths of the sides of the duct are in simple proportion, it is convenient to choose $\Delta y = \Delta x$. Hence equation V becomes

$$u_{m,n} = \tfrac{1}{4}\alpha(\Delta x)^2 + \tfrac{1}{4}(u_{m+1,n} + u_{m-1,n} + u_{m,n+1} + u_{m,n-1}) \qquad\qquad \text{VI}$$

Denoting the length of the shorter side of the rectangle by L and choosing $\Delta x = \tfrac{1}{8}L$ will produce a network of 7×15 points inside the rectangular duct.

For numerical convenience αL^2 is given the value 256 so that the constant in equation VI becomes 1·0 thus

$$u_{m,n} = 1\cdot0 + \tfrac{1}{4}(u_{m+1,n} + u_{m-1,n} + u_{m,n+1} + u_{m,n-1}) \qquad \text{VII}$$

The above numerical value for αL^2 means that the total force per unit length causing the flow will be given by

$$-2L^2\frac{\partial p}{\partial z} = 2\mu\alpha L^2 = 512\mu \qquad \text{VIII}$$

The same force applied along unit length of a cylindrical pipe of the same cross-sectional area would produce an average velocity ($\bar{u}_c$) given by

$$-2L^2\frac{\partial p}{\partial z} = 8\pi\mu\bar{u}_c = 512\mu$$

$$\therefore \quad \bar{u}_c = 64/\pi = 20\cdot37 \qquad \text{IX}$$

This is the numerical value which will be needed for comparison with the average velocity in the rectangular duct.

Because of symmetry in the duct, it is only necessary to carry out the calculations in a quarter section of the duct. Denoting the walls of the duct by $m = 0$ and $n = 0$, the planes of symmetry will be given by $m = 4$ and $n = 8$. The boundary conditions thus become

$$u_{m,0} = u_{0,n} = 0 \qquad \text{X}$$

$$u_{5,n} = u_{3,n} \qquad \text{XI}$$

and $\qquad\qquad u_{m,9} = u_{m,7} \qquad\qquad\qquad\qquad\qquad\qquad \text{XII}$

Equations X imply that there is no slip at the solid boundary, and equations XI and XII imposed planes of symmetry at $m = 4$ and $n = 8$. The logic flow diagram illustrated in Fig. 11.15 can now be constructed from which the computer program may be written. The velocity distribution obtained is given in Table 11.10.

TABLE 11.10. *Velocity Distribution in a Rectangular Duct*

m\n	0	1	2	3	4	5	6	7	8
0	0	0	0	0	0	0	0	0	0
1	0	4·963	7·990	9·926	11·186	11·999	12·501	12·774	12·861
2	0	7·863	13·070	16·529	18·819	20·308	21·233	21·737	21·896
3	0	9·419	15·900	20·303	23·253	25·185	26·388	27·044	27·252
4	0	9·912	16·810	21·530	24·707	26·792	28·092	28·801	29·026

The last part of the problem involves finding the average velocity in the duct. Simpson's rule (Section 10.7.2) can be used since it is basically more accurate than the finite difference representation used for the differential

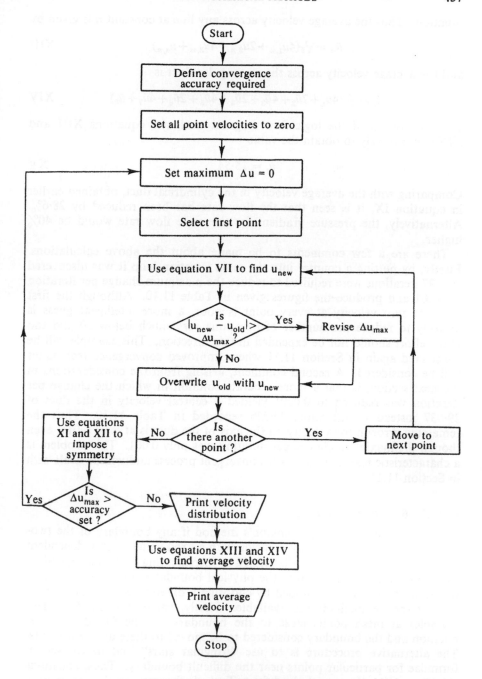

FIG. 11.15. Velocity profiles in duct

equation. Thus the average velocity across any line at constant n is given by

$$\bar{u}_n = \tfrac{1}{12}(4u_{1,n} + 2u_{2,n} + 4u_{3,n} + u_{4,n})$$ XIII

and the average velocity across the rectangular duct is

$$\bar{u}_r = \tfrac{1}{24}(4\bar{u}_1 + 2\bar{u}_2 + 4\bar{u}_3 + 2\bar{u}_4 + 4\bar{u}_5 + 2\bar{u}_6 + 4\bar{u}_7 + \bar{u}_8)$$ XIV

The last section of the logic flow diagram evaluates equations XIII and XIV consecutively to obtain the value

$$\bar{u}_r = 15\cdot54$$ XV

Comparing with the average velocity in the cylindrical duct, obtained earlier in equation IX, it is seen that the flow rate has been reduced by 28·6%. Alternatively, the pressure gradient for the same flow rate would be 40% higher.

There are a few comments to be made about the above calculations. Firstly, by putting a counter into the main iteration loop it was discovered that 97 iterations were required to reduce the maximum change per iteration to 0·001 and produce the figures given in Table 11.10. Although the first guess of zero velocity at every point is poor, a more intelligent guess is unlikely to reduce the number of iterations very much below 90 and too much effort should not be expended in this direction. This example will be considered again in Section 11.11 where improved convergence techniques will be considered. A second comment, arising from the considerations in the next section, is that a more accurate solution in which the change per iteration was reduced to 0·0001 yielded a central velocity in the duct of 29·037 instead of the value 29·026 recorded in Table 11.10. Thus the remaining error in the velocity on the axis of the duct is at least 0·011, even though the maximum change per iteration was only 0·001. This problem is a characteristic feature of a slowly convergent process and will be dealt with in Section 11.11.

11.10.2. *Boundary Conditions*

Difficulties can arise in Liebmann's method if any boundary of the two-dimensional problem does not coincide with a fixed value of an independent variable. In such a case, the cells defined by incrementing the two independent variables do not fit neatly into the physical boundaries of the problem. If this situation cannot be avoided by choosing a different coordinate system, two alternative methods are available. Firstly, values for the dependent variables at mesh points close to the boundary can be found by interpolation and the boundary considered to be moved to these adjacent points. The alternative procedure is to use "irregular stars" and derive special formulae for particular points near the difficult boundary. Thus, equations (11.40) and (11.41) were derived from Taylor's theorem on the assumption of equal intervals; without this assumption the following equations are approximately true.

$$f(x_0 + a) = f(x_0) + af'(x_0) + \tfrac{1}{2}a^2 f''(x_0) \qquad (11.88)$$

$$f(x_0 - b) = f(x_0) - bf'(x_0) + \tfrac{1}{2}b^2 f''(x_0) \qquad (11.89)$$

Solving equations (11.88) and (11.89) for the derivatives gives

$$f'(x_0) = \frac{b^2 f(x_0 + a) - a^2 f(x_0 - b) - (b^2 - a^2) f(x_0)}{ab(a + b)} \qquad (11.90)$$

$$f''(x_0) = \frac{bf(x_0 + a) + af(x_0 - b) - (a + b) f(x_0)}{\tfrac{1}{2}ab(a + b)} \qquad (11.91)$$

which can be used instead of equations (11.40) and (11.41) when a boundary is interposed between the natural points of the mesh. Equations (11.90) and (11.91) degenerate to equations (11.40) and (11.41) when $a = b = \Delta x$.

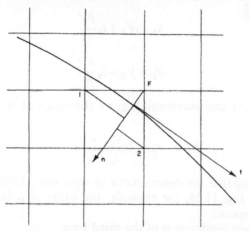

FIG. 11.16. Derivative boundary condition crossing coordinate system

So far it has been assumed that either the boundary conditions are of the first type (function specified) or that the boundary corresponds to a fixed value of an independent variable. On rare occasions, a derivative boundary condition is imposed at a boundary which runs across the coordinate system as illustrated in Fig. 11.16. The usual technique with derivative boundary conditions is to use fictitious points to extend the problem beyond the boundary. Thus consider the fictitious point F and define a temporary coordinate system with an axis (n) chosen through F and normal to the boundary with an axis (t) tangential to the boundary at the point where the n axis crosses the boundary. Choose two internal points (1 and 2 in the diagram) and calculate their coordinates (n_1, t_1) and (n_2, t_2) in the temporary system. Denoting the dependent variable by V and applying Taylor's theorem, ignoring second and higher derivatives, gives

$$V_1 = V_0 + n_1 \frac{\partial V}{\partial n} + t_1 \frac{\partial V}{\partial t} \tag{11.92}$$

$$V_2 = V_0 + n_2 \frac{\partial V}{\partial n} + t_2 \frac{\partial V}{\partial t} \tag{11.93}$$

$$V_F = V_0 + n_F \frac{\partial V}{\partial n} \tag{11.94}$$

If the boundary condition is of the second type

$$\frac{\partial V}{\partial n} = 0 \tag{11.95}$$

equations (11.92), (11.93), and (11.94) take the simple form

$$V_1 = V_F + t_1 \frac{\partial V}{\partial t} \tag{11.96}$$

$$V_2 = V_F + t_2 \frac{\partial V}{\partial t} \tag{11.97}$$

Eliminating $\partial V/\partial t$ and rearranging gives the value at the fictitious point. Thus

$$V_F = \frac{t_2 V_1 - t_1 V_2}{t_2 - t_1} \tag{11.98}$$

It is worth noting that the denominator in equation (11.98) is most unlikely to be zero. In Fig. 11.16, for example, the value of t_2 is positive and the value of t_1 is negative.

If the boundary condition is of the third type

$$\frac{\partial V}{\partial n} = \alpha(V_0 - V_s) \tag{11.99}$$

the algebra becomes quite complicated. It is necessary to solve equations (11.92), (11.93), and (11.99) for V_0 and $\partial V/\partial n$ after eliminating $\partial V/\partial t$, and substitute into equation (11.94). The result is

$$V_F = \frac{(t_2 V_1 - t_1 V_2)(1 + \alpha n_F) + \alpha V_s(n_1 t_2 - n_2 t_1 - n_F t_2 + n_F t_1)}{t_2 - t_1 + \alpha(n_1 t_2 - n_2 t_1)} \tag{11.100}$$

Geometrically, the above method involves fitting the plane surface through the values of the dependent variable at points 1 and 2 which satisfies the given boundary condition where the plane crosses the boundary.

It will be appreciated from the contents of this subsection that it is wise to chose a coordinate system which avoids the problem of derivative boundary conditions at inclined boundaries.

11.11. CONVERGENCE ACCELERATION

In all iterative processes, a recurrence relation is used repeatedly to calculate fresh estimates for an unknown quantity or a set of unknown quantities. It has been tacitly assumed so far that if the change in the estimated value is small from one iteration to the next, then the latest estimate is close to the true answer. A little thought will show that such an assumption is questionable. Suppose that an iterative process is being used to determine an unknown quantity V^* and that the consecutive estimates are plotted as shown in Fig. 11.17. There is no reason why the remaining

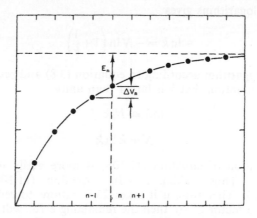

FIG. 11.17. Iterative progress to solution

error (E_n) at any stage should be smaller than the change which occurred on the last step (ΔV_n), but assume that E_n is a constant multiple of ΔV_n. Thus

$$E_n = k \Delta V_n \qquad (11.101)$$

But
$$\Delta V_n = E_{n-1} - E_n \qquad (11.102)$$

$$\therefore \quad E_n = k E_{n-1} - k E_n$$

$$\therefore \quad (1+k) E_n - k E_{n-1} = 0 \qquad (11.103)$$

Equation (11.103) is a simple, first order, linear finite difference equation which can be solved by the method of Section 9.6.1. Hence

$$E_n = A \left(\frac{k}{k+1} \right)^n \qquad (11.104)$$

This equation can be used to monitor an iterative process in the following manner. Suppose that the iterative process is stopped temporarily after n iterations with a known value for ΔV_n. If the iterative process is resumed and continued until the change per iteration is reduced to $\Delta V_n/k$, this will take a further N iterations. Since the remaining error at any stage is a constant multiple of the change per iteration, as given by equation (11.101), the remaining error after a further N iterations will also be reduced by a factor k. Thus

$$\frac{1}{k}E_n = E_{n+N} = \left(\frac{k}{k+1}\right)^N E_n \qquad (11.105)$$

$$\therefore \quad \frac{1}{k} = \left(\frac{k}{k+1}\right)^N$$

Taking natural logarithms gives

$$-\ln k = -N\ln\left(1+\frac{1}{k}\right)$$

Expanding the logarithm according to equation (3.8) and retaining only one term, on the assumption that k is larger than unity,

gives
$$\ln k \simeq N/k$$

or
$$N = k\ln k \qquad (11.106)$$

The interpretation of equation (11.106) is more easily understood in a numerical case. Thus, taking $k = 10$, equation (11.106) shows that $N = 10\ln 10 = 23$. Therefore, if it takes 23 iterations to reduce the change per iteration by a factor of 10, then the remaining error will be 10 times the change on the last iteration. If the number of iterations required is fewer than 23, the answer will clearly be more accurate.

Although it has been assumed without justification that k in equation (11.101) is a constant, this assumption implies that the approach to the solution is exponential. If the recurrence relation giving V_{n+1} in terms of V_n is linear, the approach will be exponential but in general the recurrence relation will be non-linear. However, in the late stages of any convergent iterative process, the recurrence relation usually becomes approximately linear, and the applicability of equation (11.106) improves as the solution is approached.

The validity of equation (11.106) can be demonstrated with reference to the last example in which it was stated that 97 iterations were required to reduce the change per iteration to 0·001. It is a simple matter to interrupt the iteration when the change per iteration is first below 0·01. This reveals that the last factor of 10 required 28 iterations, which is greater than $10\ln 10 = 23$. Therefore, the factor k in equation (11.106) needs to be greater than 10. Clearly, if a higher degree of accuracy is requested on resumption of the iteration, more steps will be required and a few trials show that a factor of $k = 12$ substituted into equation (11.106) may be suitable. Thus

$$N = 12 \ln 12 = 30$$

The program written to implement Fig. 11.15 was modified to keep restarting the iteration after dividing the required accuracy by 12 at each stage. The mini-computer used only worked to 7 figures and of course the program failed when the change per iteration was less than 10^{-6} because the change was recorded as zero. The last answer obtained for the centre line velocity was 29·03711 with the last non-zero change per iteration as 4×10^{-6}. Assuming that the remaining error is 12 times as large as this, the expected true answer would be 29·03716. Using this value, the remaining error at each intermediate stage of the iteration can be calculated and expressed as a multiple (k) of the maximum change recorded (rather than set) for the previous iteration. These figures are given in Table 11.11 where it can be seen that equation (11.106) with $k = 12$ and $N = 30$ applies remarkably well throughout the iteration.

TABLE 11.11. *Remaining Error as Multiple of Change per Iteration*

Stage	N	Δu_{max}	Centre line velocity	Remaining error	k
1	(38)	0·118149	27·6497	1·38740	11·743
2	31	0·009384	28·9272	0·10998	11·720
3	30	0·000809	29·0277	0·00946	11·69
4	31	0·000061	29·0364	0·00074 (5)	12·2
5	33	0·000004	29·0371	0·00004 (5)	11

11.11.1 *Aitken's δ^2 Method*

The above considerations suggest that if an iterative process is converging exponentially and slowly, the error remaining at any stage can be estimated. Furthermore, the sign of the error is also known, since the approach is exponential, and it is natural to suggest that a correction should be made to eliminate the remaining error. This is the basis of Aitken's δ^2 method which can be derived from the work of the previous section. Thus equations (11.101) and (11.104) give

$$V_n = V^* - E_n = V^* - A\alpha^n \tag{11.107}$$

where α is a constant replacing $k/(k+1)$. The estimates obtained on the two previous iterations are given by

$$V_{n-1} = V^* - A\alpha^{n-1} \tag{11.108}$$

$$V_{n-2} = V^* - A\alpha^{n-2} \tag{11.109}$$

The two unknown characteristics A and α can be eliminated from equations (11.107)–(11.109) thus

$$\frac{V^* - V_{n-1}}{V^* - V_{n-2}} = \alpha = \frac{V^* - V_n}{V^* - V_{n-1}}$$

Cross-multiplying and solving for V^* gives

$$V^{*2} - 2V^* V_{n-1} + V_{n-1}^2 = V^{*2} - (V_n + V_{n-2}) V^* + V_n V_{n-2}$$

$$\therefore \quad V^* = \frac{V_n V_{n-2} - V_{n-1}^2}{V_n - 2V_{n-1} + V_{n-2}} \tag{11.110}$$

which predicts the ultimate answer (V^*) from three consecutive iterates.

This method is useful for solving single equations or systems of equation which form a simple loop such as

$$x_2 = f_1(x_1), \quad x_3 = f_2(x_2), \quad x_4 = f_3(x_3), \quad x_1 = f_4(x_4) \tag{11.111}$$

in which all unknowns can be calculated explicitly once any one has been estimated. In more complicated problems, such as boundary value partial differential equations, it is of more limited value. This is because different unknowns converge towards the solution at different rates, it is necessary to retain solutions from three consecutive iterations, and the problem is usually non-linear. If three early iterative cycles are taken, the assumption of an exponential approach is not valid; whereas if three later iterative cycles are considered, both the numerator and denominator of equation (11.110) tend to zero with serious loss of accuracy. In such problems, Aitken's method should be applied after a predetermined number of iterative cycles in order to obtain a new guess for the solution. Normal iteration should then be allowed to proceed until the transients have decayed and the approach to the solution is again exponential. The problem is that any application of Aitken's method will amplify round-off errors or non-linear errors and these must be allowed to decay through the normal iterative process before they are disturbed again, otherwise the numerical process becomes unstable.

The following method is less ambitious but much more reliable than Aitken's method for solving multi-variable problems such as those which arise in partial differential equations.

11.11.2. Over-relaxation

If it is known that the iterative technique is converging in an approximately exponential way, as illustrated in Fig. 11.17, some acceleration towards the solution can be achieved by taking a larger value for ΔV_n than the one calculated from the recurrence relation. It is usual to multiply the calculated value of ΔV_n by a constant factor (R) at every step so that

$$V_{n+1}' = V_n + R\Delta V_n \tag{11.112}$$

where V_{n+1}' is the value to be used on the next iterative cycle. The constant R is called the over-relaxation factor if $R > 1$ and the under-relaxation factor if $R < 1$. The latter situation arises when the normal iterative cycle is oscillatory and under-relaxation can be used to damp out the oscillation. It has been found by experience that if R is chosen to be too large, the modified iteration process becomes unstable, and in most applications to the solution of partial differential equations instability arises if $R \geqslant 2$. It thus appears that if the natural iterative cycle only advances say 10% towards the solution at each step, very little is gained by putting in an over-relaxation factor which advances the solution to less than 20%. However, the acceleration is much greater than appears at first sight since the improvement occurs at every step and is cumulative. To illustrate this improvement the solution to the example in Section 11.10.1 will be repeated using various over-relaxation factors.

Example. Flow in a Rectangular Duct (continued from Section 11.10.1).
The formula used previously was

$$u_{m,n} = 1 \cdot 0 + \tfrac{1}{4}(u_{m+1,n} + u_{m-1,n} + u_{m,n+1} + u_{m,n-1}) \qquad \text{VII}$$

and at each point (m, n) the logic flow diagram (Fig. 11.15) indicated that equation VII should be used to calculate u_{new}. The next step involves calculating $(u_{\text{new}} - u_{\text{old}}) = \Delta u$ which corresponds to the calculated value of ΔV_n in equation (11.112). When the over-relaxation technique is applied, equation (11.112) takes the form

$$u_{\text{new}}' = u_{\text{old}} + R(u_{\text{new}} - u_{\text{old}}) \qquad \text{XVI}$$

or

$$u_{\text{new}}' = R u_{\text{new}} + (1 - R) u_{\text{old}} \qquad \text{XVII}$$

Therefore, the only alteration necessary in the logic flow diagram (Fig. 11.15) is to overwrite u_{old} with u_{new}' calculated from equation XVII instead of overwriting with u_{new} calculated from equation VII.

TABLE 11.12. *Effect of Over-Relaxation on Convergence Rate*

R	1·0	1·2	1·4	1·6	1·65	1·7	1·8	2·0
N	97	70	51	36	34	38	87	∞

In order to compare the effects of using different values for R, it is necessary to count the number of iterations used in the solution, and ensure that each solution is obtained to the same degree of accuracy by iterating until Δu_{max} falls below the same value, where Δu_{max} is calculated from u_{new} and not from u_{new}'. The number of iterations (N) required to reproduce Table 11.10 from an empty array is given in Table 11.12 as a function of the over-relaxation factor R. The ordinary iterative process, given by $R = 1$, took 97 iterations whereas the best, over-relaxed iterative process with $R = 1 \cdot 65$ only took 34 iterations thus saving over 60% of the computer time at a cost of two multiplications and two additions to evaluate equation XVII per iteration.

Of course, the major problem remaining is how to choose the best value for R. In the above example, the problem had to be solved 8 times to find the best way, and in normal circumstances any one solution would be good enough. The figures presented in Table 11.12 show the typical behaviour of convergence rate with R in that significant improvements are obtained with quite small values of R (1·2, 1·4, for example), but if the value selected for R is a little too large (1·8 for example) a potential gain can easily be turned into a loss. It is therefore desirable to choose a value for R which is less than the optimum. With a new problem it is impossible to know the optimum value for R and the safest plan in an unknown situation is to choose $R = 1·3$. If a group of similar problems are to be solved, some experimentation with different values of R for each problem is justified so that favourable values can be used in later calculations.

The authors have applied over-relaxation factors to Liebmann's method, as just described, and also to the Crank–Nicolson method. It appears that the optimum over-relaxation factor varies, even within the same problem, as the space increment is varied and also as the modulus is varied in the Crank–Nicolson method. However, in the Crank–Nicolson method with updated values as described in Section 11.8.2 the optimum value for R is about 1·3 whereas if all values are used from the previous cycle, the optimum value for R is greater than 1·5 and sometimes exceeds 1·7. It has already been seen from Table 11.12 that with updated values in Liebmann's method the optimum is above 1·5, and using all old values from the previous cycle the optimum value for R may be very close to 2.

11.12. PARTIAL DIFFERENTIAL EQUATIONS WITH THREE INDEPENDENT VARIABLES

The extra independent variable presents no new conceptual difficulties but the amount of calculation required for a solution is an order of magnitude greater. If time is one of the variables, the problem is still classified as an initial value problem which can be solved step-by-step or preferably by an extension of the Crank–Nicolson method. In the latter method, the calculating equation relates values at five points in a plane at the new time level to values at the same five points at the old time level.

Extra problems arise in a three-dimensional boundary value problem if the boundaries are not all of simple shape or if computer core storage is limited. The calculating equation for each point will relate seven point values rather than five and might require six different coefficients. Therefore it may be necessary to store seven values for each point. Taking twenty space increments in each dimension could therefore require $7 \times 20^3 = 56\,000$ variables to be stored. This is still within the capacity of large computers but is well beyond the capacity of mini-computers. An efficient way of using file storage to overcome this problem is to pack the data in files, one plane after another, and define three two-dimensional arrays in core to contain three consecutive planes. Using FORTRAN for the purpose of describing an efficient procedure, it is necessary to write a subroutine with three two-dimensional arrays, A, B, C in its argument list; the purpose of the subroutine being to update the values in the central plane B. Assuming that the calculation is already in progress, the sequence of operations in the main program would be:

```
      ......
  ┌→Read A from old file
  │ Call Sub (A,B,C)
  │ Write B to new file
  │ Read C from old file
  │ Call Sub (C,A,B)
  │ Write A to new file
  │ Read B from old file
  │ Call Sub (B,C,A)
  │ Write C to new file
```

This sequence involves minimum file transfers and no copying of values from one array to another.

Chapter 12

MATRICES

12.1. INTRODUCTION

IN earlier chapters it has been shown that two variables can be combined and treated as a single complex variable, and also that three variables can be treated as components of a single vector variable. Each of these techniques has resulted in the simplification of special types of problems. A further generalization of this idea leads to the consideration of greater numbers of variables as a single quantity called a "matrix". There is frequently a fundamental relationship between the various constituent parts of a matrix which leads to the variables being arranged as a rectangular array to draw attention to these links. Thus in Section 7.2.3, the stress tensor was written

$$\mathbf{T}_{mn} = \begin{bmatrix} P_{xx} & t_{xy} & t_{xz} \\ t_{yx} & P_{yy} & t_{yz} \\ t_{zx} & t_{zy} & P_{zz} \end{bmatrix} \tag{7.6}$$

Elements in a particular row refer to forces acting on one plane whereas elements in a particular column refer to forces acting in one direction. In general, the nine quantities are independent.

Students in engineering and science are introduced to the properties of determinants in their initial courses in mathematics, and it will be recalled that the determinant is a square array of numbers or symbols that has a specific value. Consideration of the above stress tensor should establish the fundamental difference between a matrix and the determinant which has the same elements in corresponding positions. It is that a matrix cannot be reduced to a specific value by multiplications between its constituent elements, whereas a determinant is evaluated in this manner.

The rules for addition and multiplication of matrices will be enunciated in this chapter and the subject will be developed to include matrix algebra. In addition, the rules of matrix calculus will be introduced, and the types of chemical engineering problems for which matrix methods are suitable will be indicated.

448

12.2. THE MATRIX

The array of symbols

$$\begin{bmatrix} a_{11} & a_{12} & a_{13} & \cdots & a_{1n} \\ a_{21} & a_{22} & a_{23} & \cdots & a_{2n} \\ \cdot & \cdot & \cdot & \cdot & \cdot \\ \cdot & \cdot & \cdot & \cdot & \cdot \\ \cdot & \cdot & \cdot & \cdot & \cdot \\ a_{m1} & a_{m2} & a_{m3} & \cdots & a_{mn} \end{bmatrix}$$

is called a matrix. Each symbol such as a_{11}, a_{2n} or a_{mn} is called an element of the matrix, and the square brackets $\begin{bmatrix} \\ \end{bmatrix}$ enclosing the array signifies that it is a matrix, just as the two straight lines $\begin{vmatrix} \\ \end{vmatrix}$ enclosing a square array signify that to be a determinant. The square brackets are thus a special part of the notation. For operations concerning the matrix rather than its elements, the above array can also be written in the short forms

$$\mathbf{A} = [a_{ij}]$$

where the size of the array has to be separately indicated.

The matrix represented above contains m rows and n columns. It is therefore called an "m by n matrix", or alternatively that matrix is said to be of order "$m \times n$". It should be noted that the matrix is always described in terms of "rows by columns". That is, the number of rows is always stated first and this is followed by the number of columns. For example, the array

$$\begin{bmatrix} 2 & 1 & 6 \\ 3 & 2 & 5 \end{bmatrix}$$

is a 2 by 3 matrix implying that there are two rows and three columns. The committing to memory of the phrase "rows by columns" will be of great help later on when matrices have to be manipulated in the solution of engineering problems.

12.2.1. *The Row Matrix*

A matrix having n elements arranged in a single row is a $(1 \times n)$ matrix, called a "row matrix" or "row vector". Thus $[a_{11}\ a_{12}\ a_{13}\ \cdots\ a_{1n}]$ is a row matrix or row vector.

The position vector of a point (Section 7.3.2) is usually described in terms of its coordinates as $[x\ y\ z]$ which is a particular (1×3) row matrix.

12.2.2. The Column Matrix

A matrix having m elements arranged in a single column is called a "column matrix" or "column vector". To adhere to convention it would be written

$$\begin{bmatrix} a_{11} \\ a_{21} \\ \cdot \\ \cdot \\ \cdot \\ a_{m1} \end{bmatrix}$$

but because a vertical array of elements would occupy a larger space, such a matrix would be written $\{a_{11}\ a_{21}\ \ldots\ a_{m1}\}$. The brackets $\{\ \}$ signify that the elements are to be arranged in a vertical column.

12.2.3. The Unit Column Matrix

A column matrix in which all elements are zero, except one element which is unity, is called a unit vector or unit column matrix. Such a vector fulfils an analogous role in matrix algebra that the three-dimensional unit vector fulfilled in vector analysis in Chapter 7. The unit column matrix will be denoted by $\mathbf{e}_i$ where the ith element is unity.

$$\mathbf{e}_1 = \{1 \quad 0 \quad 0 \quad \ldots \quad 0\} \tag{12.1}$$

12.2.4. The Unit Matrix

A matrix in which the number of rows of elements is equal to the number of columns of elements is called a "square matrix" and if all the elements except those in the diagonal from the top left-hand corner to the bottom right-hand corner are zero, the matrix is called a "diagonal matrix". A diagonal matrix in which the diagonal elements are all unity is called the "unit matrix". Thus

$$\mathbf{I} = \begin{bmatrix} 1 & 0 & 0 \\ 0 & 1 & 0 \\ 0 & 0 & 1 \end{bmatrix} = [\mathbf{e}_1 \quad \mathbf{e}_2 \quad \mathbf{e}_3] \tag{12.2}$$

is the unit matrix of order three. Furthermore, the non-zero elements are situated on the "principal diagonal". The significance of the unit matrix will be discussed later in Section 12.3.3.

12.2.5. The Null Matrix

A matrix whose elements are all zero is called a "null matrix" and is denoted by $\mathbf{0}$.

12.3. MATRIX ALGEBRA

In order to be able to use matrices in the solution of engineering problems, the manner in which the operations of addition, subtraction, multiplication, and division are performed must be defined. These mathematical processes will be considered in the following subsections.

12.3.1. *Matrix Addition*

This operation can only be carried out on matrices of the same order. Thus the sum of two matrices is obtained by adding together the corresponding elements of each matrix. For example, the sum of the matrices

$$\begin{bmatrix} 1 & 2 & 3 \\ 4 & 5 & 2 \\ 6 & 1 & 0 \end{bmatrix} \quad \text{and} \quad \begin{bmatrix} 1 & 2 & 6 \\ 3 & 8 & 4 \\ 2 & 2 & 0 \end{bmatrix} \quad \text{is:}$$

$$\begin{bmatrix} 1 & 2 & 3 \\ 4 & 5 & 2 \\ 6 & 1 & 0 \end{bmatrix} + \begin{bmatrix} 1 & 2 & 6 \\ 3 & 8 & 4 \\ 2 & 2 & 0 \end{bmatrix} = \begin{bmatrix} 2 & 4 & 9 \\ 7 & 13 & 6 \\ 8 & 3 & 0 \end{bmatrix}$$

Similarly, subtraction is negative addition. Thus

$$\begin{bmatrix} 1 & 2 & 3 \\ 4 & 5 & 2 \\ 6 & 1 & 0 \end{bmatrix} - \begin{bmatrix} 1 & 2 & 6 \\ 3 & 8 & 4 \\ 2 & 2 & 0 \end{bmatrix} = \begin{bmatrix} 0 & 0 & -3 \\ 1 & -3 & -2 \\ 4 & -1 & 0 \end{bmatrix}$$

These operations can be extended to any number of matrices of the same order, irrespective of whether the elements consist of real or complex numbers or symbols. The sum or difference does not exist if the matrices are of different order.

12.3.2. *Scalar Multiplication*

If any matrix is multiplied by a number, the elements of the matrix so formed are the products of the number and the corresponding elements of the original matrix. Thus

$$k\begin{bmatrix} 1 & 2 \\ 4 & 6 \end{bmatrix} + m\begin{bmatrix} 3 & 7 \\ 5 & -2 \end{bmatrix} = \begin{bmatrix} k+3m & 2k+7m \\ 4k+5m & 6k-2m \end{bmatrix} \tag{12.3}$$

Both k and m are scalar quantities and each operates on a matrix. For this reason such an operation is called "scalar multiplication". It should be noted from the above illustration that the coefficients k and m multiply each element within their respective matrix, and in this respect matrices differ from determinants.

12.3.3. *Matrix Multiplication*

The product of two matrices will exist only when the number of columns of the first matrix is equal to the number of rows in the second matrix. When this condition is satisfied the two matrices are said to be "comformable" and they yield a product. Thus if A is a matrix of order $(m \times n)$ and B is a matrix of order $(n \times p)$ the product AB will exist and will be a matrix of order $(m \times p)$. However the product BA will not exist since $(n \times p)$ is not conformable with $(m \times n)$. The following example will make this point clear.

Example 1. Evaluate the product

$$
\begin{bmatrix} 3 & -4 & -1 & 2 \\ 1 & 2 & 0 & 3 \\ 6 & 5 & 6 & 7 \end{bmatrix}
\begin{bmatrix} 1 & 2 \\ 4 & 8 \\ 0 & 4 \\ 7 & 9 \end{bmatrix}
\qquad \text{I}
$$

$$
\begin{bmatrix} 3 & -4 & -1 & 2 \\ 1 & 2 & 0 & 3 \\ 6 & 5 & 6 & 7 \end{bmatrix}
\begin{bmatrix} 1 & 2 \\ 4 & 8 \\ 0 & 4 \\ 7 & 9 \end{bmatrix}
=
\begin{bmatrix}
(3 \times 1) + (-4 \times 4) + (-1 \times 0) + (2 \times 7) \\
(1 \times 1) + (2 \times 4) + (0 \times 0) + (3 \times 7) \\
(6 \times 1) + (5 \times 4) + (6 \times 0) + (7 \times 7)
\end{bmatrix}
$$

$$
\begin{bmatrix}
(3 \times 2) + (-4 \times 8) + (-1 \times 4) + (2 \times 9) \\
(1 \times 2) + (2 \times 8) + (0 \times 4) + (3 \times 9) \\
(6 \times 2) + (5 \times 8) + (6 \times 4) + (7 \times 9)
\end{bmatrix}
$$

$$
=
\begin{bmatrix} 1 & -12 \\ 30 & 45 \\ 75 & 139 \end{bmatrix}
\qquad \text{II}
$$

In the above example it will be seen that the first element in the final matrix is obtained by multiplying corresponding elements from the first row of the first matrix and the first column of the second matrix and summing the products; thus

$$
[3 \quad -4 \quad -1 \quad 2]\{1 \quad 4 \quad 0 \quad 7\} = 3 - 16 - 0 + 14 = 1 \qquad (12.4)
$$

This product of a row and a column is called an "inner product".

It can also be seen that the first matrix in the above example is of order (3×4) whereas the second matrix is of order (4×2) and the product matrix is of order (3×2). That is

$$
(3 \times 4) \cdot (4 \times 2) = (3 \times 2) \qquad (12.5)
$$

No product would exist if the (4×2) matrix was multiplied by the (3×4) matrix since they are not conformable. Hence the order of terms in a matrix product is important.

In certain cases the product of two matrices will exist irrespective of the order in which the multiplication is carried out. For example, if the matrix **A** is of order $(m \times n)$ and the matrix **B** is of order $(n \times m)$ the product **C** of

$$AB = C \tag{12.6}$$

is a matrix of order $(m \times n).(n \times m) = (m \times m)$. On the other hand, the product **D** of

$$BA = D \tag{12.7}$$

is a matrix of order $(n \times m).(m \times n) = (n \times n)$. In particular, this condition will exist for square matrices, but even with these, the product will not be the same in general. It is thus necessary to use terminology which specifies the order of multiplication. Equation (12.6) can alternatively be described as:

      (i) **C** is obtained by pre-multiplying **B** by **A**

     (ii) **C** is obtained by post-multiplying **A** by **B**.

In the exceptional case when the products of two square matrices are equal, i.e.

$$AB = BA \tag{12.8}$$

the matrices are said to commute, or be commutable.

In Section 12.2.4 the unit matrix was shown to be a diagonal matrix in which the elements were unity. The significance of unity is that when it is multiplied by a number or function, the value of the number or function remains unchanged. Therefore the definition of the unit matrix must satisfy this condition. Thus consider the unit matrix of order three to operate on a third order column vector

$$\begin{bmatrix} 1 & 0 & 0 \\ 0 & 1 & 0 \\ 0 & 0 & 1 \end{bmatrix} \begin{bmatrix} x_1 \\ x_2 \\ x_3 \end{bmatrix} = \begin{bmatrix} (1 \times x_1)+(0 \times x_2)+(0 \times x_3) \\ (0 \times x_1)+(1 \times x_2)+(0 \times x_3) \\ (0 \times x_1)+(0 \times x_2)+(1 \times x_3) \end{bmatrix} = \begin{bmatrix} x_1 \\ x_2 \\ x_3 \end{bmatrix} \tag{12.9}$$

Hence the unit matrix conforms to a fundamental law of mathematics.

The unit matrix is usually given the symbol **I** and can be written into an equation as such, thereby conserving space. For example equation (12.9) would be written:

$$Ix = x \tag{12.10}$$

The reader can check that the matrices in equation (12.10) are commutable. Further evidence that the unit matrix conforms to the fundamental laws of mathematics is that

$$I^p = I \tag{12.11}$$

and again the reader is asked to check equation (12.11).

It is also important to note that the product of two matrices can be the null matrix without either of the original matrices being null.

Example 2. Find the two results of multiplying **A** and **B** together where

$$\mathbf{A} = \begin{bmatrix} 1 & 3 & 5 \\ -2 & 4 & 0 \\ 0 & 2 & 2 \end{bmatrix} \quad \text{and} \quad \mathbf{B} = \begin{bmatrix} 2 & 0 & 6 \\ 1 & 0 & 3 \\ -1 & 0 & -3 \end{bmatrix}$$

$$\mathbf{AB} = \begin{bmatrix} 0 & 0 & 0 \\ 0 & 0 & 0 \\ 0 & 0 & 0 \end{bmatrix} \quad \text{but} \quad \mathbf{BA} = \begin{bmatrix} 2 & 18 & 22 \\ 1 & 9 & 11 \\ -1 & -9 & -11 \end{bmatrix}$$

The above property gives rise to the same difficulties when defining a division process as were encountered with vectors in Chapter 7. However, division by a square matrix is possible, as will be shown in Section 12.7.

12.4. DETERMINANTS OF SQUARE MATRICES AND MATRIX PRODUCTS

The determinant of a square matrix is an array of elements which are identical with the corresponding elements of the matrix with respect to both magnitude and position. The determinant of a square matrix **A** would be written $|\mathbf{A}|$. Obviously matrices that are not square do not possess a corresponding determinant.

It has been shown in texts on determinants that the product of two determinants $|\mathbf{A}|$ and $|\mathbf{B}|$ of the same order $(n \times n)$ is also a determinant of order $(n \times n)$. Furthermore, the product

$$|\mathbf{A}| \times |\mathbf{B}| = |\mathbf{B}| \times |\mathbf{A}| = |\mathbf{C}| \tag{12.12}$$

From equation (12.12) it can be seen that the elements in $|\mathbf{C}|$ will be identical with the corresponding elements in the matrix product **AB**. Hence the determinant of the product of two square matrices is equal to the product of their determinants. That is,

$$|\mathbf{A}||\mathbf{B}| = |\mathbf{AB}| \quad \text{and} \quad |\mathbf{B}||\mathbf{A}| = |\mathbf{BA}| \tag{12.13}$$

so that

$$|\mathbf{AB}| = |\mathbf{BA}| \tag{12.14}$$

even though in general $\mathbf{AB} \neq \mathbf{BA}$.

Finally if **A** is a square matrix and $|\mathbf{A}| = 0$, the matrix is called a "singular matrix". On the other hand, if $|\mathbf{A}| \neq 0$, the matrix is "non-singular".

12.5. THE TRANSPOSE OF A MATRIX

If the rows of an $(n \times m)$ matrix are written in the form of columns, a new matrix of order $(m \times n)$ will be formed. This new matrix is called the "transpose" of the original matrix. If the original matrix is denoted by **A**, the transpose will be denoted by **A′**, the accent in the top right-hand corner signifying that **A** has been transposed. Thus if

$$\mathbf{A} = \begin{bmatrix} 2 & 0 \\ 3 & -1 \\ 4 & 6 \end{bmatrix} \quad \mathbf{A}' = \begin{bmatrix} 2 & 3 & 4 \\ 0 & -1 & 6 \end{bmatrix} \tag{12.15}$$

In a similar manner it can be seen that the transpose of a row vector is a column vector. A useful convention is to treat all row vectors as transposed column vectors and define a general column vector as **x** and a general row vector as **x**'. A matrix and its transpose are always conformable for multiplication and if

$$\mathbf{AA}' = \mathbf{I} \tag{12.16}$$

the matrix **A** is said to be "orthogonal".

12.5.1. *The Transpose of the Product of Matrices*

Let **A** be a matrix of order $(m \times n)$ and let **B** be a matrix of order $(n \times p)$. The product of these two matrices will be

$$\mathbf{AB} = \mathbf{C} \tag{12.17}$$

where **C** is a matrix of order $(m \times p)$ whose typical element in the ith row and jth column will be

$$c_{ij} = \sum_{k=1}^{n} a_{ik} b_{kj} \tag{12.18}$$

but the typical element in the transposed matrix will be c_{ij}' where

$$c_{ij}' = c_{ji} = \sum_{k=1}^{n} a_{jk} b_{ki} = \sum_{k=1}^{n} a_{kj}' b_{ik}' = \sum_{k=1}^{n} b_{ik}' a_{kj}' \tag{12.19}$$

Therefore extending equation (12.19) to all the elements in the product $(\mathbf{AB})'$ it follows that

$$(\mathbf{AB})' = \mathbf{B}'\mathbf{A}' \tag{12.20}$$

That is, the transpose of the product of two matrices is equal to the product of their transposes taken in the reverse order. This idea can be extended to any number of matrices that are conformable. Thus

$$(\mathbf{ABC})' = \mathbf{C}'\mathbf{B}'\mathbf{A}' \tag{12.21}$$

12.6. ADJOINT MATRICES

If a_{ij} is any element of a square matrix of order $(n \times n)$, and the cofactor of a_{ij} in the determinant $|\mathbf{A}|$ is A_{ij}, the transpose of the matrix whose elements are made up of all the A_{ij}s is called the "adjoint of **A**"; and is written

$$\text{adj } \mathbf{A} = [A_{ji}] \tag{12.22}$$

Example. Evaluate the adjoint of the matrix

$$\begin{bmatrix} 3 & 0 & 3 \\ -2 & 1 & 1 \\ 4 & 2 & 5 \end{bmatrix}$$

$$A = \begin{bmatrix} 3 & 0 & 3 \\ -2 & 1 & 1 \\ 4 & 2 & 5 \end{bmatrix}, \quad \text{adj } A = \begin{bmatrix} 3 & 6 & -3 \\ 14 & 3 & -9 \\ -8 & -6 & 3 \end{bmatrix}$$

12.7. Reciprocal of a Square Matrix

Consider a square matrix A to be non-singular and of order $(n \times n)$. The product of this matrix with its adjoint is

$$A \operatorname{adj} A = \left| \sum_{r=1}^{n} a_{ir} A_{rj} \right| \tag{12.23}$$

Since the properties of determinants show that

$$\sum_{r=1}^{n} a_{ir} A_{rj} = 0 \quad (i \neq j) \tag{12.24}$$

it follows that all the elements in the matrix product $A \operatorname{adj} A$ except those on the principal diagonal will be zero, hence

$$A \operatorname{adj} A = \begin{bmatrix} |A| & 0 & 0 & 0 & \dots & 0 \\ 0 & |A| & 0 & 0 & \dots & 0 \\ 0 & 0 & |A| & 0 & \dots & 0 \\ . & . & . & . & . & . \\ . & . & . & . & . & . \\ . & . & . & . & . & . \\ 0 & 0 & 0 & 0 & \dots & |A| \end{bmatrix} = |A| \begin{bmatrix} 1 & 0 & 0 & \dots & 0 \\ 0 & 1 & 0 & \dots & 0 \\ 0 & 0 & 1 & \dots & 0 \\ . & . & . & . & . \\ . & . & . & . & . \\ . & . & . & . & . \\ 0 & 0 & 0 & \dots & 1 \end{bmatrix}$$

or

$$A \operatorname{adj} A = |A| I \tag{12.25}$$

Since A is non-singular, equation (12.25) can be divided by $|A|$; and if a new matrix A^{-1} is defined by

$$A^{-1} = \frac{\operatorname{adj} A}{|A|} \tag{12.26}$$

equation (12.25) becomes

$$AA^{-1} = I \tag{12.27}$$

A^{-1} is thus seen to be the reciprocal of the non-singular square matrix A. Each element of this reciprocal matrix is obtained thus

$$a_{ij} = \frac{A_{ji}}{|A|} \tag{12.28}$$

Hence matrix division is accomplished by multiplying the one matrix by the reciprocal of the other. As with all multiplications, the order of terms must be specified, but any matrix will commute with its own reciprocal.

$$AA^{-1} = I = A^{-1}A \tag{12.29}$$

12.7.1. *The Reciprocal of the Product of Square Matrices*

Consider A and B to be two square non-singular matrices of the same order. Their reciprocals A^{-1} and B^{-1} will exist and thus the product of each is

$$(AB)(B^{-1}A^{-1}) = AIA^{-1} = AA^{-1} = I \tag{12.30}$$

Therefore $B^{-1}A^{-1}$ is the reciprocal of the product AB, or

$$(AB)^{-1} = B^{-1}A^{-1} \tag{12.31}$$

That is, the reciprocal of the product of two non-singular conformable square matrices is equal to the product of their reciprocals in the reverse order.

This concept can be extended to any number of conformable square matrices, and in general

$$(ABC \ldots XYZ)^{-1} = Z^{-1}Y^{-1}X^{-1} \ldots C^{-1}B^{-1}A^{-1} \tag{12.32}$$

12.7.2. *The Reciprocal of a Transposed Matrix*

In a manner very similar to that given above it can be shown that the reciprocal of a transpose of a matrix is equal to the transpose of its reciprocal; or

$$(A^{-1})' = (A')^{-1} \tag{12.33}$$

It is left to the reader to verify this.

12.8. THE RANK AND DEGENERACY OF A MATRIX

These two important terms are employed extensively in matrix calculus and algebra, and they must be understood in order to make use of matrices in the solution of simultaneous equations. However, the reader must be reminded of other definitions, which arise elsewhere in elementary mathematics, in order to appreciate the significance of "degeneracy" and "rank". These will be considered first.

12.8.1. *Linear Dependence*

Let $x_1, x_2, x_3, \ldots, x_n$ be initially a set of scalar variables that satisfy the equation

$$p_1 x_1 + p_2 x_2 + p_3 x_3 + \ldots + p_n x_n = 0 \tag{12.34}$$

where $p_1, p_2, \ldots, p_n$ are constants, some but not all of which may be zero. Since these scalar variables satisfy equation (12.34) they are linearly dependent.

This idea can be extended to matrices in the following way. Thus let x_1, x_2, x_3, etc. be a set of vectors of the same order, and the zero on the right of equation (12.34) be a null matrix of order corresponding to the vectors. Since p_1, p_2, p_3, etc. are not all zero the vectors are linearly dependent. However, if equation (12.34) was satisfied only when the constants were zero the vectors would be "linearly independent".

The vectors x_1, x_2, etc. could be the n columns of an $(m \times n)$ matrix X, and if there exists a set of n scalar constants, p_1, p_2, etc. not all zero so that equation (12.34) is satisfied, the columns of the matrix are linearly dependent. In particular if the matrix X is square, its columns will be linearly dependent only if the determinant $|X| = 0$. Furthermore, if $|X| = 0$

$$|X| = |X'| = 0 \qquad (12.35)$$

so that the rows of X will also be linearly dependent. To sum up, if the rows or columns of a matrix are linearly dependent the matrix is singular.

12.8.2. The Minor of a Matrix

Let A be a matrix of order $(m \times n)$ in which m is less than n. The determinants formed from the m rows and any m of the n columns are called "minors" of the matrix A. Furthermore, if r is an integer which is less than m and all the minors of order r are zero, it follows that all the minors of order greater than r will also be zero. Thus the value of the determinant of any minor of order $(r+1)$ is the sum of the products of each element in the $(r+1)$th column and its cofactor. Since these cofactors are minors of order r which are all zero, the minors of order $(r+1)$ must all be zero. By repeating this argument to the minors of order m it can be seen that all the minors of order greater than r will be zero.

Minors of a matrix are often referred to as the first minor, second minor, and so on. By this is meant that the minor of highest order is the first minor, the next highest minor is the second minor, and so on to the minor that will contain a single element only. For a square matrix of order $(n \times n)$ the first minor will be the determinant of the matrix and of order n. The second minor will be of order $(n-1)$, and so on to the minor of order 1.

12.8.3. Degeneracy

A matrix of order $(m \times n)$ is said to have a degeneracy of r when at least one of its $(r+1)$th minors does not vanish, whereas all of its rth minors are equal to zero.

For a square matrix of order $(n \times n)$ and degeneracy r this means that there are r linear relationships between the n columns of the matrix. Thus the columns of a singular matrix are related by a single equation of the type (12.34), where x_1, x_2, etc. are the columns of the matrix and such a matrix is "singly degenerate", or has a degeneracy of one.

Example. Show that the matrix below has a degeneracy of one.

$$\begin{bmatrix} 2 & 4 & -2 \\ 3 & 0 & 4 \\ 5 & 4 & 2 \end{bmatrix}$$

The determinant of this matrix is

$$\begin{vmatrix} 2 & 4 & -2 \\ 3 & 0 & 4 \\ 5 & 4 & 2 \end{vmatrix} = 0 \qquad\qquad\qquad\qquad \text{I}$$

Hence the matrix is singular. However, all the minors of order two are non-vanishing so that the degeneracy is one, and therefore the columns of the matrix are related by the single relationship

$$8\{2 \quad 3 \quad 5\} - 7\{4 \quad 0 \quad 4\} - 6\{-2 \quad 4 \quad 2\} = \mathbf{0} \qquad\qquad \text{II}$$

Similarly the rows are related by a single relationship

$$1[2 \quad 4 \quad -2] + 1[3 \quad 0 \quad 4] - 1[5 \quad 4 \quad 2] = 0 \qquad\qquad \text{III}$$

12.8.4. *The Rank of a Matrix*

The rank of a matrix is equal to the order of the highest order non-vanishing minor. In the above example, the rank of the matrix

$$\begin{bmatrix} 2 & 4 & -2 \\ 3 & 0 & 4 \\ 5 & 4 & 2 \end{bmatrix}$$

is two. For a square matrix of order n and degeneracy r, the rank is equal to $(n-r)$.

From the above definition it will be apparent that all matrices except the null matrix have a rank equal to or greater than one. Thus the matrix

$$\begin{bmatrix} 0 & 0 & 0 & 0 \\ 0 & 0 & 0 & 0 \\ 0 & 0 & 0 & 0 \\ 0 & 0 & 0 & 1 \end{bmatrix}$$

is of order 4, degeneracy 3, and rank 1.

12.9. THE SUB-MATRIX

It is frequently convenient in matrix calculations to divide a matrix of high order by dotted vertical or horizontal lines to form what are termed "sub-matrices". Thus the matrix of order 4 below is divided into four second order matrices.

$$\begin{bmatrix} a_{11} & a_{12} & a_{13} & a_{14} \\ a_{21} & a_{22} & a_{23} & a_{24} \\ \hdashline a_{31} & a_{32} & a_{33} & a_{34} \\ a_{41} & a_{42} & a_{43} & a_{44} \end{bmatrix} = \begin{bmatrix} A_1 & A_2 \\ A_3 & A_4 \end{bmatrix}$$

where

$$A_1 = \begin{bmatrix} a_{11} & a_{12} \\ a_{21} & a_{22} \end{bmatrix}, \quad A_2 = \begin{bmatrix} a_{13} & a_{14} \\ a_{23} & a_{24} \end{bmatrix}, \quad \text{etc.}$$

The dotted lines are said to "partition" the matrix.

This technique is especially useful for handling large matrices in digital computers, because it allows the matrix to be considered as a combination of smaller matrices.

12.10. Solution of Linear Algebraic Equations

Matrices are often used to express a system of linear algebraic equations in concise form. Thus consider the set of linear algebraic equations

$$\left. \begin{aligned} 2x_1 - 3x_2 - 2x_4 &= 8 \\ 3x_2 + 2x_3 + x_4 &= 5 \\ x_1 - 2x_2 - 4x_3 + 2x_4 &= 2 \\ 2x_1 + x_2 - 3x_3 - x_4 &= 6 \end{aligned} \right\} \tag{12.36}$$

This set can be written in matrix notation thus

$$\begin{bmatrix} 2 & -3 & 0 & -2 \\ 0 & 3 & 2 & 1 \\ 1 & -2 & -4 & 2 \\ 2 & 1 & -3 & -1 \end{bmatrix} \begin{bmatrix} x_1 \\ x_2 \\ x_3 \\ x_4 \end{bmatrix} = \begin{bmatrix} 8 \\ 5 \\ 2 \\ 6 \end{bmatrix}$$

or more concisely

$$Ax = b \tag{12.37}$$

where A represents a square matrix of order 4 whilst x and b are each column vectors of order 4. The solution of equation (12.37) is

$$x = A^{-1}b \tag{12.38}$$

which is possible only if the reciprocal of the matrix A can be evaluated.

The inverse can be obtained by evaluating the adjoint as described in Section 12.6, but for matrices which are larger, or not square, the methods developed below are very much superior. Consequently, the evaluation of the numerical solution to the above system will be deferred to Section 12.10.4.

12.10.1. *Interchange of Columns or Rows*

Consider the pre-multiplication of a third order matrix **A** by the modified unit matrix

$$\begin{bmatrix} 1 & 0 & 0 \\ 0 & 0 & 1 \\ 0 & 1 & 0 \end{bmatrix}$$

which has its second and third columns interchanged. The result is

$$\begin{bmatrix} 1 & 0 & 0 \\ 0 & 0 & 1 \\ 0 & 1 & 0 \end{bmatrix} \begin{bmatrix} a_{11} & a_{12} & a_{13} \\ a_{21} & a_{22} & a_{23} \\ a_{31} & a_{32} & a_{33} \end{bmatrix} = \begin{bmatrix} a_{11} & a_{12} & a_{13} \\ a_{31} & a_{32} & a_{33} \\ a_{21} & a_{22} & a_{23} \end{bmatrix} \qquad (12.39)$$

where it can be seen that the second and third rows of the matrix **A** have been interchanged. It will also be readily seen that if the matrix **A** had been post-multiplied by the same modified unit matrix the effect would have been to interchange the second and third columns of **A**.

It should be observed that the modified unit matrix is the same as its transpose and that

$$\begin{bmatrix} 1 & 0 & 0 \\ 0 & 0 & 1 \\ 0 & 1 & 0 \end{bmatrix} \begin{bmatrix} 1 & 0 & 0 \\ 0 & 0 & 1 \\ 0 & 1 & 0 \end{bmatrix} = \begin{bmatrix} 1 & 0 & 0 \\ 0 & 1 & 0 \\ 0 & 0 & 1 \end{bmatrix} \qquad (12.40)$$

The modified unit matrix has many of the properties of the unit matrix and is equivalent to the unit matrix. These operations can be extended to matrices of any order.

12.10.2. *Addition of any Column or Row, or a Multiple of any Column or Row, to any other Column or Row*

Consider the product

$$\begin{bmatrix} 1 & \beta & 0 \\ 0 & 1 & 0 \\ 0 & 0 & 1 \end{bmatrix} \begin{bmatrix} a_{11} & a_{12} & a_{13} \\ a_{21} & a_{22} & a_{23} \\ a_{31} & a_{32} & a_{33} \end{bmatrix} = \begin{bmatrix} a_{11}+\beta a_{21} & a_{12}+\beta a_{22} & a_{13}+\beta a_{23} \\ a_{21} & a_{22} & a_{23} \\ a_{31} & a_{32} & a_{33} \end{bmatrix}$$

$$(12.41)$$

The matrix which operates on **A** is a unit matrix to which has been added the element β in its first row and second column. Its determinant is unity. Pre-multiplication of the matrix **A** by this operator as in equation (12.41) has the effect of adding β times the second row to the first row. Post-multiplication of **A** by this matrix would have caused β times the first column to be added to the second column.

12.10.3. *Multiplication of any Column or Row by a Scalar*

Consider the product

$$
\begin{bmatrix} 1 & 0 & 0 \\ 0 & \gamma & 0 \\ 0 & 0 & 1 \end{bmatrix}
\begin{bmatrix} a_{11} & a_{12} & a_{13} \\ a_{21} & a_{22} & a_{23} \\ a_{31} & a_{32} & a_{33} \end{bmatrix} =
\begin{bmatrix} a_{11} & a_{12} & a_{13} \\ \gamma a_{21} & \gamma a_{22} & \gamma a_{23} \\ a_{31} & a_{32} & a_{33} \end{bmatrix}
\qquad (12.42)
$$

The result of this operation is that the second row of **A** has been multiplied by the scalar γ. The operating matrix was formed from the unit matrix by replacing the second unity element on the leading diagonal with the scalar γ. If the matrix **A** had been post-multiplied by this matrix the effect would have been to multiply the second column by the scalar γ.

The determinant of the converted unit matrix is γ.

12.10.4. *Application to the Solution of Equations*

As with any other branch of algebra, the form of an equation can be changed without altering its validity provided that the same operation is performed on both sides of the equation. Using the symbol T_i to denote any matrix of the three types just described, equation (12.37) can be altered to a succession of equivalent forms, thus

$$
T_n T_{n-1} \cdots T_2 T_1 A x = T_n T_{n-1} \cdots T_2 T_1 b \qquad (12.43)
$$

where the operating matrices T_i are selected to convert the square matrix **A** to the unit matrix **I**. That is

$$
T_n T_{n-1} \cdots T_2 T_1 A = I \qquad (12.44)
$$

Because the same operations are being performed on the vector **b** in equation (12.43) it is convenient to append the column vector **b** to the square matrix **A** and apply the operators T_i as premultipliers to the combined matrix [**Ab**] called the "augmented matrix" of the equation. If the reciprocal matrix A^{-1} is required, inspection of equation (12.44) shows that it is readily obtainable because

$$
A^{-1} = T_n T_{n-1} \cdots T_2 T_1 = T_n T_{n-1} \cdots T_2 T_1 I \qquad (12.45)
$$

Therefore, if the unit matrix is also appended to the augmented matrix, the reciprocal matrix can be generated. Thus, the steps proceed

$$
\left[\begin{array}{cccc|c|cccc}
2 & -3 & 0 & -2 & 8 & 1 & 0 & 0 & 0 \\
0 & 3 & 2 & 1 & 5 & 0 & 1 & 0 & 0 \\
1 & -2 & -4 & 2 & 2 & 0 & 0 & 1 & 0 \\
2 & 1 & -3 & -1 & 6 & 0 & 0 & 0 & 1
\end{array} \right]
$$

Divide the first row by 2 and then add (-1) times the new first row to the third row and subtract twice the new first row from the last row.

$$\begin{bmatrix} 1 & -1\cdot5 & 0 & -1 & 4 & 0\cdot5 & 0 & 0 & 0 \\ 0 & 3 & 2 & 1 & 5 & 0 & 1 & 0 & 0 \\ 0 & -0\cdot5 & -4 & 3 & -2 & -0\cdot5 & 0 & 1 & 0 \\ 0 & 4 & -3 & 1 & -2 & -1 & 0 & 0 & 1 \end{bmatrix}$$

Add the last row to the first row, subtract the last row from the second row and then subtract three times the last row from the third row.

$$\begin{bmatrix} 1 & 2\cdot5 & -3 & 0 & 2 & -0\cdot5 & 0 & 0 & 1 \\ 0 & -1 & 5 & 0 & 7 & 1 & 1 & 0 & -1 \\ 0 & -12\cdot5 & 5 & 0 & 4 & 2\cdot5 & 0 & 1 & -3 \\ 0 & 4 & -3 & 1 & -2 & -1 & 0 & 0 & 1 \end{bmatrix}$$

Reverse the signs in the second row, and eliminate the elements from the first, third, and fourth rows in the second column.

$$\begin{bmatrix} 1 & 0 & 9\cdot5 & 0 & 19\cdot5 & 2 & 2\cdot5 & 0 & -1\cdot5 \\ 0 & 1 & -5 & 0 & -7 & -1 & -1 & 0 & 1 \\ 0 & 0 & -57\cdot5 & 0 & -83\cdot5 & -10 & -12\cdot5 & 1 & 9\cdot5 \\ 0 & 0 & 17 & 1 & 26 & 3 & 4 & 0 & -3 \end{bmatrix}$$

Finally, use similar operations determined by the elements in the third column.

$$\begin{bmatrix} 1 & 0 & 0 & 0 & 5\cdot704 & 0\cdot347 & 0\cdot438 & 0\cdot165 & 0\cdot068 \\ 0 & 1 & 0 & 0 & 0\cdot261 & -0\cdot130 & 0\cdot087 & -0\cdot087 & 0\cdot174 \\ 0 & 0 & 1 & 0 & 1\cdot452 & 0\cdot174 & 0\cdot217 & -0\cdot017 & -0\cdot165 \\ 0 & 0 & 0 & 1 & 1\cdot313 & 0\cdot043 & 0\cdot304 & 0\cdot296 & -0\cdot191 \end{bmatrix}$$

Thus equation (12.43) becomes

$$Ix = T_n T_{n-1} \cdots T_2 T_1 b \tag{12.46}$$

and therefore

$$x_1 = 5\cdot704, \quad x_2 = 0\cdot261, \quad x_3 = 1\cdot452, \quad x_4 = 1\cdot313$$

and

$$A^{-1} = \begin{bmatrix} 0\cdot347 & 0\cdot438 & 0\cdot165 & 0\cdot068 \\ -0\cdot130 & 0\cdot087 & -0\cdot087 & 0\cdot174 \\ 0\cdot174 & 0\cdot217 & -0\cdot017 & -0\cdot165 \\ 0\cdot043 & 0\cdot304 & 0\cdot296 & -0\cdot191 \end{bmatrix} \tag{12.47}$$

The latter can be checked by substitution into equation (12.38).

12.10.5. *Extension to Non-square Matrices. Vector Space.*

The above calculations can be interpreted in an entirely different way by considering that each column represents an individual vector. Equation (12.37) can be written

$$b = Ib = Ax \qquad (12.48)$$

and it has been shown already in equation (12.2) that I can be considered to be a collection of unit vectors e_i. Defining the matrix A to be a collection of column vectors a_i where a_i consists of the elements of the ith column of A, equation (12.48) takes the form

$$b = [e_1 \quad e_2 \quad e_3 \quad e_4]b = [a_1 \quad a_2 \quad a_3 \quad a_4]x \qquad (12.49)$$

or more specifically

$$b = b_1 e_1 + b_2 e_2 + b_3 e_3 + b_4 e_4 = x_1 a_1 + x_2 a_2 + x_3 a_3 + x_4 a_4 \qquad (12.50)$$

In this form it is seen that b_i are the components of the vector b when it is resolved in terms of the unit vectors e_i and the unknown elements x_i are the components of the vector b when it is resolved in terms of the given vectors a_i.

It is worth noting at this point that there are strong similarities between the above expansion in terms of unit vectors e_i and the corresponding three-dimensional unit vectors defined in Section 7.3.1 in vector analysis. However, these two branches of mathematics (matrices, vector analysis) are quite distinct and have developed separately. For instance, the pre-multiplication of a column matrix with the transpose of another column matrix consists of a single element which is analogous to the dot product (Section 7.4.1) but there is no matrix operation which corresponds to the cross-product (Section 7.4.2). The reader is advised not to confuse the two subjects.

Returning to equation (12.50), it is clear that if b can be resolved in terms of the unit vectors, a_i can also be resolved in terms of the same unit vectors. The problem is to invert the process, that is, to expand the unit vectors in terms of a_i. Since this has just been done in the above calculations, it is only necessary to put a useful interpretation on the calculations. Thus the original extended augmented matrix can be labelled

$$
\begin{array}{c}
\quad\quad a_1 \quad a_2 \quad a_3 \quad a_4 \quad\ b \quad e_1 \ e_2 \ e_3 \ e_4 \\
\begin{array}{c} e_1 \\ e_2 \\ e_3 \\ e_4 \end{array}
\left[
\begin{array}{cccc|c|cccc}
2 & -3 & 0 & -2 & 8 & 1 & 0 & 0 & 0 \\
0 & 3 & 2 & 1 & 5 & 0 & 1 & 0 & 0 \\
1 & -2 & -4 & 2 & 2 & 0 & 0 & 1 & 0 \\
2 & 1 & -3 & -1 & 6 & 0 & 0 & 0 & 1
\end{array}
\right]
\end{array}
$$

and each column can be interpreted as the expansion of the vector in the heading of the column in terms of the vectors noted against each row. Thus

$$b = 8e_1 + 5e_2 + 2e_3 + 6e_4 \qquad (12.51)$$

Throughout the solution process, the labels on the columns remain fixed, but the last version of the extended augmented matrix must be relabelled on the rows, thus

	a_1	a_2	a_3	a_4	b	e_1	e_2	e_3	e_4
a_1	1	0	0	0	5·704	0·347	0·438	0·165	0·068
a_2	0	1	0	0	0·261	−0·130	0·087	−0·087	0·174
a_3	0	0	1	0	1·452	0·174	0·217	−0·017	−0·165
a_4	0	0	0	1	1·313	0·043	0·304	0·296	−0·191

This is necessary to ensure that the first four columns represent the correct identical expressions. The fifth column can also be read as the correct expansion according to equation (12.50) to give the desired values of x_i and the last four columns are simply the expansions of the unit vectors in terms of the given vectors a_i.

It is clear now how the intermediate parts of the solution should be interpreted. The labels on the rows are obtained from the labels on the columns which have the appearance of unit vectors. The first step for example involved generating the unit vector e_1 in the column headed a_1 and destroying it in its previous column. The first row thus needs to be relabelled a_1 after this first step. The whole process can be interpreted as altering the group of four vectors (labelling the rows) in terms of which the system of vectors (labelling the columns) is expanded.

With this new interpretation, there is no limit to the number of vectors which can be written down as columns and added to the system. Suppose that a vector

$$a_5 = \{1 \quad 0 \quad 2 \quad 1\}$$

is added. Since

$$a_5 = e_1 + 2e_3 + e_4 \tag{12.52}$$

the last matrix will contain the extra column

$$\begin{bmatrix} 0·347 + 0·330 + 0·068 \\ −0·130 − 0·174 + 0·174 \\ 0·174 − 0·034 − 0·165 \\ 0·043 + 0·592 − 0·191 \end{bmatrix} = \begin{bmatrix} 0·745 \\ −0·130 \\ −0·025 \\ 0·444 \end{bmatrix}$$

The addition of this vector to equation (12.49) means that x must contain an extra element (x_5), but the original solution is still valid with x_5 having the value zero. The original set of equations (12.36) has become

$$\left. \begin{aligned} 2x_1 - 3x_2 - 2x_4 + x_5 &= 8 \\ 3x_2 + 2x_3 + x_4 &= 5 \\ x_1 - 2x_2 - 4x_3 + 2x_4 + 2x_5 &= 2 \\ 2x_1 + x_2 - 3x_3 - x_4 + x_5 &= 6 \end{aligned} \right\} \tag{12.53}$$

which is a system of four equations in five unknowns. Such a system in general will permit any one unknown to be given an arbitrary value, say zero, and the system to be solved for the other four variables. Suppose it is desired to take $x_1 = 0$ in equations (12.53). It is only necessary to remove $\mathbf{a}_1$ from the row labels and replace it with $\mathbf{a}_5$ by generating the unit vector $\mathbf{e}_1$ in the new column labelled $\mathbf{a}_5$. Thus

	$\mathbf{a}_1$	$\mathbf{a}_2$	$\mathbf{a}_3$	$\mathbf{a}_4$	$\mathbf{a}_5$	$\mathbf{b}$	$\mathbf{e}_1$	$\mathbf{e}_2$	$\mathbf{e}_3$	$\mathbf{e}_4$
$\mathbf{a}_1$	1	0	0	0	0·745	5·704	0·347	0·438	0·165	0·068
$\mathbf{a}_2$	0	1	0	0	−0·130	0·261	−0·130	0·087	−0·087	0·174
$\mathbf{a}_3$	0	0	1	0	−0·025	1·452	0·174	0·217	−0·017	−0·165
$\mathbf{a}_4$	0	0	0	1	0·444	1·313	0·043	0·304	0·296	−0·191

becomes

	$\mathbf{a}_1$	$\mathbf{a}_2$	$\mathbf{a}_3$	$\mathbf{a}_4$	$\mathbf{a}_5$	$\mathbf{b}$	$\mathbf{e}_1$	$\mathbf{e}_2$	$\mathbf{e}_3$	$\mathbf{e}_4$
$\mathbf{a}_5$	1·342	0	0	0	1	7·656	0·466	0·588	0·221	0·091
$\mathbf{a}_2$	0·174	1	0	0	0	1·256	−0·069	0·163	−0·058	0·186
$\mathbf{a}_3$	0·034	0	1	0	0	1·643	0·186	0·232	−0·011	−0·163
$\mathbf{a}_4$	−0·596	0	0	1	0	−2·086	−0·164	0·043	0·198	−0·231

which gives an alternative solution

$$x_1 = 0, \quad x_2 = 1·256, \quad x_3 = 1·643, \quad x_4 = -2·086, \quad x_5 = 7·656$$

(After so much arithmetic this answer is now inaccurate and should be recalculated, keeping another decimal place throughout.)

These ideas of moving between alternative solutions will be exploited in Section 13.4.

12.11. MATRIX SERIES

Matrices can form series in much the same way as scalar variables, with each term in the series containing powers of a matrix. Hence before discussing series it is proposed to consider the properties of powers of matrices.

12.11.1. *Powers of Matrices*

If $\mathbf{A}$ is a matrix of order $(n \times m)$, the product $\mathbf{AA}$ will only exist if the components of the product are conformable. That is if $m = n$. Thus the continued product of a matrix will exist only if the matrix is square. When this is the case the continued product can be written in the form

$$\mathbf{A}_1 \mathbf{A}_2 \mathbf{A}_3 \mathbf{A}_4 \dots \mathbf{A}_n = \mathbf{A}^n \tag{12.54}$$

Raising a matrix to a power n signifies that the matrix is to be multiplied by itself n times and not that each element in the matrix is raised to the power n.

For example:

$$\begin{bmatrix} a_1 & a_3 \\ a_2 & a_4 \end{bmatrix}^2 = \begin{bmatrix} a_1 & a_3 \\ a_2 & a_4 \end{bmatrix} \begin{bmatrix} a_1 & a_3 \\ a_2 & a_4 \end{bmatrix} = \begin{bmatrix} a_1^2 + a_2 a_3 & a_1 a_3 + a_3 a_4 \\ a_1 a_2 + a_2 a_4 & a_2 a_3 + a_2^4 \end{bmatrix}$$

The usual algebraic rules for integer indices apply to matrices. Thus

$$\mathbf{A}^n \times \mathbf{A}^m = \mathbf{A}^{n+m} \tag{12.55}$$

and

$$(\mathbf{A}^n)^p = \mathbf{A}^{np} \tag{12.56}$$

If the exponent of a matrix is fractional implying that the nth root of a matrix is sought, the resulting matrix will be of the same order as, and commutable with, the parent. However, it is possible that the number of roots may be infinite, but this will have to be established for each particular matrix whose roots are required.

12.11.2. *Matrix Polynomials*

Matrix series can be formed from different powers of a square matrix, and these series have similar properties to the familiar scalar series. For example,

$$a_0 \mathbf{X}^n + a_1 \mathbf{X}^{n-1} + a_2 \mathbf{X}^{n-2} + \ldots + a_{n-1} \mathbf{X} + a_n \mathbf{I} = \mathbf{0} \tag{12.57}$$

is typical. In this series $\mathbf{X}$ is a square matrix and the coefficients $a_0, a_1, a_2, \ldots, a_n$ are scalar constants. It will be noticed that the coefficient a_n has been multiplied by the unit matrix of the same order as the matrix $\mathbf{X}$ so that all terms in equation (12.57) are conformable for addition.

Series of the type expressed by equation (12.57) can contain a finite or an infinite number of terms; and the summation and tests for convergence of such series are best made by the use of the Cayley–Hamilton theorem and Sylvester's theorem which will be treated in Sections 12.14.2 and 12.15. However, certain properties will be dealt with here.

It is possible to factorize matrix polynomials in the same manner as scalar polynomials as illustrated by the following example.

Example. Evaluate the matrix polynomial

$$\mathbf{X}^3 - 9\mathbf{X}^2 + 26\mathbf{X} - 24\mathbf{I}$$

where

$$\mathbf{X} = \begin{bmatrix} 2 & 1 \\ 3 & 4 \end{bmatrix}$$

The polynomial

$$\mathbf{X}^3 - 9\mathbf{X}^2 + 26\mathbf{X} - 24\mathbf{I} \qquad\qquad\qquad \text{I}$$

can be factorized:

$$(\mathbf{X} - 2\mathbf{I})(\mathbf{X} - 3\mathbf{I})(\mathbf{X} - 4\mathbf{I}) \qquad\qquad\qquad \text{II}$$

The first factor is

$$\begin{bmatrix} 2 & 1 \\ 3 & 4 \end{bmatrix} - 2 \begin{bmatrix} 1 & 0 \\ 0 & 1 \end{bmatrix} = \begin{bmatrix} 0 & 1 \\ 3 & 2 \end{bmatrix} \qquad\qquad \text{III}$$

and the second and third factors are

$$\begin{bmatrix} 2 & 1 \\ 3 & 4 \end{bmatrix} - \begin{bmatrix} 3 & 0 \\ 0 & 3 \end{bmatrix} = \begin{bmatrix} -1 & 1 \\ 3 & 1 \end{bmatrix}$$

and

$$\begin{bmatrix} 2 & 1 \\ 3 & 4 \end{bmatrix} - \begin{bmatrix} 4 & 0 \\ 0 & 4 \end{bmatrix} = \begin{bmatrix} -2 & 1 \\ 3 & 0 \end{bmatrix} \qquad \text{IV}$$

Hence the value of the polynomial is

$$\begin{bmatrix} 0 & 1 \\ 3 & 2 \end{bmatrix} \begin{bmatrix} -1 & 1 \\ 3 & 1 \end{bmatrix} \begin{bmatrix} -2 & 1 \\ 3 & 0 \end{bmatrix} = \begin{bmatrix} -3 & 3 \\ 9 & 3 \end{bmatrix} = 3 \begin{bmatrix} -1 & 1 \\ 3 & 1 \end{bmatrix} \qquad \text{V}$$

Finally, it is possible for matrices to be exponents of the scalar quantity or function, and the well-known exponential function is written

$$e^X = I + \frac{X}{1!} + \frac{X^2}{2!} + \frac{X^3}{3!} + \cdots \qquad (12.58)$$

where X is a square matrix.

The matrix exponent X in e^X behaves in a similar manner to a scalar exponent. Thus if Y is a second square matrix of the same order as X

$$e^X e^Y = e^Y e^X = e^{(X+Y)} \qquad (12.59)$$

Similarly

$$e^X e^{-X} = I \qquad (12.60)$$

where the unit matrix I must be of the same order as X.

12.12. Differentiation and Integration of Matrices

The elements of a matrix may be functions of an independent variable, or derivatives. In both cases the matrix can be differentiated with respect to the independent variable and integrated. Each of these operations will now be discussed.

12.12.1. *Differentiation*

Consider a square matrix Y whose elements are functions of an independent variable x. A small change in the value of x will bring about a corresponding change in the value of the matrix Y, and the limit of the change in Y for an infinitesimal change in x is the derivative of the matrix with respect to x. This is written in the normal way (dY/dx).

The differential coefficient of the product of two or more matrices with respect to an independent variable is obtained in a manner similar to that of the differential coefficient of the product of scalar functions, but in the case of the derivative of a matrix product the order of the matrices in the product must be maintained. Thus

$$\frac{d}{dx}(XYZ) = \frac{dX}{dx}YZ + X\frac{dY}{dx}Z + XY\frac{dZ}{dx} \qquad (12.61)$$

Finally, the differentiation of the exponential function of a matrix is similar to the differentiation of a scalar exponential function. That is,

$$\frac{d}{dt}(e^{xt}) = X + \frac{X^2 t}{1!} + \frac{X^3 t^2}{2!} + \ldots = Xe^{xt} \qquad (12.62)$$

as would be expected.

12.12.2. *Integration*

Integration of a matrix whose elements are derivatives cr functions of an independent variable is accomplished by integrating each element with respect to the independent variable. Thus if the typical element $y_{ij}(x)$ of the matrix Y is a function of x, then the new element in the integrated matrix will be

$$\int Y\,dx = \left[\int \int y_{ij}(x)\,dx\right] \qquad (12.63)$$

An integration series that arises frequently in the solution of systems of linear differential equations is known as "the matrizant" which by definition is

$$M_0^x(A) = I + \int_0^x A(x_1)\,dx_1 + \int_0^x A(x_1)\int_0^{x_1} A(x_2)\,dx_2\,dx_1$$

$$+ \int_0^x A(x_1)\int_0^{x_1} A(x_2)\int_0^{x_2} A(x_3)\,dx_3\,dx_2\,dx_1 \ldots \qquad (12.64)$$

where $M_0^x(A)$ is the matrizant of the matrix A. The form of the matrizant given in equation (12.64) is of little value in calculations because of its complexity. However, if the number of integration terms in equation (12.64) is large, the value of the matrizant over specified limits approximates to

$$M_0^x(A) = \lim_{n\to\infty} \prod_1^n \left[I + A(x_{s-1})\frac{x}{n}\right] \qquad (12.65)$$

where n is the number of integrals in equation (12.64). If all the elements in the matrix A are constants the matrizant becomes

$$M_0^x(A) = I + \frac{A(x-x_0)}{1!} + \frac{A^2(x-x_0)^2}{2!} + \ldots = e^{A(x-x_0)} \qquad (12.66)$$

12.13. LAMBDA-MATRICES

Matrices of this type arise in many fields of applied mathematics where systems of linear differential equations are to be solved. Therefore they deserve consideration. A λ-matrix is a square matrix in which the elements are functions of a scalar parameter λ. Such a matrix is completely defined by its order, rank, and degree. Its order depends on the number of rows or columns as described above. Its rank is the order of the highest non-vanishing minor irrespective of the value assigned to λ, and finally the degree of a λ-matrix is the highest degree of λ in any of the elements of such a matrix.

Example. What is the order, rank and degree of the λ-matrix

$$\begin{bmatrix} \lambda^2+1 & -1 & \lambda-3 \\ \lambda^2+\lambda & -\lambda^2 & \lambda+1 \\ 2\lambda^2-3\lambda+1 & 1-\lambda & \lambda^2 \end{bmatrix}$$

and expand this matrix into a matrix series?

When $\lambda = 1$ the determinant of the matrix $\Delta(\lambda)$ is zero, and hence this matrix is singular. However, its minors of order two are non-vanishing, therefore the λ-matrix is of rank 2.

The highest degree of λ in any element is 2 and therefore the λ-matrix is of degree 2.

The expansion of this matrix is

$$\begin{bmatrix} \lambda^2+1 & -1 & \lambda-3 \\ \lambda^2+\lambda & -\lambda^2 & \lambda+1 \\ 2\lambda^2-3\lambda+1 & 1-\lambda & \lambda^2 \end{bmatrix}$$

$$= \begin{bmatrix} 1 & 0 & 0 \\ 1 & -1 & 0 \\ 2 & 0 & 1 \end{bmatrix}\lambda^2 + \begin{bmatrix} 0 & 0 & 1 \\ 1 & 0 & 1 \\ -3 & -1 & 0 \end{bmatrix}\lambda + \begin{bmatrix} 1 & -1 & -3 \\ 0 & 0 & 1 \\ 1 & 1 & 0 \end{bmatrix} \qquad \text{I}$$

The rank of the λ-matrix can be easily verified from equation I.

The addition, multiplication, and division of λ-matrices is similar to other matrices. However, the reciprocal of a λ-matrix is normally independent of λ and is not therefore a λ-matrix.

12.13.1. *The Determinantal Equation*

A λ-matrix is usually represented by the symbol $f(\lambda)$, its adjoint by $F(\lambda)$ and its determinant by $\Delta(\lambda)$ or $\det(\lambda)$. These are related by the equation

$$f(\lambda)\,F(\lambda) = F(\lambda)\,f(\lambda) = \Delta(\lambda)\,I \qquad (12.67)$$

Equation (12.67) can be differentiated with respect to λ to give

$$\frac{d}{d\lambda}[f(\lambda)\,F(\lambda)] = \frac{d}{d\lambda}[F(\lambda)\,f(\lambda)] = \frac{d}{d\lambda}[\Delta(\lambda)]\,I \qquad (12.68)$$

and the differentiation can be repeated until the derivatives vanish.

The equation formed by equating (12.67) to zero, that is

$$\Delta(\lambda) = 0 \qquad (12.69)$$

is called the "determinantal equation" and its roots which may be distinct or multiple, are of great importance in matrix analysis. Thus let λ_p be a root of the determinantal equation; then the matrix formed by substitution of λ_p into the original matrix is $f(\lambda_p)$ and it will be singular. However, if λ_p is a multiple root, $f(\lambda_p)$ need not have a degeneracy greater than one. Thus the λ-matrix

$$\begin{bmatrix} (\lambda - a) & b & 0 \\ 0 & (\lambda - a) & 0 \\ 0 & 0 & (\lambda - b) \end{bmatrix}$$

has a root $\lambda = b$, and a double root $\lambda = a$, but is of degeneracy one. When $f(\lambda_p)$ is of degeneracy one the adjoint $F(\lambda_p)$ is of unit rank.

When $f(\lambda_p)$ has a degeneracy of q, at least q roots of the determinantal equation are equal. Furthermore, the adjoint matrix and its derivatives up to and including $F^{q-2}(\lambda_p)$ are all null.

12.14. THE CHARACTERISTIC EQUATION

Let A be a square matrix of order n whose elements are all constants, and let x be a column vector whose elements are related by the equation

$$Ax = \lambda x \qquad (12.70)$$

where λ is a scalar coefficient. Equation (12.70) may be written in the matrix form thus

$$(A - \lambda I)x = 0 \qquad (12.71)$$

where $(A - \lambda I)$ is a square λ-matrix of order n. The solution to equation (12.71) will be non-trivial if and only if

$$|A - \lambda I| = 0 \qquad (12.72)$$

Hence λ can only take values that will satisfy equation (12.72) and for this reason equation (12.72) is called the "characteristic equation" of the matrix A. Its roots λ_p are called the "latent roots" or the "eigenvalues" (see Section 8.7.1) of A.

When $|A - \lambda I|$ is zero, $|(A - \lambda I)'|$ is zero, and since $(A - \lambda I)' = A' - \lambda I$, both A and its transpose have the same latent roots.

12.14.1. *Diagonal Canonical Form*

For each root of equation (12.72) there will be a corresponding set of values for the elements in the column vector x, and each set of values constitutes a vector of the latent root. For this reason the vector corresponding to the latent root is called a "latent column vector" and there will be a latent column vector for each latent root. Furthermore, if the latent roots are distinct, the latent column vectors are linearly independent. Therefore consider a non-singular square matrix A of order n, and let the latent roots of A be distinct. In addition let $k_1, k_2, k_3, ..., k_n$ be the latent column vectors corresponding to the latent roots of A, and K the square matrix comprising all the column vectors. K will be of order n and non-singular. Also

$$Ak_1 = \lambda_1 k_1; \quad Ak_2 = \lambda_2 k_2; \quad ...; \quad Ak_n = \lambda_n k_n \quad (12.73)$$

and

$$AK = K\Lambda \quad (12.74)$$

where

$$\Lambda = \begin{bmatrix} \lambda_1 & 0 & ... & 0 \\ 0 & \lambda_2 & ... & 0 \\ \cdot & \cdot & \cdot & \cdot \\ \cdot & \cdot & \cdot & \cdot \\ \cdot & \cdot & \cdot & \cdot \\ 0 & 0 & ... & \lambda_n \end{bmatrix} \quad (12.75)$$

Therefore

$$K^{-1}AK = K^{-1}K\Lambda = \Lambda \quad (12.76)$$

showing that the matrix A can be reduced to what is known as "diagonal canonical form".

In a similar manner, latent row vectors can be determined which, in turn, can be used to reduce A to the diagonal canonical form.

Example 1. Evaluate the latent roots and reduce the following matrix to diagonal form.

$$\begin{bmatrix} 2 & -3 & 1 \\ 3 & 1 & 3 \\ -5 & 2 & -4 \end{bmatrix}$$

The characteristic equation can be expressed in the form

$$|I\lambda - A| = 0 \qquad\qquad\qquad I$$

in order to eliminate the negative signs from in front of the latent roots. Thus in matrix form

$$\begin{bmatrix} \lambda-2 & 3 & -1 \\ -3 & \lambda-1 & -3 \\ 5 & -2 & \lambda+4 \end{bmatrix} = 0 \qquad\qquad II$$

expanding the determinant of equation II

$$(\lambda-1)[(\lambda-2)(\lambda+4)+8] = 0 \qquad\qquad III$$

from which $\lambda_1 = 0, \quad \lambda_2 = 1, \quad \lambda_3 = -2$

When $\lambda = \lambda_1 = 0$, the adjoint is

$$
\begin{bmatrix}
-10 & -10 & -10 \\
-3 & -3 & -3 \\
11 & 11 & 11
\end{bmatrix}
=
\begin{bmatrix}
-10 \\
-3 \\
11
\end{bmatrix}
[1 \quad 1 \quad 1]
\qquad\qquad \text{IV}
$$

When $\lambda = \lambda_2 = 1$, the adjoint is

$$
\begin{bmatrix}
-6 & -13 & -9 \\
0 & 0 & 0 \\
6 & 13 & 9
\end{bmatrix}
=
\begin{bmatrix}
-1 \\
0 \\
1
\end{bmatrix}
[6 \quad 13 \quad 9]
\qquad\qquad \text{V}
$$

and when $\lambda = \lambda_3 = -2$, the adjoint is

$$
\begin{bmatrix}
-12 & -4 & -12 \\
-9 & -3 & -9 \\
21 & 7 & 21
\end{bmatrix}
=
\begin{bmatrix}
-4 \\
-3 \\
7
\end{bmatrix}
[3 \quad 1 \quad 3]
\qquad\qquad \text{VI}
$$

From equations IV, V and VI the **K** matrix is

$$
\mathbf{K} =
\begin{bmatrix}
-10 & -1 & -4 \\
-3 & 0 & -3 \\
11 & 1 & 7
\end{bmatrix}
\quad \text{and} \quad
\mathbf{K}^{-1} = -\frac{1}{6}
\begin{bmatrix}
3 & 3 & 3 \\
-12 & -26 & -18 \\
-3 & -1 & -3
\end{bmatrix}
\qquad \text{VII}
$$

so that

$$
\begin{bmatrix}
2 & -3 & 1 \\
3 & 1 & 3 \\
-5 & 2 & -4
\end{bmatrix}
$$

$$
= -\frac{1}{6}
\begin{bmatrix}
-10 & -1 & -4 \\
-3 & 0 & -3 \\
11 & 1 & 7
\end{bmatrix}
\begin{bmatrix}
0 & 0 & 0 \\
0 & 1 & 0 \\
0 & 0 & -2
\end{bmatrix}
\begin{bmatrix}
3 & 3 & 3 \\
-12 & -26 & -18 \\
-3 & -1 & -3
\end{bmatrix}
\quad \text{VIII}
$$

12.14.2. *The Cayley–Hamilton Theorem*

This theorem can be stated as follows: Any square matrix satisfies its own characteristic equation.

The proof of this theorem will not be given. Here it will suffice to illustrate its application to the above third order matrix.

Example 2. Show that the square matrix

$$\begin{bmatrix} 2 & -3 & 1 \\ 3 & 1 & 3 \\ -5 & 2 & -4 \end{bmatrix}$$

satisfies its own characteristic equation.

The characteristic equation of the given matrix is

$$|I\lambda - A| = \lambda(\lambda - 1)(\lambda + 2) = 0 \qquad \text{I}$$

By the Cayley–Hamilton theorem

$$A(A - I)(A + 2I) = 0 \qquad \text{II}$$

where **0** in equation II is the null matrix. That is,

$$\begin{bmatrix} 2 & -3 & 1 \\ 3 & 1 & 3 \\ -5 & 2 & -4 \end{bmatrix} \begin{bmatrix} 1 & -3 & 1 \\ 3 & 0 & 3 \\ -5 & 2 & -5 \end{bmatrix} \begin{bmatrix} 4 & -3 & 1 \\ 3 & 3 & 3 \\ -5 & 2 & -2 \end{bmatrix} = \begin{bmatrix} 0 & 0 & 0 \\ 0 & 0 & 0 \\ 0 & 0 & 0 \end{bmatrix}$$

$$\text{III}$$

The Cayley–Hamilton theorem can be used to calculate powers of matrices. Thus since a square matrix satisfies its own characteristic equation, the characteristic equation containing the matrix in place of λ can be expanded and rearranged to give the desired power of the matrix as illustrated below.

Example 3. Calculate A^4 if $A = \begin{bmatrix} -1 & 3 \\ -2 & 4 \end{bmatrix}$

The characteristic equation is

$$|I\lambda - A| = \lambda^2 - 3\lambda + 2 = 0 \qquad \text{I}$$

Replacing λ by A and rearranging gives

$$A^2 = 3A - 2I \qquad \text{II}$$

Multiplying by A gives

$$A^3 = 3A^2 - 2A = 9A - 6I - 2A$$

$$= 7A - 6I \qquad \text{III}$$

Repeating the process of multiplying by A and using equation II gives

$$A^4 = 15A - 14I \qquad \text{IV}$$

Hence using equation I,

$$A^4 = \begin{bmatrix} -29 & 45 \\ -30 & 46 \end{bmatrix} \qquad \text{V}$$

This process can be repeated to any positive or negative integer power of a square matrix.

12.15. SYLVESTER'S THEOREM

From the above section it will be seen that in general

$$\mathbf{A}^n = p_1 \mathbf{A}^{n-1} + p_2 \mathbf{A}^{n-2} + p_3 \mathbf{A}^{n-3} + \ldots + p_n \mathbf{I} \tag{12.77}$$

where $\mathbf{A}$ is a square matrix and $p_1, p_2, \ldots, p_n$ are scalar coefficients. Let the series on the right of equation (12.77) be represented by $P(\mathbf{A})$, then applying Lagrange's interpolation formula (Section 9.4.3)

$$P(\mathbf{A}) = P(\lambda_1) \frac{(\mathbf{A} - \mathbf{I}\lambda_2)(\mathbf{A} - \mathbf{I}\lambda_3)(\ldots)(\mathbf{A} - \mathbf{I}\lambda_n)}{(\lambda_1 - \lambda_2)(\lambda_1 - \lambda_3)(\ldots)(\lambda_1 - \lambda_n)}$$

$$+ P(\lambda_2) \frac{(\mathbf{A} - \mathbf{I}\lambda_1)(\mathbf{A} - \mathbf{I}\lambda_3)(\ldots)(\mathbf{A} - \mathbf{I}\lambda_n)}{(\lambda_2 - \lambda_1)(\lambda_2 - \lambda_3)(\ldots)(\lambda_2 - \lambda_n)} + \ldots$$

$$= \sum_{i=1}^{n} P(\lambda_1) \left(\prod_{j \neq i}^{n} (\mathbf{A} - \mathbf{I}\lambda_j) \Big/ \prod_{j \neq i}^{n} (\lambda_i - \lambda_j) \right) \tag{12.78}$$

where $\lambda_1, \lambda_2, \ldots, \lambda_n$ can be chosen as the roots of the characteristic equation which are assumed to be distinct in the case under consideration. Equation (12.78) is an expression of "Sylvester's theorem" but the extended products in the numerator and denominator are sometimes inconvenient computationally. However, they can be simplified as follows.

Denoting the λ-matrix

$$\mathbf{A} - \mathbf{I}\lambda = \mathbf{f}(\lambda) \tag{12.79}$$

equation (12.67) shows that

$$(\mathbf{A} - \mathbf{I}\lambda_i) \mathbf{F}(\lambda_i) = \Delta(\lambda_i) \mathbf{I} = 0 \tag{12.80}$$

since λ_i is one of the distinct roots of the characteristic equation. Furthermore,

$$\mathbf{A} - \mathbf{I}\lambda_i \neq 0 \quad \text{and} \quad \mathbf{F}(\lambda_i) \neq 0 \tag{12.81}$$

Similarly,

$$(\mathbf{A} - \mathbf{I}\lambda_i) \prod_{j \neq i}^{n} (\mathbf{A} - \mathbf{I}\lambda_j) = \prod_{j=1}^{n} (\mathbf{A} - \mathbf{I}\lambda_j) = 0 \tag{12.82}$$

but

$$\prod_{j \neq i}^{n} (\mathbf{A} - \mathbf{I}\lambda_j) \neq 0 \tag{12.83}$$

Hence it can be shown that

$$\prod_{j \neq i}^{n} (\mathbf{A} - \mathbf{I}\lambda_j) = \mathbf{CF}(\lambda_i) \tag{12.84}$$

where $\mathbf{C}$ can be any diagonal matrix with constant elements. The matrices in equations (12.80) and (12.82) are commutative in each case; therefore $\mathbf{C}$ must be some constant (α) times the unit matrix. By selecting square matrices of different orders in which all elements are unity it can be shown that $\alpha = (-1)^{n-1}$. Thus

$$\prod_{j\neq i}^{n} (A - I\lambda_j) = (-1)^{n-1} F(\lambda_i) = \prod_{j\neq i}^{n} f(\lambda_j) \tag{12.85}$$

where n is the order of the square matrix.

The denominator of equation (12.78) can be simplified as follows

$$\Delta(\lambda) = \prod_{j=1}^{n} (\lambda_j - \lambda) = (-1)^{n} \prod_{j=1}^{n} (\lambda - \lambda_j) \tag{12.86}$$

but

$$\frac{d}{d\lambda}[\Delta(\lambda)]|_{\lambda=\lambda_i} = (-1)^{n} \prod_{j\neq i}^{n} (\lambda_i - \lambda_j) = \Delta'(\lambda_i) \tag{12.87}$$

Hence Sylvester's theorem becomes

$$P(A) = -\sum_{i=1}^{n} \frac{P(\lambda_i) F(\lambda_i)}{\Delta'(\lambda_i)} \tag{12.88}$$

If $f(\lambda)$ is defined as $I\lambda - A$ instead of by equation (12.79) the only effect is to change some of the signs. The negative sign in front of equation (12.88) has to be removed, but equations (12.78) and (12.85) are not changed.

Finally, when the characteristic equation has multiple roots Sylvester's theorem takes a much more complicated form due to $\Delta'(\lambda_i)$ vanishing for the repeated roots. Equations (12.78) and (12.88) can be modified by using L'Hôpital's rule until the indeterminate fractions are resolved, but the work involved shows no significant advantage over evaluating the original polynomial (12.77) directly.

12.16. Quadratic Form

An expression of the type

$$[x_1 \ x_2 \ x_3 \ \cdots \ x_n] \begin{bmatrix} a_{11} & a_{12} & a_{13} & \cdots & a_{1n} \\ a_{21} & a_{22} & a_{23} & \cdots & a_{2n} \\ a_{31} & a_{32} & a_{33} & \cdots & a_{3n} \\ \cdot & \cdot & \cdot & \cdot & \cdot \\ \cdot & \cdot & \cdot & \cdot & \cdot \\ \cdot & \cdot & \cdot & \cdot & \cdot \\ a_{n1} & a_{n2} & a_{n3} & \cdots & a_{nn} \end{bmatrix} \begin{bmatrix} x_1 \\ x_2 \\ x_3 \\ \cdot \\ \cdot \\ \cdot \\ x_n \end{bmatrix} \tag{12.89}$$

is called a "quadratic form. It is a homogeneous function of the second degree. The coefficient of $x_i x_j$ in the quadratic form where $i \neq j$ is $(a_{ij} + a_{ji})$. For example, the conic

$$ax^2 + 2hxy + by^2 = [x \ y] \begin{bmatrix} a & h \\ h & b \end{bmatrix} \begin{bmatrix} x \\ y \end{bmatrix} = c \tag{12.90}$$

can be written more concisely in the quadratic form.

$$\mathbf{x}'\mathbf{A}\mathbf{x} = c \tag{12.91}$$

whilst the equation

$$a_1 x_1{}^2 + a_2 x_2{}^2 + a_3 x_3{}^2 + \ldots + a_n x_n{}^2 = 0 \tag{12.92}$$

would be written

$$[x_1 \quad x_2 \quad x_3 \quad \ldots \quad x_n] \begin{bmatrix} a_1 & 0 & 0 & \ldots & 0 \\ 0 & a_2 & 0 & \ldots & 0 \\ 0 & 0 & a_3 & \ldots & 0 \\ \cdot & \cdot & \cdot & \cdot & \cdot \\ \cdot & \cdot & \cdot & \cdot & \cdot \\ \cdot & \cdot & \cdot & \cdot & \cdot \\ 0 & 0 & 0 & \ldots & a_n \end{bmatrix} \begin{bmatrix} x_1 \\ x_2 \\ x_3 \\ \cdot \\ \cdot \\ \cdot \\ x_n \end{bmatrix} = \mathbf{x}'[A_{ii}]\,\mathbf{x} \tag{12.93}$$

where the symbol $[A_{ii}]$ indicates that the matrix $\mathbf{A}$ is a diagonal matrix.

Example. Determine the nature of the quadric

$$6x^2 + 3y^2 - 2z^2 + 4xy = 14$$

and evaluate the lengths of the semi-axes and equations of the principal planes.

The equation $\qquad 6x^2 + 3y^2 - 2z^2 + 4xy = 14 \qquad\qquad$ I

can be written in matrix form

$$[x \quad y \quad z] \begin{bmatrix} 6 & 2 & 0 \\ 2 & 3 & 0 \\ 0 & 0 & -2 \end{bmatrix} \begin{bmatrix} x \\ y \\ z \end{bmatrix} = 14 = \mathbf{x}'\mathbf{A}\mathbf{x} \qquad\qquad \text{II}$$

The latent roots of $\mathbf{A}$ are obtained from

$$\begin{vmatrix} 6-\lambda & 2 & 0 \\ 2 & 3-\lambda & 0 \\ 0 & 0 & -(2+\lambda) \end{vmatrix} = 0 \qquad\qquad \text{III}$$

or $\qquad\qquad (\lambda+2)(\lambda-2)(\lambda-7) = 0 \qquad\qquad\qquad$ IV

and the latent roots are $\lambda_1 = -2$, $\lambda_2 = 2$, $\lambda_3 = 7$. The latent column vector $\mathbf{k}_1$ corresponding to the root $\lambda_1 = -2$ is

$$\mathbf{A}\mathbf{k}_1 = \lambda_1 \mathbf{k}_1 = -2\mathbf{k}_1 \qquad\qquad \text{V}$$

or $$\begin{bmatrix} 6 & 2 & 0 \\ 2 & 3 & 0 \\ 0 & 0 & -2 \end{bmatrix} \begin{bmatrix} k_{11} \\ k_{21} \\ k_{31} \end{bmatrix} = -2 \begin{bmatrix} k_{11} \\ k_{21} \\ k_{31} \end{bmatrix} \qquad\qquad \text{VI}$$

or

$$6k_{11}+2k_{21} = -2k_{11}$$
$$2k_{11}+3k_{21} = -2k_{21}$$
$$-2k_{31} = -2k_{31}$$

VII

from which $k_{11} = k_{21} = 0$ and k_{31} is arbitrary. It is normal to choose orthogonal column vectors which satisfy equation (12.16) and thus $k_{31} = 1$. Therefore, the latent column vector corresponding to the latent root $\lambda_1 = -2$ is $\{0\ 0\ 1\}$.

In a similar manner, the latent column vector corresponding to the latent root $\lambda_2 = 2$ is $\{1/\sqrt{5}\ \ -2/\sqrt{5}\ \ 0\}$, and corresponding to $\lambda_3 = 7$ is $\{2/\sqrt{5}\ \ 1/\sqrt{5}\ \ 0\}$. Then by equation (12.76),

$$\begin{bmatrix} 0 & 0 & 1 \\ 1/\sqrt{5} & -2/\sqrt{5} & 0 \\ 2/\sqrt{5} & 1/\sqrt{5} & 0 \end{bmatrix} \begin{bmatrix} 6 & 2 & 0 \\ 2 & 3 & 0 \\ 0 & 0 & -2 \end{bmatrix} \begin{bmatrix} 0 & 1/\sqrt{5} & 2/\sqrt{5} \\ 0 & -2/\sqrt{5} & 1/\sqrt{5} \\ 1 & 0 & 0 \end{bmatrix} = \begin{bmatrix} -2 & 0 & 0 \\ 0 & 2 & 0 \\ 0 & 0 & 7 \end{bmatrix}$$

VIII

Applying the orthogonal transformation $\mathbf{x} = \mathbf{KX}$ where

$$\mathbf{X} = \{X\ \ Y\ \ Z\}$$

IX

then

$$\mathbf{K'x} = \mathbf{K'KX} = \mathbf{X}$$

X

or

$$\begin{bmatrix} 0 & 0 & 1 \\ 1/\sqrt{5} & -2/\sqrt{5} & 0 \\ 2/\sqrt{5} & 1/\sqrt{5} & 0 \end{bmatrix} \begin{bmatrix} x \\ y \\ z \end{bmatrix} = \begin{bmatrix} X \\ Y \\ Z \end{bmatrix}$$

XI

or

$$z = X, \quad \frac{x}{\sqrt{5}} - \frac{2y}{\sqrt{5}} = Y, \quad \frac{2x}{\sqrt{5}} + \frac{y}{\sqrt{5}} = Z$$

XII

Furthermore, from equation II

$$(\mathbf{KX})'\mathbf{A}(\mathbf{KX}) = 14$$

XIII

and by Section 12.5.1

$$\mathbf{X'K'AKX} = \mathbf{X'\Lambda X} = 14$$

XIV

or

$$-2X^2+2Y^2+7Z^2 = 14$$

XV

or in standard form

$$\frac{-X^2}{7}+\frac{Y^2}{7}+\frac{Z^2}{2} = 1$$

XVI

which is a hyperboloid of one sheet with semi-axes $\sqrt{7} : \sqrt{7} : \sqrt{2}$.

The principal planes are $X = 0$, $Y = 0$, and $Z = 0$ when referred to the new axes. With respect to the original axes, the principal planes are

$$x-2y = 0, \quad 2x+y = 0, \quad z = 0$$

12.17. Application to the Solution of Differential Equations

Matrices can be applied to the solution of systems of differential equations, and in this respect they are of great utility in the solution of engineering problems. Hence the essential features of the application technique will be presented in this section.

Consider a system of linear differential equations expressing relationships between a number of dependent variables $y_1, y_2, y_3, \ldots, y_n$ and an independent variable x. These equations can be written in conventional form

$$\left.\begin{array}{c} f_{11}(D)\,y_1 + f_{12}(D)\,y_2 + \ldots + f_{1n}(D)\,y_n = \phi_1(x) \\ f_{21}(D)\,y_1 + f_{22}(D)\,y_2 + \ldots + f_{2n}(D)\,y_n = \phi_2(x) \\ \cdot\;\cdot \\ f_{n1}(D)\,y_1 + f_{n2}(D)\,y_2 + \ldots + f_{nn}(D)\,y_n = \phi_n(x) \end{array}\right\} \quad (12.94)$$

or written in condensed matrix form

$$\mathbf{f}(D)\{y\} = [f_{ij}(D)]\{y\} = \{\phi(x)\} \quad (12.95)$$

In equations (12.94) and (12.95) $f_{ij}(D)$ is a polynomial of the differential operator having constant coefficients. The matrix of differential operators $\mathbf{f}(D)$ is analogous to the λ-matrix and in the determination of the complementary function the technique depends upon the substitution of λ for D. However, certain nomenclature are required and in addition to the terms for λ-matrices given in Section 12.13 the following are necessary.

$\mathbf{f}(D)$ is called the D-matrix

$|\mathbf{f}(D)|$ is called the D-determinant and is frequently expressed $\Delta(D)$.

The highest degree of D in any of the elements of $\mathbf{f}(D)$ is said to be the order of the system.

12.17.1. *Solution by Conversion to an Equivalent System*

In some systems of differential equations it is possible to transform the D-matrix into such a form that will permit ready solution of the differential equations. This is accomplished by pre-multiplying the D-matrix by another non-singular D-matrix of the same order whose D-determinant is free from the differential operator D. The following example will clarify the method.

Example. Solve the system of equations

$$D^3 y_1 + (D^2 + 3D - 2)\,y_2 = x$$

$$D^2 y_1 - D y_2 = \sin x$$

These equations can be put in matrix form thus

$$\begin{bmatrix} D^3 & (D^2 + 3D - 2) \\ D^2 & -D \end{bmatrix} \begin{bmatrix} y_1 \\ y_2 \end{bmatrix} = \begin{bmatrix} x \\ \sin x \end{bmatrix} \qquad \text{I}$$

Pre-multiplying both sides of equation I by

$$\begin{bmatrix} 1 & -D \\ -D & D^2+2 \end{bmatrix} \qquad \text{II}$$

gives

$$\begin{bmatrix} 1 & -D \\ -D & D^2+2 \end{bmatrix} \begin{bmatrix} D^3 & (D^2+3D-2) \\ D^2 & -D \end{bmatrix} \begin{bmatrix} y_1 \\ y_2 \end{bmatrix} = \begin{bmatrix} 1 & -D \\ -D & D^2+2 \end{bmatrix} \begin{bmatrix} x \\ \sin x \end{bmatrix}$$

$$\text{III}$$

$$\therefore \quad \begin{bmatrix} 0 & (2D^2+3D-2) \\ 2D^2 & -(2D^3+3D^2) \end{bmatrix} \begin{bmatrix} y_1 \\ y_2 \end{bmatrix} = \begin{bmatrix} x-D\sin x \\ -Dx+(D^2+2)\sin x \end{bmatrix} \qquad \text{IV}$$

$$\therefore \quad (2D^2+3D-2)y_2 = x - \cos x \qquad \text{V}$$

$$2D^2 y_1 - 2D^3 y_2 - 3D^2 y_2 = -1 + \sin x \qquad \text{VI}$$

Solving equation V by the methods of Chapter 2 gives

$$y_2 = A e^{-2x} + B e^{\frac{1}{2}x} - \tfrac{1}{2}x - \tfrac{3}{4} + \tfrac{1}{25}(4\cos x - 3\sin x) \qquad \text{VII}$$

Integrating equation VI twice and then substituting for y_2 from equation VII gives

$$y_1 = C + Ex - \tfrac{1}{4}x^2 - \tfrac{1}{2}A e^{-2x} + 2B e^{\frac{1}{2}x} + \tfrac{1}{25}(3\cos x - 21\sin x) \qquad \text{VIII}$$

where A, B, C, and E are the four arbitrary constants required by this third order system of equations.

In the above example the matrix

$$\begin{bmatrix} 1 & -D \\ -D & D^2+2 \end{bmatrix}$$

was obtained by trial, but guided by the aim that all but one of the dependent variables were to be eliminated from one of the equations. This enabled that variable to be evaluated and inserted into the other equation which was subsequently solved by conventional methods. When there are a large number of dependent variables it is necessary to convert the D-matrix into triangular form by pre-multiplication by another D-matrix whose determinant is a constant. The transformed system must be of the type

$$\begin{bmatrix} f_{11}(D) & f_{12}(D) & f_{13}(D) & \cdots & f_{1n}(D) \\ 0 & f_{22}(D) & f_{23}(D) & \cdots & f_{2n}(D) \\ 0 & 0 & f_{33}(D) & \cdots & f_{3n}(D) \\ \cdot & \cdot & \cdot & \cdot & \cdot \\ \cdot & \cdot & \cdot & \cdot & \cdot \\ \cdot & \cdot & \cdot & \cdot & \cdot \\ 0 & 0 & 0 & \cdots & f_{nn}(D) \end{bmatrix} \begin{bmatrix} y_1 \\ y_2 \\ y_3 \\ \cdot \\ \cdot \\ \cdot \\ y_n \end{bmatrix} = \begin{bmatrix} \phi_1(x) \\ \phi_2(x) \\ \phi_3(x) \\ \cdot \\ \cdot \\ \cdot \\ \phi_n(x) \end{bmatrix} \qquad (12.96)$$

In equation (12.96) it can be seen that the last row is a function of y_n only, whilst the row above is a function of y_n and y_{n-1} only. Hence the equation of the last row will give an expression for y_n which will be inserted into the equation of the row above to give an expression for y_{n-1} and so on until the system has been solved.

12.18. Solutions of Systems of Linear Differential Equations

Systems of linear differential equations can be solved in a similar manner to single linear differential equations by establishing the complementary function and the particular integral. However, with systems of equations the process will be much more complicated because of the inter-relationship between the dependent variables and the effect of one solution on the particular integrals of the others. Each part of the complete solution will now be considered separately.

12.18.1. *The Complementary Function*

A system of linear differential equations has been written in matrix form in equation (12.95) and the complementary function of this equation will be obtained by solving when $\{\phi(x)\}$ is zero. That is by solving the equation

$$\mathbf{f}(D)\{y\} = \mathbf{0} \tag{12.97}$$

The solution of equation (12.97) will depend on the latent roots of the D-matrix, and each latent root is said to contribute a "constituent solution" to the complementary function. Hence the solution of equation (12.97) depends upon whether the latent roots of $\mathbf{f}(D)$ are distinct or multiple. These will be considered separately.

(i) *Distinct latent roots.* Let λ_r by any distinct root of $\Delta(\lambda) = 0$ obtained by substituting λ for D in the D-matrix. Then since

$$\mathbf{f}(D)\, e^{\lambda_r x}\, \mathbf{F}(\lambda_r) = e^{\lambda_r x}\, \mathbf{f}(\lambda_r)\, \mathbf{F}(\lambda_r) = 0 \tag{12.98}$$

$e^{\lambda_r x}\, \mathbf{F}(\lambda_r)$ is a solution of equation (12.97) but as λ_r is a simple root $\mathbf{F}(\lambda_r)$ can be written in the form

$$\mathbf{F}(\lambda_r) = \{k_r\}\,[\kappa_r] \tag{12.99}$$

so that the constituent solution corresponding to the distinct root λ_r is an arbitrary multiple of $e^{\lambda_r x}\, k_r$.

(ii) *Multiple latent roots.* Let λ_s be any one of s repeated latent roots. Then if the matrices

$$\mathbf{W}_0(x, \lambda_s) = e^{\lambda_s x}\, \mathbf{F}(\lambda_s)$$

and
$$\mathbf{W}_p(x, \lambda_s) = \left[\frac{\partial^p}{\partial \lambda_s^p} e^{\lambda_s x}\, \mathbf{F}(\lambda_s)\right] = e^{\lambda_s x}\left(\frac{\partial}{\partial \lambda_s} + x\right)^p \mathbf{F}(\lambda_s)$$

it can be shown by a similar procedure to that given for distinct roots that

$$f(D)\{y\} = f(D)\,W_p(x, \lambda_s) \tag{12.100}$$

so that $W_0(x, \lambda_s)$; $W_1(x, \lambda_s)$; ...; $W_p(x, \lambda_s)$ all satisfy the system of differential equations. In all the above $p = (s-1)$ for obvious reasons.

It is convenient to write the matrix $W_p(x, \lambda_s)$ in the form $e^{\lambda_s x} Z_p(x, \lambda_s)$ where

$$Z_p(x, \lambda_s) = F^p(\lambda_s) + \frac{px F^{p-1}(\lambda_s)}{1!} + \frac{p(p-1)x^2 F^{p-2}(\lambda_s)}{2!} + \ldots + x^p\, F(\lambda_s) \tag{12.101}$$

and there will be the s matrices of Z corresponding to the set of repeated roots. Hence since each matrix will have n columns (the order of $f(D)$) there will be ns constituent solutions, but of these only s solutions will be linearly independent. Hence for a root λ_s from s repeated roots the constituent solution will be

$$\{y\} = \{k_{1s}(x) \quad k_{2s}(x) \quad \ldots \quad k_{ns}(x)\} e^{\lambda_s x} \tag{12.102}$$

and the s columns of the type represented by equation (12.102) corresponding to the s equal roots may be chosen to be proportional to any s linearly independent columns of the matrices.

$$Z_0(x, \lambda_s) = F(\lambda_s)$$

$$Z_1(x, \lambda_s) = F'(\lambda_s) + xF(\lambda_s)$$

. . . .

$$Z_p(x, \lambda_s) = F^p(\lambda_s) + \frac{px F^{p-1}(\lambda_s)}{1!} + \frac{p(p-1)x^2 F^{p-2}(\lambda_s)}{2!} + x^p\, F(\lambda_s)$$

Since $p = s-1$ the elements in the column vector of equation (12.102) will be polynomials in x of degree $(s-1)$. These column vectors are called "modal columns" and the matrix formed from all the n roots of $f(D)$ is

$$K(x) = \begin{bmatrix} k_{11}(x) & k_{12}(x) & \ldots & k_{1n}(x) \\ k_{21}(x) & k_{22}(x) & \ldots & k_{2n}(x) \\ \cdot & \cdot & \cdot & \cdot \\ \cdot & \cdot & \cdot & \cdot \\ \cdot & \cdot & \cdot & \cdot \\ k_{m1}(x) & k_{m2}(x) & \ldots & k_{mn}(x) \end{bmatrix} \tag{12.103}$$

and is known as the "modal matrix". Those columns obtained from distinct roots will have constant elements.

Finally if the exponential functions $e^{\lambda_n x}$ are collected into a diagonal matrix $\mathbf{M}(x)$ where

$$\mathbf{M}(x) = \begin{bmatrix} e^{\lambda_1 x} & 0 & 0 & \ldots & 0 \\ 0 & e^{\lambda_2 x} & 0 & \ldots & 0 \\ 0 & 0 & e^{\lambda_3 x} & \ldots & 0 \\ . & . & . & . & . \\ . & . & . & . & . \\ . & . & . & . & . \\ 0 & 0 & 0 & \ldots & e^{\lambda_n x} \end{bmatrix} \tag{12.104}$$

and $\{C\}$ is a column of n arbitrary constants, the complementary function of the system is

$$\{y\} = \mathbf{K}(x)\,\mathbf{M}(x)\,\{C\} \tag{12.105}$$

Example. Solve the following system of differential equations.

$$(D^2 - 3D + 2)\,y_1 + (D^2 - 1)\,y_2 = 0$$

$$(D^2 + 4D + 3)\,y_2 - (D^2 + D + 2)\,y_1 = 0$$

The above equations can be written in matrix form thus

$$\begin{bmatrix} (D^2 - 3D + 2) & (D^2 - 1) \\ -(D^2 + D + 2) & (D^2 + 4D + 3) \end{bmatrix} \begin{bmatrix} y_1 \\ y_2 \end{bmatrix} = \mathbf{0} \qquad \text{I}$$

The λ-matrix obtained from the D-matrix is

$$\begin{bmatrix} (\lambda - 1)(\lambda - 2) & (\lambda - 1)(\lambda + 1) \\ -(\lambda^2 + \lambda + 2) & (\lambda + 1)(\lambda + 3) \end{bmatrix}$$

from which $\qquad \Delta(\lambda) = (\lambda - 1)^2\,(\lambda + 1)\,(\lambda + 2) \qquad \text{II}$

The latent roots are

$$\lambda_1 = -1, \quad \lambda_2 = -2, \quad \lambda_3 = \lambda_4 = 1$$

The respective adjoints for λ_1 and λ_2

$$\mathbf{F}(\lambda_1) = \begin{bmatrix} 0 & 0 \\ 2 & 6 \end{bmatrix} = \begin{bmatrix} 0 \\ 2 \end{bmatrix} [1 \quad 3] \qquad \text{III}$$

$$\mathbf{F}(\lambda_2) = \begin{bmatrix} -1 & -3 \\ 4 & 12 \end{bmatrix} = \begin{bmatrix} -1 \\ 4 \end{bmatrix} [1 \quad 3] \qquad \text{IV}$$

and the adjoints for λ_3 and λ_4 are

$$F(\lambda_3) = \begin{bmatrix} 8 & 0 \\ 4 & 0 \end{bmatrix} = \begin{bmatrix} 2 \\ 1 \end{bmatrix} [4 \quad 0] \qquad\qquad \text{V}$$

$$F'(\lambda_3) = \begin{bmatrix} 2\lambda+4 & -2\lambda \\ 2\lambda+1 & 2\lambda-3 \end{bmatrix}_{\lambda=1} = \begin{bmatrix} 6 & -2 \\ 3 & -1 \end{bmatrix} = \begin{bmatrix} 2 \\ 1 \end{bmatrix} [3 \quad -1] \qquad \text{VI}$$

and for the latent roots λ_3 and λ_4 the model columns are

$$\mathbf{k}_3(x) = \{2 \quad 1\} \qquad\qquad \text{VII}$$

$$\mathbf{k}_4(x) = \{2 \quad 1\} + \{2 \quad 1\} x \qquad\qquad \text{VIII}$$

and the complementary function is

$$\begin{bmatrix} y_1 \\ y_2 \end{bmatrix} = \begin{bmatrix} 0 & -1 & 2 & 2(1+x) \\ 2 & 4 & 1 & (1+x) \end{bmatrix} \begin{bmatrix} e^{-x} & 0 & 0 & 0 \\ 0 & e^{-2x} & 0 & 0 \\ 0 & 0 & e^{x} & 0 \\ 0 & 0 & 0 & e^{x} \end{bmatrix} \begin{bmatrix} C_1 \\ C_2 \\ C_3 \\ C_4 \end{bmatrix} \qquad \text{IX}$$

or the conventional result is

$$y_1 = 2[C_3 + C_4(1+x)] e^{x} - C_2 e^{-2x} \qquad\qquad \text{X}$$

and $\qquad\qquad y_2 = 2C_1 e^{-x} + 4C_2 e^{-2x} + [C_3 + C_4(1+x)] e^{x} \qquad \text{XI}$

12.18.2. *The Particular Solution*

To complete the solution of the system of linear differential equations it is necessary to obtain a particular solution of equation (12.95). Essentially this solution will be

$$\{y_p\} = \mathbf{f}(D)^{-1}\{\phi(x)\} = \frac{F(D)}{\Delta(D)}\{\phi(x)\} \qquad\qquad (12.106)$$

Hence it is necessary to be able to evaluate the last term in equation (12.106). Therefore consider the case $\phi(x) = R e^{\theta x}$ where R is a column of constants. Then

$$\{y_p\} = \frac{F(D)}{\Delta(D)} R e^{\theta x} = e^{\theta x} \frac{F(\theta)}{\Delta(\theta)} R \qquad\qquad (12.107)$$

which is the particular solution when $\phi(x)$ is exponential provided that $\Delta(\theta) \neq 0.$.

When $\Delta(\theta) = 0$ the following procedure must be carried out. Let the particular solution be

$$\{y_p\} = \mathbf{b}[F'(\theta) + xF(\theta)] e^{\theta x} \qquad\qquad (12.108)$$

where **b** is constant. Then

$$\mathbf{f}(D)\,\mathbf{b}[\mathbf{F}'(\theta)+x\mathbf{F}(\theta)]\,e^{\theta x} = \mathbf{R}\,e^{\theta x} \qquad (12.109)$$

from which

$$\mathbf{b}\,e^{\theta x}\mathbf{f}(D+\theta)\,[\mathbf{F}'(\theta)+x\mathbf{F}(\theta)] = \mathbf{R}\,e^{\theta x} \qquad (12.110)$$

Using Taylor's theorem to expand $\mathbf{f}(D+\theta)$, and operating with the powers of D upon x gives

$$\mathbf{b}\,e^{\theta x}[\mathbf{f}(\theta)\,\mathbf{F}'(\theta)+\mathbf{f}'(\theta)\,\mathbf{F}(\theta)+x\mathbf{f}(\theta)\,\mathbf{F}(\theta)] = \mathbf{R}\,e^{\theta x} \qquad (12.111)$$

Since

$$\mathbf{f}(\theta)\,\mathbf{F}(\theta) = \Delta(\theta) = 0$$

and

$$\mathbf{f}'(\theta)\,\mathbf{F}(\theta)+\mathbf{f}(\theta)\,\mathbf{F}'(\theta) = \Delta'(\theta)$$

equation (12.111) gives the value

$$\mathbf{b} = \mathbf{R}/\Delta'(\theta) \qquad (12.112)$$

Hence the particular solution is

$$\{y_p\} = \frac{\mathbf{R}}{\Delta'(\theta)}\,[\mathbf{F}'(\theta)+x\mathbf{F}(\theta)]\,e^{\theta x} \qquad (12.113)$$

Other forms for $\boldsymbol{\phi}(x)$ can be treated in a similar manner to the corresponding forms of $\boldsymbol{\phi}(x)$ in the inverse operator method (Section 2.5.5).

Finally, the complete solution is the sum of the complementary function and the particular integral as for linear differential equations.

Example. A battery of N stirred tank reactors is arranged to operate isothermally in series. Each reactor has a volume of V m³ and is equipped with a perfect agitator so that the composition of the reactor effluent is the same as the tank contents. If initially the tanks contain pure solvent only, and at a time designated t_0, q m³/s of a reactant A of concentration C_0 kg moles/m³ are fed to the first tank, estimate the time required for the concentration of A leaving the Nth tank to be C_N kg moles/m³. The stoichiometry of the reaction is represented by the equation

$$A \underset{k_1'}{\overset{k_1}{\rightleftarrows}} B \underset{k_2'}{\overset{k_2}{\rightleftarrows}} C$$

All the reactions are first order and the feed to the reactor does not contain any product B or C, but is assumed to contain a catalyst that initiates the reaction as soon as the feed enters the first reactor.

Solution

The concentration of A will be written C_A.

The concentration of B will be written C_B.

The concentration of C will be written C_C.

Then a mass balance on any tank n will be

$$V\frac{dC}{dt} = qC_{n-1} - qC_n - rV \qquad\qquad \text{I}$$

where r is the rate of chemical reaction.

For component A this gives

$$V\frac{dC_{A,n}}{dt} = qC_{A,n-1} - (qC_{A,n} + Vk_1 C_{A,n} - Vk_1' C_{B,n}) \qquad\qquad \text{II}$$

for component B,

$$V\frac{dC_{B,n}}{dt} = qC_{B,n-1} - (qC_{B,n} + Vk_1' C_{B,n} + Vk_2 C_{B,n} - Vk_1 C_{A,n} - Vk_2' C_{C,n})$$

$$\text{III}$$

and for component C,

$$V\frac{dC_{C,n}}{dt} = qC_{C,n-1} - (qC_{C,n} + Vk_2' C_{C,n} - Vk_2 C_{B,n}) \qquad\qquad \text{IV}$$

Dividing equations II, III, and IV by V and calling (V/q), θ the nominal holding time, gives

$$\frac{dC_{A,n}}{dt} = \frac{C_{A,n-1}}{\theta} - \left(\frac{1}{\theta} + k_1\right) C_{A,n} + k_1' C_{B,n} \qquad\qquad \text{V}$$

$$\frac{dC_{B,n}}{dt} = \frac{C_{B,n-1}}{\theta} - \left(\frac{1}{\theta} + k_2 + k_1'\right) C_{B,n} + k_1 C_{A,n} + k_2' C_{C\,n} \qquad\qquad \text{VI}$$

and

$$\frac{dC_{C,n}}{dt} = \frac{C_{C,n-1}}{\theta} - \left(\frac{1}{\theta} + k_2'\right) C_{C,n} + k_2 C_{B,n} \qquad\qquad \text{VII}$$

Equations V, VI, and VII can be written in matrix form as follows

$$\frac{d}{dt}\begin{bmatrix} C_{A,n} \\ C_{B,n} \\ C_{C,n} \end{bmatrix} = \frac{1}{\theta}\begin{bmatrix} C_{A,n-1} \\ C_{B,n-1} \\ C_{C,n-1} \end{bmatrix} - \begin{bmatrix} (1/\theta + k_1) & -k_1' & 0 \\ -k_1 & (1/\theta + k_2 + k_1') & -k_2' \\ 0 & -k_2 & (1/\theta + k_2') \end{bmatrix}\begin{bmatrix} C_{A,n} \\ C_{B,n} \\ C_{C,n} \end{bmatrix}$$

$$\text{VIII}$$

or more concisely

$$\frac{dC_n}{dt} = \frac{C_{n-1}}{\theta} - GC_n \qquad\qquad \text{IX}$$

where C_n and C_{n-1} are column vectors and G is a third order square matrix with constant elements. Equation IX can be applied to the first reactor thus

$$\frac{dC_1}{dt} = \frac{C_0}{\theta} - GC_1 \qquad\qquad \text{X}$$

The solution of equation X is

$$C_1 = \frac{G^{-1} C_0}{\theta} + e^{-Gt} K_1 \qquad\qquad \text{XI}$$

where K_1 is the constant of integration. When

$$t = 0, \quad C_1 = 0$$

$$\therefore \quad K_1 = -G^{-1} C_0 / \theta \qquad\qquad \text{XII}$$

$$\therefore \quad C_1 = \frac{1}{\theta}(I - e^{-Gt}) G^{-1} C_0 \qquad\qquad \text{XIII}$$

Applying equation IX to the second tank and substituting the expression for C_1 given by equation XIII gives

$$C_2 = \frac{G^{-2} C_0}{\theta^2} - \frac{G^{-1} C_0}{\theta^2} t e^{-Gt} + e^{-Gt} K_2 \qquad\qquad \text{XIV}$$

When $t = 0$, $C_2 = 0$

$$\therefore \quad K_2 = G^{-2} C_0 / \theta^2$$

and

$$C_2 = \frac{1}{\theta^2} (I - e^{-Gt} - Gt\,e^{-Gt}) G^{-2} C_0 \qquad\qquad \text{XV}$$

Applying equation IX to the third tank, substituting the expression for C_2 and evaluating the constant of integration gives

$$C_3 = \frac{1}{\theta^3}\left[I - e^{-Gt} - \frac{(Gt)}{1!} e^{-Gt} - \frac{(Gt)^2}{2!} e^{-Gt}\right] G^{-3} C_0 \qquad\qquad \text{XVI}$$

Extending this to tank N by comparing equations XIII, XV, and XVI the concentration of the effluent from tank N is

$$C_N = \frac{G^{-N} C_0}{\theta^N} - \frac{e^{-Gt}}{\theta^N}\left[I + \frac{Gt}{1!} + \frac{(Gt)^2}{2!} + \dots + \frac{(Gt)^{N-1}}{(N-1)!}\right] G^{-N} C_0 \qquad \text{XVII}$$

By Sylvester's theorem (Section 12.15)

$$e^{-Gt} = -\sum_{j=1}^{3} e^{-\lambda_j t} \frac{F(\lambda_j)}{\Delta'(\lambda_j)} \qquad\qquad \text{XVIII}$$

on the assumption that all latent roots of G are distinct, and therefore equation XVII can be written in the very useful form that

$$C_N = \frac{-1}{\theta^N} \sum_{j=1}^{3} \left(\frac{1}{\lambda_j}\right)^N \left[1 - e^{-\lambda_j t} \sum_{i=0}^{N-1} \frac{(\lambda_j t)^i}{i!}\right] \frac{F(\lambda_j)}{\Delta'(\lambda_j)} C_0 \qquad\qquad \text{XIX}$$

Equation XIX was initially derived by Acrivos and Amundson† who also present equations for the transient behaviour of plate extractors, absorbers, and distillation equipment. The reader interested in the further application of matrix analysis to the solution of stagewise processes is recommended to read this article.

12.19. CONCLUSIONS

This chapter has attempted to present the elements of matrix algebra and calculus which has had to be concise because of the limited space available for this topic. The reader may draw the conclusion that the arithmetic involved will be exorbitant. This is true, but when one considers that these methods can be used in conjunction with high-speed digital computers it will be appreciated that the arithmetic need not prevent these methods being applied. Equation XIX in the above example looks quite formidable, but when used in conjunction with a computer the value of C_N for different process conditions and a large number of tanks is quickly obtained.

The methods described are applicable to the numerical solution of differential equations. They provide an alternative solution to the sets of algebraic equations which arise after the differential equation has been put into finite difference form. The choice of method is often a matter of personal preference and the authors prefer the methods presented in Chapter 11. Matrix methods offer a good alternative for ordinary differential equations of boundary value type and linear partial differential equations (without chemical reaction terms). If the problem has non-linearities, the matrix has variable coefficients and may have to be inverted repeatedly as the solution proceeds. Even with linear equations there are two further problems. Firstly, the size of the matrix may be large and, secondly, the effect of accumulating errors may be to destroy the solution. In a two-dimensional boundary value problem, if 20 increments are taken in each direction, the system of algebraic equations involves about 400 unknown point values and the matrix will have 160 000 elements, nearly 99% of which will be zero. Obviously, a matrix of this size cannot be stored and inverted, and various techniques are available for packing the non-zero elements more efficiently and taking advantage of the majority of zero elements to shorten the inversion. However, it only needs one suffix to be set incorrectly, or one element to be given the wrong value, to ruin the solution and it is very difficult to trace the source of the error. In the methods presented in Chapter 11, a complete dumping of the 400 point values usually gives a good indication of the location of the error. The matrix operations usually require many subtractions between nearly equal quantities with consequent loss of accuracy. The accumulation of these errors can sometimes dominate the solution and destroy it, whereas in the methods of Chapter 11, the errors tend to remain distributed.

† Acrivos, A. and Amundson, N. R. *I.E.C.* **47**, 1533 (1955).

Chapter 13

OPTIMIZATION

13.1. INTRODUCTION

OPTIMIZATION in the chemical process industries infers the selection of equipment and operating conditions for the production of a given material so that the profit will be a maximum. This could be interpreted as meaning the maximum output of a particular substance for a given capital outlay, or the minimum investment for a specified production rate. The former is a mathematical problem of evaluating the appropriate values of a set of variables to maximize a dependent variable, whereas the latter may be considered to be one of locating a minimum value. However, in terms of the profit, both types of problem are maximization problems, and the solution of both is generally accomplished by means of an economic balance between capital and operating costs. Such a balance can be represented as shown in Fig. 13.1 in which the capital, operating and total costs are plotted

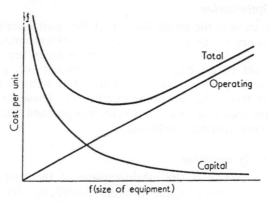

FIG. 13.1. Economic balance of process

against f (size of equipment), where f (size of equipment) implies some function of the size of the equipment. It could be the actual size of the equipment; number of pieces of equipment such as number of stirred tank reactors in a reactor battery or frames in a filter press; or some parameter related to the size of the equipment such as the reflux ratio in a distillation unit, or the solvent to feed ratio in a solvent extraction process.

In certain circumstances, it is unnecessary to make a complete economic balance involving all the variables of a process. Thus, in Section 9.6.3, the

number of stirred tank reactors for the maximum output of a product B
for the chemical reaction

$$A \longrightarrow B \longrightarrow C \tag{13.1}$$

was evaluated without cost data. This was possible because in the example
quoted the size of the reaction vessels was specified. If vessel size was
included as a parameter in the analysis, an economic balance would have
been necessary. That does not mean to say that in the example quoted a
"non-problem" was solved because in the chemical industry it is possible
that the equipment will already exist, and the engineer is expected to find
the optimum operating conditions. Hence problems involving two variables
are worth considering initially, even though in the majority of problems
encountered more than two variables will be required to describe the process
or operation adequately. However, before attempting this, it is necessary
to classify different types of optimization.

13.2. Types of Optimization

Optimization in the chemical field can be divided into two classes which
have been defined by Himsworth† as
  (i) Static Optimization.
  (ii) Dynamic Optimization.

13.2.1. *Static Optimization*

Static optimization is the establishment of the most suitable steady-state
operation conditions of the process. These include optimum size of equipment and production levels, in addition to temperatures, pressures, and flow
rates. They would be established by setting up the best possible mathematical model of the process which is maximized by some suitable technique
to give the most favourable operating conditions. These conditions would
be nominal conditions and would not take into account fluctuations in the
process about these nominal conditions.

13.2.2. *Dynamic Optimization*

Dynamic optimization is the establishment of the best procedure for
correcting the fluctuations in the process conditions about the nominal
conditions established in the static optimization analysis. It requires a
knowledge of the dynamic characteristics of the equipment and also
necessitates predicting the best way in which a change in the process conditions can be corrected. In reality it is an extension of the automatic
control analysis of the process.

In addition to the above types of optimization there are the operational
research aspects, which have been applied extensively in the chemical
industry in recent years and will be considered later in this chapter.

† Himsworth, F. R. *Trans. Inst. Chem. Eng.* **40**, 345 (1962).

13.3. ANALYTICAL PROCEDURES

When a process can be described by a small number of variables, it is possible to find the maxima or minima by differentiating and equating the resulting expressions to zero. These are then solved in order to find the optimum conditions. However as the number of variables increases, the solution of the resulting equations becomes prohibitive. Therefore analytical optimization procedures have limited applicability, but they deserve brief consideration.

13.3.1. *Problems Involving Two Variables*

On rare occasions it is possible to establish the optimum conditions of a process or operation by determining the maxima of an equation containing one independent and one dependent variable. For example, the optimum number of stirred tanks was evaluated by solution of such an equation in Section 9.6.3 and has been referred to above. The reader will be familiar with the technique of differentiating and equating to zero to find the maxima or minima and it need not be repeated here. However, it must be emphasized that such an optimum will in all probability be a "curtailed optimum", and should be treated cautiously because one or two parameters have been specified arbitrarily.

13.3.2. *Problems Involving Three Variables*

Many engineering operations of a simple type can be optimized by solution of an equation containing two independent variables. Thus let an operation be represented by the equation

$$z = f(x, y) \tag{13.2}$$

where x and y are the two independent variables. Equation (13.2) can be expressed in three dimensions by a surface in which the maximum value of z will be represented by the peak of a mountain. Similarly the minimum will be the lowest point of the bowl in the surface. Mathematically this means that for z to be a maximum at a point (x, y) it is obvious that the function $f(x, y)$ must be a maximum for displacements away from the point in any direction. That is, the conditions

$$\frac{\partial z}{\partial x} = 0 \quad \text{and} \quad \frac{\partial z}{\partial y} = 0 \tag{13.3}$$

must be satisfied simultaneously, which means that the tangent plane to the surface expressed by equation (13.2) is horizontal. In the case of functions of one independent variable, the first derivative vanishes at a maximum, a minimum, or a point of inflexion; these being resolved by reference to the second derivative. Similarly in the case of two independent variables, there may be a true maximum in the surface where all sections exhibit a maximum, a true minimum, or the two orthogonal sections may be any combination of maximum, minimum, and point of inflexion. The special combination of a maximum with respect to x and a minimum with respect to y is termed a "saddle point" and is illustrated in Fig. 13.2. To ensure that a particular point is actually a maximum or minimum, it is necessary to fulfil the conditions stated below.

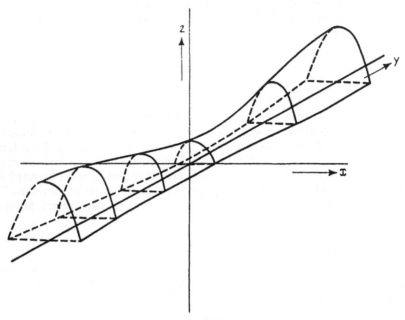

FIG. 13.2. Saddle point

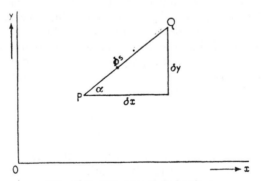

FIG. 13.3. Directional derivatives

Consider Fig. 13.3 in which z is represented as a function of position over the xy plane. Let P be any point in the plane and let the value of z change at different rates as the function moves away from P in different directions.

If PQ is one direction, the derivative of z in that direction may be defined as

$$\lim_{Q \to P} \left[\frac{f(x,y)_Q - f(x,y)_P}{PQ} \right] \qquad (13.4)$$

where PQ is the absolute length between the points P and Q. Let this be δs and, in addition, let

$$f(x,y)_Q - f(x,y)_P = \delta z \qquad (13.5)$$

Then the derivative of z in the direction PQ is

$$\lim_{Q \to P} \frac{\delta z}{\delta s}$$

But

$$\delta z = \frac{\partial z}{\partial x} \delta x + \frac{\partial z}{\partial y} \delta y$$

$$\therefore \quad \frac{dz}{ds} = \lim_{Q \to P} \frac{\delta z}{\delta s} = \frac{\partial z}{\partial x} \frac{\delta x}{\delta s} + \frac{\partial z}{\partial y} \frac{\delta y}{\delta s} \qquad (13.6)$$

but from Fig. 13.3

$$\frac{\delta x}{\delta s} = \cos \alpha \quad \text{and} \quad \frac{\delta y}{\delta s} = \sin \alpha \qquad (13.7)$$

hence

$$\frac{dz}{ds} = \frac{\partial z}{\partial x} \cos \alpha + \frac{\partial z}{\partial y} \sin \alpha \qquad (13.8)$$

where $\partial z / \partial x$ and $\partial z / \partial y$ are the derivatives in the x and y directions. The condition for a maximum or minimum is that equation (13.8) is zero whatever the value of α. Differentiating equation (13.8) with respect to s gives

$$\frac{d^2 z}{ds^2} = \frac{d}{ds} \left(\frac{dz}{ds} \right) = \frac{d}{ds} \left(\cos \alpha \frac{\partial z}{\partial x} + \sin \alpha \frac{\partial z}{\partial y} \right) \qquad (13.9)$$

$$= \cos \alpha \frac{\partial^2 z}{\partial x^2} \frac{\partial x}{\partial s} + \cos \alpha \frac{\partial^2 z}{\partial x \partial y} \frac{\partial y}{\partial s} + \sin \alpha \frac{\partial^2 z}{\partial x \partial y} \frac{\partial x}{\partial s} + \sin \alpha \frac{\partial^2 z}{\partial x^2} \frac{\partial y}{\partial s}$$

since $(d/ds)(dz/ds)$ is the total differential of (dz/ds) with respect to s. But equations (13.7) give the derivatives of x and y with respect to s in terms of α, therefore

$$\frac{d^2 z}{ds^2} = \cos^2 \alpha \frac{\partial^2 z}{\partial x^2} + 2 \sin \alpha \cos \alpha \frac{\partial^2 z}{\partial x \partial y} + \sin^2 \alpha \frac{\partial^2 z}{\partial y^2} \qquad (13.10)$$

For any value of α, $d^2 z / ds^2 > 0$ for z to be a minimum and $d^2 z / ds^2 < 0$ for z to be a maximum. That is $d^2 z / ds^2$ must have the same sign for all values of α.

Thus letting

$$z_{xx} = \frac{\partial^2 z}{\partial x^2}, \quad z_{xy} = \frac{\partial^2 z}{\partial x\,\partial y}, \quad z_{yy} = \frac{\partial^2 z}{\partial y^2}$$

substituting into equation (13.10) and dividing by z_{xx} gives

$$\frac{d^2 z}{ds^2} = \frac{1}{z_{xx}}[(\cos\alpha\, z_{xx} + \sin\alpha\, z_{xy})^2 + (z_{xx} z_{yy} - z_{xy}{}^2)\sin^2\alpha] \qquad (13.11)$$

It can be seen from equation (13.11) that if $z_{xx} z_{yy} > z_{xy}{}^2$ the terms inside the square bracket will be greater than zero for all values of α, and $d^2 z/ds^2$ will have the same sign as z_{xx}.

If z_{xx} is negative, z will be a maximum.

If z_{xx} is positive, z will be a minimum.

In the special case $z_{xx} z_{yy} = z_{xy}{}^2$, there will be one value for α which exhibits a point of inflexion whereas all other values of α give a consistent maximum or minimum. If $z_{xx} z_{yy} < z_{xy}{}^2$, the surface has a saddle point.

The above principles of maxima and minima have been used by Jenson and Jeffreys† for the evaluation of the optimum number of stages in solvent extraction processes. The analysis for counter-current processes is as follows.

Consider a feed solution containing X_0 kg solute per kg of solute-free liquor to be fed to stage 1 of a counter-current extraction process at the rate of C kg/h on a solute-free basis. Let the process contain N stages and let an extracting solvent be fed into stage N at the rate of B kg/h on a solute-free basis. The solvent passes counter-current to the feed liquor through the plant, finally emerging as the final extract from stage 1. The spent feed will be discharged from stage N with a solute concentration X_N. The carrier liquor of the feed will be assumed to be immiscible in the extracting solvent.

Assume that the feed rate and the solute content of the feed are constant and let the unit cost of the feed in any system of currency be γ per kg of solute. Let the total hourly cost of each stage including operation, fixed charges, and the depreciation of capital be α. Finally, let the cost of the solvent, including recovery from the final extract and make-up of spillage losses, be β per kg of solvent. Then on the basis of 1 hour's operation:

$$\text{Cost of feed is} \quad \gamma C X_0 \qquad (13.12)$$

$$\text{Cost of solvent is} \quad \beta B \qquad (13.13)$$

Similarly if λ is the value of 1·0 kg of extract product, the total value of the product is

$$\lambda C(X_0 - X_N) \qquad (13.14)$$

and the hourly profit from the process is given by

$$P = \lambda C(X_0 - X_N) - (\alpha N + \beta B + \gamma C X_0) \qquad (13.15)$$

† Jenson, V. G. and Jeffreys, G. V. *Brit. Chem. Eng.* **6**, 676 (1960).

But X_N is related to X_0 by

$$X_N = X_0\left(\frac{S^{N+1} - S^N}{S^{N+1} - 1}\right) \quad \text{(see Section 9.6.3)} \tag{13.16}$$

where $S = C/mB$ and m is the distribution ratio.

Substitution of equation (13.16) into (13.15) gives

$$P = \lambda C X_0\left(\frac{S^N - 1}{S^{N+1} - 1}\right) - (\gamma C X_0 + \alpha N + \beta B) \tag{13.17}$$

in which the variables are N and B. The maximum value of P is obtained by the procedure presented above. Thus

$$\frac{\partial P}{\partial N} = \lambda C X_0 S^N\left[\frac{S-1}{(S^{N+1}-1)^2}\right]\ln S - \alpha = 0 \tag{13.18}$$

From which

$$\frac{\alpha}{\lambda C X_0} = \frac{S^N(S-1)\ln S}{(S^{N+1}-1)^2} \tag{13.19}$$

and

$$\frac{\partial P}{\partial B} = \lambda C X_0\left[\frac{N(S^{N+1}-1)S^{N-1} - (N+1)(S^N-1)S^N}{(S^{N+1}-1)^2}\right]\left(\frac{-mS^2}{C}\right) - \beta = 0 \tag{13.20}$$

or

$$\frac{\beta}{\lambda m X_0} = S^{N+1}\left[\frac{S^{N+1} + N - NS - S}{(S^{N+1}-1)^2}\right] \tag{13.21}$$

In equations (13.19) and (13.21) the cost terms have been separated from the process terms. They are two complex simultaneous algebraic equations that will be difficult to solve analytically, but a solution may be obtained by one of the following methods

   (i) Graphically, as in the publication.

   (ii) Look for an optimum in P.

   (iii) Look for the roots of equations (13.19) and (13.21) by Newton's method.

Since the graph is available, (i) is the most simple but least accurate. It is not very attractive to differentiate equations (13.19) and (13.21) again in order to apply Newton's method and one of the optimum seeking methods to be presented later in this chapter should be applied. However, in the present stagewise process advantage should be taken of the fact that N is an integer when either equation (13.21) can be solved to determine S for each of the range of values of N, or equation (13.17) can be searched for optimum P at each of the same range of values of N. The former choice will be made here. Thus, denoting $(\beta/\lambda m X_0)$ by k and cross-multiplying equation (13.21) gives

$$k(S^{N+1} - 1)^2 = S^{N+1}(S^{N+1} + N - NS - S) \qquad \text{I}$$

Rearranging

$$f(S) = (1-k)S^{2N+2} - (N+1)S^{N+2} + (2k+N)S^{N+1} - k = 0 \qquad \text{II}$$

Differentiating

$$f'(S) = (2N+2)(1-k)S^{2N+1} - (N+2)(N+1)S^{N+1} + (N+1)(2k+N)S^N$$

$$= (N+1)S^N[2(1-k)S^{N+1} - (N+2)S + (2k+N)] \qquad \text{III}$$

Newton's method has been applied, as presented in the logic flow diagram in Fig. 13.4, to the following process conditions.

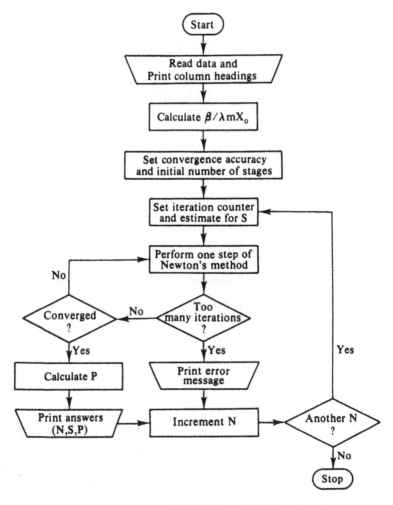

FIG. 13.4. Logic flow diagram of Newton's method for extraction process

4000 kg/h of a 5% by weight solution of acetaldehyde in toluene is to be treated with water to extract the acetaldehyde. If the annual capital and operating costs per stage are estimated to be £3200, the cost of supplying, pumping, and regenerating the solvent from the final extract is estimated to be 0·8 p/kg and the value of the acetaldehyde is taken to be 27·6 p/kg; what is the optimum number of stages and quantity of solvent required for the process? The distribution ratio of weight of acetaldehyde in water to acetaldehyde in toluene may be considered constant at 2·2.

The above data enable the constants to be evaluated as follows.

$$X_0 = 0·0526, \quad CX_0 = 200 \text{ kg/h}, \quad m = 2·2$$

$$\alpha = 40\text{p/h}, \quad \beta = 0·8\text{p/kg}, \quad \lambda = 27·6 \text{ p/kg}$$

$$\therefore \quad k = \frac{\beta}{\lambda m X_0} = 0·251$$

The solution for a range of values of N is presented in Table 13.1 from which it can be seen that the optimum number of stages is 10, corresponding to $S = 0·885$. Hence the required solvent feed rate is 1950 kg/h.

TABLE 13.1. *Optimization of Extraction Process*

N	S	$P + \gamma CX_0$ (£)
6	0·844	31·91
7	0·856	32·46
8	0·867	32·79
9	0·877	32·97
10	0·885	33·03
11	0·892	33·01
12	0·899	32·93
13	0·905	32·81
14	0·910	32·64
15	0·914	32·45

13.3.3. *Problems Involving Four Variables*

The analytical procedure presented above can be extended to four variables. That is, the maximum or minimum values of

$$w = f(x, y, z) \tag{13.22}$$

can be located by differentiating equation (13.22) and equating to zero. Thus

$$dw = \frac{\partial f}{\partial x}dx + \frac{\partial f}{\partial y}dy + \frac{\partial f}{\partial z}dz = 0 \tag{13.23}$$

when the conditions

$$\frac{\partial f}{\partial x} = 0, \quad \frac{\partial f}{\partial y} = 0, \quad \frac{\partial f}{\partial z} = 0 \qquad (13.24)$$

are satisfied.

In problems containing four or more variables, the analytical solution becomes increasingly complex and impracticable. Therefore, other techniques of optimization must be applied, and these will be considered in later sections of this chapter.

13.3.4. *Optimization with a Restrictive Condition*

In some optimization problems a relationship must be satisfied by the independent variables. In vapour–liquid equilibria, for example, the temperature, pressure, and compositions of liquid and vapour are related by the boiling point condition. One variable, say temperature, can always be eliminated analytically, but this may not be algebraically convenient, and an alternative method is available.

Thus consider equation (13.22) with the restrictive condition

$$g(x, y, z) = 0 \qquad (13.25)$$

Differentiating equation (13.25) gives

$$\frac{\partial g}{\partial x}dx + \frac{\partial g}{\partial y}dy + \frac{\partial g}{\partial z}dz = 0 \qquad (13.26)$$

Equation (13.26) restricts the free choice of dx, dy and dz so that only two of these increments, say dy and dz, can be chosen freely. The other derivative (dx) is then determined by equation (13.26). In general, the unknown increment dx can be eliminated between equations (13.23) and (13.26) by multiplying the latter by a parameter λ and adding. Thus

$$\left(\frac{\partial f}{\partial x} + \lambda\frac{\partial g}{\partial x}\right)dx + \left(\frac{\partial f}{\partial y} + \lambda\frac{\partial g}{\partial y}\right)dy + \left(\frac{\partial f}{\partial z} + \lambda\frac{\partial g}{\partial z}\right)dz = 0 \qquad (13.27)$$

where the parameter λ is chosen to satisfy the condition that

$$\frac{\partial f}{\partial x} + \lambda\frac{\partial g}{\partial x} = 0 \qquad (13.28)$$

since equation (13.27) must be satisfied by any choice of dy and dz it follows that the sum of the terms in each of the brackets must be zero, or

$$\frac{\partial f}{\partial y} + \lambda\frac{\partial g}{\partial y} = 0 \qquad (13.29)$$

and

$$\frac{\partial f}{\partial z} + \lambda\frac{\partial g}{\partial z} = 0 \qquad (13.30)$$

Equations (13.28), (13.29), and (13.30) can be obtained by optimizing the function

$$W = f(x, y, z) + \lambda g(x, y, z) \tag{13.31}$$

when the optimum point is given by the solution of the equations

$$\left.\begin{aligned}
\frac{\partial f}{\partial x} + \lambda \frac{\partial g}{\partial x} &= 0 \\[2mm]
\frac{\partial f}{\partial y} + \lambda \frac{\partial g}{\partial y} &= 0 \\[2mm]
\frac{\partial f}{\partial z} + \lambda \frac{\partial g}{\partial z} &= 0 \\[2mm]
g(x, y, z) &= 0
\end{aligned}\right\} \tag{13.32}$$

which contain the four unknowns (x, y, z, λ). λ is termed the "Lagrange multiplier", and is a very powerful parameter in many types of optimization problem, as will be seen below. This procedure can be extended to two or more restrictive conditions by introducing two or more arbitrary multipliers.

13.4. LINEAR PROGRAMMING

A very important class of optimization problem with restrictive conditions is the Linear Program problem. In this type of optimization a linear function is to be optimized with the restrictive conditions or constraints expressed by linear functions. These constraints define a plane area with straight sides in space, which is known as the feasible area, and the optimization program consists of an evaluation of the function to be optimized around the periphery of this area.

The basis of linear programming is found in the analysis of sets inequalities, and the solution of series of linear equations. The principles involved may be illustrated by the following example.

Example 1. Evaluate the maximum value of k in the function

$$k = y + 0 \cdot 5x \qquad\qquad \text{I}$$

subject to the following constraints

$$x \geqslant 0, \quad y \geqslant 0, \quad y \leqslant x + 2, \quad y \leqslant 4 - x \qquad\qquad \text{II}$$

The inequalities representing the constraints are plotted in Fig. 13.5 to enclose the shaded area shown. This area is called the polygonal convex set or simply the feasible area and the coordinates forming the edges of this shaded area include all the constraints of the system. All points within this area are feasible points.

Equation I is generally called the objective function when expressed in the form

$$k = y + 0 \cdot 5x \qquad\qquad\qquad \text{I}$$

and the object is to maximize k. Different values of k corresponding to different values of x and y are presented in Table 13.2.

TABLE 13.2. *Evaluation of Objective Function Inside the Feasible Area*

x	y	k
0	0	0
4	0	2
1	3	3·5
0	2	2
3	1	2·5
1	1	1·5
2	1	2
2	2	3

These coordinates form a family of lines of slope $-0 \cdot 5$ with intercept k, and it is obvious from the figure and Table 13.2 that the maximum value of k is 3·5. The table and the figure constitute the linear program to optimize k, and emphasizes the first theorem in linear programming. That is: The

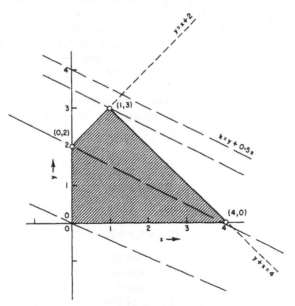

FIG. 13.5. Graphical solution of linear program

optimum value (maximum or minimum) of the objective function of a linear program is a feasible point located at one of the vertices of the polygonal convex set determined by the constraints of the problem.

In the above example the maximum value of k was easily obtained graphically because there were only two variables, and the polygonal convex set was most conveniently represented in the x–y plane. However, the feasible points could be located in an N-dimensional space and the above theorem would still apply when the optimum value of the objective function will again be found at a vertex point. In these circumstances the linear program would be solved by application of matrix algebra. Thus, the general linear program could be expressed by the objective function

$$P = p_1 x_1 + p_2 x_2 + \ldots + p_n x_n = \mathbf{p}'.\mathbf{x} \qquad (13.33)$$

with the general constraints

$$\left.\begin{array}{l} a_{11}x_1 + a_{12}x_2 + \ldots a_{1n}x_n \leqslant b_1 \\ a_{21}x_1 + a_{22}x_2 + \ldots a_{2n}x_n \leqslant b_2 \\ \cdot \quad \cdot \quad \cdot \quad \cdot \quad \cdot \quad \cdot \quad \cdot \quad \cdot \\ a_{m1}x_1 + a_{m2}x_2 + \ldots a_{mn}x_n \leqslant b_m \end{array}\right\} \qquad (13.34)$$

or

$$\mathbf{Ax} \leqslant \mathbf{b} \qquad (13.35)$$

with

$$x_i \geqslant 0, \quad i = 1, 2, 3, \ldots, n \qquad (13.36)$$

The linear program problem defined by the expressions (13.33), (13.34), and (13.36) is called the "primal problem". The inequality in the matrix form (13.35) has the interpretation that corresponding elements satisfy the inequality. For every primal problem there exists a "dual problem" which may be defined as

$$\min C = \min(\mathbf{b}'\mathbf{y}) \qquad (13.37)$$

with the constraints

$$\mathbf{A}'\mathbf{y} \geqslant \mathbf{p} \qquad (13.38)$$

and where

$$\max P = \min C \qquad (13.39)$$

so that, in the sense of equation (13.39), the primal problem and the dual problem have the same solution, and it is a matter of experience to assess which is the more convenient to solve. This will be illustrated later, but nevertheless the calculation procedure to apply once the problem has been enunciated is similar, irrespective of whether the primal or dual linear program is analysed.

13.4.1. *Elimination of Inequalities in the Constraints*

The constraints in the linear program have been expressed in (13.34) as inequalities. They can be converted to equations by the introduction of "slack variables" z_i $(i = 1, 2, 3, ..., m)$ which are also all positive or zero thus

$$\left.\begin{aligned}
a_{11}x_1 + a_{12}x_2 + ... + a_{1n}x_n + z_1 &= b_1 \\
a_{21}x_2 + a_{22}x_2 + ... + a_{2n}x_n + z_2 &= b_2 \\
\cdot \quad \cdot \quad \cdot \quad \cdot \quad \cdot \quad \cdot \quad \cdot \quad \cdot \\
a_{m1}x_1 + a_{m2}x_2 + ... + a_{mn}x_n + z_m &= b_m
\end{aligned}\right\} \tag{13.40}$$

Removing the distinction between the variables x and z by denoting all variables by f_j where

$$f_j = \begin{cases} x_j & \text{for } 1 \leqslant j \leqslant n \\ z_{j-n} & \text{for } n+1 \leqslant j \leqslant n+m = N \end{cases}$$

the constraints can be written

$$\left.\begin{aligned}
a_{11}f_1 + a_{12}f_2 + ... + a_{1N}f_N &= b_1 \\
a_{21}f_1 + a_{22}f_2 + ... + a_{2N}f_N &= b_2 \\
\cdot \quad \cdot \quad \cdot \quad \cdot \quad \cdot \quad \cdot \quad \cdot \quad \cdot \\
a_{m1}f_1 + a_{m2}f_2 + ... + a_{mN}f_N &= b_m
\end{aligned}\right\} \tag{13.41}$$

or

$$\mathbf{a}_1 f_1 + \mathbf{a}_2 f_2 + ... + \mathbf{a}_N f_N = \mathbf{b}$$

where $\mathbf{a}_j$ for $n+1 \leqslant j \leqslant N$ are the unit vectors $\mathbf{e}_i$ $(i = j - n)$.

In the same way, the objective function becomes

$$P = p_1 f_1 + p_2 f_2 + ... + p_N f_N = \mathbf{p'f} \tag{13.42}$$

where $p_{n+1} = p_{n+2} = ... = p_N = 0$.

The linear program can then be summarized as

$$\max(\mathbf{p'f}) \quad \text{with } \mathbf{f} \geqslant 0$$

and

$$\sum_{j=1}^{N} \mathbf{a}_j f_j = \mathbf{b} \tag{13.43}$$

All the points of the polygonal convex set are feasible solutions and in order to obtain the optimal feasible solution by performing a reasonable number of calculations it is necessary to devise a method that will reduce the number of possibilities.

At this stage the problem consists of solving equations (13.43) for the unknowns (f_j). There are m equations in the set and $N = m+n$ unknowns and the method previously described in Section 12.10.5 shows that, in general, n unknowns can be given arbitrary values and the remaining m unknowns can be evaluated from the m equations. A method was also described for interchanging variables and choosing values for different groups of n variables. This was interpreted as resolving the vector $\mathbf{b}$ into components in terms of the vectors $\mathbf{a}_j$ which already include the unit vectors. It is only necessary, in addition, to ensure that all components are positive or zero to obtain feasible solutions. The optimal feasible solution is known to be at a vertex of the feasible area and the second theorem of linear programming ensures that the method described in Section 12.10.5 moves between vertices of the feasible area when all surplus variables have the value zero. The formal statement of the second theorem is:

An extreme point $\mathbf{f} = \{f_1 \quad f_2 \quad \ldots \quad f_N\}$ is one for which the $\mathbf{a}_j$s with positive weights $(f_j > 0)$ form a linearly independent set amongst all of the $\mathbf{a}_j$s. The other $\mathbf{a}_j$s will have zero weight $(f_j = 0)$.

This implies that the expansion of $\mathbf{b}$ in terms of a selection of the $\mathbf{a}_j$s is unique in terms of that particular set of $\mathbf{a}_j$s, which is always true in the method of Section 12.10.5.

The third theorem of linear programming simply states that if $\mathbf{x}$ and $\mathbf{y}$ are feasible solutions of the primal and dual linear programs respectively, then

$$\mathbf{p}'\mathbf{x} \leqslant \mathbf{b}'\mathbf{y}$$

13.4.2. The Simplex Method of Solution

This calculation procedure is due to Dantzig and consists essentially of finding an initial feasible solution, usually the origin of the original set of coordinates. In this case

$$\mathbf{f} = \{0 \quad 0 \quad \ldots \quad 0 \quad b_1 \quad b_2 \quad \ldots \quad b_m\} \tag{13.44}$$

which satisfies the equations representing the constraints

$$\mathbf{b} = \sum_{j=1}^{N} \mathbf{a}_j f_j \tag{13.43}$$

and makes the objective function zero, since

$$P = \mathbf{p}'\mathbf{f} = [p_1 \quad p_2 \quad \ldots \quad p_n \quad 0 \quad \ldots \quad 0]\{0 \quad 0 \quad \ldots \quad 0 \quad b_1 \quad \ldots \quad b_m\} = 0 \tag{13.45}$$

The basis of the expansion of the set of vectors $\mathbf{a}_j$ is changed one vector at a time in such a way that the solution remains feasible and the objective function increases. After a finite number of iterations of this kind the objective function will reach its optimum when no change of vectors in the basis will lead to an increase. The change of a vector in the basis involves two choices: firstly, which vector to introduce to the basis and, secondly, which vector should be replaced. The method of making these choices will now be described.

Thus, suppose at some stage in the iterations, that the feasible solution is expanded in the basis $\mathbf{a}_i$ $(1 \leqslant i \leqslant m)$ where the vectors $\mathbf{a}_i$ may be any selection from the original $\mathbf{a}_j$ or $\mathbf{e}_j$. Denote the remaining vectors (not in the basis) by $\mathbf{a}_s (1 \leqslant s \leqslant n)$ which will also be a mixture of the original $\mathbf{a}_j$s or $\mathbf{e}_j$s. The constraints are represented by the expansion

$$\mathbf{b} = \sum_{i=1}^{m} f_i \mathbf{a}_i \tag{13.46}$$

where the other n elements of $\mathbf{f}$ are zero. As in Section 12.10.5, the remaining vectors $\mathbf{a}_s$ also possess expansions in the current basis, thus

$$\mathbf{a}_s = \beta_{1s}\mathbf{a}_1 + \beta_{2s}\mathbf{a}_2 \ldots + \beta_{ms}\mathbf{a}_m$$

or in compact form

$$\mathbf{a}_s = \sum_{i=1}^{m} \beta_{is}\mathbf{a}_i \qquad (13.47)$$

The objective function now takes the form

$$P = \sum_{i=1}^{m} f_i p_i \qquad (13.48)$$

where the f_i are as defined in equation (13.46) and each p_i has the value associated with the original $\mathbf{a}_i$; that is, if the particular $\mathbf{a}_i$ started as an $\mathbf{e}_i$, the corresponding p_i would be zero.

Now define a set of "penalty functions" z_s by

$$z_s = \sum_{i=1}^{m} \beta_{is} p_i \qquad (13.49)$$

where equation (13.49) is associated with equation (13.47) in the same way that equation (13.48) is related to equation (13.46).

Symbolically, $\mathbf{a}_s$ can be introduced into the basis and satisfy the constraints thus

$$\mathbf{b} = \sum_{i=1}^{m} f_i\mathbf{a}_i - \theta\mathbf{a}_s + \theta\mathbf{a}_s \qquad (13.50)$$

where the value of θ will be determined later. Substituting equation (13.47)

$$\therefore \quad \mathbf{b} = \sum_{i=1}^{m} f_i\mathbf{a}_i - \theta\sum_{i=1}^{m}\beta_{is}\mathbf{a}_i + \theta\mathbf{a}_s \qquad (13.50)$$

$$\therefore \quad \mathbf{b} = \sum_{i=1}^{m} (f_i - \theta\beta_{is})\mathbf{a}_i + \theta\mathbf{a}_s \qquad (13.51)$$

With an appropriate choice of θ, equation (13.51) represents the constraints in the new basis. One vector has to be removed from the previous basis and if θ is chosen to satisfy

$$\theta = f_j/\beta_{js}$$

it can be seen that $\mathbf{a}_j$ will be removed. However, the solution will only be feasible if all of the other coefficients in equation (13.51) are positive or zero. Therefore, θ must be positive, and j must be chosen to give the minimum positive value for θ. Equation (13.51) becomes

$$\mathbf{b} = \sum_{i=1}^{m} \left(f_i - \frac{f_j}{\beta_{js}}\beta_{is} \right)\mathbf{a}_i + \frac{f_j}{\beta_{js}}\mathbf{a}_s \qquad (13.52)$$

The coefficients in equation (13.52) are the new values for $\mathbf{f}$ in the new basis.

It is now necessary to look at the effect of the above change of basis on the objective function. Its new value P' will be given by

$$P' = \sum_{i=1}^{m} \left(f_i - \frac{f_j}{\beta_{js}} \beta_{is} \right) p_i + \frac{f_j}{\beta_{js}} p_s$$

$$= P + \frac{f_j}{\beta_{js}} (p_s - z_s) \qquad (13.53)$$

using equations (13.48) and (13.49). The objective function will thus increase if $p_s - z_s > 0$.

The method is now seen to consist of the following steps:

(i) for each vector which is not in the basis, evaluate $p_s - z_s$ and choose the largest to determine s.

(ii) of the positive elements β_{js}, choose the one for which f_j/β_{js} is the smallest; this determines j.

(iii) Replace $\mathbf{a}_j$ with $\mathbf{a}_s$ in the basis of the expansion.

This procedure can be set up in tabular form as illustrated in the examples below. The layout is a natural extension of that used in Section 12.10.5.

Example 2. Find the maximum value of the function

$$P = 3x_1 + x_2 + 2x_3$$

where x_1, x_2, x_3, are restricted by

$$x_1 - x_2 \leqslant 1$$
$$4x_1 + 2x_2 - 3x_3 \leqslant 10$$
$$5x_1 + x_2 + 3x_3 \leqslant 14$$
$$x_1, x_2, x_3 \geqslant 0$$

Introducing the slack variables x_4, x_5, x_6, the constraints become

$$x_1 - x_2 + x_4 = 1$$
$$4x_1 + 2x_2 - 3x_3 + x_5 = 10$$
$$5x_1 + x_2 + 3x_3 + x_6 = 14$$

The first feasible solution, which is the expansion of the vector {1 10 14} in terms of the unit vectors is set out thus

	$\mathbf{a}_1$	$\mathbf{a}_2$	$\mathbf{a}_3$	$\mathbf{a}_4$	$\mathbf{a}_5$	$\mathbf{a}_6$	$\mathbf{b}$	p_i
$\mathbf{a}_4$	1	−1	0	1	0	0	1	0
$\mathbf{a}_5$	4	2	−3	0	1	0	10	0
$\mathbf{a}_6$	5	1	3	0	0	1	14	0
p_s	3	1	2	0	0	0	—	
z_s	0	0	0	0	0	0	0	
$p_s - z_s$	3	1	2	0	0	0		

The vector expansion, enclosed in the square box, is exactly the same as that described in Section 12.10.5, and all of the calculations performed within this area subsequently are also the same. However, calculations are also made below the expansion as illustrated. The first extra row (labelled p_s) contains the coefficients of the objective function P which is constant throughout the calculation. The basis of the above expansion is a_4, a_5, a_6 and their coefficients in the objective function (all zeros at this point) are transferred to an extra column p_i. The second extra row (labelled z_s) can be calculated from equation (13.49) and the last value, which is calculated from equation (13.48) in the column headed b, is the value of the objective function at the current feasible solution. Finally, the third extra row (labelled $p_s - z_s$) is obtained by subtraction from the previous two rows.

The largest element in the row $(p_s - z_s)$ is 3 in the first column. Every value of β_{j1} is positive in this column and the values of f_j/β_{j1} are $1/1 = 1$, $10/4 = 2.5$, $14/5 = 2.8$. Since f_1/β_{11} is the smallest of these, a_4 must be replaced by a_1 giving the second tableau.

	a_1	a_2	a_3	a_4	a_5	a_6	b	p_i
a_1	1	−1	0	1	0	0	1	3
a_5	0	6	−3	−4	1	0	6	0
a_6	0	6	3	−5	0	1	9	0
p_s	3	1	2	0	0	0	—	
z_s	3	−3	0	3	0	0	3	
$p_s - z_s$	0	4	2	−3	0	0		

The maximum $(p_s - z_s)$ occurs in column 2 and the minimum f_j/β_{j2}, for positive β_{j2}, is the second row. Therefore, the next step is to replace a_5 with a_2, thus

	a_1	a_2	a_3	a_4	a_5	a_6	b	p_i
a_1	1	0	$-\frac{1}{2}$	$\frac{1}{3}$	$\frac{1}{6}$	0	2	3
a_2	0	1	$-\frac{1}{2}$	$-\frac{2}{3}$	$\frac{1}{6}$	0	1	1
a_6	0	0	6	−1	−1	1	3	0
p_s	3	1	2	0	0	0	—	
z_s	3	1	−2	$\frac{1}{3}$	$\frac{2}{3}$	0	7	
$p_s - z_s$	0	0	4	$-\frac{1}{3}$	$-\frac{2}{3}$	0		

There is only one positive value for $(p_s - z_s)$ now, and this is in column 3. There is also only one positive value of β_{j3}, therefore a_6 must be replaced by a_3.

	a_1	a_2	a_3	a_4	a_5	a_6	b	p_i
a_1	1	0	0	$\frac{1}{4}$	$\frac{1}{12}$	$\frac{1}{12}$	$\frac{9}{4}$	3
a_2	0	1	0	$-\frac{3}{4}$	$\frac{1}{12}$	$\frac{1}{12}$	$\frac{5}{4}$	1
a_3	0	0	1	$-\frac{1}{6}$	$-\frac{1}{6}$	$\frac{1}{6}$	$\frac{1}{2}$	2
p_s	3	1	2	0	0	0	—	
z_s	3	1	2	$-\frac{1}{3}$	0	$\frac{2}{3}$	9	
$p_s - z_s$	0	0	0	$\frac{1}{3}$	0	$-\frac{2}{3}$		

It might be thought that having removed all of the slack variables from the basis, and achieving a solution which is at the limit of all of the constraints, that the optimum value of P has been obtained. This is not true! There is still a positive value in the row $(p_s - z_s)$ in column 4, and β_{14} is also positive. It is therefore necessary to replace a_1 with a_4.

	a_1	a_2	a_3	a_4	a_5	a_6	b	p_i
a_4	4	0	0	1	$\frac{1}{3}$	$\frac{1}{3}$	9	0
a_2	3	1	0	0	$\frac{1}{3}$	$\frac{1}{3}$	8	1
a_3	$\frac{2}{3}$	0	1	0	$-\frac{1}{9}$	$\frac{2}{9}$	2	2
p_s	3	1	2	0	0	0	—	
z_s	$\frac{13}{3}$	1	2	0	$\frac{1}{9}$	$\frac{7}{9}$	12	
$p_s - z_s$	$-\frac{1}{3}$	0	0	0	$-\frac{1}{9}$	$-\frac{7}{9}$		

There are no longer any positive values in the row $(p_s - z_s)$ and the maximum value of the objective function has been found. The complete solution appears in the column headed b, thus

$$b = 9a_4 + 8a_2 + 2a_3$$

$$\therefore \quad x_1 = 0, \quad x_2 = 8, \quad x_3 = 2 \quad \text{and} \quad x_4 = 9$$

The maximum value of the function P is 12.

The last step in the above problem actually reversed the first step from the original expansion. This is not unusual and cannot lead to a closed loop of calculation because the value of the objective function always increases (in this example 0, 3, 7, 9, 12) thus preventing the possibility of a return to a previous feasible solution.

Example 3. A chemical company can produce four products A, B, C, and D by a batch process. The raw material requirements, space needed for storage, production rates, and profits are given in the table below. The total amount of raw material available per day is 18 000 kg and the total storage space for all products is 47·5 m². A maximum of 7 h per day is available for production and the total production is shipped to the storage area at the end of each day. Assuming that the four products must share the total available storage space, production time, and raw materials, estimate the number of drums of each chemical product that should be produced in order to maximize the total profit.

Product	A	B	C	D
Raw materials (kg/drum)	200	200	150	250
Storage space (m²/drum)	0·4	0·5	0·4	0·3
Production rate (drums/h)	30	60	20	30
Profit (£/drum)	10	13	10	11

Solution. Let the number of drums of A, B, C, D produced be x_1, x_2, x_3, x_4. The objective function is

$$P = 10x_1 + 13x_2 + 10x_3 + 11x_4$$

The constraints are

$$200x_1 + 200x_2 + 150x_3 + 250x_4 \leqslant 18\,000$$

$$0.4x_1 + 0.5x_2 + 0.4x_3 + 0.3x_4 \leqslant 47.5$$

$$2x_1 + \quad x_2 + 3x_3 + 2x_4 \quad \leqslant 420$$

where the last constraint has been expressed in minutes to avoid fractions. Dividing the first constraint by the common factor 50, multiplying the second by 10, and introducing the slack variables, the first tableau is

	a_1	a_2	a_3	a_4	a_5	a_6	a_7	b	p_i
a_5	4	4	3	5	1	0	0	360	0
a_6	4	5	4	3	0	1	0	475	0
a_7	2	1	3	2	0	0	1	420	0
p_s	10	13	10	11	0	0	0	—	
z_s	0	0	0	0	0	0	0		0
$p_s - z_s$	10	13	10	11	0	0	0		

The maximum value of $p_s - z_s$ is in column 2, the values of β_{j2} are all positive, and the minimum value of f_j / β_{j2} is in the first row. Therefore replace a_5 with a_2 to obtain the second tableau.

	a_1	a_2	a_3	a_4	a_5	a_6	a_7	b	p_i
a_2	1	1	$\frac{3}{4}$	$\frac{5}{4}$	$\frac{1}{4}$	0	0	90	13
a_6	-1	0	$\frac{1}{4}$	$-\frac{13}{4}$	$-\frac{5}{4}$	1	0	25	0
a_7	1	0	$\frac{9}{4}$	$\frac{3}{4}$	$-\frac{1}{4}$	0	1	330	0
p_s	10	13	10	11	0	0	0	—	
z_s	13	13	$9\frac{3}{4}$	$16\frac{1}{4}$	$3\frac{1}{4}$	0	0	1170	
$p_s - z_s$	-3	0	$\frac{1}{4}$	$-5\frac{1}{4}$	$-3\frac{1}{4}$	0	0		

The only positive value for $(p_s - z_s)$ is in column 3 and the minimum value of f_j / β_{j3} is in the second row. Therefore, a_6 is replaced with a_3, thus

	a_1	a_2	a_3	a_4	a_5	a_6	a_7	b	p_i
a_2	4	1	0	11	4	-3	0	15	13
a_3	-4	0	1	-13	-5	4	0	100	10
a_7	10	0	0	30	11	-9	1	105	0
p_s	10	13	10	11	0	0	0	—	
z_s	12	13	10	13	2	1	0	1195	
$p_s - z_s$	-2	0	0	-2	-2	-1	0		

Because there are no longer any positive values in the last row, the optimum profit has been reached. The company should produce 15 drums of B and 100 drums of C each day. This will use all of the raw materials and all of the storage space but will leave 105 min or $1\frac{3}{4}$ h of productive capacity available.

In general, the answers to problems similar to the last one will not be expressed as round numbers and some adjustment would have to be made to obtain suitable batch sizes. If so, the production schedule should vary day by day to obtain the theoretical balance of products on average.

13.4.3. *The Dual Problem*

Equations (13.37) and (13.38) define the dual problem as

$$\min C = \min(\mathbf{b}'\mathbf{y}) \tag{13.37}$$

with the constraints

$$\mathbf{A}'\mathbf{y} \geqslant \mathbf{p} \tag{13.38}$$

Because the inequalities in (13.38) are now reversed, the slack variables have to be subtracted instead of added and the origin is no longer a feasible solution; also, a minimum is being sought instead of a maximum. The latter difference is easily removed by looking for $\max(-\mathbf{b}'\mathbf{y})$, but the former difference presents bigger problems. These are resolved by performing a preliminary search for a feasible solution before tackling the main problem. The equations containing the slack variables consist of an expansion in terms of the unit vectors $\mathbf{e}_i$ which are not yet included in the system, so "auxiliary variables" are introduced to bring them in and the preliminary problem is to remove the auxiliary variables from the basis. This can be done by defining an objective function consisting of the sum of the auxiliary variables with reversed sign. The procedure will be illustrated by solving Example 1 again in its dual form.

Example 4. Evaluate the maximum value of the function

$$P = 0.5x_1 + x_2$$

subject to the constraints

$$x_1 + x_2 \leqslant 4$$

$$-x_1 + x_2 \leqslant 2$$

by formulating the dual problem and solving it.

The dual problem is thus

$$\min C = 4y_1 + 2y_2$$

with the constraints

$$y_1 - y_2 \geqslant 0.5$$

$$y_1 + y_2 \geqslant 1$$

Introducing slack variables (y_3, y_4) and auxiliary variables (y_5, y_6), the constraints become

$$y_1 - y_2 - y_3 + y_5 = 0.5$$

$$y_1 + y_2 - y_4 + y_6 = 1$$

and the preliminary objective function is

$$P = \max(-y_5 - y_6)$$

The first tableau is thus

	a_1	a_2	a_3	a_4	a_5	a_6	b	p_i
a_5	1	-1	-1	0	1	0	0·5	-1
a_6	1	1	0	-1	0	1	1	-1
p_s	0	0	0	0	-1	-1	—	
z_s	-2	0	1	1	-1	-1	$-1·5$	
$p_s - z_s$	2	0	-1	-1	0	0		

Replace a_5 with a_1 to obtain the second tableau

	a_1	a_2	a_3	a_4	a_5	a_6	b	p_i
a_1	1	-1	-1	0	1	0	0·5	0
a_6	0	2	1	-1	-1	1	0·5	-1
p_s	0	0	0	0	-1	-1	—	
z_s	0	-2	-1	1	1	-1	$-0·5$	
$p_s - z_s$	0	2	1	-1	-2	0		

Replace a_6 with a_2 to obtain the third tableau

	a_1	a_2	a_3	a_4	a_5	a_6	b	p_i
a_1	1	0	$-0·5$	$-0·5$	0·5	0·5	0·75	0
a_2	0	1	0·5	$-0·5$	$-0·5$	0·5	0·25	0

There is no need to complete this tableau because both p_i are zero, the auxiliary variables have been removed from the basis, and a feasible solution to the dual problem has been obtained. The columns corresponding to the auxiliary variables (headed a_5 and a_6) are now deleted and the true objective function is inserted thus

	a_1	a_2	a_3	a_4	b	p_i
a_1	1	0	$-0·5$	$-0·5$	0·75	-4
a_2	0	1	0·5	$-0·5$	0·25	-2
p_s	-4	-2	0	0	—	
z_s	-4	-2	1	3	$-3·5$	
$p_s - z_s$	0	0	-1	-3		

Because all of the values of $(p_s - z_s)$ are zero or negative, this first tableau for the dual problem itself is the optimal solution. It shows that the maximum value of $-C$ is $-3·5$, or that $\min(C) = 3·5$ which was the maximum value found for P (or k) in Example 1. This verifies the statement made in equation (13.39).

It has been assumed so far that all elements of **b** and **p** are positive or zero. If one of the b_is is negative, the signs throughout that particular inequality are reversed, and so is the inequality itself. Slack variables are introduced as always, and one auxiliary variable is also needed to correspond to this particular b_i. Again a preliminary problem has to be solved to remove the one auxiliary variable, following the technique just used for a dual problem. Alternatively, if one element of **p** is negative, the dual problem needs one fewer auxiliary variable. In other words, the distinction between the linear programming problem and its dual becomes slightly blurred when the elements of **b** and **p** are not all positive or zero.

13.5. The Calculus of Variations
(and Pontryagin's Maximum Principle)

Consider a chemical reaction of the type

$$A \longrightarrow B \longrightarrow C$$

to take place in a batch reaction process and it is required that the reaction temperature be programmed as a function of time in order to maximize the production of the product B. That is, the integral

$$I = \int_0^t r_B \, dt \tag{13.54}$$

is to be maximized. The rate of each reaction contributing to r_B may possibly be expressed by the Arrhenius equation that

$$r = r_0 \exp \frac{-E}{RT} \tag{13.55}$$

and the temperature T must be expressed as a function of time

$$T = f(t) \tag{13.56}$$

in order to maximize the integral I. This is a variational problem and would be solved using the calculus of variations. It involves the selection of an unknown function (in the example above equation (13.56) is the integrand) so that the integral is optimized.

In general terms an optimization problem can be set up in variational form as follows

$$I = \int_{x_1}^{x_2} F[y(x), y_x(x), x] \, dx \tag{13.57}$$

where, in this situation, I is called the "functional" since its value depends on the choice of the function $y(x)$ inserted into the "integrand" F in order to make the integral a maximum or a minimum. However $y(x)$ must satisfy certain conditions. Suppose that $y(x)$ is the particular function that makes

I in equation (13.57) a minimum. Now let $y^*(x)$ be another function which is only infinitesimally different from $y(x)$ at every point within the interval $x_1 \rightarrow x_2$. Let

$$\delta y = y^*(x) - y(x) = \varepsilon\phi(x) \qquad (13.58)$$

where δ, in the context used here, is called "the variational symbol". Equation (13.58) emphasizes that δy is an infinitesimal change in the value of the function at a particular value of x, and is *not* the change in the value of $y(x)$ resulting from a change δx in the value of the independent variable x. Here $\delta x = 0$. Also, in equation (13.58), $\phi(x)$ is an arbitrary continuous function with a continuous first and second derivative and ε is a parameter tending to zero. The variational symbol is an operator which commutes in both the operations of differentiation and integration. That is

$$\frac{d}{dx}(\delta y) = \delta\frac{dy}{dx} \quad \text{and} \quad \int_{x_1}^{x_2} \delta y \, dx = \delta \int_{x_1}^{x_2} y \, dx$$

The search for maxima or minima in conventional calculus is well known as summarized in Section 13.3.2. The principles that apply in variational calculus are very similar. That is I in equation (13.57) is considered to be "stationary" if

$$\delta I = 0 \qquad (13.59)$$

However, the change in δI must be investigated with respect to changes in the functional relationships. That is $y(x)$ is subjected to small variations rather than x. Now

$$\delta F = F(y^*, y_x^*, x) - F(y, y_x, x)$$

$$= F[(y + \varepsilon\phi), (y_x + \varepsilon\phi_x), x] - F(y, y_x, x) \qquad (13.60)$$

Expanding δF by Taylor series (Section 3.3.6) and taking only first terms, since ε^n (where n is greater than 2) is negligible, gives

$$\delta F = \varepsilon\left(\frac{\partial F}{\partial y}\phi + \frac{\partial F}{\partial y_x}\phi_x\right) \qquad (13.61)$$

If I is stationary, equation (13.59) applies, or

$$\delta I = \varepsilon \int_{x_1}^{x_2}\left(\frac{\partial F}{\partial y}\phi + \frac{\partial F}{\partial y_x}\phi_x\right)dx = 0 \quad \text{for all } \phi(x) \qquad (13.62)$$

The second term in equation (13.62) can be integrated by parts to give

$$\int_{x_1}^{x_2}\frac{\partial F}{\partial y_x}\phi_x \, dx = \left[\phi\frac{\partial F}{\partial y_x}\right]_{x_1}^{x_2} - \int_{x_1}^{x_2}\phi\frac{d}{dx}\left(\frac{\partial F}{\partial y_x}\right)dx \qquad (13.63)$$

The first term in equation (13.63) is zero because no variation is permitted at the limits x_1 and x_2.

Insertion of equation (13.63) into (13.62) gives

$$\delta I = \varepsilon \int_{x_1}^{x_2} \left[\frac{\partial F}{\partial y} - \frac{d}{dx} \left(\frac{\partial F}{\partial y_x} \right) \right] \phi(x)\, dx = 0 \qquad (13.64)$$

since $\phi(x)$ is arbitrary, equation (13.64) is zero only if the sum of the terms within the square brackets is zero. That is

$$\frac{\partial F}{\partial y} - \frac{d}{dx} \left(\frac{\partial F}{\partial y_x} \right) = 0 \qquad (13.65)$$

Equation (13.65) is independent of the parameter ε and the arbitrary function $\phi(x)$ and depends on the choice of the function inserted in the integrand F. It is known as the "Euler–Lagrange" equation and is a necessary, but not a sufficient condition that the functional is stationary. Confirmation of the optimum must be made by testing the second variation in an example to ascertain the optimum. That is, if

$\delta^2 I > 0$ for all possible admissable variations, the stationary value is a minimum.

$\delta^2 I < 0$ for all possible admissable variations, the stationary value is a maximum.

$\delta^2 I$ is positive for some variations and negative for others, the stationary value is not an "extremal".

An alternative form of the Euler–Lagrange equation, when the integrand of the variational integral does not depend on the independent variable explicitly (only through $y(x)$ and $y_x(x)$), is

$$\frac{d}{dx} \left(F - y_x \frac{\partial F}{\partial y_x} \right) = 0 \qquad (13.66)$$

which surprisingly finds extensive application in the solution of variational problems in engineering.

Example. A particle slides from rest down a smooth curve in a vertical plane joining two points A and B. If the particle's motion is sustained by gravity evaluate the path taken by the particle if the time interval between A and B is to be a minimum.

Let the point A be the origin where the initial velocity is zero. Then the equation of motion is

$$v^2 = U^2 + 2gy = 2gy \qquad \text{I}$$

where y is the vertical component of the distance from the origin.

Let s be the length of the optimum curve between A and B. Then an element of this curve

$$ds = [(dy)^2 + (dx)^2]^{\frac{1}{2}}$$

or

$$ds = (1 + y_x^2)^{\frac{1}{2}}\, dx \qquad \text{II}$$

and

$$I = \int_A^B \frac{ds}{v} = \frac{1}{\sqrt{2g}} \int_A^B \left(\frac{1+y_x^2}{y} \right)^{\frac{1}{2}} dx \qquad \text{III}$$

Then the Euler–Lagrange equation is, for this example,

$$\frac{d}{dx} \left(F - y_x \frac{\partial F}{\partial y_x} \right) = 0$$

$$\therefore \quad \frac{d}{dx} \left[\left(\frac{1+y_x^2}{y} \right)^{\frac{1}{2}} - \frac{y_x^2}{\sqrt{(y)}(1+y_x^2)^{\frac{1}{2}}} \right] = 0 \qquad \text{IV}$$

Integrating

$$y(1+y_x^2) = 2R \quad \text{(a constant)} \qquad \text{V}$$

Rearranging and integrating again gives

$$y = R(1+\cos 2\theta) \qquad \text{VI}$$

and

$$x = A - R(2\theta + \sin 2\theta) \qquad \text{VII}$$

where

$$\tan \theta = \frac{dy}{dx} = y_x$$

and A and R are arbitrary constants.

Equations VI and VII are equations of a cycloid where 2θ is the angular parameter and R is the radius of the generating circle.

The second variation can be obtained from Legendre's test which will be stated but not proved here. Thus, "If the Euler–Lagrange equation is satisfied, the range of integration is small, and the sign of $(\partial^2 F/\partial y_x^2)$ is constant throughout the range, I is a maximum or minimum according to the sign of $(\partial^2 F/\partial y_x^2)$ being negative or positive". Then

$$\delta^2 I = \frac{\partial^2}{\partial y_x^2} \left(\frac{1+y_x^2}{y} \right)^{\frac{1}{2}} = \frac{1}{y^{\frac{1}{2}}(1+y_x^2)^{\frac{1}{2}}} \qquad \text{VIII}$$

Taking the positive value of each root in equation VIII results in $\delta^2 I$ being a minimum.

Problems of this kind are called "Brachistochrome problems".

13.5.1. *Functions of Several Dependent Variables*

The principles developed above to obtain the extremal of a functional of a single dependent variable can be extended to a functional embodying many dependent variables. For example; for a functional

$$I = \int_{x_1}^{x_2} F[y(x), y_x(x), z(x), z_x(x), x] dx \qquad (13.67)$$

to be stationary at an extremal there will be an Euler–Lagrange equation for each dependent variable, thus

$$\frac{\partial F}{\partial y} - \frac{d}{dx}\frac{\partial F}{\partial y_x} = 0 \quad \text{and} \quad \frac{\partial F}{\partial z} - \frac{d}{dx}\frac{\partial F}{\partial z_x} = 0 \tag{13.68}$$

must be satisfied. Each is a necessary, but not a sufficient condition for the functional to be an extremal. Tests on the second variation are necessary to confirm that conditions are an optimum.

13.5.2. *Functions of Several Independent Variables*

In many engineering problems the optimum is required for a function containing more than one independent variable. The necessary and sufficient conditions for the functional

$$I = \int_V F[u(x,y,z), u_x(x,y,z), u_y(x,y,z), u_z(x,y,z), x, y, z]\,dV \tag{13.69}$$

to be an extremal is obtained by the method described above where, in this case, the integration is throughout a volume V bounded by a surface S. That is, no variation is permitted on the surface S. Insertion of the variation

$$\delta u = u^* - u = \varepsilon\eta(x,y,z) \tag{13.70}$$

into equation (13.69) and following the mathematical procedures presented for a function of a single independent variable leads to the following extended form of the Euler–Lagrange equation

$$\frac{\partial F}{\partial u} - \frac{\partial}{\partial x}\left(\frac{\partial F}{\partial u_x}\right) - \frac{\partial}{\partial y}\left(\frac{\partial F}{\partial u_y}\right) - \frac{\partial}{\partial z}\left(\frac{\partial F}{\partial u_z}\right) = 0 \tag{13.71}$$

which is the necessary condition for the existence of an optimum. The sufficient conditions would be established by testing the second variations.

The forms of the Euler–Lagrange equations presented in Section 13.5.1 and 13.5.2 have not been proved and it is left for the reader to undertake this task since the steps to follow are identical with that presented initially.

13.5.3. *Isoperimetric Problems—Constrained Extremals*

In many engineering problems it is frequently necessary to evaluate optimum conditions within specified boundaries when the objective function is constrained by some additional condition. For example, it is well known that in solvent extraction processes the rate of mass transfer is a maximum from oscillating drops having the maximum surface area but that the phase separation is controlled by the size of the drops in the dispersed phase. This variational problem then takes the form of establishing the function $y(x)$ that will make the integral

$$I = \int_{x_1}^{x_2} F[y(x), y_x(x), x]\,dx \tag{13.72}$$

a maximum (or a minimum) with the auxiliary condition that $y(x)$ must also satisfy the equation

$$J = \int_{x_1}^{x_2} G[y(x), y_x(x), x] \, dx = \text{constant} \tag{13.73}$$

In this type of problem it is not necessary to test all possible functions in equation (13.72) to obtain the extremal since an admissible function must first satisfy equation (13.73) before it is necessary to consider the function as a possible extremalizing function.

Let the optimizing function be $y(x)$ and again define the variation by $y^*(x)$ where again

$$\delta y = y^*(x) - y(x) = \varepsilon \phi(x) \tag{13.74}$$

with the previous conditions applying to $\phi(x)$ and ε. Also

$$\delta I = \varepsilon \int_{x_1}^{x_2} \left(\frac{\partial F}{\partial y} \phi + \frac{\partial F}{\partial y_x} \phi_x \right) dx \tag{13.75}$$

Since J is a constant, similar treatment of equation (13.73) leads to

$$\varepsilon \int_{x_1}^{x_2} \left(\frac{\partial G}{\partial y} \phi + \frac{\partial G}{\partial y_x} \phi_x \right) dx = 0 \tag{13.76}$$

Again the second term of each of these integrals is integrated by parts and since $\phi(x_1) = \phi(x_2) = 0$ equations (13.75) and (13.76) respectively become

$$\delta I = \varepsilon \int_{x_1}^{x_2} \phi(x) \left[\frac{\partial F}{\partial y} - \frac{d}{dx} \left(\frac{\partial F}{\partial y_x} \right) \right] dx \tag{13.77}$$

and

$$\varepsilon \int_{x_1}^{x_2} \phi(x) \left[\frac{\partial G}{\partial y} - \frac{d}{dx} \left(\frac{\partial G}{\partial y_x} \right) \right] dx = 0 \tag{13.78}$$

From equation (13.78) it is evident that in this case ε and $\phi(x)$ are not independent of each other and therefore equations (13.77) and (13.78) must be expressed through the Lagrange multiplier (see Section 13.3.4) thus

$$\delta I = \varepsilon \left[\int_{x_1}^{x_2} \phi(x) \left(\frac{\partial F}{\partial y} - \frac{d}{dx} \frac{\partial F}{\partial y_x} \right) dx - \lambda \int_{x_1}^{x_2} \phi(x) \left(\frac{\partial G}{\partial y} - \frac{d}{dx} \frac{\partial G}{\partial y_x} \right) dx \right] \tag{13.79}$$

If I is to be stationary the sign of δI must be independent of the choice of ε and $\phi(x)$. This result follows if λ is a constant and $y(x)$ is chosen so as to satisfy the second order differential equation

$$\frac{\partial}{\partial y} (F - \lambda G) - \frac{d}{dx} \left[\frac{\partial}{\partial y_x} (F - \lambda G) \right] = 0 \tag{13.80}$$

It may appear to be sufficient to choose $y(x)$ so that it satisfies the equation

$$\frac{\partial F}{\partial y} - \frac{d}{dx} \left(\frac{\partial F}{\partial y_x} \right) = 0 \tag{13.81}$$

since it would follow by equation (13.77) that the sign of δI is independent of ε and $\phi(x)$ but this need not satisfy equation (13.73). In the case of equation (13.80) λ can be considered to be a third arbitrary constant in addition to the two of the solution. With an appropriate choice of λ it is possible to find a solution which passes through the correct end-points x_1 and x_2, and also satisfies equation (13.73). Frequently equation (13.80) is written in the form

$$\left[\frac{\partial F}{\partial y} - \frac{d}{dx}\left(\frac{\partial F}{\partial y_x}\right)\right] - \lambda\left[\frac{\partial G}{\partial y} - \frac{d}{dx}\left(\frac{\partial G}{\partial y_x}\right)\right] = 0 \qquad (13.82)$$

thereby separating the functions F and G.

Example. It is postulated that a fluid flows within a cylindrical pipe such that the surface integral of the square of the vorticity across a cross-section is a minimum. If, in addition, there is no slip at the walls, the flow is always parallel to the axis of the pipe and the average velocity is U_0, what is the velocity distribution?

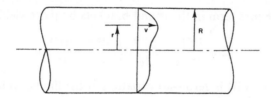

FIG. 13.6. Fluid flow with vorticity in a cylindrical pipe

Solution. The flow conditions are represented in Fig. 13.6.

The vorticity is expressed as ζ (Section 7.6.2) and in cylindrical polar coordinates

$$\zeta = \nabla \wedge \mathbf{V} = \begin{vmatrix} \mathbf{i} & \mathbf{j} & \mathbf{k} \\ \dfrac{\partial}{\partial r} & 0 & \dfrac{\partial}{\partial z} \\ 0 & 0 & V \end{vmatrix} \qquad \text{I}$$

$$\therefore \quad \zeta = -\mathbf{j}\frac{\partial V}{\partial r}$$

so that the variational form of the integral is

$$I = 2\pi \int_0^R r\left(\frac{dV}{dr}\right)^2 dr \qquad \text{II}$$

and the constraint

$$J = 2\pi \int_0^R rV\,dr = \text{constant} \quad \text{for incompressible liquid flow} \qquad \text{III}$$

so that

$$I_s = 2\pi \int_0^R r\left(\frac{dV}{dr}\right)^2 dr + 2\pi\lambda \int_0^R rV dr \qquad \text{IV}$$

where λ is the Lagrange multiplier.

Let the variation in velocity be

$$V(r)^* = V(r) + \varepsilon\phi(r) \qquad \text{V}$$

Then

$$\frac{dV^*}{dr} = \frac{dV}{dr} + \varepsilon\frac{d\phi}{dr} \qquad \text{VI}$$

and insertion of equation VI into IV gives that

$$\delta I_s = 2\pi\varepsilon \int_0^R \left(2r\frac{dV}{dr}\frac{d\phi}{dr} + \lambda r\phi\right) dr \qquad \text{VII}$$

Integrating the first term in the round brackets by parts yields

$$\left[2r\phi\frac{dV}{dr}\right]_0^R - 2\int_0^R \phi\left(r\frac{d^2V}{dr^2} + \frac{dV}{dr}\right) dr. \qquad \text{VIII}$$

The first term in VIII is zero—no variation at the limits, so that insertion of VIII in equation VII gives

$$\delta I_s = 2\pi\varepsilon \int_0^R \left[r\lambda - \frac{d}{dr}\left(2r\frac{dV}{dr}\right)\right]\phi\, dr = 0 \qquad \text{IX}$$

or the Euler–Lagrange equation is

$$r\lambda - \frac{d}{dr}\left(2r\frac{dV}{dr}\right) = 0 \qquad \text{X}\dagger$$

Integrating

$$\frac{\lambda r^2}{2} - 2r\frac{dV}{dr} = A$$

or

$$\int dV = \int\left(\frac{\lambda r}{4} - \frac{A}{2r}\right) dr \qquad \text{XI}$$

which gives

$$V = \frac{\lambda r^2}{8} - \frac{A}{2}\ln r + B \qquad \text{XII}$$

† Equation X could be obtained directly from equation (13.82).

at $r = 0$, V must be finite so that $A = 0$ and

$$2\pi \int_0^R rV\,dr = \pi R^2 U_0$$

$$\therefore \quad R^2 U_0 = \int_0^R \left(\frac{\lambda r^3}{4} + 2Br\right) dr = \left[\frac{\lambda r^4}{16} + Br^2\right]_0^R \qquad \text{XIII}$$

so that

$$U_0 = \frac{\lambda R^2}{16} + B$$

at $r = R$, $V = 0$ so that

$$0 = \frac{\lambda R^2}{8} + B \quad \text{or} \quad B = -\frac{\lambda R^2}{8}$$

and

$$U_0 = -\frac{\lambda R^2}{16} \quad \text{or} \quad \lambda = -\frac{16U_0}{R^2} \quad \text{and} \quad B = 2U_0$$

substitution of A, B and U_0 into equation XII gives the velocity distribution:

$$V = 2U_0\left(1 - \frac{r^2}{R^2}\right) \qquad \text{XIV}$$

13.5.4. *The Pontryagin Maximum Principle*

A most important problem in chemical reaction engineering is to evaluate the optimum temperature profile in a reaction system, and such an analysis necessitates the application of variable Lagrange multipliers. Thus consider the following rate equation to represent the reactions taking place in either a batch or plugflow tubular system having a residence time t.

$$\frac{dx_i}{dt} = f_i(x_1, x_2, \ldots, x_n, T) \qquad (13.83)$$

and let the objective function to be optimized be

$$I = \int_0^\theta f_0(x_1, x_2, \ldots x_n, T)\,dt \qquad (13.84)$$

The problem is to obtain the optimum temperature profile during the residence time 0 to θ. This will be accomplished using the method proposed by Katz.†

Let x_0 be a hypothetical component such that

$$\frac{dx_0}{dt} = f_0 \qquad (13.85)$$

† S. Katz. *Ann. N.Y. Acad. Sci.* **84**, 441 (1960).

with the boundary condition that at $t = 0$, $x_0 = 0$. Now introduce the variable Lagrange multipliers of time by defining

$$J = \int_0^\theta \frac{d}{dt} \sum_{i=0}^n \lambda_i x_i \, dt \tag{13.86}$$

where $\lambda_0 = 1\cdot0$ at $t = \theta$ and $\lambda_i = 0$ at $t = \theta$ for $i = 1, 2, 3, ..., n$.
   Integrating equation (13.86) gives

$$J = \sum_{i=0}^n \lambda_i x_i \Big|_{t=\theta} - \sum_{i=0}^n \lambda_i x_i \Big|_{t=0} = x_0 - \sum_{i=0}^n \lambda_i(0) x_i(0) \tag{13.87}$$

or with equations (13.84) and (13.85) equation (13.87) becomes

$$J - I = - \sum_{i=0}^n \lambda_i(0) x_i(0) = \text{constant} \tag{13.88}$$

$(J-I)$ is a constant for any specified set of initial conditions which are the conditions in all reaction engineering problems. Therefore, maximization of J will maximize I, and here it is proposed to maximize J. Now let the variation in the temperature at any time t be

$$\delta T = T^* - T = \varepsilon \phi(t) \tag{13.89}$$

and the corresponding variation in composition of x_i be

$$\delta x_i = x_i^* - x_i = \varepsilon \mu_i(t) \tag{13.90}$$

where, as throughout this section, ε is a small constant and $\phi(t)$ and $\mu_i(t)$ are continuous functions of time. Then the variation in J is δJ where

$$\delta J = \varepsilon \int_0^\theta \frac{d}{dt} \sum_{i=0}^n \lambda_i \mu_i \, dt \tag{13.91}$$

since the multipliers are independent of temperature. From the kinetic equation (13.83) and equations (13.89) and (13.90) it follows that

$$\frac{dx_i^*}{dt} = \frac{dx_i}{dt} + \varepsilon \frac{d\mu_i}{dt} = \varepsilon \sum_{j=0}^n \frac{\partial f_i}{\partial x_j} \mu_j + \varepsilon \frac{\partial f_i}{\partial T} \phi + f_i \tag{13.92}$$

by expanding $f_i[(x_j + \varepsilon \mu_j)...(T + \varepsilon \phi)]$ using Taylor's theorem and taking only first terms.
   From equation (13.92) it follows that

$$\frac{d\mu_i}{dt} = \sum_{j=0}^n \frac{\partial f_i}{\partial x_j} \mu_j + \frac{\partial f_i}{\partial T} \phi \tag{13.93}$$

Now expanding equation (13.91) gives

$$\delta J = \varepsilon \int_0^\theta \frac{d}{dt}(\lambda_0 \mu_0 + \lambda_1 \mu_1 + \ldots + \lambda_n \mu_n)\, dt$$

$$= \varepsilon \int_0^\theta \left[\left(\lambda_0 \frac{d\mu_0}{dt} + \mu_0 \frac{d\lambda_0}{dt}\right) + \ldots + \left(\lambda_n \frac{d\mu_n}{dt} + \mu_n \frac{d\lambda_n}{dt}\right)\right] dt$$

$$\therefore \quad \delta J = \varepsilon \int_0^\theta \left(\sum_{i=0}^n \lambda_i \frac{d\mu_i}{dt} + \sum_{i=0}^n \mu_i \frac{d\lambda_i}{dt}\right) dt \tag{13.94}$$

Inserting equation (13.93) gives

$$\delta J = \varepsilon \int_0^\theta \left[\sum_{j=0}^n \sum_{i=0}^n \lambda_i \frac{\partial f_i}{\partial x_j} \mu_j + \phi \sum_{i=0}^n \frac{\partial f_i}{\partial T} \lambda_i + \sum_{i=0}^n \mu_i \frac{d\lambda_i}{dt}\right] dt \tag{13.95}$$

Restricting the multipliers λ_i to those functions that satisfy the set of coupled differential equations

$$\frac{d\lambda_j}{dt} = -\sum_{i=0}^n \frac{\partial f_i}{\partial x_j} \lambda_i \tag{13.96}$$

with the boundary conditions stated above, leads to

$$\delta J = \varepsilon \int_0^\theta \phi \left(\sum_{i=0}^n \lambda_i \frac{\partial f_i}{\partial T}\right) dt = 0 \tag{13.97}$$

since $\delta J = 0$ for all arbitrary functions ϕ it follows that

$$\sum_{i=0}^n \lambda_i \frac{\partial f_i}{\partial T} = 0 \quad \text{for all } 0 \leqslant t \leqslant \theta. \tag{13.98}$$

Equation (13.98) states that the temperature profile which is necessary for the integral

$$I = \int_0^\theta f_0\, dt$$

to be a maximum is that which satisfies equation (13.98) where the λ_i are related to the components of the reaction x_i by the equations

$$\frac{d\lambda_j}{dt} = -\sum_{i=0}^n \frac{\partial f_i}{\partial x_j} \lambda_i$$

which are known as the "adjoint functions" of the problem, with the boundary conditions

$$x_i(0) = \text{known for } i = 1, 2, 3, \ldots, n$$

$$x_0(0) = 0$$

$$\lambda_0(\theta) = 1{\cdot}0$$

$$\lambda_i(\theta) = 0 \text{ for } i = 1, 2, 3, \ldots, n$$

Equation (13.98) and the adjoint functions are the necessary, but not the sufficient conditions for the objective function to be an optimum. However, suppose that there are a number of possible roots to equation (13.98). It is then possible that there are several temperatures that satisfy equation (13.98) and it is necessary to evaluate the optimum temperatures at specific time intervals during the course of the reaction. Alternatively, it is possible that the constraints on the reaction temperature prevent operating the process at the true optimum. In such cases the best choice of temperature may be made through the use of Pontryagin's Maximum Principle which may be stated as follows: "The Hamiltonian function H expressed in the form

$$H = f_0 + \sum_{i=1}^{n} \lambda_i f_i$$

must be a maximum for the value of an independent variable to render the objective function a maximum. Alternatively the Hamiltonian must be a minimum if the objective function is to be a minimum."

Consider the summation in the Hamiltonian. That is $\sum_{i=1}^{n} \lambda_i f_i$ in the context of the reaction system analysed above. Now

$$\frac{d}{dt} \sum_{i=1}^{n} \lambda_i f_i = \sum_{i=1}^{n} \lambda_i \frac{df_i}{dt} + \sum_{i=1}^{n} f_i \frac{d\lambda_i}{dt} \tag{13.99}$$

Now for the above reaction system

$$\frac{df_i}{dt} = \sum_{j=1}^{n} \frac{\partial f_i}{\partial x_j} \frac{dx_j}{dt} + \frac{\partial f_i}{\partial T} \frac{dT}{dt} \tag{13.100}$$

which on substituting into equation (13.99) gives

$$\frac{d}{dt} \sum_{i=1}^{n} \lambda_i f_i = \sum_{i=1}^{n} \sum_{j=1}^{n} \lambda_i \frac{\partial f_i}{\partial x_j} \frac{dx_j}{dt} + \sum_{i=1}^{n} \lambda_i \frac{\partial f_i}{\partial T} \frac{dT}{dt} + \sum_{i=1}^{n} \frac{dx_i}{dt} \frac{d\lambda_i}{dt} \tag{13.101}$$

where $f_i = dx_i/dt$. Insertion of equation (13.96) and (13.98) in equation (13.101) results in

$$\frac{d}{dt} \sum_{i=1}^{n} \lambda_i f_i = 0 \tag{13.102}$$

Integration of equation (13.102) yields

$$\sum_{i=1}^{n} \lambda_i f_i = \text{constant} \tag{13.103}$$

and since f_0 is also a constant at the optimum, the Hamiltonian must be constant at the optimum.

Example. The chemical reaction

$$A \longrightarrow B \longrightarrow C$$

is to be carried out in a tubular reactor. If all the reactions are first order and the rate constants are related to temperature through the Arrhenuis Law $k_i = k_{0i}\exp(-E_i/RT)$ and the flow through the reactor is considered to be plug flow, derive the set of equations to evaluate the temperature profile that will maximize the production of the product B.

Solution. The rate equations at any point in the reactor are

$$\frac{dx_1}{dt} = -k_1 x_1 = f_1 \quad \text{(say)} \qquad\qquad \text{I}$$

and

$$\frac{dx_2}{dt} = k_1 x_1 - k_2 x_2 = f_2 \quad \text{(say)} \qquad\qquad \text{II}$$

where $t = l/v$ and l is the distance from the entrance of the reactor to the plane in the reactor where the compositions are x_1 and x_2; v is the superficial velocity of the reaction mixture. The variation of the reaction rate constants with temperature

$$k_1 = k_{01}\exp\left(-\frac{E_1}{RT}\right) \quad \text{and} \quad k_2 = k_{02}\exp\left(-\frac{E_2}{RT}\right)$$

It is required to maximize the yield of B. That is $x_2(\theta)$ must be a maximum. Then the objective function becomes

$$I = x_2(\theta) - x_2(0) = \int_0^\theta \frac{dx_2}{dt}\,dt = \int_0^\theta f_2\,dt \qquad\qquad \text{III}$$

where $\theta = L/v$ and L is the reactor length.
  The adjoint functions are

$$\frac{d\lambda_1}{dt} = -\frac{\partial f_0}{\partial x_1} - \sum_{j=1}^n \frac{\partial f_j}{\partial x_1}\lambda_j \qquad\qquad \text{IV}$$

where in this problem $f_0(x, T) = f_2$. Then

$$\frac{\partial f_0}{\partial x_1} = k_1 \quad \text{and} \quad \frac{\partial f_1}{\partial x_1}\lambda_1 = -k_1\lambda_1, \quad \text{also} \quad \frac{\partial f_2}{\partial x_1}\lambda_2 = k_1\lambda_2$$

and

$$\frac{\partial f_0}{\partial x_2} = -k_2, \quad \frac{\partial f_1}{\partial x_2}\lambda_1 = 0 \quad \text{and} \quad \frac{\partial f_2}{\partial x_2}\lambda_2 = -k_2\lambda_2$$

so that from equation IV

$$\frac{d\lambda_1}{dt} = k_1(\lambda_1 - \lambda_2) - k_1 \qquad\qquad \text{V}$$

and

$$\frac{d\lambda_2}{dt} = k_2(\lambda_2 + 1) \qquad\qquad \text{VI}$$

Now the temperature T must be chosen to maximize the Hamiltonian

$$H = f_0 + \sum_{i=1}^{n} f_i \lambda_i$$

or

$$\frac{\partial H}{\partial T} = \frac{\partial f_0}{\partial T} + \frac{\partial}{\partial T} \sum_{i=1}^{n} f_i \lambda_i$$

where

$$\frac{\partial f_0}{\partial T} = x_1 \frac{\partial k_1}{\partial T} - x_2 \frac{\partial k_2}{\partial T} \quad \text{(at any plane in the reactor)}$$

$$= \frac{k_{01} x_1 E_1}{RT^2} \exp\left(-\frac{E_1}{RT}\right) - \frac{k_{02} x_2 E_2}{RT^2} \exp\left(-\frac{E_2}{RT}\right)$$

and

$$\lambda_1 \frac{\partial f_1}{\partial T} = -\frac{k_{01} x_1 E_1 \lambda_1}{RT^2} \exp\left(-\frac{E_1}{RT}\right)$$

with a similar expression for $\lambda_2 \partial f_2/\partial T$. Therefore,

$$\frac{\partial H}{\partial T} = \frac{1}{RT^2}\left[(1 - \lambda_1 + \lambda_2) k_{01} x_1 E_1 \exp\left(-\frac{E_1}{RT}\right) - (1 + \lambda_2) k_{02} x_2 E_2 \exp\left(-\frac{E_2}{RT}\right)\right]$$

VII

Because the factor $\lambda_2 + 1$ occurs frequently, it is convenient to change to the variable $\bar{\lambda}_2$ defined by

$$\bar{\lambda}_2 = \lambda_2 + 1 \qquad \text{VIII}$$

The complete problem is now posed as the solution of the following equations

$$\frac{dx_1}{dt} = -k_1 x_1 \qquad\qquad \text{I}$$

$$\frac{dx_2}{dt} = k_1 x_1 - k_2 x_2 \qquad\qquad \text{II}$$

$$\frac{d\lambda_1}{dt} = k_1(\lambda_1 - \bar{\lambda}_2) \qquad\qquad \text{V}$$

$$\frac{d\bar{\lambda}_2}{dt} = k_2 \bar{\lambda}_2 \qquad\qquad \text{VI}$$

with the boundary conditions that

$$\text{at} \quad t = 0, \quad x_1 = 1{\cdot}0, \quad x_2 = 0$$

and

$$\text{at} \quad t = \theta, \quad \lambda_1 = 0, \quad \bar{\lambda}_2 = 1$$

The optimum temperature profile along the reactor is that which maximizes the Hamiltonian, or satisfies

$$\frac{\partial H}{\partial T} = \frac{1}{RT^2}[(\bar{\lambda}_2 - \lambda_1)k_1 x_1 E_1 - \bar{\lambda}_2 k_2 x_2 E_2] = 0$$

There are many ways of solving the above system of equations, as described by Rosenbrock and Storey,† and one method is illustrated in the computer flow diagram, Fig. 13.7. Here it is shown that a standard Runge–Kutta subroutine can be used twice, once to integrate equations I and II from $t = 0$ to $t = \theta$, and again to integrate equations V and VI in the reverse direction from $t = \theta$ to $t = 0$. All of these profiles are stored so that an optimizing method can be used at each value of t to adjust the temperature T in such a way as to increase the Hamiltonian. The gradient method was selected for this step, thus

$$T_{new} = T_{old} + \alpha \frac{\partial H}{\partial T} \qquad\qquad \text{IX}$$

where α is an arbitrary scale factor relating the adjustment in temperature to the gradient of the Hamiltonian. Clearly, if the desired turning value in H is reached, $\partial H/\partial T$ will be zero and the optimum temperature profile will have been reached.

The following numerical values were used in the program of Fig. 13.7 to obtain the optimum temperature and concentration profiles illustrated in Fig. 13.8.

$$k_{01} = 65 \cdot 60 \text{ s}^{-1} \quad E_1 = 10 \text{ kcals/mole}$$

$$k_{02} = 1970 \text{ s}^{-1} \quad E_2 = 16 \text{ kcals/mole}$$

$$\theta = 12 \cdot 5 \text{ s} \qquad \alpha = 4 \times 10^5 \text{ (°K)}^2 \text{ s}$$

From an initially assumed constant temperature profile of 700 °K, the first ten iterations gave the following succession of concentrations of the desired product in the outlet from the reactor.

$$0 \cdot 41247, \quad 0 \cdot 46896, \quad 0 \cdot 47465, \quad 0 \cdot 47706, \quad 0 \cdot 47810$$

$$0 \cdot 47834, \quad 0 \cdot 47843, \quad 0 \cdot 47844, \quad 0 \cdot 47845, \quad 0 \cdot 47845$$

13.6. HILL CLIMBING OPTIMIZATION PROCEDURES

Hill-climbing optimization techniques have developed very rapidly over the past 10–20 years, and are nowadays the most extensively applied optimization procedures when the problem involves the analysis of a large number of variables. The principles involved are not new and date back to Cauchy; but from an optimization point of view, most of the development work was

† Rosenbrock, H. H. and Storey, C. "Computational Techniques for Chemical Engineers", Pergamon Press, London (1966).

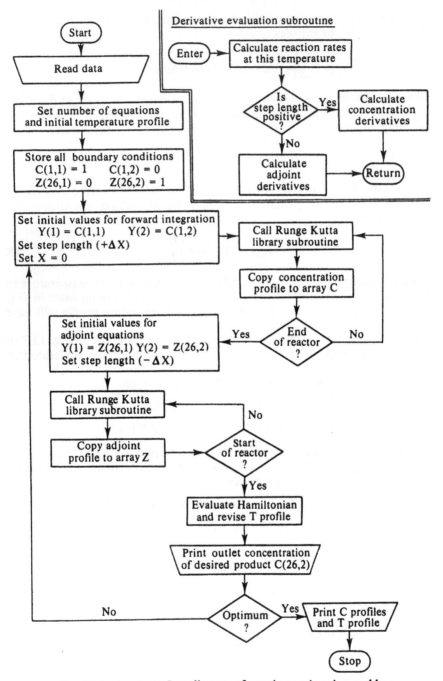

FIG. 13.7. Computer flow diagram of reaction engineering problem

carried out by Box and his coworkers.† Essentially all the techniques developed embody a search procedure. That is, an initial estimate of the solution of the problem to be optimized is made and tested. Depending on this result a search is performed in order to improve the initial estimate, and

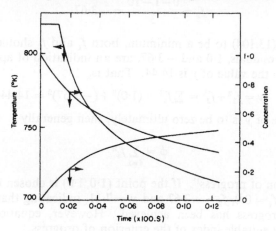

FIG. 13.8. Temperature and concentration profiles

this procedure is repeated until the optimum is found. Thus, let the problem to be optimized be finally expressed by the set of simultaneous non-linear equations,

$$f_i(x_1, x_2, ..., x_n) = 0 \quad (1 \leqslant i \leqslant m)$$

where it is necessary to evaluate $x_1, x_2, ..., x_n$ for each of the equations. An indication of the progress to the evaluation of these equations is to estimate the "residuals" and examine the "criterion of progress". By residual is meant the difference in the value of the functions from zero at each iteration. For example, consider the quadratic test function suggested by Rosenbrock and Storey

$$y = (x_1 - x_2)^2 + \left(\frac{x_1 + x_2 - 10}{3}\right)^2 \tag{13.104}$$

with components

$$f_1(x_1, x_2) = (x_1 - x_2)$$

and

$$f_2(x_1, x_2) = \left(\frac{x_1 + x_2 - 10}{3}\right)$$

Equation (13.104) is the equation of an ellipse with centre at the point $(5, 5)$ as shown in Fig. 13.10. Let it be assumed that a search is to be initiated to

† Box, G. E. P. and Wilson, K. B. *J. Statist. Soc.* 1, B13 (1951).

locate the minimum of this function starting at the point $(0, -1)$. For these coordinates:

$$f_1 = (0 - (-1 \cdot 0)) = 1 \cdot 0 \tag{13.105}$$

and

$$f_2 = \frac{0 - 1 - 10}{3} = -3 \cdot 67 \tag{13.106}$$

For equation (13.104) to be a minimum, both f_1 and f_2 should be zero and therefore the residuals, $1 \cdot 0$ and $-3 \cdot 67$, are an indication of approach to the solution. Also the value of y is $14 \cdot 44$. That is,

$$y = f_1^2 + f_2^2 = \Sigma f_i^2 = (1 \cdot 0)^2 + (-3 \cdot 67)^2 = 14 \cdot 44$$

since the value of y is to be zero ultimately, then generally

$$\phi = \sum_{i=1}^{n} f_i^2 \tag{13.107}$$

is the "criterion of progress". If the point $(1 \cdot 0, 1 \cdot 0)$ is chosen for the second search point, $f_1 = 0 \cdot 0$, $f_2 = -2 \cdot 67$ and $y = 7 \cdot 11$, indicating that only a small amount of progress has been achieved. However, equation (13.107) is confirmed as a suitable index of the criterion of progress.

In the above analysis, the progress from the initial base point $(0, -1 \cdot 0)$ to the second point $(1 \cdot 0, 1 \cdot 0)$ appeared to have been taken in a random manner, and this can be very time consuming computationally with problems involving a larger number of variables. Hence to make the search for the optimum more systematic it is appropriate to identify

(i) The Search Vector.
(ii) The Step Length.

13.6.1. *The Search Vector*

The search vector is a vector indicating the direction in which the hill climb should proceed. Mathematically this can be expressed

$$\mathbf{x}_{m+1} = \mathbf{x}_m + p\mathbf{D}_m \tag{13.108}$$

where $\mathbf{D}_m$ is the vector indicating the direction of the hill climb and p is the step length. For unit step length in the above search the search vector would be

$$\begin{bmatrix} 0 \\ -1 \cdot 0 \end{bmatrix} + 1 \cdot 0 \begin{bmatrix} d_1 \\ d_2 \end{bmatrix} = \begin{bmatrix} 1 \cdot 0 \\ 1 \cdot 0 \end{bmatrix} \tag{13.109}$$

or $\mathbf{D}_m = \begin{bmatrix} 1 \cdot 0 \\ 2 \cdot 0 \end{bmatrix}$. The evaluation of the search vector has proved to be a

great difficulty in hill-climbing optimization problems and many techniques have been developed to reduce the number of iterations necessary to obtain the optimum. Probably the first method was the "successive variation of parameter method".

13.6.1.1. *Successive variation of parameters.* Essentially this method involves varying each parameter in turn. For example, if the maximum of the function $f(x_1, x_2)$ is sought the parameter x_1 is varied stepwise until no further improvement is achieved. Thereafter the parameter x_2 is varied in a similar manner until the maximum value in this direction is obtained. Then the search is continued in the x_1 direction again, and this procedure is repeated until the maximum of the objective function is obtained. The search takes the form shown in Fig. 13.9 where it can be seen that, for the function presented,

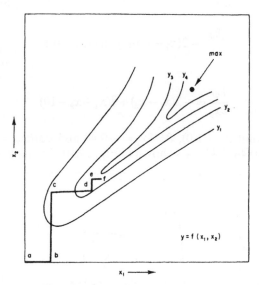

FIG. 13.9. Hill climbing by successive variation of parameters

progress to the optimum will be very slow and probably will not locate the maximum irrespective of the number of iterations attempted. A further, and equally formidable, difficulty of this method arises with multivariable problems when each parameter is varied in sequence $x_1, x_2, ..., x_n$. With such problems attempts have been made to reduce the number of iterations by evaluating $(\partial f/\partial x_i)$ after each step and then varying the parameter x_i corresponding to the largest value of the derivative $(\partial f/\partial x_i)$. This reduces the computing time but does not always provide a method of overcoming obstacles encountered with functions of the kind shown in the contour plot in Fig. 13.9. However, the calculation of derivatives in the hill-climbing search leads to the method of steepest ascent.

13.6.1.2. *The method of steepest ascent or descent.* The method of steepest ascent is one of the best known optimization procedures to apply when the number of variables involved is large. The method was developed by Box and his coworkers in 1951. The search direction is evaluated at each step. In this case the search vector $\mathbf{D}_m$ is

$$\mathbf{D}_m = \pm \nabla\phi \qquad (13.110)$$

where the sign depends on whether a maximum or a minimum is sought, and each component of $\mathbf{D}_m$ is evaluated. Thus in the example given above

$$\phi = (x_1 - x_2)^2 + \left(\frac{x_1 + x_2 - 10}{3}\right)^2$$

and

$$\frac{\partial\phi}{\partial x_1} = 2(x_1 - x_2) + \tfrac{2}{9}(x_1 + x_2 - 10)$$

also

$$\frac{\partial\phi}{\partial x_2} = -2(x_1 - x_2) + \tfrac{2}{9}(x_1 + x_2 - 10)$$

so that at the point $(0, -1\cdot0)$, $\partial\phi/\partial x_1 = -0\cdot44$ and $\partial\phi/\partial x_2 = -4\cdot44$ and the search vector is $\{0\cdot44, 4\cdot44\}$ as is shown in Fig. 13.10. In many real engineering

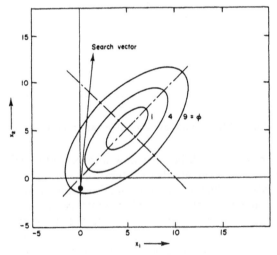

FIG. 13.10. Contour plot of $\phi = (x_1 - x_2)^2 + [\tfrac{1}{3}(x_1 + x_2 - 10)]^2$ showing search vector

problems these partial derivatives cannot be obtained analytically and must be estimated numerically by choosing a small step length, say h, so that

$$\frac{\partial f}{\partial x_i} = \frac{f[x_1, x_2, \dots, (x_i + h), \dots, x_n] - f[x_1, x_2, \dots, x_n]}{h} \qquad (13.111)$$

The search vector of steepest descent is then computed.

13.6.2. *The Step Length*

In all the above stepwise searches it is advisable to vary the step length. Rosenbrock recommends adjusting the step length as follows. If evaluation of the objective function is successful the step length is trebled for the next iteration, but if unsuccessful the step length is reduced by half. The effect of this procedure is that a successful step in a particular direction is accelerated threefold whereas a series of unsuccessful trials causes the step length to decrease steadily in magnitude. If successive steps are successful the step length grows rapidly, accelerating the solution of the problem. In any case this procedure compels the step length to adjust to the desired length very rapidly and, in a limited sense, is built into the method of Hooke and Jeeves described below.

13.7. THE SEQUENTIAL SIMPLEX METHOD

The method is an extension of the steepest ascent method and was proposed by Himsworth.† It differs from the above method in that no attempt is made to find the lines of steepest ascent, but rather that a rapid determination is made of a direction which is steep although not necessarily the steepest, so that moves are made frequently in a favourable direction.

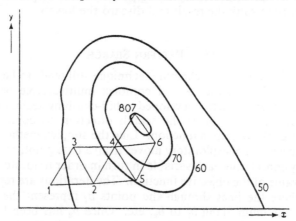

FIG. 13.11. The sequential simplex method of optimization

For a process involving two variables the procedure is as follows. In Fig. 13.11, three points are selected fairly close together in such a manner that they form the vertices of an equilateral triangle. Let these be points 1, 2, and 3. Following this, three further points are selected such that each one forms an equilateral triangle with any two of the first three points. If the response surface is locally a plane or nearly so, one of the possible new points will give a higher result than the other two, and furthermore the point 4 will be a

† Himsworth, F. R. *Trans. Inst. Chem. Eng.* **40**, 345 (1962).

"mirror image" to the lowest point in the first triangle. Hence after the first three trials the point giving the lowest results is rejected and replaced by the mirror image point forming the second triangle. This construction process is repeated until the summit is reached. Thus as long as the surface is sloping and reasonably plane over the area of a particular triangle, each step in the construction leads to a more favourable region and the path taken from the starting point will zig-zag about the line of steepest ascent.

For a system containing more than two variables, the procedure will be similar except that $(k+1)$ trial points will be made, where k is the number of variables. The most favourable of these points is retained to replace the appropriate member of the original set, thus completing a move towards the optimum. The $(k+1)$ points correspond to the vertices of a regular simplex in k dimensions. Hence the name of the optimizing technique.

It is claimed that this method has the following advantages:

   (i) The calculations are elementary requiring no mathematical or statistical knowledge.

  (ii) Each move is determined by the previous result and therefore human judgement is eliminated.

 (iii) It is possible to add a variable at any time.

 (iv) A numerical measure of the response is not required. It is only necessary to rank the result and discard the lowest.

13.8. Pattern Search

The sequential simplex hill-climbing technique still tends to be extravagant in computation time in the case of complex multivariable problems and therefore Hooke and Jeeves† developed a hill-climbing technique in which an attempt is made to induce the search direction to coincide with the principal axis of the objective function. Like the simplex method, the search is started with a series of exploratory moves from a base point, as illustrated in Fig. 13.12, by considering each variable in turn in order to locate an improvement in the value of the objective function. The iterations are represented on the figure where the lines through the points b_i represent the progress in locating the minimum along b_1 to b_2, etc. Once b_2 has been located, b_3 is found in the direction $b_1 \rightarrow b_2$ in accordance with Hooke and Jeeves assumption—that the direction $(b_1 \rightarrow b_2)$ approximates the direction of the local principal axis of the objective function. Consequently a pattern search is made to the point $2(b_2 - b_1)$. At this point (b_3) the value of the objective function is not compared with $f_2(x_1, x_2)$ but a search is started to establish a new base point. This procedure is continued until the minimum of the objective function is located. This procedure was applied by Jeffreys, Mumford and Herridge‡ to optimize a solvent extraction process. Details of

† Hooke, R. and Jeeves, T. A. *J. Ass. Comput. Mech.* **8**, 212 (1961).
‡ Jeffreys, G. V., Mumford, C. J. and Herridge, M. H. *J. App. Chem. Biotechnol.* **22**, 319 (1972).

their analysis are presented below where the Hooke and Jeeves method is compared with a direct cost analysis.

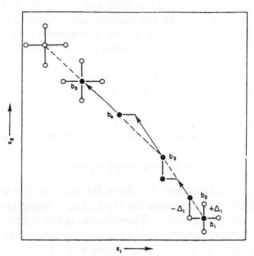

FIG. 13.12. Pattern search method of Hooke and Jeeves

13.8.1. *Basic Equations for the Optimization of the Process*

The solvent extraction process under consideration is illustrated in Fig. 13.13. It will be assumed that the system has been decided (i.e. the solvent is known) and that the type of extractor has been selected. The operation will be counter-current because this is by far the most efficient. The feed rate and the concentration of the solute in the feed will be considered constant and

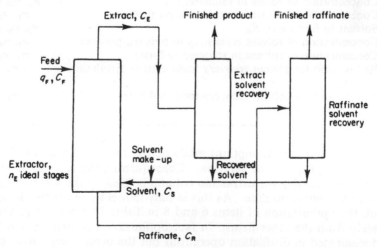

FIG. 13.13. Process flow diagram

independent of time, and the products and the solvent will be recovered by distillation. The following major assumptions will apply: (i) the extracting solvent and the mother liquor in the feed are insoluble; (ii) the distribution coefficient of the solute between the mother liquor and extracting solvent is constant; (iii) only one solute is to be extracted; (iv) constant relative volatility between the solvent and extracted solute; (v) losses of solvent in the distillation processes are negligible.

The objective function, defined as the total annual cost of extracting the solute, C_T, can be expressed in terms of the following cost items: π_1, annual cost of multistage extractor; π_2, annual cost of unextracted solute; π_3, annual cost of solvent recovery from extract; π_4, annual cost of solvent recovery from raffinate; π_5, annual cost of solvent; π_6, annual cost of labour. Then

$$C_T = \pi_1 + \pi_2 + \pi_3 + \pi_4 + \pi_5 + \pi_6 \qquad (13.113)$$

Whilst the annual labour cost has been included in the objective function it will be considered constant, since the type of equipment involved is unlikely to introduce variable labour costs. Therefore π_6 is not subject to optimization. However, each of the cost items are influenced by a number of factors which may be set to produce an optimum arrangement. These are discussed below and summarized in Table 13.3.

TABLE 13.3

Quantity	Cost item affected
1. Design of extractor stages	π_1
2. Concentration of solute in raffinate, C_R	π_1, π_2, π_3
3. Concentration of solute in recycled solvent, C_S	π_1, π_3
4. Solvent to feed ratio, R_E	π_1, π_3, π_5
5. Concentration of solvent remaining in finished product	π_3, π_5
6. Concentration of solvent in stripped raffinate	π_4, π_5
7. Reflux ratio for solvent recovery from extract distillation column	π_3
8. Reflux ratio for solvent recovery from raffinate distillation column	π_4

On the basis of the assumptions made above, the concentration of solute in the raffinate, C_R, will be very low. Consequently, the amount of solvent in the raffinate leaving the extraction column is fixed by the solubility of the solvent in the non-consulate. As this is only indirectly under the designer's control, the optimization of items 6 and 8 in Table 13.3 may be performed separately from the other items. In fact optimization of items 6 and 8 are well documented in distillation operations but the other components of the cost items need to be considered.

13.8.2. *Relation between Component Costs*

13.8.2.1. *The annual cost of the extraction column π_1.* The annual cost of the extraction column is expressed by the equation

$$\pi_1 = n_E C_E/E_0 \tag{13.114}$$

where n_E is the number of ideal stages. For the assumptions specified above, n_E can be estimated from the equation proposed by Treybal:[†]

$$n_E = \left\{ \ln\left[\frac{C_E - C_S/m}{C_R - C_S/m} \left(1 - \frac{1}{mR_E} \right) + \frac{1}{mR_E} \right] \right\} \Big/ \ln mR_E \tag{13.115}$$

The term C_E is the annual cost per stage and is given by

$$C_E = C_{VE}(p/Y+b)_E + P \tag{13.116}$$

where $p = 1 +$ (cost of ancillary equipment/cost of major equipment) and P is the total power cost. Values of p have been given by Aries and Newton.[‡] C_{VE} is the uninstalled cost of the major extraction equipment and is related to the capacity of the equipment. That is,

$$C_{VE} = f(Q) \tag{13.117}$$

where Q is the total volume rate of flow of the phases. Since this is proportional to the cross-sectional area of the column C_{VE} can be expressed as

$$C_{VE} = Ka^x \tag{13.118}$$

where a is the column cross-section and K and x are constants. In terms of the solvent-to-feed ratio and the volumetric flow rates of the dispersed phase V_D and the continuous phase V_C equation (13.118) becomes

$$C_{VE} = K[q_F(R_E+1)/(V_C+V_D)]^x \tag{13.119}$$

Substitution of equation (13.119) into equation (13.116) gives

$$C_E = K[q_F(R_E+1)/(V_C+V_D)]^x (R/Y+b) + P \tag{13.120}$$

Thus the annual cost per stage can be expressed in terms of the process variables.

Finally E_0 in equation (13.114) is the overall efficiency which is dependent on the type of extractor and the system being treated. Expressions for different types of extractor are presented by Treybal and for this analysis a perforated plate column will be considered. The overall stage efficiency is expressed as

$$E_0 = 89\,500 H_C^{0.5}/\sigma R_E^{0.42} \tag{13.121}$$

[†] Treybal, R. E. "Liquid Extraction", McGraw-Hill, New York (1963).
[‡] Aries, R. S. and Newton, R. D. "Chemical Engineering Cost Estimation", McGraw-Hill, New York (1955).

13.8.2.2. Annual value of the unextracted solute, π_2. Since the feed is essentially insoluble in the solvent, the quantity of the solute in the raffinate is $q_F C_R$. Therefore

$$\pi_2 = F q_F C_R H, \tag{13.122}$$

where F is the solute cost.

13.8.2.3. Annual cost of solvent recovery from extract by distillation. The annual cost of a distillation operation is given by Happel† and can be summarized by the equation

$$\pi_3 = C_D n_D O(1+r)(p/Y+b)_D/E_D G_D$$
$$+ C_h O(1+r)(p/Y+b)_D/G_h + C_{hC} O(1+r) H \tag{13.123}$$

The three terms in equation (13.123) represent the cost of distillation equipment, cost of heat exchange equipment, and the cost of reboiler heat and condenser coolant.

The number of ideal trays n_D required for the recovery of the solvent may be expressed in terms of the minimum number of ideal trays at total reflux $n_{D\,\text{min}}$ by

$$n_D = \gamma n_{D\,\text{min}} \tag{13.124}$$

where γ is a coefficient obtainable from Gilliland's correlation‡ and $n_{D\,\text{min}}$ is obtained from the Fenske equation.§

$$n_{D\,\text{min}} = \frac{\ln[x_D/(1-x_D)]}{\ln \alpha} + \frac{\ln[(1-x_w)x_w]}{\ln \alpha} - 1 \tag{13.125}$$

Restricting the analyses to the case in which the solvent is the volatile component in the extract, and therefore becomes the distillate in the recovery system, and assuming that the relative volatility α between the solvent and extracted solute is high, this becomes

$$n_{D\,\text{min}} = \frac{\ln[x_D/(1-x_D)]}{\ln \alpha} + C_1 \tag{13.126}$$

where C_1 is a constant of value near 1.0 since x_w approaches zero. Therefore

$$x_D/(1-x_D) = [q_F R_E \rho_S/M_S O][M_E O/q_F R_E C_S] = \rho_S M_E/C_S M_S \tag{13.127}$$

Substituting equation (13.127) into equation (13.126) gives

$$n_{D\,\text{min}} = \ln \frac{(\rho_S M_E/C_S M_S)}{\ln \alpha} + C_1 \tag{13.128}$$

The distillation rate O can be expressed as

$$O = R_E q_F \rho_S/M_S \tag{13.129}$$

† Happel, J. "Chemical Process Economics", McGraw-Hill, New York (1958).
‡ Gilliland, E. R. *Ind. Eng. Chem.* **32**, 1220 (1940).
§ Fenske, M. R. *Ind. Eng. Chem.* **24**, 482 (1932).

Then if U is the number of kg mol solute/h entering the extract still, then

$$U = q_F(C_F - C_R)/M_E \qquad (13.130)$$

so that the mole fraction of solvent in the feed to the extract still z will be

$$z = O/(U+O) = R_E q_F \rho_S/(U M_S + R_E q_F \rho_S) \qquad (13.131)$$

Furthermore, for a volatile solvent, x_D and $y_D \simeq 1 \cdot 0$ and since at minimum reflux r_m, the value of y in equilibrium with z will be y_m, where

$$y_m = \alpha z/[1 + z(\alpha - 1)] \qquad (13.132)$$

since

$$r_m/(1 + r_m) = (y_D - y_m)/(x_D - z) \qquad (13.133)$$

so that for these conditions by equations (13.132) and (13.133)

$$r_m = 1/z(\alpha - 1) \qquad (13.134)$$

Now the optimum reflux ratio is related to the minimum reflux ratio as follows

$$r = \beta r_m. \qquad (13.135)$$

Then the vapour rate B is related to the solvent recovery rate by

$$B = O(1+r) = \frac{R_E q_F \rho_S}{M_S}\left(1 + \frac{\beta}{\alpha - 1}\right) + \frac{\beta U}{\alpha - 1} \qquad (13.136)$$

Equation (13.123) can be rearranged to

$$\pi_3 = \left[\left(\frac{C_D \gamma n_{D\,\min}}{E_D\,G_D} + \frac{C_h}{G_h}\right)\left(\frac{p}{Y} + b\right) + C_{hC}H\right]O(1+r) \qquad (13.137)$$

$$\pi_3 = AB \qquad (13.138)$$

where both A and B are obtainable in terms of the process variables.

13.8.2.4. *Annual cost of solvent recovery from raffinate by distillation.* As already stated, this item can be treated separately.

13.8.2.5. *Annual cost of lost solvent.* For the plate type of extraction column the loss of solvent is proportional to the recirculation rate, so that

$$\pi_5 = q_F R_E\, lS.H. \qquad (13.139)$$

where l is fractional loss and S is solvent cost.

Equations (13.114), (13.122), and (13.128) are the fundamental equations for the optimization of a solvent extraction process. However, in order to apply these equations it is necessary to examine how the quantities in Table 13.3 affect the component costs.

13.8.3. *Effects of π_1 on Component Costs*

13.8.3.1. *Effect of solute concentration in the raffinate, C_R.* Table 13.3 indicates that π_1, π_2, and π_3 are influenced. Item π_3 is considered to be affected because the extract concentration is influenced by the material balance. However, C_R is usually so small that the effect of its variation on item π_3 is negligible. Therefore considering items π_1 and π_2 which, for a minimum, give

$$\frac{\partial C_T}{\partial C_R} = \frac{\partial \pi_1}{\partial C_R} + \frac{\partial \pi_2}{\partial C_R} = \frac{C_E}{E} \frac{\partial n_E}{\partial C_R} + F q_F H = 0 \qquad (13.140)$$

Differentiating equation (13.115) and substituting into equation (13.140) gives

$$\frac{C_E}{E_0} \left(\frac{1 - 1/M_E R_E}{\ln M_E R_E} \right) \left(C_F - \frac{C_S}{M_E} \right)$$

$$\times \left\{ \left[\frac{(C_F - C_S/M_E)(1 - M_E R_E)^{-1}}{C_R - C_S/M_E} + \frac{1}{M_E R_E} \right] \frac{(C_R - C_S)^2}{M_E} \right\} - 1 = F q_F H$$

$$(13.141)$$

13.8.3.2. *Effect of solute concentration in recovered solvents, C_S.* Table 13.3 indicates that π_1 and π_3 are affected. Then

$$\frac{\partial C_T}{\partial C_S} = \frac{\partial \pi_1}{\partial C_S} + \frac{\partial \pi_3}{\partial C_S} = 0 \qquad (13.142)$$

For insoluble solvents the amount of solvent returned from the raffinate recovery still will be very low. Furthermore assuming competent operation, the losses of solvent and therefore make-up will be low. Consequently it may be assumed that C_S is the same as the concentration of solute leaving with the solvent from the extract still. In practice this will mean that there will be increasing amounts of solute in the recycle stream. Hence a purge of this stream may be necessary when it should then be possible to keep the value of C_S near the calculated optimum. Substitution of the appropriate equations into equation (13.142) gives

$$\frac{C_E}{E_0} \left(\frac{1 - 1/M_E R_E}{\ln M_E R_E} \right) \left[\left(C_F - \frac{C_S}{M_E} \right) - \left(C_R - \frac{C_S}{M_E} \right) \right]$$

$$\times \left\{ M_E \left[\frac{(C_F - C_S/M_E)(1 - 1/M_E R_E)}{C_R - C_S/M_E} + \frac{1}{M_E R_E} \right] \left[C_R - \frac{C_S}{M_E} \right]^2 \right\}^{-1}$$

$$= \frac{C_D \gamma (p/Y + b)_D B}{C_S E_D G_D \ln \alpha} \qquad (13.143)$$

Equations (13.141) and (13.143) can readily be solved simultaneously if it is assumed that in equation (13.143) the term $(C_R - C_S/M_E)$ is negligible in comparison with the term $(C_F - C_S/M_E)$. The solution is that

$$\text{the optimum } C_S = C_{S \text{ opt}} = \frac{M_E \gamma C_D (p/Y + b)_D B}{E_D G_D F q_F H \ln \alpha} \qquad (13.144)$$

and

the optimum $C_R = C_{R \text{ opt}} = (C_S - C_{S \text{ opt}}/M_E)\left[(1 - M_E R_E)/2\right](1 \pm \sqrt{J})$

$$+ C_{S \text{ opt}}/M_E \qquad (13.145)$$

where

$$J = 1 + \frac{4C_E}{(C_F - C_{S \text{ opt}}/E) Fq_F HE_0(M_E R_E - 1) \ln M_E R_E}$$

for $M_E R_E > 1$, then $-\sqrt{J}$ is applicable and $M_E R_E < 1$, $+\sqrt{J}$ should be used. Substituting the above values of C_R and C_S into equation (13.114) gives the optimum value of n_E as

$$n_{E \text{ opt}} = \left\{\frac{\ln\left[1 - 2/(1 + \sqrt{J})\right]}{\ln M_E R_E}\right\} - 1 \qquad (13.146)$$

The above values of $C_{S \text{ opt}}$, $C_{R \text{ opt}}$, and $n_{E \text{ opt}}$ are the true optimum values only when the optimum value of R_E is used. If not, the values estimated give the minimum total cost at the particular solvent to feed ratio R_E. The effect of this ratio will now be considered.

13.8.3.3. *Effect of solvent/feed ratio.* Table 13.3 shows that π_1, π_3 and π_5 are affected. That is:

$$\frac{\partial C_T}{\partial R_E} = \frac{\partial(n_E C_E/E_0)}{\partial R_E} + \frac{\partial \pi_3}{\partial R_E} + \frac{\partial \pi_5}{\partial R_E} = 0 \quad \text{for a minimum} \quad (13.147)$$

Inspection of equation (13.147) shows that differentiation of the expressions involving n_E, C_E and E_0 with respect to R_E is formidable. Therefore it is more convenient to plot the sum $(\pi_1 + \pi_3 + \pi_5)$ against R_E by inserting the optimum values of n_E, C_R and C_S when the minimum of this curve locates $R_{E \text{ opt}}$.

The estimation of minimum total cost is most conveniently achieved by computer and the flow chart of the program is presented in Fig. 13.14. This optimization procedure was compared with the Hooke and Jeeves method.

The Hooke and Jeeves algorithm is summarized in the flow chart Fig. 13.15 based on equations (13.144), (13.120), (13.121), (13.136), and (13.128).

13.8.4. *Numerical example.* The optimization procedures were compared by analysing the solvent extraction process described by Treybal.†

An organic solute valued at \$0.07/kg in a 10% solution in water is to be recovered by extraction to give a product of purity 99·95% mole solute. The feed solution flow rate will be 5·5 m³/h and its density is 993 kg/m³. A perforated plate extractor is to be used for this process with a plate spacing of 0·305 m. For this plate spacing the uninstalled cost is estimated to be \$45.5$a^{0.7}$ per plate where a is the cross-sectional area of the plate in m². The

† Treybal, R. E. "Liquid Extraction", McGraw-Hill, New York (1963).

liquid flowing in the larger volume rate will be dispersed and the allowable liquid handling capacity $(V_c + V_d)$ is $30 \cdot 5$ m/h were $V_c + V_d$ represent the superficial velocities of the continuous and dispersed phases respectively. The distribution coefficient is $2 \cdot 0$ and the interfacial tension is $13 \cdot 0 \times 10^{-2}$ N/m. The molecular weight of the solute is $60 \cdot 0$ and that of the solvent is 102. The solvent costs \$164.20 per m³.

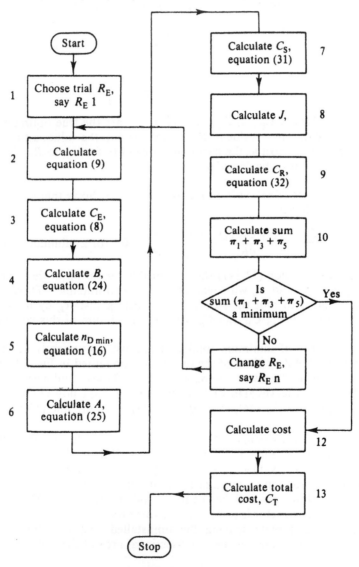

FIG. 13.14. Logic flow diagram of extraction process

Solvent recovery from the extract is to be by distillation in a perforated plate column for which the uninstalled cost per tray is $538/m². An allowable vapour velocity of 73·3 kg mole/m².h and an overall efficiency of 80% is anticipated. The solvent is the more volatile and the relative volatility is 2·5. The optimum reflux ratio is 1·25 times the minimum. The vapour-handling capacity of the still heat exchangers is 0·5 kg mole/m².h; the cost of the heat exchange surface is $16.0/m² and the cost of steam and coolant is $0·022/kg mole of distillate. The "on stream" time is 7200 h/annum and the payout

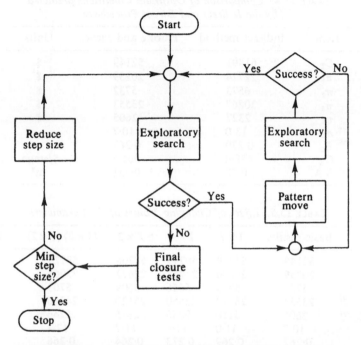

FIG. 13.15. Logic flow diagram of Hooke and Jeeves method

time is 3 years. Cost of installation, instruments, piping, etc. is estimated to be 1·3 times the uninstalled cost. Solvent losses have been estimated to be 0·05% of the circulation rate.

In terms of the analysis the relevant quantities are:

$$q_F = 5·5 \qquad S = 164·20 \qquad \beta = 1·5 \qquad Y = 3·0$$
$$m_E = 2·0 \qquad C_D = 538 \qquad \gamma = 2·5 \qquad b = 0·1$$
$$p_s = 993 \qquad G_D = 73·3 \qquad C_{hC} = 0·022 \qquad \alpha = 2·5$$
$$M_S = 102 \qquad E_0 = 0·8 \qquad l = 0·0005 \qquad C_F = 99·3$$
$$M_E = 60 \qquad G_h = 0·5 \qquad H = 7200$$
$$F = 0·07 \qquad C_h = 16·0 \qquad P = 1492$$

The above data were fed into the programs represented by the block diagrams in Figs. 13.13 and 13.14 and a comparison of the optimum conditions obtained by the two procedures are summarized in Table 13.4.

The effect of changing the most important parameters is presented in Table 13.5. These were estimated by the Hooke and Jeeves method and if a detailed interpretation of the results are required reference should be made to the original paper.

TABLE 13.4. *Comparison of Optimum Conditions predicted by the Indirect and Direct Procedures*

Item	Indirect method	Hooke and Jeeves	Units
C_T	61891	52149	\$
π_1	22724	20259	\$
π_2	6573	5732	\$
π_3	30267	23553	\$
π_5	2327	2605	\$
n_E	13·0	10·7	
E_0	0·279	0·267	
C_E	488·8	505·5	\$/stage
C.S.A.	0·32	0·331	m²

TABLE 13.5. *Effect of Changing Values of the Parameters*

Item	Basic results	1×2	1×4	$F \times 2$	$(1 \times 2)(F \times 2)$	Units
C_T	52149	54729	59749	58706	58725	\$
π_1	20259	20600	21258	22672	23066	\$
π_2	5732	5839	6056	5609	5703	\$
π_3	23553	28810	22550	25120	24710	\$
π_5	2605	5110	9885	2675	5246	\$
n_E	10·7	11·0	11·6	11·7	12·1	
E_0	0·267	0·269	0·273	0·264	0·266	
C_E	505·5	502·6	497·6	509·7	506·6	\$/stage
C.S.A.	0·331	0·328	0·323	0·334	0·332	m²
$C_{S \text{ opt}}$	0·216	0·213	0·207	0·111	0·109	kg/m³
C_R	0·425	0·433	0·449	0·208	0·212	kg/m³
$R_{E \text{ opt}}$	0·778	0·763	0·738	0·799	0·783	

13.9. DYNAMIC PROGRAMMING

This method is probably the most powerful method of optimization and has been developed by Bellman.† It finds considerable application in the optimization of multi-stage processes with no feedback such as stirred tank reactor batteries, cross-flow extractors and whole chemical processes. Thus dynamic programming formulation reduces a classical problem containing

† Bellman, R. "Dynamic Programming", Princetown University Press (1957).

N variables to N problems containing one variable, and when this technique is used in conjunction with high-speed computers it offers a powerful method of solving "multi-dimensional" constrained optimization problems encountered in chemical engineering.

In this section it is only possible to introduce the elements of dynamic programming, and therefore any reader wishing to acquire a greater knowledge of this optimization technique should consult Bellman's textbook referred to above, and also "The Optimal Design of Chemical Reactors" by Aris.† However, for an understanding of the basis of this technique certain terms and definitions must be understood and these will be considered initially. Thus consider a multi-stage chemical engineering process as, for example, a cross-flow extraction plant into which a feed of known composition is fed continuously at a constant rate. This feed enters stage 1 and mass transfer occurs to produce an extract and raffinate. The raffinate will pass to the next stage for further treatment and will subsequently emerge from stage N with considerably reduced concentration of solute, it is hoped. Extraction will occur in each stage and the extent of extraction will depend on the interfacial area (a function of agitation), on the solubility, and the temperature, etc., which are variables selected by the chemical engineer.

That is the selection of the process variables calls for decisions to be taken at each stage, and these decisions affect the change in composition of the raffinate produced by any one stage. Since the raffinate passes through each stage in turn these decisions affect the state of the final raffinate from the process. The set of all decisions taken is called the "operating policy" or simply the policy and the set of decisions that optimizes the extraction process is called "the optimal policy".

Bellman‡ stated that the essential condition of an optimal policy was that whatever the transformation that took place in the first stage of an N-stage process, the remaining stages must make use of an optimal $(N-1)$-stage policy with respect to the state of the product from the first stage, if the whole process is to operate at optimum conditions. Furthermore, by varying the operating conditions of the first stage systematically and employing the optimal $(N-1)$-stage policy for the remaining stages it will be possible to find the optimum policy for all N stages of the process. By applying these concepts to a one-stage process, then a two-stage process and so on until all the N stages have been considered, it is possible to establish the optimal policy for any number of stages in a process provided that there is no feedback.

The advantage of analysing an optimization problem in this way is that it is only necessary to vary the operating parameters of one stage at a time. For example, if there are p operating variables to each stage of a process, and the difficulty of evaluating the state of the product from a stage for one set of conditions of the p operating variables is measured by a constant A, the determination of the optimum operating conditions for the whole process for

† Aris, R. "The Optimal Design of Chemical Reactors", Academic Press (1960).
‡ Bellman, R. "Dynamic Programming", Princetown University Press (1957).

a given feed state will involve NA^p mathematical operations by dynamic programming, whereas the classical solution of the same problem would require A^{Np} mathematical operations. This constitutes considerable saving of mathematical effort and computing time.

Aris, Bellman and Kalaba[†] illustrated the principles of dynamic programming in a simple manner by evaluating the minimum energy required for the compression of a gas from P_0 to P_N in an N-stage compressor with perfect interstage cooling. Thus consider the operation of a gas compressor containing N stages with perfect intercooling so that the temperature of the gas is the same at the entrance to each stage. Let the fixed charges on the compressor depend on the number of stages, but be independent of the interstage pressures employed. Furthermore, let the cost of cooling be a fixed charge included as part of the stage cost. Then the problem becomes the determination of the interstage pressure ratio for which the energy consumption is a minimum. Thus the total energy consumed by an N-stage compressor is

$$E_N = \frac{nRT\gamma}{\gamma-1}\left[\left(\frac{p_1}{p_0}\right)^{(\gamma-1)/\gamma} + \ldots + \left(\frac{p_N}{p_{N-1}}\right)^{(\gamma-1)/\gamma} - N\right] \qquad (13.148)$$

Let

$$K = \frac{nRT\gamma}{\gamma-1}, \quad \alpha = \frac{\gamma-1}{\gamma}, \quad r_i = \frac{p_i}{p_{i-1}}$$

which on substitution into equation (13.148) gives

$$E_N = K\left(\sum_{i=1}^{N} r_i^\alpha - N\right) \qquad (13.149)$$

Also

$$r_1 r_2 r_3 \ldots r_N = \prod_{i=1}^{N} r_i = \frac{p_N}{p_0} = r \quad \text{(say)} \qquad (13.150)$$

For one stage

$$r_{1,1} = p_N/p_0$$

and

$$E_1 = K(r_1^\alpha - 1) \qquad (13.151)$$

which is the only solution.

For two stages

$$r_{2,1} = \frac{p_1}{p_0}, \quad r_{2,2} = \frac{p_N}{p_1}$$

and

$$E_2 = K\left[\left(\frac{p_1}{p_0}\right)^\alpha + \left(\frac{p_N}{p_1}\right)^\alpha - 2\right] \qquad (13.152)$$

[†] Aris, R., Bellman, R. and Kalaba, R. "C.E.P. Symposium Series", **56**, No. 31, 95 (1960).

The only variable is p_1, hence

$$\frac{dE_2}{dp_1} = K\left[\frac{\alpha}{p_1}\left(\frac{p_1}{p_0}\right)^\alpha - \frac{\alpha}{p_1}\left(\frac{p_N}{p_1}\right)^\alpha\right] = 0 \qquad (13.153)$$

$$\therefore \quad \frac{p_1}{p_0} = \frac{p_N}{p_1} \qquad (13.154)$$

The optimal policy is to choose $r_{2,1} = r_{2,2}$ and this gives an optimum energy consumption denoted by $f_2(r)$ where r from equation (13.150) is the pressure ratio across the two stages.

For three stages

$$r_{3,1} = \frac{p_1}{p_0}, \quad r_{3,2} = \frac{p_2}{p_1}, \quad r_{3,3} = \frac{p_N}{p_2}$$

The energy consumption in the first stage will be

$$E_{3,1} = K(r_{3,1}{}^\alpha - 1) \qquad (13.155)$$

but the optimal policy for the other two stages has already been determined, and the optimal energy consumption is

$$f_2\left(\frac{p_N}{p_1}\right) = f_2\left(\frac{r}{r_{3,1}}\right)$$

The optimal three-stage policy will be obtained from

$$f_3(r) = \min\left[K(r_{3,1}{}^\alpha - 1) + f_2(r/r_{3,1})\right] \qquad (13.156)$$

where $r_{3,1}$ is the only variable.

For k stages. The optimal policy for $(k-1)$ stages will have been determined and the optimal energy consumption for the latter $(k-1)$ stages will be $f_{k-1}(r/r_{k,1})$. The energy consumed in the first stage will be

$$E_{k,1} = K(r_{k,1}{}^\alpha - 1) \qquad (13.157)$$

hence the optimal k-stage policy will be determined by

$$f_k(r) = \min_{r_{k,1}}\left[K(r_{k,1}{}^\alpha - 1) + f_{k-1}(r/r_{k,1})\right] \qquad (13.158)$$

By repeated application of equation (13.158) the N-dimensional problem with no feedback has been converted into a sequence of N one-dimensional problems. Generally at this point, solution of the equations would be made by means of a digital computer, but in this particular problem an analytical solution is possible which can be found in most texts on chemical engineering thermodynamics.† The solution will show that the pressure ratio will be the same for each stage.

† Dodge, B. "Chemical Engineering Thermodynamics", McGraw-Hill, New York (1944).

To conclude this introduction to dynamic programming, consider the optimization of a cross-flow extraction process.

Example. In a cross-flow extraction process, a solute B is to be extracted from its solution in a parent liquid C by means of a solvent A which is completely immiscible with C. The feed rate to the process is to be q kg/h and W kg/h of solvent are to be distributed between the N stages of the process in order to obtain the maximum amount of extract product. The cost of pumping, regenerating, and providing make-up solvent may be taken to be a fraction (λ) of the value of the extract.

The equilibrium relation of the solute B between the phases is

$$y_n = mx_n$$

Solution.

A mass balance over the stage n in the process is

$$q(x_{n-1} - x_n) = w_n y_n \qquad\qquad \text{I}$$

Since the phases are immiscible q is constant; therefore for any stage n let

$$\frac{w_n}{q} = v_n \qquad\qquad \text{II}$$

so that $$x_{n-1} - x_n = v_n y_n \qquad\qquad \text{III}$$

Since the cost of pumping etc. of the solvent is to be a fraction λ of the value of the extract, the problem can be treated as one in which the quantity of solvent fed to each stage is such that the profit is a maximum. That is, the profit is $P_N(x_0)$

$$P_N(x_0) = q(x_0 - x_n) - \lambda \sum_1^N w_n \qquad\qquad \text{IV}$$

$$= q \sum_1^N [(x_{n-1} - x_n) - \lambda v_n] \qquad\qquad \text{V}$$

$$= q \sum_1^N v_n(y_n - \lambda) \qquad\qquad \text{VI}$$

Now let $$P_N(x_0) = \max_{v_n} \sum_1^N v_n(y_n - \lambda) \qquad\qquad \text{VII}$$

Then the functional equation corresponding to the formalism of dynamic programming is that

$$P_N(x_0) = \max_{v_1} [P_{N-1}(x_1) + v_1(y_1 - \lambda)] \qquad\qquad \text{VIII}$$

where
$$x_1 = x_0 - v_1 y_1 = \frac{y_1}{m} \qquad\qquad \text{IX}$$

and
$$P_0(x_0) = 0 \qquad\qquad \text{X}$$

From equation IX,
$$x_1 = \frac{x_0}{mv_1 + 1} \quad \text{and} \quad y_1 = \frac{mx_0}{mv_1 + 1} \qquad\qquad \text{XI}$$

Substituting into equation VII gives for stage 1,
$$P_1(x_0) = \max_{v_1}\left[v_1\left(\frac{mx_0}{mv_1 + 1} - \lambda\right)\right] \qquad\qquad \text{XII}$$

Differentiating with respect to v_1 gives
$$\frac{dP_1}{dv_1} = \left(\frac{mx_0}{mv_1 + 1} - \lambda\right) - v_1\frac{m^2 x_0}{(mv_1 + 1)^2} = 0 \qquad\qquad \text{XIII}$$

or
$$mx_0 - \lambda(mv_1 + 1)^2 = 0 \qquad\qquad \text{XIV}$$

from which, with the aid of equation XI,
$$v_1 = \sqrt{\left(\frac{x_0}{m\lambda}\right)} - \frac{1}{m}, \quad x_1 = \sqrt{\left(\frac{x_0\lambda}{m}\right)}, \quad y_1 = \sqrt{(\lambda mx_0)} \qquad\qquad \text{XV}$$

Substituting the terms in equations XV into equation VII with $N = 1$ gives
$$P_1(x_0) = x_0 - 2\sqrt{(x_0\lambda/m)} + \lambda/m \qquad\qquad \text{XVI}$$

Using the functional equation VIII on stage 2 gives
$$P_2(x_0) = \max_{v_1}\left[\left(v_1\frac{mx_0}{mv_1 + 1} - \lambda\right) + P_1(x_1)\right] \qquad\qquad \text{XVII}$$

whence, by differentiating equation XVII in a manner similar to that already presented,
$$v_1 = \sqrt[3]{\left(\frac{x_0}{m^2\lambda}\right)} - \frac{1}{m}, \quad x_1 = \sqrt[3]{\left(\frac{x_0^2\lambda}{m}\right)}, \quad y_1 = \sqrt[3]{(m^2 x_0^2 \lambda)} \qquad \text{XVIII}$$

and
$$P_2(x_0) = x_0 - 3\sqrt[3]{\left(\frac{x_0\lambda^2}{m^2}\right)} + \frac{2\lambda}{m} \qquad\qquad \text{XIX}$$

The flow ratio v_2 is obtained from the functional equation VIII by considering the second stage to be using an optimal one-stage policy with a feed of composition x_1. Then by repeating the procedure given above, it can be shown that
$$v_2 = v_1 = \sqrt[3]{\left(\frac{x_0}{m^2\lambda}\right)} - \frac{1}{m} \qquad\qquad \text{XX}$$

That is, for the two-stage policy the ratio of solvent to raffinate rate to each stage must be the same, or extending this concept to N stages, the optimal policy is obtained by dividing the total quantity of solvent available equally between the N stages.

The above problem was presented by Aris, Rudd and Amundson† and extended to more complex cross-flow extractions in which the solvent and the parent liquid in the feed to the process are partially miscible, and the equilibrium is not represented by a straight line. Solution of such problems can only be solved by computer and these authors did in fact use a Univac Scientific Computer Model 1103.

† Aris, R., Rudd, D. F. and Amundson, N. R. *Chem. Eng. Sci.* **12**, 88 (1960).

PROBLEMS

1. It is desired to produce a substance B from a raw material A in a continuous stirred tank reactor of effective volume V m³. If Q m³/s of a solution of A of concentration C_0 is fed to the empty reactor, and the chemical reaction of A is represented by

$$A \underset{k_2}{\overset{k_1}{\rightleftharpoons}} B \xrightarrow{k_3} C$$

in which all the reactions are first order, show that the number of moles of B in the initial discharge from the reactor is given by the solution of the differential equation

$$\frac{d^2 N_B}{dt^2} + P\frac{dN_B}{dt} + RN_B = C$$

where
$$P = k_1 + k_2 + k_3$$
$$R = k_1 k_3$$
$$C = QC_0 k_1$$

2. Find the general solution of the differential equation

$$\frac{dy}{dx} + 2y \tan x = 4 \sin x$$

If $y = 0$ when $x = \pi/3$, show that y has a maximum value of $\frac{1}{2}$.

3. Solve the differential equation

$$x^3 \frac{d^2 y}{dx^2} + x^2 \left(\frac{dx}{dy}\right)^2 - y^2 = 0$$

with the help of the boundary conditions:

$$\text{at} \quad x = 1, \quad y = 2, \quad \frac{dy}{dx} = -1$$

4. Solve the equation

$$\frac{d^2 y}{dx^2} + 2\frac{dy}{dx} + 2y = e^{-x} \cos x$$

(*Birmingham*, 1962.)

549

5. Solve the equation

$$x^2 \frac{d^2 y}{dx^2} + 2x \frac{dy}{dx} + 2y = x + \ln x$$

(*Birmingham*, 1960.)

6. It is proposed to produce X, Y, and Z from A and B by chemical reaction on a semi-batch basis. That is, 20 moles of pure A is charged into the reactor and B is added at the rate of 0·75 moles/min. If all the reactions are second order and are represented stoichiometrically by the equations

$$A + B \longrightarrow X \tag{1}$$

$$X + B \longrightarrow Y \tag{2}$$

$$Y + B \longrightarrow Z \tag{3}$$

prepare a plot of the composition of the reaction mixture as a function of the number of moles of B reacted. The specific reaction rate constants are:

for reaction 1, $k_1 = 1\cdot8 \times 10^{-4}$ m³/mole.min

for reaction 2, $k_2 = 0\cdot3 \times 10^{-4}$ m³/mole.min

for reaction 3, $k_3 = 0\cdot1 \times 10^{-3}$ m³/mole.min

7. Product from a plant flows at a rate of 2 kg/s into a storage vessel holding 80 000 kg of material and is then pumped to another process. Owing to a mishap in the plant, the material flowing to storage contains 1% of a contaminant.

(*a*) If the flow from storage is 2 kg/s and the contamination of supply lasts for 1 h, what is the peak contamination in the discharge?

(b) What is the peak contamination in (a) if the discharge is reduced to 1 kg/s?

(c) For how long would the 1% contamination of supply have to continue before the discharge contamination became 0·8% with flows as in (a)?

(d) In (a), if the supply contamination ceased completely after 1 h, what further time would elapse before the discharge contamination fell to 0·01%?

(e) In (a), what supply contamination could be tolerated for 1 h if discharge contamination could be allowed up to 1%?

Assume complete mixing in the storage vessel.

8. A component A is absorbed from a gas phase by a liquid containing a constituent B with which it reacts chemically by a pseudo first order reaction. The rate of absorption can be assumed to depend on both the rate of chemical reaction and the rate of molecular diffusion through the

laminar film of liquid of thickness δ. Show that when the concentration of A at a liquid depth δ is zero, the concentration profile of A through the film is

$$C_A = \frac{C_{A,i}\sinh[\alpha(\delta-x)]}{\sinh(\alpha\delta)}$$

where C_A is the concentration at depth x, and $\alpha = \sqrt{(k/D)}$.

Show also that the rate of mass transfer through the surface is

$$N_A = \frac{DC_{A,i}}{\delta}\left[\frac{\alpha\delta}{\tanh(\alpha\delta)}\right]$$

where $[\alpha\delta/\tanh(\alpha\delta)]$ is a dimensionless quantity known as the Hatta number.

9. (a) Two thin wall pipes of 2 cm external diameter have flanges 1 cm thick and 10 cm diameter on the ends joining them together. If the conductivity of the flange metal is k W/m °K, and the exposed surfaces of the flange lose heat to the surroundings which are at T_1 °K by means of a heat transfer coefficient h W/m² °K, show that the equation giving the temperature distribution in the flange is

$$100k\left(r\frac{d^2 T}{dr^2}+\frac{dT}{dr}\right) = hr(T-T_1)$$

where r is the radial distance coordinate in cm.

(b) If, in addition, the circular faces of the flanges are thermally insulated (i.e. $h = 0$) and the flanges only lose heat through the rim where $h = 20$, solve the simplified equation and determine the rim temperature if the pipe temperature is 370 °K, $T_1 = 290$ °K and $k = 60$.

10. A second order irreversible chemical reaction between two materials A and B takes place in a single stirred tank reactor according to

$$A+B \longrightarrow C+D$$

Two feed streams enter the tank each at a rate of R kg mole/h, the one stream contains $4C_0$ mole fraction of A, and the other contains $2C_0$ mole fraction of B. Material leaves the tank continuously at a rate $2R$ kg mole/h so that the tank always contains M kg moles and V m³ of material. The specific reaction rate constant is k m³/kg mole h.

When the reactor is closed down, both feeds are stopped, the vessel remains full, and the reaction goes to completion leaving C_0 mole fraction of A unreacted. If the reactor is started from this condition, by turning on both feed streams simultaneously, perform the following calculations:

(a) Show that the difference in concentration between the two reactants A and B in the tank remains constant during start-up.

(b) Determine how the concentration of B in the tank varies with time during start-up.

(*Note*: In part (b) a quadratic equation should arise which will not factorize; denote the roots by α and β and continue the calculation in terms of α and β.)

(*M.Sc. Birmingham*, 1962.)

11. As liquid flows across any plate of a distillation column its composition (x) changes from the entry concentration (x_0) to the exit concentration (x_1). The composition at any point on the plate is influenced by the passage of the stripping gas at a rate G, the bulk flow of the liquid at a rate L, and mixing on the plate which can be expressed in terms of an eddy diffusivity (D_E). Assuming a constant Murphree point efficiency given by

$$E_{MV}{}^* = \frac{y - y_1}{y^* - y_1}$$

and a straight-line equilibrium curve given by $y^* = mx + b$ show that the liquid composition satisfies the equation

$$D_E \frac{d^2 x}{dz^2} - V \frac{dx}{dz} - \frac{mGV}{Lz_1} E_{MV}{}^*(x - x^*) = 0$$

where z is the distance measured along the plate from the inlet weir,

      z_1 is the distance between the weirs,

      V is the linear velocity of the liquid,

      x^* is the liquid composition in equilibrium with the entering gas which is constant across the plate.

Allowing for diffusion in the entry section where no gas flows, find the liquid composition at all points on the plate.

(*Andrew, S. P. S., Hutchinson, R. C. and Mackay, R. W.*   Chem. Eng. Sci. **17**, 541, 1962.)

12. A surge tank of 2 m³ capacity initially contains 1 m³ of liquid. There are two inlet streams to the tank and one outlet stream. The first inlet stream is steady at 0·2 m³/min with a temperature of 50 °C, whereas the second inlet stream at a temperature of 100 °C is subject to surging on a 2 min cycle. During the first minute, the flow rate is steady at 0·2 m³/min; and during the second minute the flow stops completely; it then continues to start and stop at 1 min intervals. The outlet stream is always steady at 0·3 m³/min. If the tank is well stirred and the temperature in the outlet stream is always the same as that in the tank, over what range does this outlet temperature vary when the system has settled to its equilibrium cycle?

13. Two vertical cylindrical tanks of 3 m diameter and 2 m diameter are joined at the base by a horizontal pipe 0·75 m long and 1 cm diameter. The 3 m diameter tank also has an outlet at the base consisting of a horizontal tube 0·6 m long and 1 cm diameter.

   If the larger tank is filled with oil to a depth of 6 m whilst the smaller tank is empty, and both tubes are open simultaneously, prove that the maximum depth of oil occurring in the smaller tank will be 2·4 m. Assume that the flow in both tubes is laminar and neglect kinetic losses.

14. Derive an expression giving the true average temperature difference, in terms of the terminal temperatures and fluid properties, which is applicable to a split flow heat exchanger.

   (*This problem has been solved by D. Q. Kern and C. LeRoy Carpenter, Chem. Eng. Prog. 47, 211, 1951.*)

15. Is the series

$$\frac{1}{2} + \frac{1.3}{2.4} + \frac{1.3.5}{2.4.6} + \dots$$

   convergent or divergent?

16. Although the series $\sum 1/n$ is divergent, show that the series obtained from it by removing all terms in which the digit 0 occurs is convergent, i.e.

$$1 + \tfrac{1}{2} + \tfrac{1}{3} + \dots + \tfrac{1}{9} + \tfrac{1}{11} + \tfrac{1}{12} + \dots + \tfrac{1}{19} + \tfrac{1}{21} + \dots$$

17. "Log-mean" temperature differences are frequently used and defined by

$$T_{\mathrm{m}} = \frac{T_2 - T_1}{\ln(T_2/T_1)}$$

   What is T_{m} when $T_2 = T_1$?
   Show further that the log-mean value tends to the arithmetic mean value as $T_2/T_1 \to 1$.

18. Given the differential equation

$$x\frac{\mathrm{d}^2 y}{\mathrm{d}x^2} + k\frac{\mathrm{d}y}{\mathrm{d}x} + y = 0$$

   (a) Use the method of Frobenius to find the roots of the indicial equation and determine the recurrence relation.
   (b) State without further calculation what form the complete solution takes for all real values of k.
   (c) Complete the solution for the case when $k = \tfrac{1}{2}$.

   (*Birmingham, 1960.*)

19. Use the method of Frobenius to solve the equations:

(a)
$$x\frac{d^2 y}{dx^2} + (1-2x)\frac{dy}{dx} - 2y = 0$$

(b)
$$x(1-x)\frac{d^2 y}{dx^2} + (3-7x)\frac{dy}{dx} - 9y = 0$$

20. A constant temperature vessel is supported on a truncated cone which is in turn resting on another surface which is at constant temperature. The small end of the support is of radius 4 cm and is at 50 °C, the temperature of the vessel; whilst the wide end of the support is of radius 5 cm and is at 10 °C, the temperature of the surface. The vertical height of the support is 10 cm and its thermal conductivity is 70 W/m °K.

The curved surface of the support loses heat to the surroundings at a temperature 10 °C according to a heat transfer coefficient of 10 W/m² °K.

By assuming that the temperature is constant across any circular cross-section of the support and by measuring all temperatures relative to the temperature of the base, calculate the first five terms in the series giving the temperature distribution within the support.

21. (a) Use the substitution $x = a e^{it}$ to solve

$$\frac{d^2 y}{dt^2} - a^2 e^{2it} y = 0$$

(b) Use the substitution $y = wx^{-\frac{1}{2}}$ to solve

$$\frac{d^2 y}{dx^2} + \frac{2}{x}\frac{dy}{dx} + k^2 y = \frac{p^2}{x^2} y$$

22. Show that the nth roots of unity can be expressed as $1, z, z^2, \ldots, z^{n-1}$, where $z = \cos(2\pi/n) + i\sin(2\pi/n)$. Hence, or otherwise, prove that

$$z^a + z^{2a} + \ldots + z^{na} = \begin{cases} 0 & \text{if } n \text{ is not a factor of } a \\ n & \text{if } n \text{ is a factor of } a \end{cases}$$

where a is any integer.

23. By using the known trigonometrical identities, and equations (4.42), (4.43), and (4.44),
(a) express $\tanh(a+b)$ in terms of $\tanh a$ and $\tanh b$,
(b) express $\sinh a$ in terms of $\cosh 2a$,
(c) prove that

$$\sinh^{-1} a + \sinh^{-1} b = \sinh^{-1}[a\sqrt{(1+b^2)} + b\sqrt{(1+a^2)}]$$

24. By putting $z = e^{i\theta}$, show that

$$\int \frac{dz}{\sqrt{z}} = -4e^{i\phi/2}$$

where the integral is taken around a unit circle centred at the origin, starting from the point $e^{i\phi}$.

If two circuits of the origin are permitted by allowing θ to vary from ϕ to $\phi + 4\pi$, show that the integral is zero for all values of ϕ.

(This example illustrates the importance of making $z^{-\frac{1}{2}}$ single valued by taking a branch cut at $\theta = \phi$.)

25. Integrate the following functions around any finite curve enclosing the origin.

(a) $\dfrac{\sin z}{z^2}$,   (b) $\dfrac{\cos z}{z^2}$

26. By integrating $e^{imz}/(1+z^2)$ around a suitable contour, show that

$$\int_0^\infty \frac{\cos mx}{1+x^2} dx = \frac{\pi}{2} e^{-m}$$

for all positive real values of m.

27. Evaluate the residue of $e^z/(z \sin z)$ at the origin.

28. Integrate the function $e^{iz} z^{a-1}$ around a contour consisting of the parts of two circles centred at the origin which lie in the first quadrant, and the parts of the positive real and imaginary axes joining the two arcs. By allowing the one circle to increase in radius indefinitely and the other to shrink to the origin, show that

$$\int_0^\infty x^{a-1} \cos x \, dx = \Gamma(a) \cos \tfrac{1}{2}\pi a$$

provided that $0 < a < 1$.

29. By raising the trigonometrical identity, $\sin 2\theta = 2\sin\theta\cos\theta$, to the $(2n-1)$th power and integrating with respect to θ over the range 0 to $\frac{1}{2}\pi$, prove the duplication formula for the gamma function.

$$\Gamma(2n) = 2^{2n-1} \pi^{-\frac{1}{2}} \Gamma(n) \Gamma(n+\tfrac{1}{2})$$

30. Prove by substitution that the function

$$I(x) = \int_0^\pi \cos(n\phi - x\sin\phi) \, d\phi$$

satisfies Bessel's equation of order n provided that n is an integer or zero. Hence prove that

$$I(x) = \pi J_n(x)$$

31. Show that

$$\int_0^\infty \frac{\cos ax - \cos bx}{x}\,dx = \ln\left(\frac{a}{b}\right)$$

(Introduce the function $e^{-\alpha x}$ and use the method of Section 5.4.4.)

32. Show that the Laplace transform of:

(i)
$$J_0(t) = \frac{1}{(s^2+1)^{\frac{1}{2}}}$$

(ii)
$$\cos \omega t = \frac{s}{s^2+\omega^2}$$

(iii)
$$e^{-at}\sin \omega t = \frac{\omega}{(s+a)^2+\omega^2}$$

33. Invert the following Laplace transforms:

(i) $\dfrac{s-1}{s(s+2)}$, (ii) $\dfrac{sA+B}{(s+i\omega)(s-i\omega)}$, (iii) $\dfrac{1}{(s^2+4)(s^2+1)}$

34. Prove by integration that $a^n/s(s+a)^n$ is the Laplace transform of

$$1 - e^{-at}\left[1 + at + \frac{a^2 t^2}{2!} + \ldots + \frac{a^{n-1} t^{n-1}}{(n-1)!}\right]$$

35. The diagram illustrates a hydraulic device used in automatic control mechanisms. It operates as follows. The movable cylinder A is completely filled with oil, and when a sudden change in pressure occurs at

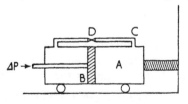

the piston B, oil is forced via the line C through the valve D which has a very narrow aperture so that the flow of oil is restricted. Consequently the cylinder A moves against the spring S, but as the oil passes through the valve the compressed spring forces the cylinder to return to its original position.

For a pressure change ΔP and corresponding displacement x show that if the cross-sectional area of the piston is a, the spring constant is k and the flow of oil through the valve is laminar, the differential equation describing the system is

$$\frac{d^2 x}{dt^2} + \left(\frac{a}{mK}\right)\frac{dx}{dt} + \frac{k}{m}x = 0$$

where K is the pressure change-flow proportionality constant,
  m is the effective mass of the system.
Solve this equation using the initial conditions that the displacement of the cylinder is initially x_0.

36. A stirred tank reactor of effective volume V m^3 containing a solution of a reactant A, of concentration $C_{A,0}$ is heated to such a temperature that the pseudo first order reaction

$$A \longrightarrow B$$

takes place. When the chemical reaction is initiated a solution of A of concentration $C_{A,I}$ is fed continuously to the reactor at the rate of q m^3/min. If the specific reaction rate is k min^{-1} and the vessel is assumed to contain a perfect stirrer, develop an expression giving the concentration of A in the effluent as a function of time.

37. A liquid feed stream of heat capacity C_p is passed through a small bore cylindrical pipe of radius a m at the rate of R kg/h. The pipe wall is maintained at a temperature of T_1 °C and the entering liquid at T_0 °C. Determine the temperature of the liquid as a function of time and distance from the inlet if the pipe is initially full of hot liquid at T_1 °C when the flow is suddenly started.
    The heat transfer coefficient is a function of distance x m along the pipe from the inlet and is expressed as

$$U = kx^{-\frac{1}{2}}$$

where k is constant.
    Conduction within the liquid can be neglected and radial variation of temperature can be assumed to be zero.

38. The temperature of a liquid flowing at a rate q through an agitated electrically heated vessel of volume V is to be controlled by placing a temperature-detecting element in the liquid leaving the vessel. The detector is connected to a controller which regulates the supply of energy to an immersion heater in such a way that the heat supplied is proportional to the temperature difference $(T_m - T)$; where T_m is the constant safe maximum temperature of the heating element, and T is the temperature of the liquid. The immersion heater supplies heat to the liquid at a rate $Q_c = U_c A(T_c - T)$, where A is the heat transfer area, U_c is the overall heat transfer coefficient and T_c is the surface temperature of the coil.

If the temperature of the entering liquid suddenly falls from T_0 to a new steady-state temperature T_1, develop an equation to show the variation of liquid temperature as a function of time, during the period of transient behaviour.

The effective mass of the immersion heater can be taken to be m and the specific heat of the heater material s, the heat capacity of the liquid C_p and its density ρ. Agitation in the vessel can be assumed to be perfect.

39. The denominator of the complete transfer function of a control system can be represented by

$$6s^4 + 5(\alpha - 1)s^3 + 5\alpha s^2 + (\alpha^2 - 2\alpha + 1)s - (\alpha - 6) = 0$$

For what range of values of α is the system stable?

40. Plot the Nyquist diagram and discuss the stability of the system

$$\bar{G}_1(s)\,\bar{G}_2(s) = \frac{-2}{(s-1)(4s+1)}$$

41. Prove the following by vector methods:

(a) In any skew quadrilateral, the mid-points of the sides are at the corners of a parallelogram.

(b) A cube has twelve edges, six of which form the corners at the ends of one diagonal. Prove that the mid-points of the other six edges form the corners of a regular hexagon.

42. (a) Find the unit vector perpendicular to the vectors $2i + 3j + 4k$, $i + 2j - 2k$. Also determine the angle between these two vectors.

(b) Prove that the vectors $i + 3j - 2k$, $i - 5j + 3k$, and $2i - 2j + k$ are coplanar.

43. (a) Show that

$$(A \wedge B).(A \wedge C) = A^2(B.C) - (A.B)(A.C)$$

(b)   If $(A \wedge B) \wedge (A \wedge D) = A$, what is $(A \wedge B) \wedge (B \wedge D)$?

44. If A, B, and C are three independent vectors, find the values of x, y, and z which make

$$M \wedge N = xA + yB + zC$$

Hence show that

$$M \wedge N = \begin{vmatrix} A & A.M & A.N \\ B & B.M & B.N \\ C & C.M & C.N \end{vmatrix} \; [A, B, C]$$

45. If r is the position vector, show that:

(a) $\operatorname{div} r = 3$,   (c) $\operatorname{div} r^n r = (n+3) r^n$

(b) $\operatorname{curl} r = 0$,   (d) $\operatorname{curl} r^n r = 0$

46. Use the theorems connecting line, surface, and volume integrals to establish the results

(i) $$\int_c (A \wedge \phi) \cdot dr = \int_s (\nabla \cdot \phi) A \cdot dS - \int_s (A \cdot \nabla) \phi \cdot dS$$

(ii) $$\int_s U\phi \cdot dS = \int_\sigma U\nabla \cdot \phi \, d\sigma + \int_\sigma \phi \cdot \nabla U \, d\sigma$$

(iii) $$\int_\sigma \psi \cdot \nabla \wedge \phi \, d\sigma = \int_s (\phi \wedge \psi) \cdot dS + \int_\sigma \phi \cdot \nabla \wedge \psi \, d\sigma$$

where A is a constant vector, ϕ and ψ are variable vectors and U is a variable scalar.

(*Birmingham (Maths)*, 1953.)

47. In a porous medium, the continuity equation can be written

$$\varepsilon \frac{\partial \rho}{\partial t} = -\nabla \cdot \rho u$$

where ε is the porosity; and the equation of motion (Darcy's Law) can be written

$$u = -\frac{\kappa}{\mu} \nabla p$$

where gravity is neglected, κ is the permeability, and μ the viscosity.
(a) Show that for an incompressible fluid, $\nabla^2 p = 0$.
(b) Show that for isothermal flow of a compressible gas,

$$\frac{2\varepsilon\mu\rho_0}{\kappa} \frac{\partial \rho}{\partial t} = \nabla^2 \rho^2$$

48. Show that the equation of fluid motion (7.126) can be expressed in the alternative forms:

(a) $$\frac{\partial u}{\partial t} - u \wedge \zeta = -\nabla\chi + \nu\nabla^2 u$$

(b) $$\frac{\partial u}{\partial t} - u \wedge \zeta = -\nabla\chi - \nu\nabla \wedge \zeta$$

where $$\chi = p/\rho + \tfrac{1}{2}u^2 + \Omega$$

and $$F = -\nabla\Omega$$

49. Two concentric spherical metallic shells of radii a and b cm $(a < b)$ are separated by a solid of thermal diffusivity α cm²/s. The outer surface of the inner shell is maintained at T_0 °C and the inner surface of the outer shell at T_1 °C. Derive the differential equation governing the unsteady-state temperature distribution in the solid as a function of time and radial coordinate.

Show that the solution takes the form

$$T = \frac{T_1 b - T_0 a}{b - a} + \frac{ab}{r} \frac{T_0 - T_1}{b - a} + \sum_{n=1}^{\infty} \frac{B_n}{r} \sin[\beta(r-a)]\, e^{-\beta^2 \alpha t}$$

where $\beta = n\pi/(b-a)$. Show how B_n can be determined from any initial temperature distribution.

50. A cylindrical block of metal at a uniform temperature T_0 °C has it circular faces thermally insulated. The temperature of the curved surface is suddenly raised to and maintained at T_1 °C. Prove that the temperature T of any point in the block at any time t is given by

$$\frac{T_1 - T}{T_1 - T_0} = \frac{2}{a} \sum_{n=1}^{\infty} \frac{J_0(\alpha_n r)}{\alpha_n J_1(\alpha_n a)}\, e^{-\alpha_n^2 k t}$$

where r is the radial coordinate, a is the radius of the cylinder, k is the thermal diffusivity, and α_n are the positive roots of

$$J_0(\alpha_n a) = 0$$

(*M.Sc., Birmingham*, 1962.)

51. If a liquid is flowing through a fixed bed at a rate v cm/s when a pulse of a tracer is added, show that the later distribution of the tracer is given by the solution of

$$E \frac{\partial^2 C}{\partial x^2} - v \frac{\partial C}{\partial x} = \frac{\partial C}{\partial t}$$

where E is the mixing coefficient which operates in the axial direction (x) only. Removing x by the substitution $z = x - vt$, and assuming an infinite bed length, solve the above equation.

Show that the maximum concentration of tracer passing a down-stream observation station arrives earlier than the residence time by an amount E/v^2.

(*See J. Carberry and R. H. Bretton*, A.I.Ch.E. **4**, 367, 1958.)

52. Considering two dimensional flow and neglecting the inertia terms in the Navier–Stokes equation, show that the change of variables given by

$$u = -\frac{\partial \phi}{\partial x} - \frac{\partial \psi}{\partial y}$$

$$v = -\frac{\partial \phi}{\partial y} + \frac{\partial \psi}{\partial x}$$

$$p = p_0 - g\rho y + \rho \frac{\partial \phi}{\partial t} - F_y \rho y$$

yields the equations

$$\frac{\partial^2 \phi}{\partial x^2} + \frac{\partial^2 \phi}{\partial y^2} = 0$$

$$\frac{\mu}{\rho}\left(\frac{\partial^2 \psi}{\partial x^2} + \frac{\partial^2 \psi}{\partial y^2}\right) = \frac{\partial \psi}{\partial t}$$

Show that a solution of these equations takes the form

$$\phi = A e^{-Kv+nt}\cos Kx + My$$

$$\psi = B e^{-mv+nt}\sin Kx$$

if

$$m = \sqrt{[K^2 + (\rho n/\mu)]}$$

*(This method has been used to describe flow in fluidized beds by W. J. Rice and R. H. Wilhelm, A.I.Ch.E.J. **4**, 423, 1958.)*

53. Liquid enters tangentially a swirl chamber of radius R_3 and emerges through an orifice of radius R_2, which is considerably less than R_3, after passing down a converging cone whose vertex subtends an angle 2α. Assume that the main body of the liquid is moving with velocity Ω/r in circles around an axis where r is the distance from the axis, Ω is constant, and the velocity component along the axis is negligible. Centrifugal forces will generate a pressure gradient along the wall of the cone which will produce a current in the boundary layer towards the orifice at the point of the cone. Use the Pohlhausen cubic velocity profiles to determine how the boundary layer thickness varies along the wall of the cone.

*(The solution to this problem was published by G. I. Taylor, Quart. Journal Mech & Applied Maths. **3**, 129, 1950.)*

54. The steady laminar flow of a liquid through a heated cylindrical pipe has a parabolic velocity distribution if natural convection effects and variation of physical properties with temperature are neglected. If the fluid entering the heated section is at a uniform temperature (t_1), and the wall is maintained at a constant temperature (t_w), develop Graetz's solution by neglecting the thermal conductivity in the axial direction.

*(The solution is reported by T. B. Drew, Trans. Am. Inst. Chem. Engrs. **26**, 26, 1931.)*

55. Assuming that Stokes' approximation is valid and that the interfacial tension (γ) varies according to

$$\gamma = \alpha \cos \theta + \beta$$

determine the fall velocity of a spherical droplet of a material of viscosity μ_1 through a medium of viscosity μ_2, where θ is the polar angle measured from the front of the droplet.

Show that there is only motion inside the droplet if

$$a^2 g \Delta \rho > 3\alpha$$

where a is the radius of the drop and $\Delta \rho$ is the density difference.

56. The sudden closure of a valve generates a pressure wave within the liquid flowing in the pipe leading to the valve. The passage of this wave causes compression of the liquid and expansion of the pipe. Show that the velocity of the liquid and the pressure are related by the equations

$$-\frac{\partial p}{\partial x} = \frac{\rho}{g} \frac{\partial v}{\partial t}$$

$$\frac{\partial^2 p}{\partial t^2} = c^2 \frac{\partial^2 p}{\partial x^2}$$

where c is the velocity of propagation of the pressure wave.

If it can be assumed that a uniform pipe of length L connects a reservoir at $x = 0$ to the valve at $x = L$, show that

$$p(x, t) = p_0 + \frac{4c\rho v_0}{\pi g} \sum_{n=0}^{\infty} \frac{(-1)^n}{(2n+1)} \sin(2n+1)\frac{\pi x}{2L} \sin(2n+1)\frac{\pi ct}{2L}$$

(*Rich, G. R.* Trans. A.S.M.E. **67**, 361, 1945.)

57. A liquid at a uniform temperature falls as a thin film down a flat vertical surface. Determine the velocity distribution and the steady-state thickness of the film. If a short section of the wall is maintained at a higher temperature than the liquid, determine the average heat transfer coefficient in the heated section. The velocity distribution can be simplified in the thin layer where temperature gradients are important.

58. Given that the values for $\sinh x$ for $x = 0.1, 0.15, 0.2, 0.25, 0.30$, and 0.35 are $0.100, 0.150, 0.201, 0.253, 0.305$, and 0.357; estimate the value of $\sinh x$ when $x = 0.328$.

59. Solve the difference equations

(i) $$y_{n+3} - 6y_{n+2} + 11y_{n+1} - 6y_n = 0$$

(ii) $$y_n y_{n+2} = y_{n+1}^2$$

(iii) $$y_{n+4} - 9y_{n+3} + 30y_{n+2} - 44y_{n+1} + 24y_n = 4^n a$$

(iv) given that $y_0 = 0$, $y_1 = 3$, $y_2 = 6$, $y_3 = 36$,

solve $y_{n+2} + 2Ay_{n+1} + By_n = 0$

60. In the production of crude phosphoric acid (29% P_2O_5), phosphate rock is treated with sulphuric acid in an agitated vessel. The suspension, which is formed, is allowed to settle in a thickener of the Dorr type. The mud which is withdrawn from the thickener contains acid which must be recovered by counter-current washing with water in a stripping system consisting of a cascade of Dorr thickeners. If the concentration of the acid on the sludge discharged from the final thickener, designated number 1, is not to exceed a certain concentration C_1 kg P_2O_5 per kg water, the mud passing through the system contains f kg of liquid per kg of dry mud and the acid concentration in the liquid overflow from each thickener is the same as that attached to the mud leaving the thickener, derive an equation to predict the number of thickeners required for a given duty.

61. G kg/h of a carrier gas containing Y_0 kg solute A per kg of carrier gas is fed into the base of a plate column containing N ideal plates and contacted with L kg/h of solute-free liquid containing X_{N+1} kg solute A per kg of solute-free liquor. If the concentration of A in the exit gas and liquid streams are Y_N and X_1 respectively, show that the concentration of A in the gas phase leaving any plate n in the column is given by

$$Y_n = \left(\frac{Y_0 - HX_1}{1 - HG/L}\right)\left(\frac{HG}{L}\right)^n + \left(HX_{N+1} - \frac{HG}{L}Y_N\right)\bigg/\left(1 - \frac{HG}{L}\right)$$

where H is Henry's Law constant expressed by $Y_N = HX_N$.

62. Calculate the number of plates required to absorb 99% mole ketene from 58 kg moles/h of gas containing 4·36% ketene by volume using 59 kg moles/h of glacial acetic acid. The plate efficiency of a 1 m diameter tower, containing 0·7D weirs and therefore plate hold-up of 0·85 kg moles liquid, is 40%. The pseudo first order reaction rate constant for the chemical absorption process may be taken to be 0·075 s^{-1} and the equilibrium relationship

$$Y^* = 2X$$

Y^* is the kg moles ketene/kg mole carrier gas in equilibrium with X kg moles ketene/kg mole of acetic acid in the liquid.

(*Problem taken from* "A Problem of Chemical Engineering Design",
by G. V. Jeffreys; published by Institution of Chemical Engineers,
London, 1962.)

63. Butyl acetate is to be produced in a battery of continuous stirred tank reactors operating in series by reacting 1000 kg/h of butanol with 163 kg/h of glacial acetic acid containing sufficient sulphuric acid to catalyse the reaction at 100 °C. Under these conditions the rate of reaction can be expressed by the equation

$$r = kC_A^2$$

where k is $1\cdot05$ m³/kg mole h and C_A is the concentration of acetic acid in kg moles/m³. If the effective volume of each tank is $0\cdot3$ m³ and the density of the reaction mixture is assumed to be constant at 770 kg/m³, estimate the number of reaction vessels required if the concentration of the acetic acid in the final discharge is not to exceed 48 kg/m³.

64. A reactant A is to be converted into a product B in a battery of N continuous stirred tank reactors of total volume V m³. If the feed rate is q m³/min and the concentration of A is $C_{A,0}$ show that the production of B is a maximum when all the tanks are the same size. The rate of reaction can be taken to be first order and the battery is to operate isothermally.

65. 4000 kg/h of a 5% by weight solution of acetaldehyde in toluene is to be treated with 2500 kg/h of water in a six theoretical stage extraction column operating under counter-flow conditions. If the concentration of the feed conditions is suddenly changed to a 3% by weight solution, how long will it take for the extractor to settle down to steady-state operations under the new conditions?

The equilibrium relation is $Y = 2\cdot2X$

where Y = kg acetaldehyde per kg of water

X = kg acetaldehyde per kg of toluene.

The hold-up per stage can be assumed to be constant at

375 kg toluene,   150 kg water

66. A block of metal $30 \times 30 \times 5$ cm has its rectangular faces thermally insulated and its vertical square faces losing heat to the surroundings at 16 °C by means of a heat transfer coefficient. It can be assumed that the temperature in the block is uniform and that the properties of the metal are:

specific heat $c = 0\cdot4$ kJ/kg °K

density $\rho = 6700$ kg/m³

The following table gives the metal temperature as a function of time:

Time (min)	0	5	10	15	20	25	30	35	40
Temperature (°C)	94	86	79	72	66	61	56	52	49

Determine the heat transfer coefficient.

In addition to the experimental readings, the given values are subject to the following errors:

$c, \pm1\%$, $\rho, \pm0\cdot1\%$,   all dimensions, ± 1 mm

Assuming that the thermocouples are accurate, how accurately is the heat transfer coefficient determined?

(*Birmingham*, 1961.)

67. Use the method of averages to find the best curve of the type:

$$y = A \sin(x + B)$$

which fits the following values (A and B are constant):

x	0°	30°	60°	90°	120°	150°
y	0·944	1·242	1·208	0·850	0·264	−0·392

68. Use the method of least squares to fit the best equation of the type

$$Nu = aPr^n$$

to the following data.

Pr	Nu	Pr	Nu	Pr	Nu
0·46	24·8	10·0	84·5	32	140
0·53	26·5	17·7	115	31·6	127
0·63	28·5	18·6	115	70·3	189
0·74	30·0	25·3	150	93	245
4·2	60·3	41·0	170	185	315
5·6	69·0	37·0	165	340	380
3·0	58·4	58·5	193	590	480
5·0	70·7	95·0	245	55	195

(The information above was collected from various sources by W. L. Friend and A. B. Metzner, A.I.Ch.E.J. 4, 393, 1958.)

69. The top product leaving a batch distillation column is being continuously reprocessed so that it is difficult to measure directly the amount produced. However, the flow rate can be continuously measured and samples are taken at hourly intervals giving the following results:

Time (hours)	0	1	2	3	4	5	6	7	8
More volatile component (%)	90	92	91	91	89	87	84	80	75
Flow rate (kg/min)	60	62	63	62	62	60	59	57	57

(a) What is the total amount produced during the period of 8 h?
(b) What proportion of this total is the more volatile component?

70. The reaction rate constant for the decomposition of a substituted dibasic acid has been determined at various temperatures as follows:

T (°C)	50·0	70·1	89·4	101·0
$k \times 10^4$ (h^{-1})	1·08	7·34	45·4	138

Use the method of least squares to determine the activation energy (E) in the equation $k = A e^{-E/RT}$ where T is measured in degrees Kelvin.

71. Calculate the first four positive roots of

$$x \tan x = 2$$

to four decimal places.

72. A closed cylindrical vessel 12 cm long has a partition perpendicular to the axis 4 cm from the one end. The larger section is filled with pure nitrogen and the smaller section with pure oxygen at the same pressure. At time $t = 0$, the partition is removed to allow diffusion to commence.

If the diffusivities of both gases in any mixture of the two are equal to 0.2 cm²/s, use a numerical method to estimate the distribution of the oxygen after about half a minute.

(Neglect density differences and convection.)

(Birmingham, 1968.)

73. Oil of density 800 kg/m³ and viscosity 0.08 Ns/m² is fed into an open channel 0.12 m wide, inclined on a gradient of 1 in 98.1. Sufficient oil is fed to the channel to maintain a steady-state depth of 20 mm. It may be assumed that the oil is Newtonian and that the flow is laminar without ripples.
    (a) What would be the surface velocity of the same depth of oil if the channel were infinitely wide?
    (b) Use a numerical method to determine the surface velocity distribution in the given channel.

74. A long, thin-walled metal pipe of 0.4 m diameter is encased symmetrically in a square-section concrete pillar of side 1 m. If liquid at 100 °C flows continuously through the pipe and the surface temperature of the concrete is 30 °C, determine the rate of heat loss per unit length of the pipe as a multiple of the thermal conductivity of the concrete.

Assume steady state and no longitudinal temperature variation.

75. A mathematical model for the washing of a filter cake has been proposed by M. T. Kuo (*A.I.Ch.E.J.* 6, 566, 1960). It is assumed that the wash liquor is in plug flow through the pores of the cake and that a stagnant film of filtrate remains on the pore walls. A mass transfer coefficient governs the passage of solute across the boundary between the two liquid layers. Show that the process is described by the equations:

$$A_2 \frac{\partial c_2}{\partial t} = hS(c_1 - c_2)$$

$$A_1 \frac{\partial c_1}{\partial t} = hS(c_2 - c_1) - A_1 U \frac{\partial c_1}{\partial x}$$

for $0 < x < L$, $t > x/U$.

- U velocity of the wash liquor
- L length of the pore channel
- h mass transfer coefficient
- S transfer area per unit length of pore
- A_1 cross-sectional area occupied by wash liquor
- A_2 cross-sectional area occupied by filtrate
- C_1 concentration of solute in wash liquor
- C_2 concentration of solute in filtrate

Calculate the concentration of the exit wash liquor as a function of time for the following numerical values.

$$U = 0 \cdot 076 \text{ mm/s}, \qquad L = 28 \text{ mm}$$

$$hS/A_1 = 0 \cdot 00682 \text{ s}^{-1}, \quad hS/A_2 = 0 \cdot 00475 \text{ s}^{-1}$$

Initial concentration of filtrate is $0 \cdot 24$ kg moles/m^3.

76. An oil-bearing formation is assumed to be of uniform depth and circular in shape. A single well penetrates the formation at a point midway between the centre of the formation and its perimeter. If oil is withdrawn at constant pressure, determine how the pressure distribution in the formation changes, and the rate of production of oil as a function of time. (The governing equation can be checked with Example 4, Section 8.8.)

77. Two immiscible liquids are in laminar flow through a cylindrical pipe and the flow rates are adjusted so that each liquid occupies a semi-circular section of the pipe. If the viscosity ratio between the two liquids is 10 : 1, find the flow rate ratio.

(*Relaxation methods have been used to solve this problem by A. R. Gemmel and N. Epstein*, Canad. J. Chem. Eng. **40**, 215, 1962.)

78. If a matrix

$$A = \begin{bmatrix} 1 & 2 & 3 \\ -1 & 1 & 3 \\ 3 & 0 & 2 \end{bmatrix}$$

and a matrix

$$B = \begin{bmatrix} 1 & -3 & 0 \\ 2 & 0 & 1 \\ 4 & 1 & 3 \end{bmatrix}$$

write down

(i) $A + B$;   (ii) AB;   (iii) BA;   (iv) $B^{-1}A$.

79. Determine the latent roots of each of the following matrices.

$$(i) \begin{bmatrix} 2 & -3 & 1 \\ 3 & 1 & 3 \\ -5 & 2 & -4 \end{bmatrix} \quad \text{and} \quad (ii) \begin{bmatrix} 7 & 4 & -1 \\ 4 & 7 & -1 \\ -4 & -4 & 4 \end{bmatrix}$$

What is the rank and degeneracy of each of these matrices?

80. 1·24 kg/s of sulphuric acid (sp. heat 1·5 kJ/kg °K) is to be cooled in a three-stage counter-current cooler of the following type. Hot acid at 174 °C is fed to tank No. 1 where it is thoroughly agitated in contact with cooling coils. The continuous discharge from this tank at 88 °C flows to a second stirred tank and leaves at 45 °C. Thereafter it flows to the third stirred tank and leaves at 28 °C. Cooling water at 18 °C flows into the coil of tank No. 3 and thence to the coils of tank No. 2 and finally to the coils of the first tank. The temperature of the water leaving the coils of the hot acid tank is 88 °C. To what temperatures would the contents of each tank rise if due to trouble in the supply, the cooling water suddenly stopped for 1 h? On restoration of the water supply, water is put on the system at the rate of 1·5 kg/s. Calculate the acid discharge temperature after 1 h. The capacity of each tank is 3600 kg of acid and the overall coefficient of heat transfer in tank 1 is 1300 W/m² °C, in tank 2 is 1000 W/m² °C and in tank 3 is 650 W/m² °C which may be assumed constant.

81. A battery of N stirred tank reactors is arranged to operate isothermally in series. Each reactor has a volume of V m³ and is equipped with a perfect agitator so that the composition of the reactor effluent is the same as the tank contents. If initially the tanks contain pure solvent only, and at a time designated t_0, q m³/h of a reactant A of concentration C_0 kg moles/m³ are fed to the first tank, estimate the time required for the concentration of A leaving the Nth tank to be C_N kg moles/m³.

The reaction can be represented stoichiometrically by the equation

$$A \underset{k_2}{\overset{k_1}{\rightleftharpoons}} B \xrightarrow{k_3} C$$

and all the reactions are first order. There is no B or C in the feed but it may be assumed that the feed contains a catalyst that initiates the reaction as soon as the feed enters the first reactor.

82. 2000 kg/h of dry coal gas containing 0·2 kg light oil/kg dry gas is to be treated with 1250 kg/h of wash oil in a six-plate absorption column to remove the light oil. If the concentration of light oil in the feed suddenly

changes to 0·3 kg oil/kg dry gas show that the disturbance in the absorption column can be described by the equation

$$\frac{dX}{dt} = RX + PQ$$

where

$$X = \{X_1 \quad X_2 \quad \ldots \quad X_6\}$$

$$P = \begin{Bmatrix} A & 0 & \ldots & O_6 \\ 0 & . & \ldots & C \end{Bmatrix}$$

$$Q = \left\{ X_0 \quad \frac{Y_{N+1} - b}{m} \right\}$$

$$R = \begin{bmatrix} B & C & 0 & . & . & O_6 \\ A & B & C & . & . & . \\ 0 & A & B & . & . & . \\ & & 0 & . & & \\ . & 0 & . & & . & . \\ . & . & . & A & B & C \\ O_6 & . & . & O & A & B_6 \end{bmatrix}$$

and

$$A = \frac{L}{mH + h}, \quad B = \frac{L + mG}{mH + h}, \quad C = \frac{mG}{mH + h}$$

The symbols are:

h = liquid hold-up on each plate and may be taken to be 35 kg
H = vapour hold-up on each plate and may be taken to be 0·5 kg
L and G are the wash oil and dry gas flow rates and the equilibrium relationship of the light oil between gas and liquid may be written $Y = mX$, where m is 0·7.

Solve the transient equation and plot the composition of light oil on each plate as a function of time.

83. X_0 kg of a solute/kg solute-free liquor is fed to the first stage of an N-stage cross-flow extraction process at the rate of C kg/h on a solute-free basis. If B kg of extracting solvent free from solute is fed to each stage and the solute distribution ratio is m, calculate the optimum number of stages and solvent to feed ratio using the data given below.

Total cost of operating one stage is α including operation, fixed charges and depreciation of capital.

Cost of solvent is β per kg of solvent.

Cost of feed liquor is γ per kg of solute.

Cost of extract product is λ per kg of solute.

(*Problem taken from V. G. Jenson and G. V. Jeffreys*, Brit. Chem.
Eng. **6**, 676, 1961.)

84. 30 m³/h of an aqueous liquor containing 80 kg of diethylamine per m³ of solution is to be treated with toluene in a counter-current liquid extraction process. Estimate the optimum number of stages and the quantity of solvent required if the capital and operating cost of each stage including vessels, pumps and labour is estimated to be $7000 per annum. The cost of regenerating recycling and adding make-up solvent is estimated to be $0·5 per m³ of solvent and the value of diethylamine is estimated to be $1·2 per kg. The plant is to be operated 8000 h per annum and the distribution ratio of diethylamine between toluene and water is 0·86.

In question 83 above it will be found that the optimum number of stages for cross-flow extraction is related to the other cost and process variables by the equation

$$1 - \frac{\alpha m}{C\beta} = \left(\frac{\beta}{\lambda m X_0}\right)^{N+1} \left[1 + \frac{1}{N+1} \ln\left(\frac{\beta}{\lambda m X_0}\right)\right]$$

Therefore, compare the two methods of operation in carrying out the extraction of diethylamine.

85. Four different kinds of crude petroleum are available for purchase by an oil company; 100 000 bulk barrels (bbl) per week each of crudes 1, 2, and 3 and 200 000 bbl per week of crude 4. Crude oils 1, 2, and 3 are processed to produce gasolene, heating oil, and jet fuel whereas crude oil 4 can be processed in two ways. In the first heating oil is produced primarily while in the second lubricating oil is produced primarily. The expected yield of the different products is given in the table below together

		Crude					Product
					4		
		1	2	3	Fuel process	Lube process	Product on order barrels/wk
Yield bbl product per bbl crude	Gasoline	0·6	0·5	0·3	0·4	0·4	170 000
	Heating oil	0·2	0·2	0·3	0·3	0·1	85 000
	Lube oil	0	0	0	0	0·2	20 000
	Jet fuel	0·1	0·2	0·3	0·2	0·2	85 000
	Loss	0·1	0·1	0·1	0·1	0·1	
Crude available bbl/wk		100 000	100 000	100 000	200 000		
Profit, $/1000 bbl crude processed		100	200	70	150	250	

with the estimated profit and maximum amount of each product that can be produced each week. Estimate the amount of each crude that should be processed each week in order to obtain the maximum amount of profit.

("Foundations of Optimisation" *by D. J. Wilde and C. S. Beightler, Prentice–Hall, Inc.* 1967.)

86. Three chemical plants, 1, 2, and 3, require liquid feedstocks, benzene, and formaldehyde, that are stored in three tank farms, A, B, and C. The daily requirements are:

Plant 1   5000 kg benzene

Plant 2   10 000 kg benzene + 7000 kg formaldehyde

Plant 3   9000 kg benzene + 8000 kg formaldehyde

For either liquid the cost of pumping 1000 kg between a given tank farm and plant is as given below:

Plant	1	2	3
Tank farm			
A	1	1	3
B	2	3	1
C	3	2	1

How should the tank farms distribute the two feedstocks so as to minimize the total pumping costs on a day when the amounts available are:

	A	B	C
Benzene (kg)	8000	10 000	9000
Formaldehyde (kg)	8000	5000	5000

87. A heat exchanger is to be designed with longitudinal fins in which the mass of fins per unit base width of fin must not exceed M. Show that for maximum heat transfer rate, the height of the fin b and the profile of the fin $f(x)$ are related by

$$f(x) = a(x^2 + 2bx - 3b^2)$$

where

$$a = \frac{h}{2k}, \quad b = \left(\frac{3Mk}{\rho h}\right)^{\frac{1}{3}}$$

h is the fin surface heat transfer coefficient

k is the thermal conductivity of the fin

ρ is the density of the material of the fin

(*Problem taken from* "The Variational Method in Engineering" *by R. C. Schechter, McGraw-Hill, New York,* 1967.)

88. Estimation of the quantity of liquid held up in the form of a bridge between adjacent particles is important in the recovery of oil from porous media, and in the adsorption hysteresis in porous adsorbents. Assuming that the porous medium consists of spheres of radius R derive an expression to estimate the volume of the liquid bridge between two particles as a function of the wetted perimeter of the bridge, the distance between centres of the spherical particles, and the surface tension of the liquid.

(*Erle, M. E., Dyson, D. C., and Morrow, N. R. A.I.Ch.E.J.* **17**, 115, 1971.)

89. The chemical reaction

$$A \xrightarrow{\ k_1\ } B \xrightarrow{\ k_2\ } C$$

takes place in a tubular reactor under plug-flow conditions in order to produce the maximum excess of the desired product B above the waste product C. All of the reactions are first order and the reaction rate constants vary with temperature in accordance with the Arrhenius equation

$$k_i = k_{i0}\exp-\frac{E_i}{RT}$$

where $k_{10} = 0.535 \times 10^{11}$ min^{-1}, $E_1 = 18\,000$ k cal/kg mole

$k_{20} = 0.461 \times 10^{18}$ min^{-1}, $E_2 = 30\,000$ k cal/kg mole

If the initial feed consists of a mixture of 0.95 kg moles/m^3 of A and 0.05 kg moles/m^3 of B evaluate the optimum temperature profile using Pontriaguin's maximum principle.

("Quasilinearisation and Invariant Imbedding" *by E. S. Lee, Academic Press, New York, 1968.*)

90. The chemical reaction $A \rightarrow B$ is to be carried out in a tubular reactor, and the reaction product is passed to a separator for fractionation and recycling. It is desired to optimize the profit from the process and it is believed that the cost of maintaining the level of temperature along the reactor length is the only important operating cost and that all other costs are constant. If the reaction is nth order establish the optimum temperature profile over the reactor.

(*See* "Dynamic Programming in Chemical Engineering and Process Control" *by S. M. Roberts, Chapter* 10, *Section* 11, *Academic Press, New York, 1964.*)

91. The product P is to be produced at the rate of 20×10^6 kg/yr from raw materials A and B in the following process. The pure reactants A and B are fed separately to a water cooled C.S.T.R. in which the following

second order reactions occur:

(1) $A + B \rightarrow C$ specific reaction rate k_1

(2) $C + B \rightarrow P + E$ specific reaction rate k_2

(3) $P + C \rightarrow G$ specific reaction rate k_3

The by-products C and E have no sale value but can be disposed of as plant fuel. By-product G is an immiscible heavy tar which must be disposed of as a waste material. The products from the reactor are cooled on a heat exchanger and then pumped to a separator maintained at a temperature of 40 °C to remove the tar G. The overflow from the separator is fed to a distillation column where P is obtained as distillate. The bottom product of the still flows through a divider where part is fed to the fuel system and the remainder is recycled to the reactor. Using the following data optimize the process:

$$k_1 = 1.66 \times 10^6 \left(\exp -\frac{6649}{T} \right), \quad \text{where } k_1 \text{ is in s}^{-1} \text{ and } T \,^\circ K$$

$$k_2 = 7.21 \times 10^5 \left(\exp -\frac{8316}{T} \right), \quad \text{where } k_2 \text{ is in s}^{-1} \text{ and } T \,^\circ K$$

$$k_3 = 2.675 \times 10^6 \left(\exp -\frac{11093}{T} \right), \quad \text{where } k_3 \text{ is in s}^{-1} \text{ and } T \,^\circ K$$

and the heats of reaction are: for reaction (1), $\Delta H_1 = -290$ kJ/kg of C; reaction (2), $\Delta H_2 = -116$ kJ/kg of E+P; and reaction (3), $\Delta H_3 = -333.0$ kJ/kg of G. The reactor must operate in the temperature range between 50 °C and 105 °C and the gravity of the reaction mixture can be considered constant at 0.8. The gravity of the separated tar G is 1.10 and the molecular weights of the components in the system may be taken to be

$$A = B = P = 100, \quad C = E = 200 \quad \text{and} \quad G = 300$$

The viscosity–temperature relationship of the reaction mixture is presented in Fig. X and the vapour pressure–temperature relationship in Fig. Y. The thermal conductivity of reaction mixture $k = 0.15$ W/m °K. Latent heat is 221 kJ/kg and the overall heat transfer coefficient of reactor cooling coils is 284 W/m² °K. Heat capacity $= 0.4$. Relative volatility of product is 2.2.

The reactor after cooler is to be a shell and tube heat exchanger that is required to cool reactor products to 40 °C and the overall heat transfer coefficient of this equipment may be taken to be 468 W/m² °K. Cooling water is available at 23 °C.

The decanter is a vertical separator and is required to have a holding time of 6.0 min in order to ensure phase separation of the tarry material G.

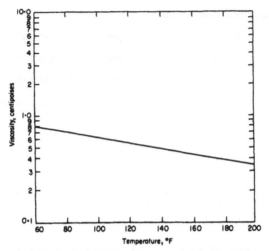

FIG. X. Viscosity relationships for reactant mixture

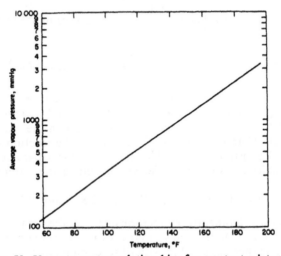

FIG. Y. Vapour pressure relationships for reactant mixture

The product P is to be purified by distillation and the vapour–liquid equilibria for the distillation system is presented in Fig. Z and the distillation column temperature must not exceed 50 °C at which temperature reaction (3) could be initiated in the reboiler or column.

Cost data are as follows:
Total sales available 20×10^6 kg/y. Excess production must be discarded.
Sale price of P = $0·66/kg.
By product (discard stream) fuel value $0·015 /kg.

Reactant A costs $0·044/kg and is available at the rate of 6600 ± 960 kg/h.
Reactant B costs $0·066/kg and is available in unlimited quantities.
Unit price of main plant components $382 000.
Total installed cost with all accessories (manufacturing): capital $1 035 000.
Steam production capacity 5000 kg/h at $40/kg h.
Electricity: 50 kW at $100/kW = $5000.
Miscellaneous costs: 1·0 of manufacturing costs.

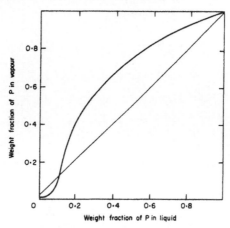

FIG. Z. Vapour–liquid equilibria

Utility costs:

Steam costs: $0·5/$M$ kgs
Water costs: $0·25/$M$ gallons
   where M is the net hourly value of sales, less raw material and waste disposal charges
Electricity costs: $0·01 kW–h
Waste disposal: $0·022/kg of G

Notes

This problem has been developed from the "General chemical processing model for the investigation of computer control" by T. J. Williams and R. E. Otto, *A.I.E.E. Trans.* **79**, 458 (1960), who also presented a steady-state optimized program based on the sequential use of an analogue and digital computer. Their numerical procedure was based on a fourth order Runge–Kutta techique.

C. W. DiBella and W. F. Stevens (*I.E.C.* (*Chem. Proc. Design*) **4**, 16, 1965) also optimized this problem using a combination of steepest descent and linear programming. A. Adelman and W. F. Stevens (*A.I.Ch.E.J.* **18**, 20, 1972) applied a constrained simplex method developed by M. J. Box (*Computer J.* **1**, 42, 1965) to obtain the optimum. R. Luns and T. H. S. Joakola (*A.I.Ch.E.J.* **19**, 645, 1973) used a direct-search technique which has great utility.

The reader is advised to consider the different methods cited here and attempt each method in order to "obtain the feel" for the different optimization procedures.

　　　(*M.Sc.—Computer Project Exercise, Process Analysis and Development,*
Aston University.)

APPENDIX

TABLE OF LAPLACE TRANSFORMS

	$\bar{f}(s)$	$f(t)$
1	$\dfrac{1}{s}$	1
2	$\dfrac{1}{s^2}$	t
3	$\dfrac{1}{s^n}$ $(n = 1, 2, \ldots)$	$\dfrac{t^{n-1}}{(n-1)!}$
4	$\dfrac{1}{\sqrt{s}}$	$\dfrac{1}{\sqrt{\pi t}}$
5	$s^{-\frac{3}{2}}$	$2\sqrt{\dfrac{t}{\pi}}$
6	$s^{-(n+\frac{1}{2})}$ $(n = 1, 2, \ldots$	$\dfrac{2^n t^{n-\frac{1}{2}}}{1 \times 3 \times 5 \ldots (2n-1)\sqrt{\pi}}$
7	$\dfrac{\Gamma(k)}{s^k}$ $(k > 0)$	t^{k-1}
8	$\dfrac{1}{s-a}$	e^{at}
9	$\dfrac{1}{(s-a)^2}$	$t\,e^{at}$
10	$\dfrac{1}{(s-a)^n}$ $(n = 1, 2, \ldots$	$\dfrac{1}{(n-1)!}t^{n-1}e^{at}$
11	$\dfrac{\Gamma(k)}{(s-a)^k}$ $(k > 0)$	$t^{k-1}e^{at}$
12*	$\dfrac{1}{(s-a)(s-b)}$	$\dfrac{1}{a-b}(e^{at} - e^{bt})$

* Here a and b represent distinct constants.

(From *Operational Mathematics* (Ed. 2) by R. V. Churchill. Copyright 1958, McGraw-Hill Book Company Inc. Used by permission.)

TABLE OF LAPLACE TRANSFORMS (*continued*)

	$\bar{f}(s)$	$f(t)$
13*	$\dfrac{s}{(s-a)(s-b)}$	$\dfrac{1}{a-b}(a\,e^{at}-b\,e^{bt})$
14*	$\dfrac{1}{(s-a)(s-b)(s-c)}$	$-\dfrac{(b-c)\,e^{at}+(c-a)\,e^{bt}+(a-b)\,e^{ct}}{(a-b)(b-c)(c-a)}$
15	$\dfrac{1}{s^2+a^2}$	$\dfrac{1}{a}\sin at$
16	$\dfrac{s}{s^2+a^2}$	$\cos at$
17	$\dfrac{1}{s^2-a^2}$	$\dfrac{1}{a}\sinh at$
18	$\dfrac{s}{s^2-a^2}$	$\cosh at$
19	$\dfrac{1}{s(s^2+a^2)}$	$\dfrac{1}{a^2}(1-\cos at)$
20	$\dfrac{1}{s^2(s^2+a^2)}$	$\dfrac{1}{a^3}(at-\sin at)$
21	$\dfrac{1}{(s^2+a^2)^2}$	$\dfrac{1}{2a^3}(\sin at-at\cos at)$
22	$\dfrac{s}{(s^2+a^2)^2}$	$\dfrac{t}{2a}\sin at$
23	$\dfrac{s^2}{(s^2+a^2)^2}$	$\dfrac{1}{2a}(\sin at+at\cos at)$
24	$\dfrac{s^2-a^2}{(s^2+a^2)^2}$	$t\cos at$
25	$\dfrac{s}{(s^2+a^2)(s^2+b^2)}(a^2\neq b^2)$	$\dfrac{\cos at-\cos bt}{b^2-a^2}$
26	$\dfrac{1}{(s-a)^2+b^2}$	$\dfrac{1}{b}e^{at}\sin bt$
27	$\dfrac{s-a}{(s-a)^2+b^2}$	$e^{at}\cos bt$

* Here a, b, and (in 14) c represent distinct constants.

TABLE OF LAPLACE TRANSFORMS (*continued*)

	$\bar{f}(s)$	$f(t)$
28	$\dfrac{3a^2}{s^3+a^3}$	$e^{-at}-e^{at/2}\times$ $\times\left(\cos\dfrac{at\sqrt{3}}{2}-\sqrt{3}\sin\dfrac{at\sqrt{3}}{2}\right)$
29	$\dfrac{4a^3}{s^4+4a^4}$	$\sin at \cosh at - \cos at \sinh at$
30	$\dfrac{s}{s^4+4a^4}$	$\dfrac{1}{2a^2}\sin at \sinh at$
31	$\dfrac{1}{s^4-a^4}$	$\dfrac{1}{2a^3}(\sinh at - \sin at)$
32	$\dfrac{s}{s^4-a^4}$	$\dfrac{1}{2a^2}(\cosh at - \cos at)$
33	$\dfrac{8a^3s^2}{(s^2+a^2)^3}$	$(1+a^2t^2)\sin at - at\cos a$
34*	$\dfrac{1}{s}\left(\dfrac{s-1}{s}\right)^n$	$L_n(t)=\dfrac{e^t}{n!}\dfrac{d^n}{dt^n}(t^n e^{-t})$
35	$\dfrac{s}{(s-a)^{\frac{3}{2}}}$	$\dfrac{1}{\sqrt{\pi t}}e^{at}(1+2at)$
36	$\sqrt{s-a}-\sqrt{s-b}$	$\dfrac{1}{2\sqrt{\pi t^3}}(e^{bt}-e^{at})$
37	$\dfrac{1}{\sqrt{s}+a}$	$\dfrac{1}{\sqrt{\pi t}}-a\,e^{a^2t}\operatorname{erfc}(a\sqrt{t})$
38	$\dfrac{\sqrt{s}}{s-a^2}$	$\dfrac{1}{\sqrt{\pi t}}+a\,e^{a^2t}\operatorname{erf}(a\sqrt{t})$
39	$\dfrac{\sqrt{s}}{s+a^2}$	$\dfrac{1}{\sqrt{\pi t}}-\dfrac{2a}{\sqrt{\pi}}e^{-a^2t}\displaystyle\int_0^{a\sqrt{t}}e^{\lambda^2}\,d\lambda$
40	$\dfrac{1}{\sqrt{s}(s-a^2)}$	$\dfrac{1}{a}e^{a^2t}\operatorname{erf}(a\sqrt{t})$
41	$\dfrac{1}{\sqrt{s}(s+a^2)}$	$\dfrac{2}{a\sqrt{\pi}}e^{-a^2t}\displaystyle\int_0^{a\sqrt{t}}e^{\lambda^2}\,d\lambda$

* $L_n(t)$ is the Laguerre polynomial of degree n.

TABLE OF LAPLACE TRANSFORMS (*continued*)

	$\bar{f}(s)$	$f(t)$
42	$\dfrac{b^2 - a^2}{(s - a^2)(b + \sqrt{s})}$	$e^{a^2 t}\left[b - a\,\mathrm{erf}(a\sqrt{t})\right]$ $ - b\,e^{b^2 t}\,\mathrm{erfc}(b\sqrt{t})$
43	$\dfrac{1}{\sqrt{s}(\sqrt{s} + a)}$	$e^{a^2 t}\,\mathrm{erfc}(a\sqrt{t})$
44	$\dfrac{1}{(s + a)\sqrt{s + b}}$	$\dfrac{1}{\sqrt{b - a}}\,e^{-at}\,\mathrm{erf}(\sqrt{b - a}\,\sqrt{t})$
45	$\dfrac{b^2 - a^2}{\sqrt{s}(s - a^2)(\sqrt{s} + b)}$	$e^{a^2 t}\left[\dfrac{b}{a}\,\mathrm{erf}(a\sqrt{t}) - 1\right]$ $ + e^{b^2 t}\,\mathrm{erfc}(b\sqrt{t})$
46*	$\dfrac{(1 - s)^n}{s^{n + \frac{1}{2}}}$	$\dfrac{n!}{(2n)!\sqrt{\pi t}}\,H_{2n}(\sqrt{t})$
47	$\dfrac{(1 - s)^n}{s^{n + \frac{3}{2}}}$	$-\dfrac{n!}{\sqrt{\pi}(2n + 1)!}\,H_{2n+1}(\sqrt{t})$
48†	$\dfrac{\sqrt{s + 2a}}{\sqrt{s}} - 1$	$a\,e^{-at}\left[I_1(at) + I_0(at)\right]$
49	$\dfrac{1}{\sqrt{s + a}\sqrt{s + b}}$	$e^{-\frac{1}{2}(a + b)t}\,I_0\!\left(\dfrac{a - b}{2}\,t\right)$
50	$\dfrac{\Gamma(k)}{(s + a)^k (s + b)^k}\,(k > 0)$	$\sqrt{\pi}\left(\dfrac{t}{a - b}\right)^{k - \frac{1}{2}} e^{-\frac{1}{2}(a + b)t}$ $ \times I_{k - \frac{1}{2}}\!\left(\dfrac{a - b}{2}\,t\right)$
51	$\dfrac{1}{(s + a)^{\frac{3}{2}}(s + b)^{\frac{1}{2}}}$	$t\,e^{-\frac{1}{2}(a + b)t}\left[I_0\!\left(\dfrac{a - b}{2}\,t\right)\right.$ $\left. + I_1\!\left(\dfrac{a - b}{2}\,t\right)\right]$
52	$\dfrac{\sqrt{s + 2a} - \sqrt{s}}{\sqrt{s + 2a} + \sqrt{s}}$	$\dfrac{1}{t}\,e^{-at}\,I_1(at)$
53	$\dfrac{(a - b)^k}{(\sqrt{s + a} + \sqrt{s + b})^{2k}}\,(k > 0)$	$\dfrac{k}{t}\,e^{-\frac{1}{2}(a + b)t}\,I_k\!\left(\dfrac{a - b}{2}\,t\right)$

* $H_n(x)$ is the Hermite polynomial, $H_n(x) = e^{x^2}\dfrac{d^n}{dx^n}(e^{-x^2})$.

† $I_n(x) = i^{-n}J_n(ix)$, where J_n is Bessel's function of the first kind.

TABLE OF LAPLACE TRANSFORMS (*continued*)

	$\bar{f}(s)$	$f(t)$
54	$\dfrac{(\sqrt{s+a}+\sqrt{s})^{-2v}}{\sqrt{s}\sqrt{s+a}}\ (v>-1)$	$\dfrac{1}{a^v}\,e^{-\frac{1}{2}at}I_v(\tfrac{1}{2}at)$
55	$\dfrac{1}{\sqrt{s^2+a^2}}$	$J_0(at)$
56	$\dfrac{(\sqrt{s^2+a^2}-s)^v}{\sqrt{s^2+a^2}}\ (v>-1)$	$a^v J_v(at)$
57	$\dfrac{1}{(s^2+a^2)^k}\ (k>0)$	$\dfrac{\sqrt{\pi}}{\Gamma(k)}\left(\dfrac{t}{2a}\right)^{k-\frac{1}{2}}J_{k-\frac{1}{2}}(at)$
58	$(\sqrt{s^2+a^2}-s)^k\ (k>0)$	$\dfrac{ka^k}{t}J_k(at)$
59	$\dfrac{(s-\sqrt{s^2-a^2})^v}{\sqrt{s^2-a^2}}\ (v>-1)$	$a^v I_v(at)$
60	$\dfrac{1}{(s^2-a^2)^k}\ (k>0)$	$\dfrac{\sqrt{\pi}}{\Gamma(k)}\left(\dfrac{t}{2a}\right)^{k-\frac{1}{2}}I_{k-\frac{1}{2}}(at)$
61	$\dfrac{e^{-ks}}{s}$	$S_k(t)=\begin{cases}0 & \text{when } 0<t<k \\ 1 & \text{when } t>k\end{cases}$
62	$\dfrac{e^{-ks}}{s^2}$	$\begin{cases}0 & \text{when } 0<t<k \\ t-k & \text{when } t>k\end{cases}$
63	$\dfrac{e^{-ks}}{s^\mu}\ (\mu>0)$	$\begin{cases}0 & \text{when } 0<t<k \\ \dfrac{(t-k)^{\mu-1}}{\Gamma(\mu)} & \text{when } t>k\end{cases}$
64	$\dfrac{1-e^{-ks}}{s}$	$\begin{cases}1 & \text{when } 0<t<k \\ 0 & \text{when } t>k\end{cases}$
65	$\dfrac{1}{s(1-e^{-ks})}=\dfrac{1+\coth\frac{1}{2}ks}{2s}$	$S(k,t)=n$ when $(n-1)k<t<nk$ (Fig. 6.4)
66	$\dfrac{1}{s(e^{ks}-a)}$	$\begin{cases}0 & \text{when } 0<t<k \\ 1+a+a^2+\ldots+a^{n-1} \\ \quad\text{when } nk<t<(n+1)k \\ \quad (n=1,2,\ldots)\end{cases}$
67	$\dfrac{1}{s}\tanh ks$	$M(2k,t)=(-1)^{n-1}$ when $2k(n-1)<t<2kn$ $(n=1,2,\ldots)$

TABLE OF LAPLACE TRANSFORMS (*continued*)

	$\bar{f}(s)$	$f(t)$
68	$\dfrac{1}{s(1+e^{-ks})}$	$\frac{1}{2}M(k,t)+\frac{1}{2}=\dfrac{1-(-1)^n}{2}$ when $(n-1)k < t < nk$
69	$\dfrac{1}{s^2}\tanh\frac{1}{2}ks$	$H(k,t)=\begin{cases} t & \text{when } 2nk < t \\ & < (2n+1)k \\ 2k-t & \text{when } (2n+1)k \\ & < t < (2n+2)k \end{cases}$
70	$\dfrac{1}{s\sinh ks}$	$F(t)=2(n-1)$ when $(2n-3)k < t < (2n-1)k$ $(t>0)$
71	$\dfrac{1}{s\cosh ks}$	$M(2k,t+3k)+1=1+(-1)^n$ when $(2n-3)k < t < (2n-1)k$ $(t>0)$
72	$\dfrac{1}{s}\coth ks$	$F(t)=2n-1$ when $2k(n-1) < t < 2kn$
73	$\dfrac{k}{s^2+k^2}\coth\dfrac{\pi s}{2k}$	$\lvert\sin kt\rvert$
74	$\dfrac{1}{(s^2+1)(1-e^{-\pi s})}$	$\begin{cases} \sin t & \text{when } (2n-2)\pi \\ & < t < (2n-1)\pi \\ 0 & \text{when } (2n-1)\pi < t < 2n\pi \end{cases}$
75	$\dfrac{1}{s}e^{-(k/s)}$	$J_0(2\sqrt{kt})$
76	$\dfrac{1}{\sqrt{s}}e^{-(k/s)}$	$\dfrac{1}{\sqrt{\pi t}}\cos 2\sqrt{kt}$
77	$\dfrac{1}{\sqrt{s}}e^{k/s}$	$\dfrac{1}{\sqrt{\pi t}}\cosh 2\sqrt{kt}$
78	$\dfrac{1}{s^{\frac{3}{2}}}e^{-(k/s)}$	$\dfrac{1}{\sqrt{\pi k}}\sin 2\sqrt{kt}$
79	$\dfrac{1}{s^{\frac{3}{2}}}e^{k/s}$	$\dfrac{1}{\sqrt{\pi k}}\sinh 2\sqrt{kt}$
80	$\dfrac{1}{s^\mu}e^{-(k/s)}\,(\mu>0)$	$\left(\dfrac{t}{k}\right)^{(\mu-1)/2}J_{\mu-1}(2\sqrt{kt})$
81	$\dfrac{1}{s^\mu}e^{k/s}\,(\mu>0)$	$\left(\dfrac{t}{k}\right)^{(\mu-1)/2}I_{\mu-1}(2\sqrt{kt})$

TABLE OF LAPLACE TRANSFORMS (*continued*)

	$\bar{f}(s)$	$f(t)$
82	$e^{-k\sqrt{s}}\,(k>0)$	$\dfrac{k}{2\sqrt{\pi t^3}}\exp\left(-\dfrac{k^2}{4t}\right)$
83	$\dfrac{1}{s}e^{-k\sqrt{s}}\,(k\geqq 0)$	$\operatorname{erfc}\left(\dfrac{k}{2\sqrt{t}}\right)$
84	$\dfrac{1}{\sqrt{s}}e^{-k\sqrt{s}}\,(k\geqq 0)$	$\dfrac{1}{\sqrt{\pi t}}\exp\left(-\dfrac{k^2}{4t}\right)$
85	$s^{-\frac{3}{2}}e^{-k\sqrt{s}}\,(k\geqq 0)$	$2\sqrt{\dfrac{t}{\pi}}\exp\left(-\dfrac{k^2}{4t}\right)-k\operatorname{erfc}\left(\dfrac{k}{2\sqrt{t}}\right)$
86	$\dfrac{a\,e^{-k\sqrt{s}}}{s(a+\sqrt{s})}\,(k\geqq 0)$	$-e^{ak}e^{a^2 t}\operatorname{erfc}\left(a\sqrt{t}+\dfrac{k}{2\sqrt{t}}\right)$ $+\operatorname{erfc}\left(\dfrac{k}{2\sqrt{t}}\right)$
87	$\dfrac{e^{-k\sqrt{s}}}{\sqrt{s}(a+\sqrt{s})}\,(k\geqq 0)$	$e^{ak}e^{a^2 t}\operatorname{erfc}\left(a\sqrt{t}+\dfrac{k}{2\sqrt{t}}\right)$
88	$\dfrac{e^{-k\sqrt{s(s+a)}}}{\sqrt{s(s+a)}}$	$\begin{cases}0 & \text{when } 0<t<k\\ e^{-\frac{1}{2}at}I_0(\tfrac{1}{2}a\sqrt{t^2-k^2}) & \text{when } t>k\end{cases}$
89	$\dfrac{e^{-k\sqrt{s^2+a^2}}}{\sqrt{s^2+a^2}}$	$\begin{cases}0 & \text{when } 0<t<k\\ J_0(a\sqrt{t^2-k^2}) & \text{when } t>k\end{cases}$
90	$\dfrac{e^{-k\sqrt{s^2-a^2}}}{\sqrt{s^2-a^2}}$	$\begin{cases}0 & \text{when } 0<t<k\\ I_0(a\sqrt{t^2-k^2}) & \text{when } t>k\end{cases}$
91	$\dfrac{e^{-k(\sqrt{s^2+a^2}-s)}}{\sqrt{s^2+a^2}}\,(k\geqq 0)$	$J_0(a\sqrt{t^2+2kt})$
92	$e^{-ks}-e^{-k\sqrt{s^2+a^2}}$	$\begin{cases}0 & \text{when } 0<t<k\\ \dfrac{ak}{\sqrt{t^2-k^2}}J_1(a\sqrt{t^2-k^2}) & \\ & \text{when } t>k\end{cases}$
93	$e^{-k\sqrt{s^2-a^2}}-e^{-ks}$	$\begin{cases}0 & \text{when } 0<t<k\\ \dfrac{ak}{\sqrt{t^2-k^2}}I_1(a\sqrt{t^2-k^2}) & \\ & \text{when } t>k\end{cases}$

TABLE OF LAPLACE TRANSFORMS (*continued*)

	$\bar{f}(s)$	$f(t)$
94	$\dfrac{a^v e^{-k\sqrt{s^2+a^2}}}{\sqrt{s^2+a^2}\,(\sqrt{s^2+a^2}+s)^v}\,(v > -1)$	$\begin{cases} 0 \quad \text{when } 0 < t < k \\ \left(\dfrac{t-k}{t+k}\right)^{\pm v} J_v(a\sqrt{t^2-k^2}) \\ \qquad\qquad \text{when } t > k \end{cases}$
95	$\dfrac{1}{s}\log s$	$\Gamma'(1) - \log t \quad [\Gamma'(1) = -0{\cdot}5772]$
96	$\dfrac{1}{s^k}\log s\,(k > 0)$	$t^{k-1}\left[\dfrac{\Gamma'(k)}{[\Gamma(k)]^2} - \dfrac{\log t}{\Gamma(k)}\right]$
97*	$\dfrac{\log s}{s-a}\,(a > 0)$	$e^{at}[\log a - \mathrm{Ei}(-at)]$
98*	$\dfrac{\log s}{s^2+1}$	$\cos t\,\mathrm{Si}(t) - \sin t\,\mathrm{Ci}(t)$
99*	$\dfrac{s\log s}{s^2+1}$	$-\sin t\,\mathrm{Si}(t) - \cos t\,\mathrm{Ci}(t)$
100*	$\dfrac{1}{s}\log(1+ks)\,(k > 0)$	$-\mathrm{Ei}\left(-\dfrac{t}{k}\right)$
101	$\log\dfrac{s-a}{s-b}$	$\dfrac{1}{t}(e^{bt} - e^{at})$
102*	$\dfrac{1}{s}\log(1+k^2s^2)$	$-2\,\mathrm{Ci}\left(\dfrac{t}{k}\right)$
103*	$\dfrac{1}{s}\log(s^2+a^2)\,(a > 0)$	$2\log a - 2\,\mathrm{Ci}(at)$
104*	$\dfrac{1}{s^2}\log(s^2+a^2)\,(a > 0)$	$\dfrac{2}{a}[at\log a + \sin at - at\,\mathrm{Ci}(at)]$
105	$\log\dfrac{s^2+a^2}{s^2}$	$\dfrac{2}{t}(1 - \cos at)$
106	$\log\dfrac{s^2-a^2}{s^2}$	$\dfrac{2}{t}(1 - \cosh at)$
107	$\arctan\dfrac{k}{s}$	$\dfrac{1}{t}\sin kt$

* The exponential-integral, cosine-integral, and sine-integral functions Ei($-t$), Ci(t), and Si(t) are defined in Section 5.5.3.

TABLE OF LAPLACE TRANSFORMS (*continued*)

	$\bar{f}(s)$	$f(t)$
108	$\dfrac{1}{s}\arctan\dfrac{k}{s}$	$\mathrm{Si}\,(kt)$
109	$e^{k^2s^2}\,\mathrm{erfc}\,(ks)\,(k>0)$	$\dfrac{1}{k\sqrt{\pi}}\exp\left(-\dfrac{t^2}{4k^2}\right)$
110	$\dfrac{1}{s}e^{k^2s^2}\,\mathrm{erfc}\,(ks)\,(k>0)$	$\mathrm{erf}\left(\dfrac{t}{2k}\right)$
111	$e^{ks}\,\mathrm{erfc}\,\sqrt{ks}\,(k>0)$	$\dfrac{\sqrt{k}}{\pi\sqrt{t}(t+k)}$
112	$\dfrac{1}{\sqrt{s}}\,\mathrm{erfc}\,(\sqrt{ks})$	$\begin{cases}0 & \text{when } 0<t<k \\ (\pi t)^{-\frac{1}{2}} & \text{when } t>k\end{cases}$
113	$\dfrac{1}{\sqrt{s}}e^{ks}\,\mathrm{erfc}\,(\sqrt{ks})\,(k>0)$	$\dfrac{1}{\sqrt{\pi(t+k)}}$
114	$\mathrm{erf}\left(\dfrac{k}{\sqrt{s}}\right)$	$\dfrac{1}{\pi t}\sin(2k\sqrt{t})$
115	$\dfrac{1}{\sqrt{s}}e^{k^2/s}\,\mathrm{erfc}\left(\dfrac{k}{\sqrt{s}}\right)$	$\dfrac{1}{\sqrt{\pi t}}e^{-2k\sqrt{t}}$
116	$K_0(ks)$	$\begin{cases}0 & \text{when } 0<t<k \\ (t^2-k^2)^{-\frac{1}{2}} & \text{when } t>k\end{cases}$
117	$K_0(k\sqrt{s})$	$\dfrac{1}{2t}\exp\left(-\dfrac{k^2}{4t}\right)$
118	$\dfrac{1}{s}e^{ks}K_1(ks)$	$\dfrac{1}{k}\sqrt{t(t+2k)}$
119	$\dfrac{1}{\sqrt{s}}K_1(k\sqrt{s})$	$\dfrac{1}{k}\exp\left(-\dfrac{k^2}{4t}\right)$
120	$\dfrac{1}{\sqrt{s}}e^{k/s}K_0\left(\dfrac{k}{s}\right)$	$\dfrac{2}{\sqrt{\pi t}}K_0(2\sqrt{2kt})$
121	$\pi e^{-ks}I_0(ks)$	$\begin{cases}[t(2k-t)]^{-\frac{1}{2}} & \text{when } 0<t<2k \\ 0 & \text{when } t>2k\end{cases}$
122	$e^{-ks}I_1(ks)$	$\begin{cases}\dfrac{k-t}{\pi k\sqrt{t(2k-t)}} & \text{when } 0<t<2k \\ 0 & \text{when } t>2k\end{cases}$

TABLE OF LAPLACE TRANSFORMS (*continued*)

	$\bar{f}(s)$	$f(t)$
123	$-e^{as}\mathrm{Ei}(-as)$	$\dfrac{1}{t+a}\,(a>0)$
124	$\dfrac{1}{a}+se^{as}\mathrm{Ei}(-as)$	$\dfrac{1}{(t+a)^2}\,(a>0)$
125	$\left(\dfrac{\pi}{2}-\mathrm{Si}\,s\right)\cos s+\mathrm{Ci}\,s\sin s$	$\dfrac{1}{t^2+1}$

Printed and bound by CPI Group (UK) Ltd, Croydon, CR0 4YY

03/10/2024

01040413-0015